Geometry, Topology and Physics

Geometry, Topology and Physics

Proceedings of the First Brazil – USA Workshop
held in
Campinas, Brazil, June 30 – July 7, 1996

Editors

Boris N. Apanasov
Steven B. Bradlow
Waldyr A. Rodrigues, Jr.
Karen K. Uhlenbeck

Walter de Gruyter · Berlin · New York 1997

Editors

Boris N. Apanasov
Dept. of Mathematics
University of Oklahoma
Norman, OK 73019
USA

Steven B. Bradlow
Dept. of Mathematics
University of Illinois
Urbana, IL 61801
USA

Waldyr A. Rodrigues, Jr.
IMECC-UNICAMP
13084-970 Campinas
Brazil

Karen K. Uhlenbeck
Math. Department
University of Texas
Austin, TX 78712-1082
USA

1991 Mathematics Subject Classification:
32Cxx, 53Cxx, 57Mxx, 58Exx, 15A66, 53B50

Keywords:
Harmonic maps, Kähler and contact geometry, gauge theory, low dimensional topology, Seiberg-Witten theory

⊗ Printed on acid-free paper which falls within the guidelines of the ANSI to ensure permanence and durability.

Library of Congress Cataloging-in-Publication-Data

Brazil–USA Workshop (1st : 1996 : Campinas, Brazil)
Geometry, topology and physics : proceedings of the first Brazil–USA Workshop, held in Campinas, Brazil, June 30–July 7, 1996 / editors Boris N. Apanasov ... [et al.].
p. cm.
Includes bibliographical references.
ISBN 3-11-015594-X (alk. paper)
1. Geometry – Congresses. 2. Topology – Congresses. 3. Mathematical physics – Congresses. I. Apanasov, B. N. (Boris Nikolaevich) II. Title
QC20.7.G44B73 1996
530.156–dc21 97-35550
CIP

Die Deutsche Bibliothek – Cataloging-in-Publication-Data

Geometry, topology and physics : proceedings of the first Brazil USA workshop, held in Campinas, Brazil, June 30–July 7, 1996 / ed. Boris N. Apanasov ... – Berlin ; New York : de Gruyter, 1997
ISBN 3-11-015594-X

Printed in Germany. Typeset using the authors' TEX files: I. Zimmermann, Freiburg.
Printing: Arthur Collignon GmbH, Berlin. Binding: Lüderitz & Bauer, Berlin.
Cover design: Thomas Bonnie, Hamburg.

Preface

This volume consists of contributions from participants in the Workshop on Geometry, Topology and Physics which took place in Campinas (Brazil) under the auspices of the Instituto de Matematica, Estatistica e Ciencia da Computacao (IMECC) at the Universidade Estadual de Campinas from June 30 to July 8, 1996.

The suggestion of such a workshop was first raised by Caio Negreiros in December of 1993. He and Valery Marenitch were the main local organizers of the workshop. That the workshop took place at all is a tribute to their perseverance and refusal to be daunted by all manner of trials and tribulations.

The workshop (initially planned as a joint USA/Brazil workshop) attracted more than 100 participants from multiple centers in Brazil, the US, Asia and Europe, and ranging in experience from graduate students to leading experts in their fields. The program covered topics in geometry, topology and physics, with special emphasis on harmonic maps, Kähler and contact geometry, gauge theory, low dimensional topology, and Seiberg–Witten theory. The presented talks included surveys of 'the state-of-the-art', as well as reports on recent progress. This mix is reflected in the contents of this Proceedings.

Acknowledgements. First and foremost, the editors thank the local organizers, Valery B. Marenich, Caio J. C. Negreiros and Waldyr A. Rodrigues Jr., for their contributions to the success of this project. They gratefully acknowledge the support of the Universidade Estadual de Campinas and its Instituto de Matematica, Estatistica e Ciencia da Computacao (IMECC), and also funding from the USA National Science Foundation, the CNPq (Brazilian Research Council), FAPESP (Sao Paulo State Research Council) and FINEP (Special Research Project Agency). They are also happy to thank the staff of IMECC, particularly Olesya Belinskaya and Leila Sanches, for their help in the organization and running the Workshop.

All contributions to this volume were refereed, and the editors thank the referees for their invaluable service. Many authors helped us by preparing their contributions in different formats of TeX, with using of macros provided by the Walter de Gruyter Verlag. We thank Dr. Manfred Karbe, the Mathematics Editor of Walter de Gruyter Verlag, for his valuable assistence.

Boris Apanasov and Steven Bradlow, for the editors, March 1997

Table of Contents

Program of the Workshop

Monday, July 1

One hour talks:

Mauricio Peixoto
Sigma decomposition in geometry, arithmetics and physics

Oscar Garcia-Prada
Seiberg–Witten invariants and vortex equations

Waldyr A. Rodrigues Jr.
Some news from Maxwell and Dirac and the lepton mass spectrum

30 minutes talks:

Peter Gilkey
The Gromov–Lawson conjecture: when do manifolds admit metrics of positive scalar curvature

Nikolai Saveliev
Invariants of homology 4-cobordism from gauge theory

Valery Marenich
Topological and metric gap-phenomenons for open manifolds of non-negative curvature

Patricio Letelier
Space-time defects

Jayme Vaz Jr.
Clifford algebras and Witten's monopole equations

Tuesday, July 2

One hour talks:

William Goldman
Ergodic theory on moduli spaces

Paul Schweitzer
Realizing Riemannian manifolds as leaves of foliations

Dmitry Alekseevsky
N-extended Poincaré Lie algebras and superalgebras and homogeneous quaternionic Kähler manifolds

Boris Apanasov
Negatively curved 4-manifolds: complex surfaces

30 minutes talks:

Nikolay Gusevskii
Intersection of bisectors in complex hyperbolic space

Frederico Palmeira
Topological methods in the study of non linear waves

Yoshinobu Kamishima
Locally conformally Kählerian structures on complex manifolds

Prem Srivastava
Light-cone coordinates and light-front quantized field theory

Rafael Herrera
Intersection numbers of moduli spaces of stable bundles and symmetries of a Verlinde formula

Dmitry Galtsov
Harmonic maps to Kähler spaces in stringy gravity

Wednesday, July 3

One hour talks:

Ron Fintushel
Geography of smooth 4 manifolds

Francis Burstall
Harmonic 2-spheres and the Bruhat decomposition of the loop group

Georgios Daskalopoulos
On the Brill–Noether problem for bundles

Ron Donagi
Strings dualities superpotentials

30 minutes talks:

Dan Pollack
Gluing constructions for noncompact geometric problems

Yan Lin Yu
Augmented spinors bi-spinors and Seiberg–Witten equation

Thursday, July 4

One hour talks:

Karen Uhlenbeck
Some interesting evolution equations related to harmonic maps

John Wood
Harmonic maps and morphisms

Steven Bradlow
A new look at vortices and dimensional reduction

Lyndon Woodward
Some geometric aspects of the 2-dimensional Toda equations

30 minutes talks:

Dave Aukley
Wall crossing in 3D Seiberg–Witten theory

John Bolton
G_2-geometry of certain submanifolds of the nearly Kähler 6-sphere

Caio Negreiros
On harmonic two dimensional spheres in the twistor space of $CP^{(n-1)}$

Armando Machado
Ruled manifolds and the geometry of the affine Grassmannian

Jose Ricardo Zeni
An elementary view of Lie group analysis of differential equations

Friday, July 5

One hour talks:

Lorenzo Sadun
Simple type is not a surface phenomenon

Jan Segert
Integrability and $SL(2C)$ invariant theory

30 minutes talks:

Jonathan Poritz
From polygons to vector bundles and back again

Takashi Kimura
Moduli spaces and algebraic structures

Antonella Marini
A topological method for finding non-absolute minima for the Yang–Mills functional with Dirichlet boundary data

Sergio Santa Cruz
Nahm's equations and hyperkähler metrics on complex adjoint orbits

Jorge Sotomayor
Stability and metric properties of umbilic points on surfaces

Cecilia Ferreira
Some embeddings of the space of partially complex structures

Shuhuong Choi
The universal cover of an affine three-manifold with holonomy of discompactedness two

Open Problems Session

Compact cohomogeneity one Riemannian manifolds of positive Euler characteristic and quaternionic Kähler manifolds

D. V. Alekseevsky and F. Podestà

Abstract. We investigate compact manifolds with positive Euler characteristic which admit a smooth action of a compact Lie group G with codimension one orbit. We give a classification for a large class of such G-manifolds up to a diffeomorphism and we apply the classification to the case of G-manifolds with an invariant quaternionic Kähler structure.

1991 Mathematics Subject Classification: 57S15, 53C30, 53C25.

Key words: G-manifolds, cohomogeneity one, Euler characteristic, quaternionic Kähler structure.

0. Introduction

A cohomogeneity one G-manifold is a manifold M with a smooth action of a compact connected Lie group G which has a codimension one orbit. Such manifolds have been investigated in some pioneer works by T. Nagano ([Na]) and P. S. Mostert ([Mo]), who studied the topology of the manifold M and of its orbit space M/G.

A new impulse to the research was given by L. Bérard Bergery in 1982 ([BeBe]), who intensively used cohomogeneity one manifolds to construct new explicit examples of Einstein metrics. Some years later R. Bryant and S. Salamon ([BrSa]) could find explicit complete metrics with exceptional holonomy on cohomogeneity one manifolds, while it is well-known that no such example can be found in the class of homogeneous spaces. These papers showed how the high degree of symmetry enjoyed by cohomogeneity one manifolds allows the construction of interesting invariant geometric structures.

In ([PaTe]), R. S. Palais and Ch. L. Terng investigated the differential geometric properties of cohomogeneity one (Riemannian) manifolds, while in series of joint papers ([A2], [AA], [AA1]) the group-theoretic description of such manifolds was given. More recently, several papers appeared on the geometry of cohomogeneity one manifolds endowed with some invariant geometric structure; among others, in view of our results, we like to refer to the work by A. Dancer and A. F. Swann, who classified cohomogeneity one hyperkähler Riemannian manifolds ([DaSw]).

This paper is aimed at studying compact cohomogeneity one manifolds with positive Euler characteristic. Some basic facts about G-manifolds are recalled in Section 1.1. Our starting point is established in [AA], namely a one to one correspondence between G-manifolds M of cohomogeneity one with orbit space $[0, \pi]$, and admissible triples (H, K, H') of subgroups of G, see Theorem 1.1.3. The subgroup K corresponds to the common stability subgroup of all interior points of a suitable geodesic segment, while the two subgroups H, H' correspond to the singular stability subgroups.

The positiveness of the Euler characteristic of M implies that at least one of the two singular orbits $G/H, G/H'$ have positive Euler characteristic and, hence, the corresponding stabilizer is a subgroup of G of maximal rank. This enable us to define the concept of *positive* triples of subgroups (H, K, H') (as an admissible triples of subgroups (H, K, H') such that the rank of H equal to the rank of G) and to reduce the classification of cohomogeneity one G-manifolds with positive Euler characteristic to classification of positive triples of subgroups of G and then to the positive triples of subalgebras of the Lie algebra $\mathfrak{g} = \mathrm{Lie}\, G$. We devide all positive triples of subgroups of a compact Lie group G into 5 types and give the classification for three of them. Some applications of these results to classification of cohomogeneity one quaternionic Kähler manifolds are given.

1. Cohomogeneity one G-manifolds with positive Euler characteristic and positive triples

1.1. Cohomogenity one G-manifolds and admissible triples of subgroups

A cohomogeneity one manifold is a manifold M together with an effective action of a compact connected Lie group G which has a codimension one orbit. If a complete G-invariant Riemannian metric g on M is chosen, we say that (M, g) is a Riemannian cohomogeneity one G-manifold. Then the orbit space $\Omega = M/G$ has a natural structure of one dimensional metrical space. Without loss of generality we may assume that it is isometric to one of the following standard metrical spaces:

(i) $\Omega = \mathbb{R}$;

(ii) $\Omega = S^1$;

(iii) $\Omega = \mathbb{R}^+$;

(iv) $\Omega = [0, \pi]$.

Throughout the following, we will refer to [AA] and [AA1] for general results on cohomogeneity one Riemannian manifolds.

In the following, we will assume that the manifold M is compact so that the orbit space Ω is either the circle S^1 or the closed interval $[0, \pi]$. In the first case, the projection $p : M \to S^1$ is a fibre bundle with fibre G/K, where K is the stabilizer of a point in M. Hence, the Euler characteristic $\chi(M) = \chi(G/K)\chi(S^1) = 0$.

In order to deal with the second case, we recall the following basic definition

Definition. A complete geodesic γ on a cohomogeneity one Riemannian manifold is called a *normal geodesic* if it crosses every orbit orthogonally.

It is well known that normal geodesics exist and that the group G acts transitively on the set of normal geodesics. Following [PaTe], [AA], we can consider the Weyl group of a normal geodesic γ as the group W of all isometries of γ which are induced by elements of G preserving γ. We note that the common stabilizer K of all regular points $x \in \gamma$ leaves γ pointwise fixed and therefore the Weyl group can be identified with a subgroup of the quotient group $N_G(K)/K$, where $N_G(K)$ denotes the normalizer of K in G.

Now, if γ is any normal geodesic, we can consider a fundamental domain δ of the Weyl group W of γ and we will call δ a *geodesic segment*. We have the following ([AA])

Proposition 1.1.1 ([Br], [AA]). *All internal points of a geodesic segment δ have the same stabilizer K. In case when $\Omega = [0, \pi]$, the stability subgroup H of a boundary point of δ contains K and the coset space H/K is homeomorphic to a sphere $S^m, m \in \{0, 1, 2, \dots\}$.*

Now, we can associate with a geodesic segment δ some set θ of invariants in the following way: when $\Omega = [0, \pi]$, we put $\theta = (H, K, H')$, where K is the common stability subgroup of regular points in δ and H, $H'$are the stability subgroups of two singular boundary points of δ, with H/K and H'/K both homeomorphic to some spheres S^m, $S^{m'}$ respectively.

Since the group G acts transitively on the set of geodesic segments, if we pass to another geodesic segment $\delta' = a\delta, a \in G$, then the elements of the set θ are transformed by conjugation associated with the element $a \in G$ in an obvious way. Therefore, we may consider the elements of θ, up to conjugation, as invariants of the Riemannian G-manifold.

Conversely, we give the following

Definition.

(1) A pair (H, K) of compact Lie groups with $K \subset H$ is called an *admissible pair* of groups if the homogeneous space H/K is diffeomorphic to a sphere S^m, $m \geq 0$. If, moreover, H acts on H/K effectively, resp. almost effectively, that is with

discrete kernel of non effectivity, the admissible pair is called *effective*, resp. *almost effective*.

(2) A triple $\theta = (H \supset K \subset H')$ of closed subgroups of a compact connected Lie group G is called an *admissible triple* if the pairs (H, K), (H', K) are admissible.
An admissible triple $\theta = (H \supset K \subset H')$ is said to be *effective*, resp. *almost effective* if any subgroup of K normal in G is trivial, resp. finite.

(3) Two admissible triples $\theta = (H, K, H')$, $\bar{\theta} = (\bar{H}, \bar{K}, \bar{H}')$ are said to be *equivalent* if there exists an element $n \in (N_G(K)/K)_0$ (where $_0$ indicates the connected component of the identity) such that the triples $(H \supset K \subset nH'n^{-1})$ and $\bar{\theta} = (\bar{H} \supset \bar{K} \subset \bar{H}')$ are conjugated by some element of G.

Remark that an almost effective admissible triple of subgroups of G becomes effective if we factorize G and subgroups of the triple by the finite subgroup $Z(G) \cap K$, where $Z(G)$ is the center of G.

The following proposition shows that in most cases one may consider only triples of connected subgroups.

Proposition 1.1.2 ([AA]). *Let (H, K, H') be an effective admissible triple of subgroups of a Lie group G. Assume that the dimension of the spheres H/K, H'/K is positive. Then the triple (H_o, K_o, H'_o), given by connected components of the identity, is admissible.*

Proof. The connected group H_o acts transitively on the sphere H/K. From the Borel classification of such groups (see Table 2.1), it follows that the stabilizer $H_o \cap K$ is connected and therefore it coincides with K_o. The same argument can be applied to H' □

Now, we define an admissible pair and an admissible triple of Lie algebras of a compact Lie algebra $\mathfrak{g}$. We will always denote Lie groups by capital Latin letters and their Lie algebras by corresponding small German letters.

Definition.

(1) The pair $(\mathfrak{h}, \mathfrak{k})$ of Lie algebras associated with an admissible (almost effective) pair (H, K) of Lie groups is called to be admissible (effective);

(2) The triple $(\mathfrak{h}, \mathfrak{k}, \mathfrak{h}')$ of subalgebras of the Lie algebras $\mathfrak{g} = \text{Lie}\, G$ associated with an admissible (almost effective) triple (H, K, H') of subgroups of a compact connected Lie group G is called an admissible (effective) triple of subalgebras.

If the Lie group G is semisimple, in the above definition one may always assume that it is simply connected.

Using this remark, we easily obtain the following intrinsic characterization of admissible triples of Lie subalgebras.

Proposition 1.1.3. *Let $\mathfrak{h}$, $\mathfrak{k}$, $\mathfrak{h}'$ be subalgebras of a compact semisimple Lie algebra $\mathfrak{g}$ such that $\mathfrak{k}$ is a proper subalgebra of $\mathfrak{h}$ and $\mathfrak{h}'$. Denote by H, K, H' the corresponding connected subgroups of the simply connected Lie group G associated with $\mathfrak{g}$. Then*

(1) *the pair $(\mathfrak{h}, \mathfrak{k})$ of Lie algebras is admissible if and only if H/K is a sphere,*

(2) *the triple $(\mathfrak{h}, \mathfrak{k}, \mathfrak{h}')$ of subalgebras of $\mathfrak{g}$ is admissible, if and only if the pairs $(\mathfrak{h}, \mathfrak{k})$, $(\mathfrak{h}', \mathfrak{k})$ are admissible.*

Remark that an admissible pair $(\mathfrak{h}, \mathfrak{k})$, resp., triple $(\mathfrak{h}, \mathfrak{k}, \mathfrak{h}')$ of subalgebras is effective iff $\mathfrak{k}$ has no non trivial ideal of $\mathfrak{h}$, resp, of $\mathfrak{g}$.

We may now state the fundamental result (see [AA]).

Theorem 1.1.3. *There exists a one to one correspondence between compact cohomogeneity one G-manifolds with orbit type iv) up to a G-diffeomorphism (resp., up to a diffeomorphism) and admissible triples of subgroups of G up to a conjugation in G (resp., up to an equivalency).*

1.2. Euler characteristic of cohomogeneity one G-manifolds and positive triples $\theta = (H, K, H')$

In this section we will investigate the topology of a compact G-manifold M of cohomogeneity one with the orbit space of type iv).

We choose a G-invariant metric g on M such that $\Omega = [0, \pi]$ and denote by γ a normal geodesic such that $\gamma|_{[0,\pi]}$ is a geodesic segment. This means that all the internal points $\gamma(t), t \in (0, \pi)]$ are regular and have the same stabilizer K and the boundary points $\gamma(0)$, $\gamma(\pi)$ are singular and have singular stabilizers H and H', respectively.

In this situation, we can consider the two G-invariant submanifolds $M_1 = G \cdot \{\gamma(t); 0 \leq t < \frac{3}{4}\pi\}$ and $M_2 = G \cdot \{\gamma(t); \frac{1}{4}\pi < t \leq \pi\}$. We have two G-equivariant projections $\pi_i : M_i \to G/H_i, i = 1, 2$ given by the projection along normal geodesics. By means of these projections, the two singular orbits G/H_1 and G/H_2 are easily seen to be deformation retracts of M_1, M_2 respectively. Moreover, the intersection $M_1 \cap M_2$ contains the regular orbit G/K as deformation retract. Since the manifold $M = M_1 \cup M_2$, we can apply the Mayer–Vietoris exact sequence in cohomology with a coefficient group Λ, to get the following

$$\cdots \to H^p(M, \Lambda) \to H^p(M_1, \Lambda) \oplus H^p(M_2, \Lambda) \to H^p(M_1 \cap M_2, \Lambda) \to$$
$$\to H^{p+1}(M, \Lambda) \to \cdots .$$

We rewrite this exact sequence as follows :

$$\cdots \to H^p(M,\Lambda) \to H^p(G/H_1,\Lambda) \oplus H^p(G/H_2,\Lambda) \to H^p(G/K,\Lambda) \to \\ \to H^{p+1}(M,\Lambda) \to \cdots . \tag{1.2.1}$$

As immediate corollary of this, we get

Proposition 1.2.1. *Let M be a cohomogeneity one G-manifold with the orbit space $[0,\pi]$ and let $\theta = (H, K, H')$ be the associated triple of subgroups of G. Then the Euler characteristic $\chi(M)$ is given by*

$$\chi(M) = \chi(G/H) + \chi(G/H') - \chi(G/K).$$

As a useful corollary, we have

Corollary 1.2.2. *If the compact cohomogeneity one G-manifold M has positive Euler characteristic, then at least one from two singular orbits G/H, G/H' has positive Euler characteristic, hence at least one singular stabilizer is a subgroup of maximal rank of G.*

Proof. Recall that the orbit space of a cohomogeneity one manifold with positive Euler characteristic has type iv), see sect. 1.1. The proof follows immediately from Proposition 1.2.1 by observing that for any compact homogeneous space G/L we have $\chi(G/L) \geq 0$ with strict inequality if and only if L has maximal rank in G. □

The last corollary suggests the following definition

Definition 1.2.3.

(1) An effective admissible triple $\theta = (H \supset K \subset H')$ is called *positive* if $\mathrm{rk}(G) = \mathrm{rk}(H) \geq \mathrm{rk}(H')$.

(2) The triple $(\mathfrak{h}, \mathfrak{k}, \mathfrak{h}')$ of Lie subalgebras of the Lie algebra $\mathfrak{g} = \mathrm{Lie}(G)$ associated with a positive triple (H, K, H') of subgroups of a compact Lie group G is called *positive*.

Theorem 1.2.4. *There exists a one to one correspondence between compact cohomogeneity one G-manifolds with positive Euler characteristic up to a G-diffeomorphism, (resp., up to a diffeomorphism) and positive admissible triples of subgroups of G up to conjugation in G (resp., up to an equivalency).*

Proof. The theorem follows immediately from Theorem 1.1.2 and Corollary 1.2.2. □

We conclude this section mentioning another topological tool which will be useful later on. With the same notations as above, if G/K is a regular orbit and G/H is a singular orbit with $K \subset H$, we have a natural fibration $\pi : G/K \to G/H$ with typical fibre H/K which is diffeomorphic to some sphere S^m; if the manifold M is orientable (hence G/K is orientable) and G/H is orientable, we may relate the cohomology of the base manifold G/H with the cohomology of the total space G/K by means of a Gysin exact sequence, given by

$$\cdots \to H^p(G/K) \xrightarrow{\pi_*} H^{p-m}(G/H) \xrightarrow{\wedge e} H^{p+1}(G/H) \xrightarrow{\pi^*} H^{p+1}(G/K) \to \cdots \quad (1.2.5)$$

where π_* denotes integration along the fibres and $\wedge e$ means multiplication by the Euler class of the sphere bundle.

1.3. The canonical decomposition of the principal stability subalgebra $\mathfrak{k}$ and the five types of admissible triples

Let $\theta = (H, K, H')$ be an effective admissible triple of subgroups of a Lie group G and $(\mathfrak{h}, \mathfrak{k}, \mathfrak{h}')$ the corresponding triple of subalgebras of $\mathfrak{g}$. Then there exist representations

$$j : H \to GL(V), \quad j' : H' \to GL(V')$$

of H and resp., H' into the spaces $V = \mathbb{R}^{m+1}$, $V' = \mathbb{R}^{m'+1}$, respectively, such that the homogeneous spheres $S^m = H/K$, $S^{m'} = H'/K$ are realized as the orbits $j(H)v$, $j(H')v'$, of some vectors v, v'. We will say that the representations j, j' are normal isotropic representations, see [Br].

Denote by $N = \ker j$, $N' = \ker j'$ the normal subgroups of K. Then we have a decomposition of the Lie algebra $\mathfrak{k}$ of K into the sum of three ideals

$$\mathfrak{k} = \mathfrak{k}_0 + \mathfrak{n} + \mathfrak{n}',$$

where $\mathfrak{n}$ and $\mathfrak{n}'$ are the Lie algebras of the ideals N, N' and $\mathfrak{k}_0$ is an ideal of $\mathfrak{k}$ complementary to the ideal $\mathfrak{n} + \mathfrak{n}'$. We call this decomposition the canonical decomposition of the subalgebra $\mathfrak{k}$.

We separate all effective admissible triples $\theta = (H, K, H')$ of subgroups of a Lie group G and the corresponding triples of subalgebras into five types:

(1) Type 0 $\quad G = H$

(For all other types we will assume that H and H' are proper subgroups of G.)

(2) Type 1 $\quad (\mathfrak{h}, \mathfrak{k})$ is an effective pair, that is $\mathfrak{n} = 0$

(3) Type 2 $\quad (\mathfrak{h}, \mathfrak{k})$ is not effective ($\mathfrak{n} \neq 0$), but $(\mathfrak{h}', \mathfrak{k})$ is effective ($\mathfrak{n}' = 0$)

(4) Type 3 $\quad \mathfrak{k} = \mathfrak{n} + \mathfrak{n}'$ with $\mathfrak{n}, \mathfrak{n}' \neq 0$

(5) Type 4 $\quad \mathfrak{k} = \mathfrak{k}_0 + \mathfrak{n} + \mathfrak{n}'$ with $\mathfrak{k}_0 \neq 0$, $\mathfrak{n} \neq 0$ and $\mathfrak{n}' \neq 0$.

2. Classification of cohomogeneity one G-manifolds with a fixed point

In this section we will classify type 0 effective admissible triples of connected subgroups (H, K, H') of a compact connected Lie group G. They correspond to cohomogeneity one G-manifolds M with a fixed point of G. In this case we have $G = H \supset H' \supset K$ and the normal isotropy representation j of $G = H$ may be identified with the isotropy representation of G at the fixed point p of G. In particular, the representation j is exact. The A. Borel classification of the transitive actions of compact connected Lie groups on spheres implies the classification of effective admissible pairs of Lie algebras. The list of such algebras is given in the Table 2.1.

$\mathfrak{h}$	$\mathfrak{k}$	$\dim \mathfrak{h}/\mathfrak{k}$	H/K	rem.
$\mathfrak{a}_n$	$\mathfrak{a}_{n-1}$	$2n+1$	SU_{n+1} / SU_n	$n \geq 1$
$\mathfrak{b}_n$	$\mathfrak{d}_n$	$2n$	SO_{2n+1} / SO_{2n}	$n \geq 1$
$\mathfrak{c}_n$	$\mathfrak{c}_{n-1}$	$4n-1$	Sp_n / Sp_{n-1}	$n \geq 2$
$\mathfrak{d}_n$	$\mathfrak{b}_{n-1}$	$2n-1$	SO_{2n} / SO_{2n-1}	$n \geq 3$
$\mathbb{R} + \mathfrak{a}_{n-1}$	$\mathbb{R} + \mathfrak{a}_{n-2}$	$2n-1$	U_n/U_{n-1}	$n \geq 2$
$\mathbb{R} + \mathfrak{c}_{n-1}$	$\mathbb{R} + \mathfrak{c}_{n-2}$	$4n-5$	$T^1 \cdot Sp_{n-1} / T^1 \cdot Sp_{n-2}$	$n \geq 3$
$\mathfrak{a}_1 + \mathfrak{c}_{n-1}$	$\mathfrak{a}_1 + \mathfrak{c}_{n-2}$	$4n-5$	$Sp_1 \cdot Sp_{n-1} / Sp_1 \cdot Sp_{n-2}$	$n \geq 2$
$\mathfrak{b}_4$	$\mathfrak{b}_3$	15	$Spin_9 / Spin_7$	
$\mathfrak{b}_3$	$\mathfrak{g}_2$	7	$Spin_7 / G_2$	
$\mathfrak{g}_2$	$\mathfrak{a}_2$	6	G_2/SU_3	
$\mathbb{R}$	0	1	$SO_2 /\{1\}$	
0	0	0	$\mathbb{Z}_2/\{1\}$	

Table 2.1. Effective admissible pairs $(\mathfrak{h}, \mathfrak{k})$ of Lie algebras and corresponding homogeneous spheres H/K

Remark that if $m = \dim H/K = \dim M - 1 > 0$, then the stabilizer K is connected. This implies that the stabilizer H' is also connected. This remark reduces the classification of effective admissible triple of subgroups of type 0 to classification of admissible triples of subalgebras $(\mathfrak{h}, \mathfrak{k}, \mathfrak{h}')$ such that

(1) $(\mathfrak{h}, \mathfrak{k})$ is an effective admissible pair;

(2) $\mathfrak{h} \supseteq \mathfrak{h}' \supseteq \mathfrak{k}$ and $(\mathfrak{h}', \mathfrak{k})$ is an admissible pair.

Using Table 2.1, we easily obtain the classification of all such triples of subalgebras and, hence, the classification of effective admissible triples of type 0. The results are presented in the Table 2.2.

$\mathfrak{h}$	$\mathfrak{k}$	$\mathfrak{h}'$
$\mathfrak{a}_n$	$\mathfrak{a}_{n-1}$	$\mathbb{R} + \mathfrak{a}_{n-1}$
$\mathfrak{c}_n$	$\mathfrak{c}_{n-1}$	$\mathbb{R} + \mathfrak{c}_{n-1}, \mathfrak{a}_1 + \mathfrak{c}_{n-1}$
$\mathbb{R} + \mathfrak{a}_{n-1}$	$\mathbb{R} + \mathfrak{a}_{n-2}$	$2\mathbb{R} + \mathfrak{a}_{n-2}$
$\mathbb{R} + \mathfrak{c}_{n-1}$	$\mathbb{R} + \mathfrak{c}_{n-2}$	$2\mathbb{R} + \mathfrak{c}_{n-2}, \mathbb{R} + \mathfrak{a}_1 + \mathfrak{c}_{n-2}$
$\mathfrak{a}_1 + \mathfrak{c}_{n-1}$	$\mathfrak{a}_1 + \mathfrak{c}_{n-2}$	$2\mathfrak{a}_1 + \mathfrak{c}_{n-2}$

Table 2.2. Admissible triples $(\mathfrak{h}, \mathfrak{k}, \mathfrak{h}')$ of Lie algebras with $\mathfrak{h} \supsetneq \mathfrak{h}' \supsetneq \mathfrak{k}$ and $(\mathfrak{h}, \mathfrak{k})$ effective admissible pair

3. Rank n extensions of compact rank n Lie algebras and maximal rank n Lie algebras

3.1. Rank n extensions of compact Lie algebras

In this section we describe all rank n extension of a compact rank n Lie algebra. We start with the following

Definition. Let $\mathfrak{h}$ be a compact Lie algebra of rank n. A compact Lie algebra $\mathfrak{g}$ of rank n will be called *rank n extension* of $\mathfrak{h}$ (or shortly *extension*) if $\mathfrak{h}$ is isomorphic to a proper subalgebra of $\mathfrak{g}$. In case, we will denote such $\mathfrak{g}$ as $\mathfrak{h}^{\uparrow}$.

Let $\mathfrak{h} = \mathbb{R}^k + \mathfrak{h}_1 + \mathfrak{h}_2 + \cdots + \mathfrak{h}_s$ the decomposition of $\mathfrak{h}$ into sum of the center and its simple ideals $\mathfrak{h}_i, i = 1, \ldots, s$. We describe now some "elementary" extensions of $\mathfrak{h}$. Any such elementary extension consists in enlarging one ideal $\mathfrak{q}$ of $\mathfrak{h}$ by its extension $\mathfrak{q}^\uparrow$ given in the following list:

(1) Extensions of a simple summand:

$$\mathfrak{d}_n \subset \mathfrak{b}_n;\ \mathbb{R} \subset \mathfrak{a}_1;\ \mathfrak{a}_2 \subset \mathfrak{g}_2;\ \mathfrak{b}_4 \subset \mathfrak{f}_4;\ \mathfrak{a}_7 \subset \mathfrak{e}_7;\ \mathfrak{a}_8, \mathfrak{d}_8 \subset \mathfrak{e}_8$$

(2) Extensions of a semisimple ideal:

$p \geq 2, q \geq 1$	$\mathfrak{d}_p + \mathfrak{b}_q \subset \mathfrak{b}_{p+q}$	$2\mathfrak{a}_1$	$\subset \mathfrak{g}_2$	$2\mathfrak{a}_2$	$\subset \mathfrak{f}_4$
$p, q \geq 1$	$\mathfrak{c}_p + \mathfrak{c}_q \subset \mathfrak{c}_{p+q}$	$\mathfrak{a}_1 + \mathfrak{c}_3$	$\subset \mathfrak{f}_4$	$\mathfrak{a}_2 + \mathfrak{a}_5$	$\subset \mathfrak{e}_7$
$p, q \geq 2$	$\mathfrak{d}_p + \mathfrak{d}_q \subset \mathfrak{d}_{p+q}$	$\mathfrak{a}_1 + \mathfrak{a}_5$	$\subset \mathfrak{e}_6$	$\mathfrak{a}_2 + \mathfrak{e}_6$	$\subset \mathfrak{e}_8$
		$\mathfrak{a}_1 + \mathfrak{d}_6$	$\subset \mathfrak{e}_7$	$2\mathfrak{a}_4$	$\subset \mathfrak{e}_8$
		$\mathfrak{a}_1 + \mathfrak{e}_7$	$\subset \mathfrak{e}_8$	$3\mathfrak{a}_2$	$\subset \mathfrak{e}_6$

(3) Extensions of a non semisimple ideal:

$\mathbb{R} + \mathfrak{a}_{n-1}$	$\subset$	$\mathfrak{a}_n, \mathfrak{b}_n, \mathfrak{c}_n, \mathfrak{d}_n$
$\mathbb{R} + \mathfrak{a}_p + \mathfrak{a}_q$	$\subset$	$\mathfrak{a}_{p+q+1}$
$\mathbb{R} + \mathfrak{b}_{n-1}$	$\subset$	$\mathfrak{b}_n$
$\mathbb{R} + \mathfrak{d}_{n-1}$	$\subset$	$\mathfrak{d}_n$
$\mathbb{R} + \mathfrak{c}_{n-1}$	$\subset$	$\mathfrak{c}_1 + \mathfrak{c}_{n-1} \subset \mathfrak{c}_n$
$\mathbb{R} + \mathfrak{a}_1$	$\subset$	$\mathfrak{g}_2$
$\mathbb{R} + \mathfrak{d}_5$	$\subset$	$\mathfrak{e}_6$
$\mathbb{R} + \mathfrak{e}_6$	$\subset$	$\mathfrak{e}_7$

We have

Proposition 3.1.1. *Any extension is a composition of elementary ones.*

Proof. It is enough to check that any rank n compact Lie algebra $\mathfrak{g}$ which contains $\mathfrak{h} = \mathbb{R}^k + \mathfrak{h}_1 + \cdots + \mathfrak{h}_s$ as a maximal rank n subalgebra can be obtained by means of elementary extensions. Let

$$\mathfrak{g} = \mathbb{R}^l + \mathfrak{g}_1 + \cdots + \mathfrak{g}_m$$

the decomposition of $\mathfrak{g}$ into sum of the center and simple ideals. Any Cartan subalgebra of $\mathfrak{h}$ has the form $\mathfrak{t} = \mathbb{R}^k + \mathfrak{t}_1 + \cdots + \mathfrak{t}_s$, where $\mathfrak{t}_i$ is a Cartan subalgebra of $\mathfrak{h}_i$. Note that $\mathfrak{t}$ is also a Cartan subalgebra of $\mathfrak{g}$. Hence any simple ideal $\mathfrak{h}_i$ is generated by root vectors and it is contained into some ideal $\mathfrak{g}_j$ of $\mathfrak{g}$. This implies that

$$\mathfrak{h} = \mathbb{R}^l + \mathfrak{h} \cap \mathfrak{g}_1 + \cdots + \mathfrak{h} \cap \mathfrak{g}_m,$$

where $\mathfrak{h} \cap \mathfrak{g}_i$ is a maximal subalgebra of maximal rank of the simple Lie algebra $\mathfrak{g}_i$. Now the claim follows from A.Borel and J.Siebenthal classification of maximal subalgebras of maximal rank in a simple Lie algebra, which is recalled in the next subsection. □

3.2. Maximal rank n subalgebras of rank n Lie algebras

In this section we recall the A. Borel and J. Siebenthal classification of maximal subalgebras $\mathfrak{h}$ of maximal rank n in a simple Lie algebra $\mathfrak{g}$ (see e.g. [GoGr], [GOV]).

Let $\Pi = \{\alpha_i\}_{i=1,\dots,n}$ be a system of simple roots of a simple Lie algebra $\mathfrak{g}$ and let $\tilde{\Pi} = \Pi \cup \{\alpha_o\}$ be the extended simple root system, where α_o is the minimal root. We then denote by $n_i \in \mathbb{N}$ the coefficients such that

$$\sum_{i=0}^{n} n_i \alpha_i = 0, \quad n_0 = 1.$$

Then any rank n maximal semisimple subalgebra $\mathfrak{h}$ of $\mathfrak{g}$ has simple root system $\tilde{\Pi}_{\alpha_i} = \tilde{\Pi} \setminus \{\alpha_i\}$, where $\alpha_i \in \tilde{\Pi}$ is such that the corresponding coefficient n_i is prime. Any non semisimple maximal subalgebra of maximal rank has simple root system $\Pi_{\alpha_j} = \Pi \setminus \{\alpha_j\}$, where $n_j = 1$.

In Table 3.1 we enumerate all rank n maximal subalgebras of simple Lie algebras. We underline all subalgebras which are Lie algebras of compact Lie groups acting transitively and effectively on some sphere.

$\mathfrak{g}$	$\mathfrak{h}$
$\mathfrak{a}_n$	$\mathbb{R} + \mathfrak{a}_p + \mathfrak{a}_q$, $n = p + q + 1, p, q > 0$; $\underline{\mathbb{R} + \mathfrak{a}_{n-1}}$
$\mathfrak{b}_n$	$\mathfrak{d}_p + \mathfrak{b}_q$, $n = p + q, p, q > 0$; $2\mathfrak{a}_1 + \mathfrak{b}_{n-2}$, $\underline{\mathfrak{d}_n}$, $\mathbb{R} + \mathfrak{b}_{n-1}$
$\mathfrak{c}_n$	$\mathfrak{c}_p + \mathfrak{c}_q$, $n = p + q$, $\underline{(\mathfrak{a}_1 + \mathfrak{c}_{n-1})}$, $\underline{\mathbb{R} + \mathfrak{a}_{n-1}}$
$\mathfrak{d}_n$	$\mathfrak{d}_p + \mathfrak{d}_q$, $n = p + q$; $\mathbb{R} + \mathfrak{d}_{n-1}$, $2\mathfrak{a}_1 + \mathfrak{d}_{n-2}$, $\underline{\mathbb{R} + \mathfrak{a}_{n-1}}$
$\mathfrak{e}_6$	$\mathfrak{a}_1 + \mathfrak{a}_5, 3\mathfrak{a}_2, \mathbb{R} + \mathfrak{d}_5$
$\mathfrak{e}_7$	$\mathfrak{a}_1 + \mathfrak{d}_6, \mathfrak{a}_2 + \mathfrak{a}_5, \underline{\mathfrak{a}_7}, \mathbb{R} + \mathfrak{e}_6$
$\mathfrak{e}_8$	$\mathfrak{a}_1 + \mathfrak{e}_7, \mathfrak{a}_2 + \mathfrak{e}_6, 2\mathfrak{a}_4, \underline{\mathfrak{a}_8}, \underline{\mathfrak{d}_8}$
$\mathfrak{f}_4$	$\underline{\mathfrak{b}_4}, 2\mathfrak{a}_2, \underline{\mathfrak{a}_1 + \mathfrak{c}_3}$
$\mathfrak{g}_2$	$\underline{\mathfrak{a}_2}, \underline{2\mathfrak{a}_1}$

Table 3.1. Rank n maximal subalgebras $\mathfrak{h}$ of simple rank n Lie algebra $\mathfrak{g}$

4. Classification of cohomogeneity one G-manifolds of type 1

In this section we classify positive triples of Lie subalgebras $\theta = (\mathfrak{h}, \mathfrak{k}, \mathfrak{h}')$ of a compact Lie algebra $\mathfrak{g}$ which are of type 1. This means that the corresponding pair $(\mathfrak{h}, \mathfrak{k})$ is effective and therefore it appears in Table 2.1.

Let $\theta = (\mathfrak{h}, \mathfrak{k}, \mathfrak{h}')$ be a positive tripe of subalgebras of a compact Lie algebra $\mathfrak{g}$. Then, by definition, $\mathrm{rk}(\mathfrak{h}) = \mathrm{rk}(\mathfrak{g}) = n$. In other words, $\mathfrak{g}$ is a rank n extension of the rank n Lie algebra $\mathfrak{h}$. All such extensions $\mathfrak{h} \subset \mathfrak{g}$, where $\mathfrak{h}$ runs through Table 2.1, can be easily classified using Theorem 3.1.1. The result is presented in Table 4.1

To obtain a classification of all type 1 positive triples of subalgebras of a rank n compact Lie algebra $\mathfrak{g}$, we subdivide them into three types:

1A) Triples of the form $(\mathfrak{h}, \mathfrak{k}, \mathfrak{k})$, where $(\mathfrak{h}, \mathfrak{k})$ is an admissible pair. They correspond to admissible triples of subgroups of the form $(H, K, \mathbb{Z}_2 \cdot K)$. The corresponding G-manifold has an exceptional orbit $G/\mathbb{Z}_2 \cdot K$.

1B) Triples $(\mathfrak{h}, \mathfrak{k}, \mathfrak{h}')$ with $\mathrm{rk}(\mathfrak{h}') = \mathrm{rk}(\mathfrak{g}) = n$.

1C) Triples $(\mathfrak{h}, \mathfrak{k}, \mathfrak{h}')$ with $\mathfrak{h}' \neq \mathfrak{k}$ and $\mathrm{rk}(\mathfrak{h}') = \mathrm{rk}(\mathfrak{k}) = \mathrm{rk}(\mathfrak{g})$-1.

The list of all triples of type 1B) can be obtained as follows.

For any admissible pair $(\mathfrak{h}, \mathfrak{k})$ from Table 2.1 which admits a proper rank n extension from Table 4.1, we determine all rank $(n-1)$ extensions $\mathfrak{h}'$ of rank $(n-1)$ Lie algebra $\mathfrak{k}$ and, using the results of Section 3 and Table 2.1, we choose those which satisfy the following two properties:

(1) $\mathfrak{h}'$ is contained in $\mathfrak{g}$;

(2) the pair $(\mathfrak{h}', \mathfrak{k})$ is an admissible pair,

We will say that an admissible triple $(\mathfrak{h}, \mathfrak{k}, \mathfrak{h}')$ is *standard* if the Lie algebras $\mathfrak{h}$, $\mathfrak{h}'$ are isomorphic (but nor necessarily conjugated by an element of G). All non standard admissible triples of type 1B are enumerated in Table 4.2.

All positive triples $(\mathfrak{h}, \mathfrak{k}, \mathfrak{h}')$ of type 1C are determined in a similar way: we consider all admissible pairs $(\mathfrak{h}, \mathfrak{k})$ from Table 2.1 such that $\mathfrak{h}$ has a proper rank n extension described in Table 4.1; we then determine all rank $(n-1)$ extensions $\mathfrak{h}'$ of rank $(n-1)$ Lie algebra $\mathfrak{k}$ and choose those $\mathfrak{h}'$ which satisfy properties (1) and (2). The result is given in the Table 4.3.

$\mathfrak{h}$	$\mathfrak{g}$
$\mathbb{R} + \mathfrak{a}_{n-1}$	$\mathfrak{a}_n, \mathfrak{b}_n, \mathfrak{c}_n, \mathfrak{d}_n, \mathfrak{a}_1 + \mathfrak{a}_{n-1}$
$\mathfrak{d}_n$	$\mathfrak{b}_n$
$\mathfrak{a}_1 + \mathfrak{c}_{n-1}$	$\mathfrak{c}_n$
$\mathbb{R} + \mathfrak{c}_{n-1}$	$\mathfrak{c}_1 + \mathfrak{c}_{n-1}, \mathfrak{c}_n$
$\mathbb{R} + \mathfrak{a}_7$	$\mathbb{R} + \mathfrak{e}_7, \mathfrak{a}_1 + \mathfrak{e}_7$
$\mathbb{R} + \mathfrak{a}_7$	$\mathfrak{e}_8$
$\mathfrak{a}_8$	$\mathfrak{e}_8$
$\mathfrak{d}_8$	$\mathfrak{e}_8$
$\mathbb{R} + \mathfrak{a}_8$	$\mathbb{R} + \mathfrak{e}_8, \mathfrak{a}_1 + \mathfrak{e}_8$
$\mathbb{R} + \mathfrak{a}_6, \mathfrak{a}_7$	$\mathfrak{e}_7$
$\mathfrak{b}_4$	$\mathfrak{f}_4$
$\mathfrak{a}_1 + \mathfrak{c}_3$	$\mathfrak{f}_4$
$\mathfrak{a}_2$	$\mathfrak{g}_2$
$\mathbb{R} + \mathfrak{a}_2$	$\mathbb{R} + \mathfrak{g}_2, \mathfrak{a}_1 + \mathfrak{g}_2$
$2\mathfrak{a}_1$	$\mathfrak{g}_2$
$\mathbb{R} + \mathfrak{a}_1$	$\mathfrak{g}_2$

Table 4.1. Proper rank n extensions $\mathfrak{g}$ of the rank n Lie algebras $\mathfrak{h}$ from Table 2.1

$\mathfrak{g}$	$\mathfrak{h}$	$\mathfrak{k}$	$\mathfrak{h}'$
$\mathfrak{a}_n$	$\mathbb{R}+\mathfrak{a}_{n-1}$	$\mathbb{R}+\mathfrak{a}_{n-2}$	$\mathbb{R}+\mathfrak{a}_1+\mathfrak{a}_{n-2}$
$\mathfrak{b}_n$	$\mathbb{R}+\mathfrak{a}_{n-1}$	$\mathbb{R}+\mathfrak{a}_{n-2}$	$2\mathbb{R}+\mathfrak{a}_{n-2}$
$\mathfrak{c}_n$	$\mathbb{R}+\mathfrak{a}_{n-1}$	$\mathbb{R}+\mathfrak{a}_{n-2}$	$2\mathbb{R}+\mathfrak{a}_{n-2}$
$\mathfrak{d}_n$	$\mathbb{R}+\mathfrak{a}_{n-1}$	$\mathbb{R}+\mathfrak{a}_{n-2}$	$2\mathbb{R}+\mathfrak{a}_{n-2}$
$\mathfrak{b}_n$	$\mathfrak{d}_n$	$\mathfrak{b}_{n-1}$	$\mathbb{R}+\mathfrak{b}_{n-1}$
$\mathfrak{c}_n$	$\mathfrak{a}_1+\mathfrak{c}_{n-1}$	$\mathfrak{a}_1+\mathfrak{c}_{n-2}$	$\mathbb{R}+\mathfrak{a}_1+\mathfrak{c}_{n-2}$
$\mathfrak{c}_n$	$\mathfrak{a}_1+\mathfrak{c}_{n-1}$	$\mathfrak{a}_1+\mathfrak{c}_{n-2}$	$2\mathfrak{a}_1+\mathfrak{c}_{n-2}$
$\mathfrak{c}_n$	$\mathbb{R}+\mathfrak{c}_{n-1}$	$\mathbb{R}+\mathfrak{c}_{n-2}$	$\mathbb{R}+\mathfrak{a}_1+\mathfrak{c}_{n-2}$
$\mathfrak{c}_n$	$\mathbb{R}+\mathfrak{c}_{n-1}$	$\mathbb{R}+\mathfrak{c}_{n-2}$	$2\mathbb{R}+\mathfrak{c}_{n-2}$
$\mathfrak{a}_1+\mathfrak{a}_{n-1}$	$\mathbb{R}+\mathfrak{a}_{n-1}$	$\mathbb{R}+\mathfrak{a}_{n-2}$	$\mathbb{R}+\mathfrak{a}_1+\mathfrak{a}_{n-2}$
$\mathfrak{c}_1+\mathfrak{c}_{n-1}$	$\mathbb{R}+\mathfrak{c}_{n-1}$	$\mathbb{R}+\mathfrak{c}_{n-2}$	$\mathbb{R}+\mathfrak{a}_1+\mathfrak{c}_{n-2}$
$\mathfrak{c}_1+\mathfrak{c}_{n-1}$	$\mathbb{R}+\mathfrak{c}_{n-1}$	$\mathbb{R}+\mathfrak{c}_{n-2}$	$2\mathbb{R}+\mathfrak{c}_{n-2}$
$\mathfrak{e}_8$	$\mathbb{R}+\mathfrak{a}_7$	$\mathbb{R}+\mathfrak{a}_6$	$\mathbb{R}+\mathfrak{a}_1+\mathfrak{a}_6$
$\mathfrak{e}_8$	$\mathfrak{a}_8$	$\mathfrak{a}_7$	$\mathfrak{a}_1+\mathfrak{a}_7$
$\mathfrak{e}_8$	$\mathfrak{a}_8$	$\mathfrak{a}_7$	$\mathbb{R}+\mathfrak{a}_7$
$\mathfrak{e}_7$	$\mathbb{R}+\mathfrak{a}_6$	$\mathbb{R}+\mathfrak{a}_5$	$\mathbb{R}+\mathfrak{a}_1+\mathfrak{a}_5$
$\mathfrak{e}_7$	$\mathbb{R}+\mathfrak{a}_6$	$\mathbb{R}+\mathfrak{a}_5$	$2\mathbb{R}+\mathfrak{a}_5$
$\mathfrak{e}_7$	$\mathfrak{a}_7$	$\mathfrak{a}_6$	$\mathbb{R}+\mathfrak{a}_6$

$\mathfrak{f}_4$	$\mathfrak{a}_1 + \mathfrak{c}_3$	$\mathfrak{a}_1 + \mathfrak{c}_2$	$2\mathfrak{a}_1 + \mathfrak{c}_2$
$\mathfrak{f}_4$	$\mathfrak{a}_1 + \mathfrak{c}_3$	$\mathfrak{a}_1 + \mathfrak{c}_2$	$\mathbb{R} + \mathfrak{a}_1 + \mathfrak{c}_2$
$\mathfrak{g}_2$	$\mathfrak{a}_2$	$\mathfrak{a}_1$	$2\mathfrak{a}_1$
$\mathfrak{g}_2$	$\mathfrak{a}_2$	$\mathfrak{a}_1$	$\mathbb{R} + \mathfrak{a}_1$
$\mathfrak{g}_2$	$\mathbb{R} + \mathfrak{a}_1$	$\mathbb{R}$	$2\mathbb{R}$
$\mathbb{R} + \mathfrak{g}_2$	$\mathbb{R} + \mathfrak{a}_2$	$\mathbb{R} + \mathfrak{a}_1$	$2\mathbb{R} + \mathfrak{a}_1$
$\mathbb{R} + \mathfrak{g}_2$	$\mathbb{R} + \mathfrak{a}_2$	$\mathbb{R} + \mathfrak{a}_1$	$\mathbb{R} + 2\mathfrak{a}_1$

Table 4.2. Non standard positive triples $(\mathfrak{h}, \mathfrak{k}, \mathfrak{h}')$ with $\mathrm{rk}(\mathfrak{h}) = \mathrm{rk}(\mathfrak{h}') = \mathrm{rk}(\mathfrak{g}) = n$

$\mathfrak{g}$	$\mathfrak{h}$	$\mathfrak{k}$	$\mathfrak{h}'$
$\mathfrak{a}_n$	$\mathbb{R} + \mathfrak{a}_{n-1}$	$\mathbb{R} + \mathfrak{a}_{n-2}$	$\mathfrak{a}_1 + \mathfrak{a}_{n-2}$
$\mathfrak{c}_n$	$\mathbb{R} + \mathfrak{c}_{n-1}$	$\mathbb{R} + \mathfrak{c}_{n-2}$	$\mathfrak{c}_1 + \mathfrak{c}_{n-2}$
$\mathfrak{a}_1 + \mathfrak{a}_{n-1}$	$\mathbb{R} + \mathfrak{a}_{n-1}$	$\mathbb{R} + \mathfrak{a}_{n-2}$	$\mathfrak{a}_1 + \mathfrak{a}_{n-2}$
$\mathfrak{c}_1 + \mathfrak{c}_{n-1}$	$\mathbb{R} + \mathfrak{c}_{n-1}$	$\mathbb{R} + \mathfrak{c}_{n-2}$	$\mathfrak{c}_1 + \mathfrak{c}_{n-2}$
$\mathfrak{e}_8$	$\mathbb{R} + \mathfrak{a}_7$	$\mathbb{R} + \mathfrak{a}_6$	$\mathfrak{a}_1 + \mathfrak{a}_6$
$\mathfrak{a}_1 + \mathfrak{e}_7$	$\mathbb{R} + \mathfrak{a}_7$	$\mathbb{R} + \mathfrak{a}_6$	$\mathfrak{a}_1 + \mathfrak{a}_6$
$\mathfrak{e}_7$	$\mathbb{R} + \mathfrak{a}_6$	$\mathbb{R} + \mathfrak{a}_5$	$\mathfrak{a}_1 + \mathfrak{a}_5$
$\mathfrak{g}_2$	$\mathbb{R} + \mathfrak{a}_1$	$\mathbb{R}$	$\mathfrak{a}_1$

Table 4.3. Type 1 positive triples $(\mathfrak{h}, \mathfrak{k}, \mathfrak{h}')$ of subalgebras of a compact rank n Lie algebra $\mathfrak{g}$ with $\mathrm{rk}(\mathfrak{h}') = \mathrm{rk}(\mathfrak{k}) = n\text{-}1 = \mathrm{rk}(\mathfrak{g})\text{-}1$ and $\mathfrak{h}' \neq \mathfrak{k}$

5. Classification of cohomogeneity one manifolds of type 2

Now we classify type 2 positive triples $(\mathfrak{h}, \mathfrak{k}, \mathfrak{h}')$ of subalgebras of a rank n compact Lie algebra $\mathfrak{g}$. Without loss of generality we will assume that $\mathrm{rk}\,\mathfrak{k} = \mathrm{rk}\,\mathfrak{h}' = n-1 = \mathrm{rk}\,\mathfrak{h} - 1$. By definition, $(\mathfrak{h}', \mathfrak{k})$ is an effective admissible pair of Lie algebras of the same rank. Table 2.1 shows that such pair is either $(\mathfrak{b}_{n-1}, \mathfrak{d}_{n-1})$ or $(\mathfrak{g}_2, \mathfrak{a}_2)$. Since the Lie algebras $\mathfrak{k} = \mathfrak{d}_{n-1}, \mathfrak{a}_2$ are simple, $\mathfrak{k} = \mathfrak{n} = \ker j$ and $\mathfrak{h} = \mathbb{R} + \mathfrak{k}$ or $\mathfrak{a}_1 + \mathfrak{k}$. Using Proposition 3.1.1., we enumerate all extensions $\mathfrak{g} = \mathfrak{h}^{\uparrow}$ of $\mathfrak{h}$ as follows :

(1) $(\mathbb{R} + \mathfrak{d}_{n-1})^{\uparrow} = \mathfrak{b}_n, \mathfrak{d}_n, \mathfrak{a}_1 + \mathfrak{d}_{n-1}, \mathfrak{a}_1 + \mathfrak{b}_{n-1}$

(2) $(\mathbb{R} + \mathfrak{d}_5)^{\uparrow} = \mathfrak{e}_6, \quad (\mathbb{R} + \mathfrak{d}_6)^{\uparrow} = \mathfrak{a}_1 + \mathfrak{d}_6, \mathfrak{e}_7, \quad (\mathfrak{a}_1 + \mathfrak{d}_6)^{\uparrow} = \mathfrak{e}_7$

(3) $(\mathbb{R} + \mathfrak{a}_2)^{\uparrow} = \mathfrak{a}_3, \mathfrak{b}_3, \mathfrak{c}_3, \mathfrak{d}_3, \mathfrak{a}_1 + \mathfrak{a}_2, \mathbb{R} + \mathfrak{g}_2, \mathfrak{a}_1 + \mathfrak{g}_2, (\mathfrak{a}_1 + \mathfrak{a}_2)^{\uparrow} = \mathfrak{a}_1 + \mathfrak{g}_2$.

Note that $\mathfrak{a}_1 + \mathfrak{d}_{n-1}$ has no extension for $n \neq 7$.Moreover, in case (1), the Lie algebra $\mathfrak{a}_1 + \mathfrak{d}_{n-1}$ does not contain $\mathfrak{b}_{n-1}$ as subalgebra and therefore has to be discarded.

Case (2) gives no positive triple, because the Lie algebra $\mathfrak{e}_6$ has no subalgebra $\mathfrak{h}' = \mathfrak{b}_5$ and $\mathfrak{a}_1 + \mathfrak{d}_6, \mathfrak{e}_7$ have no subalgebra $\mathfrak{h}' = \mathfrak{b}_6$, see [GOV], ch. 6.

In the case (3), the Lie algebras $\mathfrak{g} = \mathfrak{a}_3, \mathfrak{c}_3, \mathfrak{d}_3, \mathfrak{a}_1 + \mathfrak{a}_2$ are not admissible, because they have no subalgebra $\mathfrak{h}' = \mathfrak{g}_2$. All other Lie algebras $\mathfrak{g}$ contain appropriate triples $(\mathfrak{h}, \mathfrak{k}, \mathfrak{h}')$ of subalgebras. The final result of the classification is given in Table 5.1.

$\mathfrak{g}$	$\mathfrak{h}$	$\mathfrak{k}$	$\mathfrak{h}'$
$\mathfrak{b}_n$	$\mathbb{R} + \mathfrak{d}_{n-1}$	$\mathfrak{d}_{n-1}$	$\mathfrak{b}_{n-1}$
$\mathfrak{d}_n$	$\mathbb{R} + \mathfrak{d}_{n-1}$	$\mathfrak{d}_{n-1}$	$\mathfrak{b}_{n-1}$
$\mathfrak{b}_3$	$\mathbb{R} + \mathfrak{a}_2$	$\mathfrak{a}_2$	$\mathfrak{g}_2$
$\mathfrak{a}_1 + \mathfrak{g}_2$	$\mathbb{R} + \mathfrak{a}_2$	$\mathfrak{a}_2$	$\mathfrak{g}_2$
$\mathbb{R} + \mathfrak{g}_2$	$\mathbb{R} + \mathfrak{a}_2$	$\mathfrak{a}_2$	$\mathfrak{g}_2$
$\mathfrak{a}_1 + \mathfrak{b}_{n-1}$	$\mathbb{R} + \mathfrak{d}_{n-1}$	$\mathfrak{d}_{n-1}$	$\mathfrak{b}_{n-1}$

Table 5.1. Type 2 positive triples $(\mathfrak{h}, \mathfrak{k}, \mathfrak{h}')$ of subalgebras of a compact rank n Lie algebra $\mathfrak{g}$ with $\mathrm{rk}(\mathfrak{h}') = \mathrm{rk}(\mathfrak{k}) = n-1 = \mathrm{rk}(\mathfrak{g})-1$

6. Cohomogeneity one quaternionic Kähler G-manifolds

In this section (M, g) will be a $4n$-dimensional quaternionic-Kähler manifold, which is by definition a Riemannian manifold whose holonomy group is a subgroup of $Sp(1) \cdot Sp(n)$. We will assume that $n > 1$ throughout the following.

Quaternionic-Kähler manifolds, for brevity qk-manifolds, are automatically Einstein (see e.g. [Be]) and we will assume that M is compact with positive scalar curvature (for brevity a positive qk-manifold). It is well known that a *homogeneous* positive qk-manifold is symmetric (see [A1]), that is a Wolf space. For each compact simple Lie algebra (except $\mathfrak{su}_2$) there is exactly one compact symmetric qk-manifold, called Wolf space (see [W]) and they are given in the following table.

$$\mathbb{H}P^n = Sp(n+1)/Sp(1) \cdot Sp(n), \quad Gr_2^+(\mathbb{C}^{n+2}) = SU(n+2)/S(U(2) \times U(n)),$$

$$Gr_4^+(\mathbb{R}^{n+4}) = SO(n+4)/SO(4) \times SO(n), \quad G_2/SO(4),$$

$$F_4/Sp(1) \cdot Sp(3), \quad E_6/Sp(1) \cdot SU(6), \quad E_7/Sp(1) \cdot Spin(12), \quad E_8/Sp(1) \cdot E_7.$$

Moreover, it is known (see [PoSa]) that any 8-dimensional positive qk-manifold is symmetric. We summarize some fundamental results on positive qk-manifolds in the following Theorem, which will turn out to be essential in our classification.

Theorem 6.1 ([S],[LBS]). *Let (M, g) be a positive $4n$-dimensional qk-manifold.*

(1) M is simply connected and has vanishing odd Betti numbers. In particular M has positive Euler characteristic.

(2) The fundamental class of Marchiafava–Romani $\xi(M) \in H^2(M, \mathbb{Z}_2)$ vanishes iff (M, g) is homothetic to a quaternionic projective space with its standard metric.

(3) If $b_2 > 0$, then, up to homothety, $M = Gr_2(\mathbb{C}^{n+2})$ with its symmetric metric.

(4) The second homotopy group $\pi_2(M)$ is 0 iff M is $\mathbb{H}P^n$ and $\pi_2(M) \cong \mathbb{Z}$ iff M is $Gr_2(\mathbb{C}^{n+2})$.

Our aim will be to classify positive qk-manifolds which admit an isometric action of an compact, connected Lie group G with a codimension one orbit. Note that the associated triple of subgroups (H, K, H') is positive in view of Theorem 6.1 (1) and Corollary 1.2.2.

As a first corollary of Theorem 6.1, (1), we get the following

Lemma 6.2. *Let (M, g) be a cohomogeneity one positive qk-G- manifold. Then there are exactly two singular orbits of G and none of them is exceptional.*

Proof. Indeed, the first statement follows from the fact the $\chi(M) > 0$. The second statement follows from simply connectedness of M (see e.g. [Bre]). □

According to the results in Section 1.2, we associate to a positive qk-G-manifold (M, g) a positive triple of subgroups (H, K, H') of G and, correspondingly, a positive triple of subalgebras $(\mathfrak{h}, \mathfrak{k}, \mathfrak{h}')$ of the Lie algebra $\mathfrak{g}$ of G. We first recall the following well-known fact about transformation groups of quaternionic Kähler manifolds (see e.g.[A1]).

Lemma 6.3. *Let (M, g) be a $4n$-dimensional, compact qk-manifold which is acted on by a Lie group of isometries G. If $p \in M$ is any point and if G_p denotes the stability subgroup at p, then the isotropy representation $\iota : G_p \to O(T_pM)$ takes values into $Sp_1 \cdot Sp_n$. In particular, if G_p is simple of dimension bigger than 3, $\iota(G_p) \subset Sp_n$.*

The previous lemma gives a strong constraint on the (positive) triple of subgroups (H, K, H'), since the isotropy representations of H, K, H' must be contained in $Sp_1 \cdot Sp_n$. In the following sections, we will examine the types 0,1 and 2 separately, using the classification we obtained in Sections 2, 4 and 5.

7. Cohomogeneity one quaternionic Kähler G-manifolds with a fixed point

In this section we will consider a $4n$-dimensional ($n > 1$) positive qk-manifold (M, g) of cohomogeneity one under the action of a connected compact Lie group G with a fixed point $p \in M$ and we want to prove the following

Theorem 7.1. *Let (M, g) be a $4n$-dimensional ($n > 1$) positive qk-manifold which is of cohomogeneity one under the action of a compact Lie group G of isometries with a fixed point. Then (M, g) is homothetic to the projective space $\mathbb{H}P^n$ with its symmetric metric.*

The proof of Theorem 7.1 will be carried out through several lemmata.

We start noting that, if $\theta = (H, K, H')$ (resp., $(\mathfrak{h}, \mathfrak{k}, \mathfrak{h}')$) is the positive triple of subgroups (resp., subalgebras) of G (resp., $\mathfrak{g}$) associated to the G-manifold M, then $H = G$ (resp.,$\mathfrak{h} = \mathfrak{g}$) and the triple is of type 0. First of all we rule out the possibility $G = H'$.

Lemma 7.2. *Let (M, g) be a $4n$-dimensional ($n > 1$) positive qk-manifold which is of cohomogeneity one under the action of a compact Lie group G of isometries with a fixed point. If (H, K, H') is the associated triple of subgroups of G, then H' is a proper subgroup of G.*

Proof. Indeed if $G = H'$, then the manifold should be diffeomorphic to a sphere. Since positive qk-manifolds have $b_4 \neq 0$, we should have $n = 1$. □

We may therefore use the classification Table 2.2; in view of Lemma 6.3, we may discard those triples of subalgebras $(\mathfrak{h}, \mathfrak{k}, \mathfrak{h}')$ from Table 2.2, for which the corresponding isotropy representation is not contained into $\mathfrak{sp}_1 + \mathfrak{sp}_n$.

Lemma 7.3. *The positive triples of type* 0 $(\mathfrak{h}, \mathfrak{k}, \mathfrak{h}')$ *from Table* 2.2 *for which the isotropy representation of* $\mathfrak{h}, \mathfrak{k}, \mathfrak{h}'$ *is contained into* $\mathfrak{sp}_1 + \mathfrak{sp}_n$ *are given in the following table:*

$n.$	$\mathfrak{h}$	$\mathfrak{k}$	$\mathfrak{h}'$
1	$\mathfrak{c}_n$	$\mathfrak{c}_{n-1}$	$\mathfrak{a}_1 + \mathfrak{c}_{n-1}$
2	$\mathfrak{c}_n$	$\mathfrak{c}_{n-1}$	$\mathbb{R} + \mathfrak{c}_{n-1}$
3	$\mathfrak{a}_1 + \mathfrak{c}_n$	$\mathfrak{a}_1 + \mathfrak{c}_{n-1}$	$2\mathfrak{a}_1 + \mathfrak{c}_{n-1}$
4	$\mathbb{R} + \mathfrak{c}_n$	$\mathbb{R} + \mathfrak{c}_{n-1}$	$2\mathbb{R} + \mathfrak{c}_{n-1}$
5	$\mathbb{R} + \mathfrak{c}_n$	$\mathbb{R} + \mathfrak{c}_{n-1}$	$\mathbb{R} + \mathfrak{a}_1 + \mathfrak{c}_{n-1}$

The corresponding triples of subgroups (H, K, H') *are given by the following:*

$n.$	H	K	H'
1	Sp_n	Sp_{n-1}	$Sp_1 \cdot Sp_{n-1}$
2	Sp_n	Sp_{n-1}	$T^1 \cdot Sp_{n-1}$
3	$Sp_1 \cdot Sp_n$	$Sp_1 \cdot Sp_{n-1}$	$Sp_1 \cdot Sp_1 \cdot Sp_{n-1}$
4	$T^1 \cdot Sp_n$	$T^1 \cdot Sp_{n-1}$	$T^1 \cdot T^1 \cdot Sp_{n-1}$
5	$T^1 \cdot Sp_n$	$T^1 \cdot Sp_{n-1}$	$T^1 \cdot Sp_1 \cdot Sp_{n-1}$

Proof. It is easy to see that all the triples of subalgebras $(\mathfrak{h}, \mathfrak{k}, \mathfrak{h}')$ given above have the claimed property; on the other hand, by direct inspection of Table 2.2, the only triples of subalgebras which we have to discard are $(\mathfrak{a}_n, \mathfrak{a}_{n-1}, \mathbb{R} + \mathfrak{a}_{n-1})$ and $(\mathbb{R} + \mathfrak{a}_n, \mathbb{R} + \mathfrak{a}_{n-1}, 2\mathbb{R} + \mathfrak{a}_{n-1})$; these are easily ruled out, since the algebras $\mathfrak{h} = \mathfrak{a}_n$ and $\mathfrak{h} = \mathbb{R} + \mathfrak{a}_n$ acting on $V = \mathbb{C}^{n+1}$ do not normalize any quaternionic structure on V.

The second statement about the triples of subgroups is immediate, if we prove that H' is connected. But this follows from the fact that $G = H$ and K are connected and H'/K is a sphere. □

We now use some topological arguments in order to characterize the manifold (M, g) according to its associated triple. We have the following result.

Lemma 7.4. *Let (M, g) be a $4n$-dimensional $(n > 1)$ positive qk-manifold which is of cohomogeneity one under the action of a compact Lie group G of isometries with a fixed point. The associated triple of subgroups (H, K, H') of G with $G = H$ is one of the following:*

H	K	H'
Sp_n	Sp_{n-1}	$Sp_1 \cdot Sp_{n-1}$
$Sp_1 \cdot Sp_n$	$Sp_1 \cdot Sp_{n-1}$	$Sp_1 \cdot Sp_1 \cdot Sp_{n-1}$
$T^1 \cdot Sp_n$	$T^1 \cdot Sp_{n-1}$	$T^1 \cdot Sp_1 \cdot Sp_{n-1}$

Proof. We have to rule out the triples (2) and (4) from the table in Lemma 7.3. Indeed, in these cases, one singular orbit is a point and the other one is a copy of a complex projective space $\mathbb{C}P^{2n-1}$. We may use the Mayer–Vietoris sequence (1.2.1), noting that regular orbits are spheres and the second Betti number $b_2(\mathbb{C}P^m) = 1$. We get that $b_2(M) = 1$ and, by Theorem 6.1 (3), the manifold (M, g) should be homothetic to the symmetric space $Gr_2(\mathbb{C}^{n+2})$, hence

$$\chi(M) = \chi(Gr_2(\mathbb{C}^{n+2})) = \frac{1}{2}(n+2)(n+1).$$

But, by Prop. 1.2.1, we have $\chi(M) = 2n$ and this gives a contradiction for $n \geq 2$. □

The proof of the Theorem will be complete if we show that each of the triples enumerated in Lemma 7.4 is associated to the quaternionic projective space.

Lemma 7.5.

(1) *For any triple $\theta = (H, K, H')$ enumerated in Lemma 7.4, there exists an isometric action of the group $G = H$ on the quaternionic projective space $\mathbb{H}P^n$ which is of cohomogeneity one and has θ as associated triple.*

(2) *Let (M, g) be a $4n$-dimensional $(n > 1)$ positive qk-manifold which is of cohomogeneity one under the action of a compact Lie group G of isometries with associated triple of subgroups $(G = H, K, H')$ given in the table of Lemma 7.4;*

then (M, g) is homothetic to the quaternionic projective space $\mathbb{H}P^n$ with its symmetric metric.

Proof. (1) is easy to be verified, if we identify the group $H = G$ with a subgroup of Sp_{n+1}, acting by isometries on $\mathbb{H}P^n$.

(2) In these cases, one singular orbit is a point and the other one is a copy of $\mathbb{H}P^{n-1}$. We may use the Mayer–Vietoris exact sequence (1.2.1) with coefficient group $\mathbb{Z}_2$, using the fact that $H^2(\mathbb{H}P^{n-1}, \mathbb{Z}_2) = \{0\}$ and regular orbits are spheres. We get that $H^2(M, \mathbb{Z}_2) = \{0\}$ and it then follows that the fundamental class $\xi(M)$ vanishes; by Theorem 6.1 (2), we get our claim. □

8. Cohomogeneity one quaternionic Kähler G-manifolds of type 1

In this section we will examine the case of a cohomogeneity one, positive qk-manifold (M, g) which is acted on isometrically by a connected compact Lie group G with cohomogeneity one and associated triple of subalgebras $\theta = (\mathfrak{h}, \mathfrak{k}, \mathfrak{h}')$ of type 1. This means that the normal isotropy representation $j : \mathfrak{h} \to \mathfrak{o}(V)$ is exact and $\mathrm{rk}(\mathfrak{h}) = \mathrm{rk}(\mathfrak{g})$. Throughout the following we will refer to the discussion of type 1 positive triples given in §4. In particular we will prove the following Theorem

Theorem 8.1. *Let (M, g) be a positive qk-manifold which is of cohomogeneity one under the (almost) effective action of a connected, compact Lie group G with associated triple of subalgebras of type* 1. *Then G is simple.*

If G is not exceptional and $\dim M > 8$, *then (M, g) is homothetic to $Gr_2(\mathbb{C}^{n+2})$ or to $\mathbb{H}P^n$, endowed with their symmetric metrics.*

If G is exceptional, then its Lie algebra is $\mathfrak{g}_2$ or $\mathfrak{f}_4$. If $G = G_2$, then M is diffeomorphic to the Wolf space $SO_7 / SO_4 \times SO_3$.

Remark. The authors do not know whether the case $\mathfrak{g} = \mathfrak{f}_4$ is realizable.

The proof will be carried out through several lemmata and propositions.

We start recalling that, in Table 4.1, we classified all pairs $(\mathfrak{h}, \mathfrak{g})$ of Lie algebras with $\mathfrak{h} \subset \mathfrak{g}$, $\mathrm{rk}(\mathfrak{h}) = \mathrm{rk}(\mathfrak{g})$ and $\mathfrak{h}$ Lie algebra of a compact Lie group H acting transitively on the unit sphere S of a vector space V.

In view of Lemma 6.3, the isotropy representation of H must normalize a quaternionic structure; in terms of Lie algebra, this means that, if we choose an $\mathrm{Ad}(H)$-invariant subspace $\mathfrak{m}$ of $\mathfrak{g}$ with $\mathfrak{g} = \mathfrak{h} \oplus \mathfrak{m}$, the isotropy representation ι given by

$$ad|_{\mathfrak{m}} + j : \mathfrak{h} \to \mathfrak{o}(\mathfrak{m} \oplus V),$$

must normalize some quaternionic structure $\mathfrak{q}$ on $\mathfrak{m} \oplus V$. We now prove the following algebraic lemma.

Lemma 8.2. *Let* $(\mathfrak{h}, \mathfrak{g})$ *be a pair of compact Lie algebras with* $\mathfrak{h} \subset \mathfrak{g}$, *rk*$(\mathfrak{h})$ = *rk*$(\mathfrak{g})$ *and* $\mathfrak{h}$ *Lie algebra of a compact Lie group H acting transitively on the unit sphere S of a vector space V, via an exact representation* $j : \mathfrak{h} \to \mathfrak{o}(V)$. *For each* $(\mathfrak{h}, \mathfrak{g})$, *we choose an ad*$(\mathfrak{h})$*-invariant subspace* $\mathfrak{m}$ *of* $\mathfrak{g}$ *with* $\mathfrak{g} = \mathfrak{h} \oplus \mathfrak{m}$. *The pairs* $(\mathfrak{h}, \mathfrak{g})$ *for which the representation* $ad|_{\mathfrak{m}} + j$ *normalizes some quaternionic structure* $\mathfrak{q}$ *on* $\mathfrak{m} \oplus V$ *are given in the following table:*

$\mathfrak{h}$	$\mathfrak{g}$
$\mathbb{R} + \mathfrak{a}_{n-1}$	$\mathfrak{a}_n$
$\mathfrak{a}_1 + \mathfrak{c}_{n-1}$	$\mathfrak{c}_n$
$\mathfrak{a}_1 + \mathfrak{c}_3$	$\mathfrak{f}_4$
$\mathfrak{a}_2$	$\mathfrak{g}_2$
$2\mathfrak{a}_1$	$\mathfrak{g}_2$
$\mathbb{R} + \mathfrak{a}_2$	$\mathbb{R} + \mathfrak{g}_2$

Proof. First of all it is easy to see that all the pairs $(\mathfrak{h}, \mathfrak{g})$ enumerated in the table have the desidered property. In order to prove the lemma, we need to consider all the pairs $(\mathfrak{h}, \mathfrak{g})$ in Table 4.1 and rule out the following ($n \geq 2$)

(1) $(\mathbb{R} + \mathfrak{c}_{n-1}, \mathfrak{c}_1 + \mathfrak{c}_{n-1})$;

(2) $(\mathbb{R} + \mathfrak{c}_{n-1}, \mathfrak{c}_n)$;

(3) $(\mathbb{R} + \mathfrak{a}_7, \mathfrak{a}_1 + \mathfrak{e}_7)$, $(\mathbb{R} + \mathfrak{a}_7, \mathbb{R} + \mathfrak{e}_7)$;

(4) $(\mathfrak{a}_8, \mathfrak{e}_8)$;

(5) $(\mathbb{R} + \mathfrak{a}_8, \mathfrak{a}_1 + \mathfrak{e}_8)$, $(\mathbb{R} + \mathfrak{a}_8, \mathbb{R} + \mathfrak{e}_8)$;

(6) $(\mathbb{R} + \mathfrak{a}_6, \mathfrak{e}_7)$;

(7) $(\mathfrak{b}_4, \mathfrak{f}_4)$;

(8) $(\mathfrak{a}_7, \mathfrak{e}_7)$;

(9) $(\mathbb{R} + \mathfrak{a}_1, \mathfrak{g}_2)$;

(10) $(\mathbb{R} + \mathfrak{a}_2, \mathfrak{a}_1 + \mathfrak{g}_2)$;

(11) $(\mathbb{R} + \mathfrak{a}_{n-1}, \mathfrak{c}_n)$;

(12) $(\mathbb{R} + \mathfrak{a}_{n-1}, \mathfrak{d}_n)$ for $n \geq 4$;

(13) $(\mathbb{R} + \mathfrak{a}_{n-1}, \mathfrak{b}_n)$;

(14) $(\mathfrak{d}_n, \mathfrak{b}_n)$;

(15) $(\mathfrak{d}_8, \mathfrak{e}_8)$;

(16) $(\mathbb{R} + \mathfrak{a}_7, \mathfrak{e}_8)$.

This means that we have to show that the representation ι associated to each pair (1)–(16) does not normalize any quaternionic structure $\mathfrak{q}$ on $\mathfrak{m} \oplus V$.

Cases (1)–(10) are easily ruled out since $\dim(\mathfrak{m} \oplus V)$ is not a multiple of 4.

Case (11). The pair $(\mathbb{R} + \mathfrak{a}_{n-1}, \mathfrak{c}_n)$ is an irreducible symmetric pair with $\dim \mathfrak{m} = n^2 + n$ and $\dim V = 2n$. Let us suppose that we have a quaternionic structure $\mathfrak{q}$ on $V \oplus \mathfrak{m}$ which is normalized by $\iota(\mathfrak{h})$. We note that the $\iota(\mathfrak{a}_{n-1})$-modules V and $\mathfrak{m}$ are irreducible. Now, the simple Lie algebra $\mathfrak{a}_{n-1} \neq \mathfrak{a}_1$ since $k = n^2 + 3n$ must be a multiple of 4, hence $n \geq 3$; so by Lemma 6.3, $\iota(\mathfrak{a}_{n-1})$ centralizes $\mathfrak{q}$. But this is not possible, since V and $\mathfrak{m}$ are (complex) irreducible representations of $\mathfrak{a}_{n-1}$ which are not quaternionic and not equivalent.

Case (12) The same argument as in (1) works.

Case (13) In this case $\mathbb{R} + \mathfrak{a}_{n-1} \subset \mathfrak{d}_n$ and the space $\mathfrak{m}$ splits as a sum of irreducible $\iota(h)$-submodules

$$\mathfrak{m} = \mathfrak{m}_1 + \mathfrak{m}_2,$$

where $\mathfrak{m}_1$ is equivalent to V, while $\mathfrak{m}_2 \cong \Lambda^2(\mathbb{C}^n)$. At this point we note that the each summand $\mathfrak{m}_1$ and $\mathfrak{m}_2$ remain irreducible if we restrict the representation to $\mathfrak{a}_{n-1}$. Since $n \neq 2$ by dimensional reasons, we have that $\iota(\mathfrak{a}_{n-1})$ must centralize the quaternionic structure $\mathfrak{q}$; but this is not possible, since $\mathfrak{m}_2$ is not a quaternionic (irreducible) representation of $\mathfrak{a}_{n-1}$ and it appears with multiplicity one.

Case (14) In this case the representation ι of $\mathfrak{d}_n$ is equivalent to $2\rho_n$, where ρ_n denotes the standard representation of $\mathfrak{d}_n$ on $\mathbb{R}^{2n}$, and this is not contained in $sp_1 + sp_n$.

Case (15) The pair $(\mathfrak{d}_8, \mathfrak{e}_8)$ is an irreducible symmetric pair with $\dim \mathfrak{m} = 128$ and $\dim V = 16$. Since $(\mathfrak{d}_8)$ is simple, $\iota(\mathfrak{d}_8)$ should centralize the quaternionic structure; but this is impossible, since V and $\mathfrak{m}$ are not equivalent.

Case (16) We need to distinguish three different imbeddings of $\mathbb{R} + \mathfrak{a}_7$ into $\mathfrak{e}_8$, namely

(A) $\mathbb{R} + \mathfrak{a}_7 \subset \mathfrak{a}_8 \subset \mathfrak{e}_8$;

(B) $\mathbb{R} + \mathfrak{a}_7 \subset \mathfrak{d}_8 \subset \mathfrak{e}_8$;

(C) $\mathbb{R} + \mathfrak{a}_7 \subset \mathfrak{a}_1 + \mathfrak{a}_7 \subset a_1 + \mathfrak{e}_7 \subset \mathfrak{e}_8$.

We will now consider these three subcases separately.

Subcase (A). We write $\mathfrak{a}_8 = \mathbb{R} + \mathfrak{a}_7 + \mathfrak{n}$ with $\mathfrak{n}$ an $\mathrm{ad}(\mathbb{R} + \mathfrak{a}_7)$-invariant subspace of $\mathfrak{a}_8$ and $\mathfrak{e}_8 = \mathfrak{a}_8 + \mathfrak{p}$, with $\mathfrak{p}$ and $\mathrm{ad}(\mathfrak{a}_8)$-invariant subspace of $\mathfrak{e}_8$, so that $\mathfrak{m} = \mathfrak{n} + \mathfrak{p}$. Moreover, we have that $\mathfrak{n} \cong V$ and $\mathfrak{p}$ is an $\mathrm{ad}(\mathfrak{a}_8)$-irreducible module, corresponding to the fundamental representation $\Lambda^3(\mathbb{C}^9)$ (see e.g. [GOV]). Now, we decompose $\mathfrak{p}$ into irreducible $\mathrm{ad}(\mathfrak{a}_7)$-submodules as

$$\mathfrak{p} = \Lambda^3(\mathbb{C}^8) \oplus \Lambda^2(\mathbb{C}^8),$$

so that

$$V + \mathfrak{m} = 2V + \Lambda^3(\mathbb{C}^8) + \Lambda^2(\mathbb{C}^8).$$

But since $\Lambda^3(\mathbb{C}^8)$ is an irreducible $\mathfrak{a}_7$-submodule which is not quaternionic and which appears with multiplicity one, there is no quaternionic structure on $V + \mathfrak{m}$ which is centralized by $\mathfrak{a}_7$.

Subcase (B). We write $\mathfrak{d}_8 = \mathbb{R} + \mathfrak{a}_7 + \mathfrak{n}$ with $\mathfrak{n}$ subspace of $\mathfrak{d}_8$ and $\mathrm{ad}(\mathbb{R} + \mathfrak{a}_7)$-invariant; similarly, we decompose $\mathfrak{e}_8 = \mathfrak{d}_8 + \mathfrak{p}$, with $\mathfrak{p}$ subspace of $\mathfrak{e}_8$ and $\mathrm{ad}(\mathfrak{d}_8)$-invariant.

We have that $\mathfrak{n} \cong \Lambda^2(\mathbb{C}^8)$ and is $\mathrm{ad}(\mathfrak{a}_7)$-irreducible, while $\dim \mathfrak{p} = 128$ and $\mathrm{ad}(\mathfrak{d}_8)|_{\mathfrak{p}}$ is a spin representation. We have

$$V + \mathfrak{m} = V + \mathfrak{n} + \mathfrak{p}.$$

Now, we suppose that there exists a quaternionic structure $\mathfrak{q}$ on $V + \mathfrak{m}$ which is centralized by $\mathfrak{a}_7$. Since V and $\mathfrak{n}$ are not quaternionic, the $\mathfrak{a}_7$-module $\mathfrak{p}$ must contain at least one copy of V and $\mathfrak{n}$, so that we should write

$$\mathfrak{p} = V + \mathfrak{n} + \mathfrak{p}',$$

with $\dim \mathfrak{p}' = 56$. Since the irreducible (unitary) representations of $\mathfrak{a}_7$ of real dimension less or equal to 56 have dimensions 16 or 56 and none of them is quaternionic, we should have that the a_7-module $\mathfrak{p}'$ contains a trivial submodule of dimension at least 24; but this is excluded by looking at the restriction of the spin representation of $\mathfrak{so}(16)$ to $\mathfrak{su}(8)$.

Subcase (C). We write $\mathfrak{a}_1 + \mathfrak{a}_7 = \mathbb{R} + \mathfrak{a}_7 + \mathfrak{p}_o$ where $\mathfrak{p}_o$ is a trivial $\mathfrak{a}_7$-module of dimension 2. Moreover, we may write $\mathfrak{a}_1 + \mathfrak{e}_7 = \mathfrak{a}_1 + \mathfrak{a}_7 + \mathfrak{p}_1$, with $\mathfrak{p}_1$ an $\mathrm{ad}(\mathfrak{a}_1 + \mathfrak{a}_7)$-invariant subspace. Now, the pair $(\mathfrak{e}_7, \mathfrak{a}_7)$ is a symmetric irreducible pair, so that $\mathfrak{p}_1$ is an irreducible $\mathrm{ad}(\mathfrak{a}_7)$-module of real dimension 70. We now decompose $\mathfrak{e}_8 = (\mathfrak{a}_1 + \mathfrak{e}_7) + \mathfrak{p}_2$, with $\mathfrak{p}_2$ an $\mathrm{ad}(\mathfrak{a}_1 + \mathfrak{e}_7)$-invariant subspace of real dimension 112. Note that the pair $(\mathfrak{e}_8, \mathfrak{a}_1 + \mathfrak{e}_7)$ is an irreducible symmetric pair which corresponds to the Wolf space $E_8/Sp_1 \cdot E_7$; hence $\mathfrak{p}_2$ is $\mathrm{ad}(\mathfrak{a}_1 + \mathfrak{e}_7)$-irreducible and admits a quaternionic structure which is centralized by $\mathfrak{e}_7$.

Now we have

$$V + \mathfrak{m} = V + \mathfrak{p}_o + \mathfrak{p}_1 + \mathfrak{p}_2.$$

Suppose there exists a quaternionic structure $\mathfrak{q}$ on $V + \mathfrak{m}$ which is centralized by $\mathfrak{a}_7$. Then the $\mathfrak{a}_7$-module $\mathfrak{p}_2$ should contain at least one copy of V, $\mathfrak{p}_o$, $\mathfrak{p}_1$. But the action of $\mathfrak{a}_7$ on $\mathfrak{p}_2$ is quaternionic, while the action on $\mathfrak{p}_1$ is not quaternionic, so that $\mathfrak{p}_2$ should contain at least two copies of $\mathfrak{p}_1$. But $2 \dim \mathfrak{p}_1 = 140 > \dim \mathfrak{p}_2 = 112$ and this gives a contradiction. □

Now, for each pair $(\mathfrak{h}, \mathfrak{g})$ enumerated in Lemma 8.2, we find all possible triples $(\mathfrak{h}, \mathfrak{k}, \mathfrak{h}')$ of Lie subalgebras of $\mathfrak{g}$. We recall that in §4 we devided the classification into three subtypes; if we exclude the existence of exceptional orbits (see Lemma 6.2), we have triples of type 1B, namely with $\mathrm{rk}(\mathfrak{h}') = \mathrm{rk}(\mathfrak{g})$ and of type 1C, i.e. with $\mathrm{rk}(\mathfrak{h}') = \mathrm{rk}(\mathfrak{k}) = \mathrm{rk}(\mathfrak{g})$-1.

Using the Tables 4.2 and 4.3, we list here all possible triples $(\mathfrak{h}, \mathfrak{k}, \mathfrak{h}')$ of type 1 of Lie subalgebras of $\mathfrak{g}$ with $(\mathfrak{h}, \mathfrak{g})$ as in Lemma 8.2. Besides standard triples $(\mathfrak{h}, \mathfrak{k}, \mathfrak{h}')$ with $\mathfrak{h}'$ isomorphic to $\mathfrak{h}$, we have present Table 8.1.

$\mathfrak{g}$	$\mathfrak{h}$	$\mathfrak{k}$	$\mathfrak{h}'$
$\mathfrak{a}_n$	$\mathbb{R} + \mathfrak{a}_{n-1}$	$\mathbb{R} + \mathfrak{a}_{n-2}$	$\mathbb{R} + \mathfrak{a}_1 + \mathfrak{a}_{n-2}$
$\mathfrak{a}_n$	$\mathbb{R} + \mathfrak{a}_{n-1}$	$\mathbb{R} + \mathfrak{a}_{n-2}$	$\mathfrak{a}_1 + \mathfrak{a}_{n-2}$
$\mathfrak{c}_n$	$\mathfrak{a}_1 + \mathfrak{c}_{n-1}$	$\mathfrak{a}_1 + \mathfrak{c}_{n-2}$	$\mathbb{R} + \mathfrak{a}_1 + \mathfrak{c}_{n-2}$
$\mathfrak{c}_n$	$\mathfrak{a}_1 + \mathfrak{c}_{n-1}$	$\mathfrak{a}_1 + \mathfrak{c}_{n-2}$	$2\mathfrak{a}_1 + \mathfrak{c}_{n-2}$
$\mathfrak{c}_n$	$\mathfrak{a}_1 + \mathfrak{c}_{n-1}$	$\mathfrak{a}_1 + \mathfrak{c}_{n-2}$	$\mathfrak{c}_{n-1}$
$\mathfrak{f}_4$	$\mathfrak{a}_1 + \mathfrak{c}_3$	$\mathfrak{a}_1 + \mathfrak{c}_2$	$2\mathfrak{a}_1 + \mathfrak{c}_2$
$\mathfrak{f}_4$	$\mathfrak{a}_1 + \mathfrak{c}_3$	$\mathfrak{a}_1 + \mathfrak{c}_2$	$\mathbb{R} + \mathfrak{a}_1 + \mathfrak{c}_2$
$\mathfrak{g}_2$	$\mathfrak{a}_2$	$\mathfrak{a}_1$	$2\mathfrak{a}_1$
$\mathfrak{g}_2$	$\mathfrak{a}_2$	$\mathfrak{a}_1$	$\mathbb{R} + \mathfrak{a}_1$
$\mathbb{R} + \mathfrak{g}_2$	$\mathbb{R} + \mathfrak{a}_2$	$\mathbb{R} + \mathfrak{a}_1$	$2\mathbb{R} + \mathfrak{a}_1$
$\mathbb{R} + \mathfrak{g}_2$	$\mathbb{R} + \mathfrak{a}_2$	$\mathbb{R} + \mathfrak{a}_1$	$\mathbb{R} + 2\mathfrak{a}_1$

Table 8.1.

We now proceed by analyzing the triples in Table 8.1. We start proving the following proposition which takes care of the first three triples in Table 8.1.

Proposition 8.3. *Let (M, g) be a positive qk-manifold which is acted on isometrically by a connected compact Lie group G with cohomogeneity one and associated triple* *of subalgebras of type* 1. *If the Lie algebra* $\mathfrak{g}$ *of* G *is* $\mathfrak{a}_n$ *with* $n \geq 3$, *then the manifold* (M, g) *is homothetic to* $Gr_2(\mathbb{C}^{n+2})$ *or to* $\mathbb{H}P^n$ *endowed with their symmetric metrics.*

Conversely, the symmetric spaces $Gr_2(\mathbb{C}^{n+2})$ *and* $\mathbb{H}P^n$ *admit isometric actions of the group* $G = SU_{n+1}$ *of cohomogeneity one and associated triple of type* 1.

Proof. If we allow the action of G to be almost effective, we may suppose that $G = SU_{n+1}$. Let us consider the associated triple of subgroups (H, K, H') (resp.,subalgebras $(\mathfrak{h}, \mathfrak{k}, \mathfrak{h}')$) of G (resp., $\mathfrak{g}$). Looking at Table 8.1, we see that $\mathfrak{h} = \mathbb{R} + \mathfrak{a}_{n-1}$; the subgroup H has therefore connected component of identity $H^0 = U_n \subset SU_{n+1}$. Since the normalizer $N_{SU_{n+1}} U_n = U_n$, we get that H is connected; so $H = U_n$ and one singular orbit is diffeomorphic to $\mathbb{C}P^n$.

First of all, we prove the following lemma.

Lemma 8.4.

(1) *A regular orbit* $R = G/K$ *is simply connected and has second Betti number* $b_2 = 1$.

(2) *The singular orbit* G/H' *is simply connected.*

Proof. (1) Indeed we have a sphere bundle $S^{2n-1} \to R \to \mathbb{C}P^n$; the first assertion follows from the long exact homotopy sequence associated to this bundle and the second assertion from the Gysin sequence (1.2.5).
(2) We know that the singular orbit G/H has codimension $2n$ in M. By the general position theorem, we get that $M \setminus (G/H)$ is simply connected; our claim now follows from the fact that G/H' is a deformation retract of $M \setminus (G/H)$. □

We will now examine the second singular stability subgroup H' with its normal representation $j' : H' \to O(V')$ and prove the following

Lemma 8.5. *We have* $\dim V' \geq 3$.

Proof. Indeed $\dim V' \geq 2$, since we know from Lemma 6.2 that there are no exceptional orbits. So, we have to rule out the case $\dim V' = 2$. If this is the case, we use the Gysin sequence associated to the fibering $S^1 \to R \to G/H'$, to get

$$\{0\} = H^1(R) \to H^0(G/H') \to H^2(G/H') \to H^2(R) \to H^1(G/H').$$

Since G/H' is simply connected by Lemma 8.4, $H^1(G/H') = \{0\}$ and therefore $b_2(G/H') = 2$; but then by the Mayer–Vietoris sequence (1.2.1), we would have

$$\{0\} \to H^2(M) \to H^2(\mathbb{C}P^n) \oplus H^2(G/H') \to H^2(R) \to \{0\},$$

which yields $b_2(M) = 2$ and this is not possible, according to Theorem 6.1,(3). □

We now prove the next lemma.

Lemma 8.6. *When* $\dim V' \geq 4$, *we have* $b_2(M) = 1$ *and the manifold* (M, g) *is homothetic to the grassmannian* $Gr_2(\mathbb{C}^{n+2})$. *Conversely, the symmetric space* $Gr_2(\mathbb{C}^{n+2})$ *admits an isometric action of the group* $G = SU_{n+1}$ *of cohomogeneity one and associated triple of subalgebras* $\theta = (\mathbb{R} + \mathfrak{a}_{n-1}, \mathbb{R} + \mathfrak{a}_{n-2}, \mathbb{R} + \mathfrak{a}_1 + \mathfrak{a}_{n-2})$.

Proof. By Lemma 8.4, the singular orbit G/H' is simply connected, hence orientable and we may use the Gysin exact sequence in cohomology associated to the sphere bundle $S^k \to R \to G/H'$, where $\dim V' = k+1$. Since, by assumption, $\dim S^k \geq 3$, we obtain $H^2(G/H') \cong H^2(R)$ and, by Lemma 8.4, we obtain that $b_2(G/H') = 1$.

By the Mayer–Vietoris exact sequence (1.2.1), we have

$$\{0\} = H^1(R) \to H^2(M) \to H^2(\mathbb{C}P^n) \oplus H^2(G/H') \to H^2(R) \to \{0\}$$

hence $b_2(M) = 1$ and the manifold (M, g) is homothetic to the grassmannian $Gr_2(\mathbb{C}^{n+2})$ by Theorem 6.1, (3).

Conversely, it is easy to see that the group $G = SU_{n+1}$, viewed as a subgroup of SU_{n+2} acts isometrically on $Gr_2(\mathbb{C}^{n+2}) = SU_{n+2}/S(U_2 \times U_n)$ with cohomogeneity one and associated triple as claimed. □

We are left with the case $\dim V' = 3$. Looking at the Table 8.1, the only triple with $\mathfrak{g} = \mathfrak{a}_n$ and $\dim \mathfrak{h}' - \dim \mathfrak{k} = 2$ is

$$\theta = (\mathbb{R} + \mathfrak{a}_{n-1}, \mathbb{R} + \mathfrak{a}_{n-2}, \mathfrak{a}_1 + \mathfrak{a}_{n-2}).$$

We note here that, when $n \geq 3$, there is only one subalgebra $\mathfrak{h}'$ of $\mathfrak{a}_n$ which is isomorphic to $\mathfrak{a}_1 + \mathfrak{a}_{n-2}$ and contains $\mathfrak{k} = \mathbb{R} + \mathfrak{a}_{n-2}$.

By Lemma 8.4, the two subgroups K and H' are both connected and the triple θ of subalgebras corresponds to the triple of subgroups $\theta = (H, K, H') = (U_n, U_{n-1}, SU_2 \times SU_{n-1})$.

The proof of the proposition will be complete proving the following

Lemma 8.7.

(1) *The group* $G = SU_{n+1}$ *acts on the qk-manifold* $\mathbb{H}P^n$ *with cohomogeneity one and associated triple*

$$\theta = (U_n, U_{n-1}, SU_2 \times SU_{n-1}).$$

(2) *If the Lie group* $G = SU_{n+1}$ *acts isometrically on a qk-manifold* (M, g) *with cohomogeneity one and associated triple* θ *of type* 1, *then* (M, g) *is homothetic to* $\mathbb{H}P^n$ *endowed with its symmetric metric.*

Proof. (1) Indeed, we consider the standard embedding of G into Sp_{n+1} and let it act on $\mathbb{H}P^n = Sp_{n+1}/Sp_1 \cdot Sp_n$ (the action of G is almost effective and becomes effective after factorization by $\mathbb{Z}_2$ if n is odd). If we consider the geodesic

$$\gamma_\theta = [(\cos\theta, \sin\theta j, 0 \ldots, 0)], \theta \in [0, \pi/4],$$

it is easy to see that γ is a geodesic segment for the G-action and that the associated triple coincides with θ.
(2) By Theorem 1.2.4 and point (1), we get that M is diffeomorphic to $\mathbb{H}P^n$ and, by Theorem 6.1, (2), we have that (M, g) is homothetic to $\mathbb{H}P^n$ with its symmetric metric. □

Remark. When $n = 2$ and $\mathfrak{g} = \mathfrak{a}_2$, the manifold is eight-dimensional, hence symmetric by [PoSa]; but, the complex grassmannian is not the only example, since the group $SU(3)$ acts on the Wolf space $G_2/SO(4)$ with cohomogeneity one.

We now prove the following proposition, which takes care of the case when $\mathfrak{g} = \mathfrak{c}_n$.

Proposition 8.8. *Let (M, g) be a positive qk-manifold which is acted on isometrically by a connected compact Lie group G with cohomogeneity one and associated triple of subalgebras of type* 1. *If the Lie algebra $\mathfrak{g} = \mathfrak{c}_{n+1}$ with $n \geq 2$, then the manifold (M, g) is homothetic to the symmetric space $Gr_2(\mathbb{C}^{2n+2})$.*

Conversely, the symmetric space $Gr_2(\mathbb{C}^{2n+2})$ admits an isometric action of the group Sp_{n+1} with cohomogeneity one and associated triple of subalgebras of type 1.

Proof. We will follow the same lines as in the proof of Proposition 8.3. We denote by $\theta = (H, K, H')$ (resp., $(\mathfrak{h}, \mathfrak{k}, \mathfrak{h}')$) the associated triple of subgroups (resp., subalgebras) of G (resp., $\mathfrak{g}$) and, looking at Table 8.1, we see that $\mathfrak{h} = \mathfrak{c}_1 + \mathfrak{c}_n$. If we allow the action of G to be almost effective, we may take $G = Sp_{n+1}$, so that the connected component $H^o = Sp_1 \times Sp_n$; since the normalizer $N_G(H^o) = H^o$, we get that $H = H^o = Sp_1 \times Sp_n$.

Looking at the Table 8.1, we see that the only possible triples of subalgebras $(\mathfrak{h}, \mathfrak{k}, \mathfrak{h}')$, including the standard one, are the following:

n^o	$\mathfrak{h}$	$\mathfrak{k}$	$\mathfrak{h}'$
1	$\mathfrak{c}_1 + \mathfrak{c}_n$	$\mathfrak{c}_1 + \mathfrak{c}_{n-1}$	$\mathfrak{c}_1 + \mathfrak{c}_n$
2	$\mathfrak{c}_1 + \mathfrak{c}_n$	$\mathfrak{c}_1 + \mathfrak{c}_{n-1}$	$2\mathfrak{a}_1 + \mathfrak{c}_{n-1}$
3	$\mathfrak{c}_1 + \mathfrak{c}_n$	$\mathfrak{c}_1 + \mathfrak{c}_{n-1}$	$\mathbb{R} + \mathfrak{a}_1 + \mathfrak{c}_{n-1}$

Table 8.2.

The proof of next lemma is analogous to the one given for Lemma 8.4.

Lemma 8.9.

(1) *A regular orbit $R = G/K$ is simply connected and has $b_2 = 0$.*

(2) *The singular orbit G/H' is simply connected, hence H' is connected.*

Using some topological arguments, we may prove the following

Lemma 8.10. *Let (M, g) be a positive qk-manifold which is acted on isometrically by a connected compact Lie group G with cohomogeneity one and associated triple θ of subalgebras of type 1. If the Lie algebra $\mathfrak{g} = \mathfrak{c}_{n+1}$, then*

$$\theta = (\mathfrak{a}_1 + \mathfrak{c}_n, \mathfrak{a}_1 + \mathfrak{c}_{n-1}, \mathbb{R} + \mathfrak{a}_1 + \mathfrak{c}_{n-1}).$$

Proof. We have to rule out the cases (1) and (2) in the Table 8.2.

First of all we note that in both cases $\operatorname{codim} G/H' = \dim \mathfrak{h}' - \dim \mathfrak{k} + 1 \geq 4$ and this implies, by the general position theorem, that $\pi_2(M) \cong \pi_2(M \setminus (G/H'))$.

On the other hand $M \setminus (G/H')$ is a bundle over $G/H \cong \mathbb{H}P^n$ with contractible fibres, hence $\pi_2(M) \cong \pi_2(M \setminus (G/H')) = \{0\}$. By Theorem 6.1, part (4), cases (1) and (2) would occur iff M should be diffeomorphic to $\mathbb{H}P^{2n}$. We now show that this is not possible.

Indeed, in case (1), $\mathfrak{h}'$ is conjugate to $\mathfrak{h}$ and G/H' is diffeomorphic to $\mathbb{H}P^n$. It then follows from Prop. 1.2.1 that $\chi(M) = 2(n+1)$, while, on the other hand, $\chi(\mathbb{H}P^{2n}) = 2n + 1$ and this is a contradiction.

In case (2), it is easy to check that $\chi(G/H') = n(n+1)$ (see e.g. [GHV]), so that

$$\chi(M) = (n+1)^2 > \chi(\mathbb{H}P^{2n}) = 2n + 1,$$

which gives a contradiction. □

The proof of the theorem will be complete proving the following

Lemma 8.11. *Let (M, g) be a positive qk-manifold which is acted on isometrically by a connected compact Lie group G with cohomogeneity one. If the Lie algebra $\mathfrak{g}$ is $\mathfrak{c}_{n+1}$ and the associated triple of subalgebras of $\mathfrak{g}$ is of type 1 and it is given by*

$$\theta = (\mathfrak{a}_1 + \mathfrak{c}_n, \mathfrak{a}_1 + \mathfrak{c}_{n-1}, \mathbb{R} + \mathfrak{a}_1 + \mathfrak{c}_{n-1}),$$

then (M, g) is homothetic to the symmetric space $Gr_2(\mathbb{C}^{2n+2})$.

Conversely, the symmetric space $Gr_2(\mathbb{C}^{2n+2})$ admits an isometric action of the Lie group Sp_{n+1} with cohomogeneity one and associated triple of subalgebras θ of type 1.

Proof. (1) We will show that $b_2(M) = 1$, forcing the manifold to be homothetic to the grassmannian $Gr_2(\mathbb{C}^{2n+2})$ by Theorem 6.1, part (3).

Since $\dim \mathfrak{h}' - \dim \mathfrak{k} = 1$, the fibration $G/K \to G/H'$ has S^1 as typical fibre; using the Gysin sequence 1.2.5 and Lemma 8.9, we get

$$\{0\} = H^1(G/K) \to H^0(G/H') \to H^2(G/H') \to H^2(G/K) = \{0\},$$

which yields $b_2(G/H') = 1$; by the Mayer–Vietoris sequence 1.2.1, we get $H^2(M) \cong H^2(G/H')$ and we are done.
(2) the standard embedding of Sp_{n+1} into SU_{2n+2} induces an action of Sp_{n+1} on $Gr_2(\mathbb{C}^{2n+2}) = SU_{2n+2}/S(U_2 \times U_{2n})$, which is easily seen to be of cohomogeneity one, with associated triple θ of Lie subalgebras. □

We now come to the proof of the main theorem.

Proof of Theorem 8.1. We have to prove that G is simple. We have found in Lemma 8.2 the possible candidates for the Lie algebra $\mathfrak{g}$ of G; in particular the only non simple case is given by $\mathfrak{g} = \mathbb{R} + \mathfrak{g}_2$, with singular isotropy subalgebra $\mathfrak{h}$ given by $\mathbb{R} + \mathfrak{a}_2$. This case is excluded by the following lemma.

Lemma 8.12. *Let (M, g) be a positive qk-manifold which is acted on isometrically and effectively by a Lie group G with cohomogeneity one and associated triple of subalgebras of type* 1. *Then $\mathfrak{g}$ cannot be isomorphic to $\mathbb{R} + \mathfrak{g}_2$.*

Proof. We suppose that $\mathfrak{g} = \mathbb{R} + \mathfrak{g}_2$; by Lemma 8.2, we know that, if $(\mathfrak{h}, \mathfrak{k}, \mathfrak{h}')$ is the associated triple of subalgebras, then $\mathfrak{h} = \mathbb{R} + \mathfrak{a}_2$. Note that, since the group G_2 is centerless, we have $G = S^1 \times G_2$. We then have that one singular orbit $S = G/H$ is given by the sphere $G_2/SU_3 \cong S^6$ and $\dim M = 12$.

First of all, we exclude the possibility that M is isometric to the complex grassmannian $Gr_2(\mathbb{C}^5)$: indeed, the Lie algebra of the full isometry group of $Gr_2(\mathbb{C}^5)$ is $\mathfrak{a}_4$ and $\mathfrak{g}$ does not embed into $\mathfrak{a}_4$, since $\mathfrak{g}_2$ has no 5-dimensional complex representation.

By Theorem 6.1, (3), it follows that $b_2(M) = 0$. We now use a result by LeBrun and Salamon ([LBS]), stating that, for 12-dimensional positive qk-manifold, we have $b_6(M) = 2b_2(M)$. In our case we have therefore $b_6(M) = 0$ and

$$\chi(M) = 2(1 + b_4(M)).$$

We denote by $S' = G/H'$ the second singular orbit and note that S' is simply connecte, since the codimension of S is bigger that 3. We now have

$$\chi(M) = \chi(S^6) + \chi(S') = 2 + \chi(S'),$$

so that $\chi(S') = 2b_4(M)$.

We now consider a regular orbit R, which fibers over S^6 with typical fibre S^5. By the Gysin exact sequence in cohomology, we have tha $b_3(R) = b_4(R) = 0$. Using the Mayer–Vietoris sequence (1.2.1), we have

$$\{0\} = H^3(R) \to H^4(M) \to H^4(S') \to H^4(R) = \{0\},$$

hence $b_4(S') = b_4(M)$ and $\chi(S') = 2b_4(S')$. Since H' is of maximal rank in G, all odd Betti numbers of G/H' vanish (see [Bo]) and therefore $\chi(S') = 2b_4(S')$ implies $\dim S' = 4$. But there is no subgroup of G of dimension 11 and this contradiction proves our claim. □

So far we have proven that G must be simple. If G is not exceptional, then by the table in Lemma 8.2, we have $\mathfrak{g} = \mathfrak{a}_n$ or $\mathfrak{g} = \mathfrak{c}_n$ and our claim follows from Propositions 8.3 and 8.8.

Our last assertion is about the case $\mathfrak{g} = \mathfrak{g}_2$. We prove the following

Lemma 8.13. *Let* (M, g) *be a positive qk-manifold which is acted on isometrically by a compact Lie group* G *with Lie algebra* $\mathfrak{g}_2$ *and associated triple of subalgebras* $(\mathfrak{h}, \mathfrak{k}, \mathfrak{h}')$ *of type* 1. *Then* $\mathfrak{h} \cong \mathfrak{h}' \cong 2\mathfrak{a}_1$.

Proof. By Lemma 8.2, we have only two possibilities for $\mathfrak{h}$, namely $\mathfrak{h} = \mathfrak{a}_2$ or $\mathfrak{h} = 2\mathfrak{a}_1$. Using the same argument as in Lemma 8.12, we rule out the possibility $\mathfrak{h} = \mathfrak{a}_2$. We must prove that $\mathfrak{h}' \cong \mathfrak{h}$.

We note that, since the group G_2 is centerless, $G = G_2$ and $H = SO_4$; again we have that $\dim M = 12$. As in the proof of Lemma 8.13, we may assume that $b_6(M) = 0$ and therefore $\chi(M) = 2(1 + b_4(M))$. If $S = G_2/SO_4$ and $S' = G/H'$ are the two singular orbits, we have that S' is simply connected and has positive dimension (see §7); moreover

$$\chi(M) = 2(1 + b_4(M)) = \chi(S) + \chi(S') = 3 + \chi(S'),$$

so that $\chi(S') = 2b_4(M) - 1 > 0$ and H' is of maximal rank in G. We recall that

$$\chi(S') = \frac{w(G_2)}{w(H')} = \frac{12}{w(H')},$$

where $w(\cdot)$ denotes the order of the Weyl group. Therefore $\chi(S')$ is an odd integer which devides 12, so $\chi(S') = 3$ and $w(H') = 4$. Since H' is connected and of maximal rank, it is contained into some maximal subgroup of maximal rank in G_2, which are, up to isomorphism, only SU_3 and SO_4. But, since $w(SU_3) = 6$ and $w(H')$ does not devide 6, we have that H' is isomorphic to SO_4. □

We now consider the associated triple of subalgebras $(\mathfrak{h}, \mathfrak{k}, \mathfrak{h}') = (2\mathfrak{a}_1, \mathfrak{a}_1, \mathfrak{h}')$ with $\mathfrak{h}' \cong 2\mathfrak{a}_1$ and note that there is only one subalgebra of $\mathfrak{g}_2$ which is isomorphic to $\mathfrak{h} = 2\mathfrak{a}_1$ and contains $\mathfrak{k} = \mathfrak{a}_1$, namely $\mathfrak{h}' = \mathfrak{h}$.

It then follows from Theorem 1.1.3 and the previous lemma, that there is, up to diffeomorphism, at most one positive qk-manifold of cohomogenity one under the action of G_2 and with associated triple of subalgebras of type 1.

On the other side, the group G_2 acts on the Wolf space $M = SO_7 / SO_4 \times SO_3$ by considering the standard embedding $G_2 \hookrightarrow SO_7$; this action is easily seen to be of cohomogeneity one and with associated triple of subalgebras of type 1. Our last claim is now proved. □

9. Quaternionic Kähler manifolds of type 2

The aim of this section is to prove the following theorem.

Theorem 9.1. *There exists no positive qk-manifold (M, g) of dimension greater than eight, which is acted on isometrically by a connected, compact Lie group G with cohomogeneity one and with associated triple of subalgebras of type 2.*

The proof will rely on the results obtained in §5; in particular in Table 5.1, we enumerated all type 2 positive triples $(\mathfrak{h}, \mathfrak{k}, \mathfrak{h}')$ of a compact Lie algebra $\mathfrak{g}$.

Proof of the theorem. Let $(\mathfrak{h}, \mathfrak{k}, \mathfrak{h}')$ be the associated triple of subalgebras of $\mathfrak{g}$. We recall that by Lemma 6.3, each isotropy representation associated with $\mathfrak{h}, \mathfrak{k}, \mathfrak{h}'$ has to normalize some quaternionic structures. We recall that we may decompose $\mathfrak{g}$ as sum $\mathfrak{g} = \mathfrak{h}+\mathfrak{m}$, where $\mathfrak{m}$ is an $\mathrm{ad}(\mathfrak{h})$-invariant subspace of $\mathfrak{g}$; the isotropy representation ι of $\mathfrak{h}$ is then given by $ad|\mathfrak{m}+j$, where $j : \mathfrak{h} \to \mathfrak{o}(V)$ is the normal isotropy representation. We will now check each triple θ appearing in Table 5.1.

(1) $\mathfrak{g} = \mathfrak{b}_n$, $\theta = (\mathbb{R} + \mathfrak{d}_{n-1}, \mathfrak{d}_{n-1}, \mathfrak{b}_{n-1})$, $n > 3$. If we restrict ι to the simple summand $\mathfrak{d}_{n-1}$, we have that $\iota(\mathfrak{d}_{n-1})$ centralizes some quaternionic structure on $\mathfrak{m}\oplus V$; on the other hand, in both cases, the $\mathfrak{d}_{n-1}$-module $\mathfrak{m} \oplus V$ splits as an othogonal sum of invariant subspaces

$$\mathfrak{m} \oplus V = \mathfrak{p}_0 + 3\mathfrak{p},$$

where $\mathfrak{p}_0$ is a trivial 4-dimensional module and $\mathfrak{p} \cong \mathbb{R}^{2n-2}$ is the standard representation space of $\mathfrak{d}_{n-1}$. It would then follow that the subspace $\mathfrak{p} + \mathfrak{p} + \mathfrak{p}$ has an invariant quaternionic structure and this is not possible.

(2) $\mathfrak{g} = \mathfrak{d}_n$, $\theta = (\mathbb{R} + \mathfrak{d}_{n-1}, \mathfrak{d}_{n-1}, \mathfrak{b}_{n-1})$. In this case

$$\dim M = \dim \mathfrak{g} - \dim \mathfrak{k} + 1 = 4n - 2$$

which is not a multiple of 4.

(3) $\mathfrak{g} = \mathfrak{b}_3$; in this case we have $\dim M = \dim \mathfrak{g} - \dim \mathfrak{k} + 1 = 14$.

(4) $\mathfrak{g} = \mathfrak{a}_1 + \mathfrak{g}_2$, $\theta = (\mathbb{R} + \mathfrak{a}_2, \mathfrak{a}_2, \mathfrak{g}_2)$; in this case $\dim M = 10$;

(5) $\mathfrak{g} = \mathbb{R} + \mathfrak{g}_2$ $\theta = (\mathbb{R} + \mathfrak{a}_2, \mathfrak{a}_2, \mathfrak{g}_2)$; in this case $\dim M = 8$.

(6) $\mathfrak{g} = \mathfrak{a}_1 + \mathfrak{b}_{n-1}$ $\theta = (\mathbb{R} + \mathfrak{d}_{n-1}, \mathfrak{d}_{n-1}, \mathfrak{b}_{n-1})$. In this case, the $\iota(\mathfrak{d}_{n-1})$-module $\mathfrak{m} \oplus V$ splits as

$$\mathfrak{m} \oplus V = \mathfrak{p}_0 + \mathfrak{p}_1,$$

where $\mathfrak{p}_0$ is a trivial 4-dimensional module and $\mathfrak{p}_1 \cong \mathbb{R}^{2n-2}$ is the standard representation space of $\mathfrak{d}_{n-1}$. So, there is no quaternionic structure on $\mathfrak{m} \oplus V$ which is centralized by $\iota(\mathfrak{d}_{n-1})$. □

References

[A1] D. V. Alekseevsky, Compact quaternion spaces, Functional Anal. Appl. 2 (1968), 106–114.

[AA] A. V. Alekseevsky and D. V. Alekseevsky, G-manifolds with one dimensional orbit space, Adv. Sov. Math. 8 (1992), 1–31.

[AA1] A. V. Alekseevsky and D. V. Alekseevsky, Riemannian G-manifolds with one dimensional orbit space, Ann. Global Anal. Geom. 11 (1993), 197–211.

[Be] A. Besse, Einstein manifolds, Springer-Verlag, Berlin-Heidelberg–New York 1987.

[BeBe] L. Bérard Bergery, Sur de nouvelles variétés riemanniennes d'Einstein, Pub. Inst. E. Cartan 4 (1982), 1–60.

[Bo] A. Borel, Sur la cohomologie des espaces fibrés principaux et des espaces homogenes de groupes de Lie compacts, Ann. of Math. 57 (1953), 115–207.

[Br] G. E. Bredon, Introduction to compact transformation groups, Academic Press, New York–London 1972.

[BrSa] R. Bryant, S. Salamon, On the construction of some complete metrics with exceptional holonomy, Duke Math. J. 58 (1985), 829–850.

[DaSw] A. Dancer, A. F. Swann, Hyperkähler metrics of cohomogeneity one, J. Geom. Phys., to appear 1996.

[GHV] W. Greub, S. Halperin, R. Vanstone, Curvature, Connections and Characteristic Classes, vol. 3, Academic Press, New York 1976.

[GOV] V. V. Gorbatzevich, A. L. Onishchik, E. B. Vinberg, Structure of Lie groups and Lie algebras, Encyclopaedia. Math. Sci. 41, Springer-Verlag, 1992.

[GoGr] M. Goto, F. D. Grosshans, Semisimple Lie algebras, Lecture Notes in Pure and Appl. Math. 38, Marcel Dekker Inc., 1978.

[LBS] C. LeBrun, S. M. Salamon, Strong rigidity of positive quaternionic Kähler manifolds, Invent. Math. 118 (1994), 109–132.

[Mo] P. S. Mostert, On a compact Lie group action on manifolds, Ann. of Math. 65 (1957), 447–455.

[Na] T. Nagano, Transformation groups with $(n-1)$-dimensional orbits on non-compact manifolds, Nagoya Math. J. 14 (1959), 25–38.

[PaTe] R. S. Palais and Ch. L. Terng, A general theory of canonical forms, Trans. Amer. Math. Soc. 300 (1987), 771–789.

[PoSa] Y. S. Poon, S. M. Salamon, Eight-dimensional quaternionic Kähler manifolds with positive scalar curvature, J. Differential Geom. 33 (1991), 363–378.

[S] S. M. Salamon, Quaternionic Kähler manifolds, Invent. Math. 67 (1982), 143–171.

[W] J. A. Wolf, Complex homogeneous contact structures and quaternionic symmetric spaces, J. Math. Mech. 14 (1965), 1033–1047.

Quasiconformality and geometrical finiteness in Carnot–Carathéodory and negatively curved spaces

Boris Apanasov†

Abstract. The paper sketches a recent progress and formulates several open problems in studying equivariant quasiconformal and quasisymmetric homeomorphisms in negatively curved spaces as well as geometry and topology of noncompact geometrically finite negatively curved manifolds and their boundaries at infinity having Carnot–Carathéodory structures. Especially, the most interesting are complex hyperbolic manifolds with Cauchy–Riemannian structure at infinity, which occupy a distinguished niche and whose properties make them surprisingly different from real hyperbolic ones.

1991 Mathematics Subject Classification: 57, 55, 53, 51.

Key words: Negatively curved manifolds; nilpotent, Carnot and Heisenberg groups; discrete isometry groups; geometrical finiteness; complex hyperbolic manifolds; Cauchy–Riemannian structure; quasiconformal and quasisymmetric mappings; varieties of representations; deformations.

1. Introduction

The paper sketches a recent progress and formulates several open problems in studying quasiconformal and quasisymmetric homeomorphisms as well as geometry and topology of noncompact geometrically finite negatively curved manifolds and their boundaries at infinity having Carnot–Carathéodory structures in the sense of M. Gromov [Gr2]. Here, complex hyperbolic and Cauchy–Riemannian manifolds occupy a distinguished niche. First, due to their complex analytic nature, a broad spectrum of techniques can contribute to the study, and already obtained results show surprising differences between geometry and topology of noncompact complex and real hyperbolic manifolds, see [BS, BuM, EMM, GoM, KR1-2]. Second, these Kähler manifolds are

† Supported in part by the NSF. Research at MSRI was supported in part by NSF grant DMS-9022140.

the most known manifolds with variable pinched negative curvature [BGS, B, Mg1, MaG]. Finally, for complex analytic surfaces, one can apply Seiberg–Witten invariants, decomposition of 4-manifolds along homology 3-spheres, Floer homology and new (homology) cobordism invariants [LB, A11, Sa1-Sa4].

There are three main themes united by using of general Thurston's idea of geometric uniformization of low dimensional manifolds provided some canonical decomposition of them into geometric pieces. First, basic geometric block-manifolds of negative (variable) curvature which possess geometrical finiteness and may however have infinite volume. Second, the interaction between the negatively curved geometry of such blocks and the induced Carnot–Carathéodory structure of their boundary manifolds at infinity, especially the interaction between Kähler geometry of geometrically finite complex hyperbolic manifolds and Cauchy–Riemannian structure of their boundaries modeled on the Heisenberg group (which is a particular case of Carnot groups). And finally, deformations of negatively curved manifolds inducing automorphisms of boundaries at infinity which preserves their natural contact structures, together with their metrical (quasiconformal or quasisymmetric) properties.

Our study of geometrical finiteness and homeomorphisms in spaces with negative (variable) curvature exploits a new structural theorem about isometric actions of discrete groups on nilpotent Lie groups (in particular on the Heisenberg group $\mathcal{H}_n$). The problems there have a unique appeal both for the amount of similarity with the model situation of interactions between real hyperbolic geometrically finite manifolds, their boundaries with natural conformal structures and their quasiconformal deformations (see [A1, A11]), and for the interesting ways in which the similarity breaks down.

Besides geometrical finiteness in negatively curved spaces, one of inspiring ideas of our study is a well known theorem of D. Sullivan, which gave rise to many important results in geometry and topology of manifolds and theory of quasiconformal mappings. It states, see [Su1, TV], that homeomorphisms of quasiconformal n-manifolds, $n \neq 4$, can be approximated by quasiconformal ones, where one of important classes is the class of conformal manifolds. Here by the quasiconformality one means the boundedness of distortion with respect to the Euclidean metric. This result rises questions of approximation in the quasiconformal category in another negatively curved geometries and the corresponding sub-Riemannian Carnot–Carathéodory structures which appear at infinity of those geometries as well. We mention here a recent important development by M.Gromov on Carnot–Carathéodory spaces [Gr2], where he shown that continues mappings can be approximated by mappings that are Lipschitz with respect to the Carnot–Carathéodory metric. Another important achievement is a recent result by G. Margulis and G. Mostow [MM] that quasiconformal mappings of Carnot–Carathéodory spaces are a.e. differentiable and preserve their contact structures.

Another set of problems is related to the stability theorem proven by D. Sullivan [Su3] for planar Kleinian groups (see also [A11, MaG]). It raises questions on deformations of discrete isometry groups in spaces of negative (variable) curvature, varieties of their representations (Teichmüller spaces) and their boundaries, and quasiconformal/quasisymmetric homeomorphisms induced by isomorphisms of geometrically finite groups (in conformal category, it is known as Tukia isomorphism theorems [Tu]).

2. Complex hyperbolic and Heisenberg manifolds

The natural class of manifolds where the above problems can be considered is the class of geometrically finite locally homogeneous manifolds. That is why we start with studying such manifolds and their boundaries at infinity, especially complex hyperbolic manifolds and Cauchy–Riemannian and Heisenberg manifolds at their infinity.

We recall some facts concerning the link between nilpotent geometry of the Heisenberg group and the Kähler geometry of the complex hyperbolic space (see [Go, GP1, KR1-2]). Let $\mathbb{B}^n_{\mathbb{C}} \subset \mathbb{C}^n$ be the unit complex ball equipped with the Bergman metric d which turns the ball into the complex hyperbolic n-space $\mathbb{H}^n_{\mathbb{C}}$ whose sectional curvature is between -1 and $-1/4$. The automorphism group of $\mathbb{H}^n_{\mathbb{C}} \cong \mathbb{B}^n_{\mathbb{C}}$ is the projective unitary group $PU(n,1)$ whose elements $g \in PU(n,1)$ are biholomorphisms of $\mathbb{B}^n_{\mathbb{C}}$ of the following three types. If g fixes a point in $\mathbb{H}^n_{\mathbb{C}}$, it is called *elliptic*. If g has exactly one fixed point, and it lies in $\partial\mathbb{H}^n_{\mathbb{C}}$, g is called *parabolic*. If g has exactly two fixed points, and they lie in $\partial\mathbb{H}^n_{\mathbb{C}}$, g is called *loxodromic*. These three types exhaust all the possibilities. Since $PU(n,1)$ can be embedded in a linear group (for details, see [AX1]), any finitely generated group $G \subset PU(n,1)$ is residually finite and has a finite index torsion free subgroup.

For any point $\infty \in \partial\mathbb{B}^n_{\mathbb{C}}$, one can identify $\overline{\mathbb{B}^n_{\mathbb{C}}}\backslash\{\infty\}$ with the closure of the Siegel domain $\mathfrak{S}_n$, which is conveniently represented in horospherical coordinates as $\mathbb{C}^{n-1} \times \mathbb{R} \times [0,\infty)$ ([GP1]). Then "the boundary plane" $\mathbb{C}^{n-1} \times \mathbb{R} \times \{0\} = \partial\mathbb{H}^n_{\mathbb{C}}\backslash\{\infty\}$ and the horospheres $H_u = \mathbb{C}^{n-1} \times \mathbb{R} \times \{u\}$, $0 < u < \infty$, centered at ∞ are identified with the Heisenberg group $\mathcal{H}_n = \mathbb{C}^{n-1} \times \mathbb{R}$. It is a 2-step nilpotent group with center $\{0\} \times \mathbb{R} \subset \mathbb{C}^{n-1} \times \mathbb{R}$, and the inverse of (ξ, v) is $(\xi, v)^{-1} = (-\xi, -v)$. The Heisenberg group $\mathcal{H}_n$ isometrically acts on itself and on $\mathbb{H}^n_{\mathbb{C}}$ by left translations:

$$T_{(\xi_0, v_0)} : (\xi, v, u) \longmapsto (\xi_0 + \xi\,,\, v_0 + v + 2\,\mathrm{Im}\langle\langle \xi_0, \xi \rangle\rangle\,,\, u)\,.$$

The unitary group $U(n-1)$ acts on $\mathcal{H}_n$ and $\mathbb{H}^n_{\mathbb{C}}$ by rotations: $A(\xi, v, u) = (A\xi\,,\, v\,,\, u)$ for $A \in U(n-1)$. The semidirect product $\mathcal{H}(n) = \mathcal{H}_n \rtimes U(n-1)$ is naturally embedded in $U(n,1)$ as follows:

$$A \longmapsto \begin{pmatrix} A & 0 & 0 \\ 0 & 1 & 0 \\ 0 & 0 & 1 \end{pmatrix} \in U(n,1) \quad \text{for } A \in U(n-1)\,,$$

$$(\xi, v) \longmapsto \begin{pmatrix} I_{n-1} & \xi & \xi \\ -\bar{\xi}^t & 1 - \frac{1}{2}(|\xi|^2 - iv) & -\frac{1}{2}(|\xi|^2 - iv) \\ \bar{\xi}^t & \frac{1}{2}(|\xi|^2 - iv) & 1 + \frac{1}{2}(|\xi|^2 - iv) \end{pmatrix} \in U(n, 1)$$

where $(\xi, v) \in \mathcal{H}_n = \mathbb{C}^{n-1} \times \mathbb{R}$ and $\bar{\xi}^t$ is the conjugate transpose of ξ.

The action of $\mathcal{H}(n)$ on $\overline{\mathbb{H}^n_{\mathbb{C}}}\setminus\{\infty\}$ also preserves the Cygan metric ρ_c there, which plays the same role as the Euclidean metric does on the upper half-space model of the real hyperbolic space $\mathbb{H}^n$ and is induced by the following norm:

$$||(\xi, v, u)||_c = |\,||\xi||^2 + u - iv|^{1/2}, \quad (\xi, v, u) \in \mathbb{C}^{n-1} \times \mathbb{R} \times [0, \infty).$$

The relevant geometry on each horosphere $H_u \subset \mathbb{H}^n_{\mathbb{C}}$, $H_u \cong \mathcal{H}_n = \mathbb{C}^{n-1} \times \mathbb{R}$, is the spherical CR-geometry induced by the complex hyperbolic structure. The geodesic perspective from ∞ defines CR-maps between horospheres, which extend to CR-maps between the one-point compactifications $H_u \cup \infty \approx S^{2n-1}$. In the limit, the induced metrics on horospheres fail to converge but the CR-structure remains fixed. In this way, the complex hyperbolic geometry induces CR-geometry on the sphere at infinity $\partial\mathbb{H}^n_{\mathbb{C}} \approx S^{2n-1}$, naturally identified with the one-point compactification of the Heisenberg group $\mathcal{H}_n$.

3. Geometrical finiteness in negative curvature

Our main assumption on a negatively curved n-manifold M is the geometrical finiteness condition on its fundamental group $G \subset \operatorname{Isom} X$ acting by isometries on a simply connected space $X = \widetilde{M}$, which in particular implies (see below) that the discrete group G is finitely generated.

A subgroup $G \subset \operatorname{Isom} X$ is called *discrete* if it is a discrete subset of $\operatorname{Isom} X$. The *limit set* $\Lambda(G) \subset \partial X$ of a discrete group G is the set of accumulation points of (any) orbit $G(y)$, $y \in X$. The complement of $\Lambda(G)$ in ∂X is called the *discontinuity set* $\Omega(G)$. A discrete group G is called *elementary* if its limit set $\Lambda(G)$ consists of at most two points. An infinite discrete group G is called *parabolic* if it has exactly one fixed point $\operatorname{fix}(G)$; then $\Lambda(G) = \operatorname{fix}(G)$, and G consists of either parabolic or elliptic elements. As it was observed by many authors (c.f. [MaG]), parabolicity in the variable curvature case is not as easy a condition to deal with as it is in the constant curvature space. However the results below simplify the situation.

Due to the absence of totally geodesic hypersurfaces in a space X of variable negative curvature, we cannot use the original definition of geometrical finiteness which came from an assumption that the corresponding real hyperbolic manifold $M = \mathbb{H}^n/G$ may be decomposed into a cell by cutting along a finite number of its totally geodesic hypersurfaces, that is the group G should possess a finite-sided fundamental polyhedron, see [Ah]. However, one can use another definition of *geometrically finite* groups $G \subset \operatorname{Isom} X$ as those ones whose limit sets $\Lambda(G) \subset \partial X$ consist of only conical limit points and parabolic (cusp) points p with compact quotients $(\Lambda(G)\setminus\{p\})/G_p$

with respect to parabolic stabilizers $G_p \subset G$ of p, see [BM, Bow]. There are other definitions of geometrical finiteness in terms of ends and the minimal convex retract of the noncompact manifold M, which work well not only in the real hyperbolic spaces $\mathbb{H}^n$ (see [Md, Th, A3, A1]) but also in spaces with variable pinched negative curvature [Bo].

In the case of variable curvature, it is problematic to use geometric methods based on consideration of finite sided fundamental polyhedra. In particular, Dirichlet polyhedra $D_y(G)$ in the complex hyperbolic space $X = \mathbb{H}^n_{\mathbb{C}}$ are fundamental polyhedra for a discrete subgroup $G \subset PU(n, 1)$. They are bounded by bisectors in a complicated way (see [Mo2, GP1]), and the bisectors are not totally geodesic hypersurfaces. For discrete parabolic groups $G \subset \operatorname{Isom} X$, one may expect that the Dirichlet polyhedron $D_y(G)$ centered at a point y lying in a G-invariant subspace has finitely many sides. It is true for real hyperbolic spaces [A1] as well as for cyclic and dihedral parabolic groups in complex hyperbolic spaces $X = \mathbb{H}^n_{\mathbb{C}}$. Namely, due to [Ph], Dirichlet polyhedra $D_y(G)$ are always two sided for any cyclic group $G \subset PU(n, 1)$ generated by a Heisenberg translation. Due to [GP1], such finiteness also holds for a cyclic ellipto-parabolic group or a dihedral parabolic group $G \subset PU(n, 1)$ generated by inversions in asymptotic complex hyperplanes in $\mathbb{H}^n_{\mathbb{C}}$ if the central point y lies in a G-invariant vertical line or $\mathbb{R}$-plane (for any other center y, $D_y(G)$ has infinitely many sides). Our technique easily implies that this finiteness still holds for generic parabolic cyclic groups [AX1]:

Theorem 3.1. *For any discrete group $G \subset PU(n, 1)$ generated by a parabolic element, there exists a point $y_0 \in \mathbb{H}^n_{\mathbb{C}}$ such that the Dirichlet polyhedron $D_{y_0}(G)$ centered at y_0 has two sides.*

However, the behavior of Dirichlet polyhedra for parabolic groups $G \subset PU(n, 1)$ of rank more than one can be very bad. It is given by our construction [AX1]:

Theorem 3.2. *Let $G \subset PU(2, 1)$ be a discrete parabolic group conjugate to the following subgroup Γ of the Heisenberg group $\mathcal{H}_2 = \mathbb{C} \times \mathbb{R}$:*

$$\Gamma = \{(m, n) \in \mathbb{C} \times \mathbb{R} : m, n \in \mathbb{Z}\}.$$

Then any Dirichlet polyhedron $D_y(G)$ centered at an arbitrary point $y \in \mathbb{H}^2_{\mathbb{C}}$ has infinitely many sides.

Despite this, we are providing [AX1] a construction of fundamental polyhedra $P(G) \subset \mathbb{H}^n_{\mathbb{C}}$ for arbitrary discrete parabolic groups $G \subset PU(n, 1)$, which are bounded by finitely many hypersurfaces (different from Dirichlet bisectors). This construction may be seen as a base for extension of Apanasov's construction [A1] of finite sided pseudo-Dirichlet polyhedra in $\mathbb{H}^n$ to the case of the complex hyperbolic space $\mathbb{H}^n_{\mathbb{C}}$, which may solve the following problem:

Problem 3.3. *Given geometrically finite group $G \subset \operatorname{Isom} X$ in a negatively curved space X, is there any finitely sided fundamental polyhedron $P(G) \subset X$?*

As another tool for attacking this problem, one can use the following structural theorem for discrete groups acting on nilpotent Lie groups (lying at infinity of X), in particular on the Heisenberg group $\mathcal{H}_n$, $\mathcal{H}_n \cup \{\infty\} = \partial \mathbb{H}^n_{\mathbb{C}}$ (see [AX1–AX3]):

Theorem 3.4. *Let N be a connected, simply connected nilpotent Lie group, C a compact group of automorphisms of N, and Γ a discrete subgroup of the semidirect product $N \rtimes C$. Then there exist a connected Lie subgroup V of N and a finite index normal subgroup Γ^* of Γ with the following properties:*

(1) *There exists $b \in N$ such that $b\Gamma b^{-1}$ preserves V.*

(2) *$V/b\Gamma b^{-1}$ is compact.*

(3) *$b\Gamma^* b^{-1}$ acts on V by left translations and this action is free.*

Here we remark that

(1) Compactness of C is an essential condition because of Margulis [Mg2] construction of nonabelian free discrete subgroups Γ of $R^3 \rtimes GL(3, R)$.

(2) This theorem generalizes a result by L.Auslander [Au] who claimed those properties not for whole group Γ but only for some finite index subgroup. We also remark that an extension in [AX1] of Wolf's argument [Wo] to the complex hyperbolic case (based on Margulis Lemma [Mg1, BGS] and geometry of $\mathcal{H}_n$) does not work in general nilpotent groups. A different algebraic proof of Theorem 3.4 can be found in [AX2, AX3].

This result allows one to study

Problem 3.5. *Investigate parabolic cusp ends of negatively curved manifolds.*

To solve this problem in the complex hyperbolic case, we use the above Theorem 3.4 to define standard cusp neighborhoods of parabolic cusps. This allows us to prove the following results about cusp ends of complex hyperbolic manifolds (or, equivalently, about the structure of Heisenberg manifolds).

Namely, suppose a point $p \in \partial \mathbb{H}^n_{\mathbb{C}}$ is fixed by some parabolic element of the group G, and G_p is the stabilizer of p in G. Conjugating the group G by an element $h \in PU(n, 1)$, $h(p) = \infty$, we have that $G_\infty \subset \mathcal{H}(n)$. In particular, if p is the origin in $\mathcal{H}_n$, we can take such h as the Heisenberg inversion $\mathcal{I}$ in the hyperchain in $\mathcal{H}_n$, which preserves the unit Heisenberg sphere $S_c(0, 1)$ and acts in $\mathcal{H}_n$ as follows:

$$\mathcal{I}(\xi, v) = \left(\frac{\xi}{|\xi|^2 - iv}, \frac{-v}{v^2 + |\xi|^4} \right) \quad \text{where } (\xi, v) \in \mathcal{H}_n = \mathbb{C}^{n-1} \times \mathbb{R}. \tag{3.6}$$

Then, due to Theorem 3.4, there exists a connected Lie subgroup $\mathcal{H}_\infty \subseteq \mathcal{H}_n$ preserved by G_∞.

Definition. A set $U_{p,r} \subset \overline{\mathbb{H}^n_{\mathbb{C}}}\backslash\{p\}$ is called a *standard cusp neighborhood of radius* $r > 0$ at a parabolic fixed point $p \in \partial\mathbb{H}^n_{\mathbb{C}}$ of a discrete group $G \subset PU(n,1)$ if, for the Heisenberg inversion $\mathcal{I}_p \in PU(n,1)$ with respect to the unit sphere $S_c(p,1) = \{(\xi, v) : \rho_c(p, (\xi, v)) = 1\}$, $\mathcal{I}_p(p) = \infty$, the following conditions hold:

(1) $U_{p,r} = \mathcal{I}_p^{-1}\left(\{x \in \mathbb{H}^n_{\mathbb{C}} \cup \mathcal{H}_n : \rho_c(x, \mathcal{H}_\infty) \geq 1/r\}\right)$;

(2) $U_{p,r}$ is precisely invariant with respect to $G_p \subset G$, that is:

$$\gamma(U_{p,r}) = U_{p,r} \quad \text{for } \gamma \in G_p \quad \text{and} \quad g(U_{p,r}) \cap U_{p,r} = \emptyset \quad \text{for } g \in G\backslash G_p .$$

A parabolic point $p \in \partial\mathbb{H}^n_{\mathbb{C}}$ of $G \subset PU(n,1)$ is a *cusp point* if it has a cusp neighborhood $U_{p,r}$.

We remark that some parabolic points of a discrete group $G \subset PU(n,1)$ may not be cusp points, see [AX1, §5.4]. Applying Theorem 3.4 and [Bo], we have:

Lemma 3.7. *Let $p \in \partial\mathbb{H}^n_{\mathbb{C}}$ be a parabolic fixed point of a discrete group $G \subset PU(n,1)$. Then p is a cusp point if and only if $(\Lambda(G)\backslash\{p\})/G_p$ is compact.*

This fact and [Bo] allows us to use another equivalent definitions of geometrical finiteness. In particular, a group $G \subset PU(n,1)$ is *geometrically finite* if and only if its quotient space $M(G) = [\mathbb{H}^n_{\mathbb{C}} \cup \Omega(G)]/G$ has finitely many ends, and each of them is a cusp end, that is an end whose neighborhood can be taken as $U_{p,r}/G_p \approx (S_{p,r_0}/G_p) \times (0,1]$, where

$$S_{p,r} = \mathcal{I}_p^{-1}\left(\{x \in H^n_{\mathbb{C}} \cup \mathcal{H}_n : \rho_c(x, \mathcal{H}_\infty) = 1/r\}\right) .$$

Using the above description of discrete group actions on a nilpotent group (Theorem 3.4), one can study Carnot–Carathéodory manifolds whose fundamental groups act at infinity of a negatively curved space X as discrete parabolic groups. In particular, we establish fiber bundle structures on Heisenberg manifolds which have the form $\mathcal{H}_n/G$ where G is a discrete group freely acting on $\mathcal{H}_n$ by isometries, i.e. a torsion free discrete subgroup of $\mathcal{H}(n) = \mathcal{H}_n \rtimes U(n-1)$, see [AX1]:

Theorem 3.8. *Let $\Gamma \subset \mathcal{H}_n \rtimes U(n-1)$ be a torsion-free discrete group acting on the Heisenberg group $\mathcal{H}_n = \mathbb{C}^{n-1} \times \mathbb{R}$ with non-compact quotient. Then the quotient $\mathcal{H}_n/\Gamma$ has zero Euler characteristic and is a vector bundle over a compact manifold. Furthermore, this compact manifold is finitely covered by a nil-manifold which is either a torus or the total space of a circle bundle over a torus.*

It gives, in addition to finiteness of generators of geometrically finite groups established in [Bo], the following important finiteness:

Corollary 3.9. *The fundamental groups of Heisenberg manifolds and geometrically finite complex hyperbolic manifolds are finitely presented.*

Due to Theorem 3.4, any Heisenberg manifold $N = \mathcal{H}_n/\Gamma$ is the vector bundle $\mathcal{H}_n/\Gamma \to \mathcal{H}_\Gamma/\Gamma$ where $\mathcal{H}_\Gamma \subset \mathcal{H}_n$ is a minimal Γ-invariant subspace. As simple examples in [AX1] show, such vector bundles are non-trivial in general. However, up to finite coverings, they are trivial [AX1]:

Theorem 3.10. *Let $\Gamma \subset \mathcal{H}_n \rtimes U(n-1)$ be a discrete group and $\mathcal{H}_\Gamma \subset \mathcal{H}_n$ a connected Γ-invariant Lie subgroup on which Γ acts co-compactly. Then there exists a finite index subgroup $\Gamma_0 \subset \Gamma$ such that the vector bundle $\mathcal{H}_n/\Gamma_0 \to \mathcal{H}_\Gamma/\Gamma_0$ is trivial. In particular, any Heisenberg orbifold $\mathcal{H}_n/\Gamma$ is finitely covered by the product of a compact nil-manifold $\mathcal{H}_\Gamma/\Gamma_0$ and an Euclidean space.*

We remark that in the case when $\Gamma \subset \mathcal{H}_n \rtimes U(n-1)$ is a lattice (i.e. $\mathcal{H}_n/\Gamma$ is compact), the existence of such finite cover of $\mathcal{H}_n/\Gamma$ by a closed nilpotent manifold $\mathcal{H}_n/\Gamma_0$ is due Gromov [Gr1] and Buser–Karcher [BK] results for almost flat manifolds. As we see, this finite covering property holds for non-compact Heisenberg manifolds as well. Moreover, noncompact geometrically finite complex hyperbolic manifolds have a similar property in the sense that they are finitely covered by manifolds with "nilpotent ends" each of which is topologically the products of a nilpotent k-manifold and an Euclidean space. Algebraically this property can be reformulated as follows:

Theorem 3.11. *Let $G \subset PU(n,1)$ be a geometrically finite discrete group. Then G has a subgroup G_0 of finite index such that every parabolic subgroup of G_0 is isomorphic to a discrete subgroup of the Heisenberg group $\mathcal{H}_n = \mathbb{C}^{n-1} \times \mathbb{R}$. In particular, each parabolic subgroup of G_0 is free Abelian or 2-step nilpotent.*

It seems this property holds for discrete isometry groups in general negatively curved spaces because our proof of Theorem 3.11 in [AX1] is based on the residual finiteness of geometrically finite subgroups $G \subset PU(n,1)$ and the following two lemmas (the second of which generalizes a result for finite volume real hyperbolic manifolds ([AF]).

Lemma 3.12. *Let $G \subset \mathcal{H}_n \rtimes U(n-1)$ be a discrete group and $\mathcal{H}_G \subset \mathcal{H}_n$ a minimal G-invariant connected Lie subgroup (given by Theorem 4.1). Then G acts on $\mathcal{H}_G$ by translations if G is either Abelian or 2-step nilpotent.*

Lemma 3.13. *Let $G \subset \mathcal{H}_n \rtimes U(n-1)$ be a torsion free discrete group, F a finite group and $\phi : G \longrightarrow F$ an epimorphism. Then the rotational part of* $\ker(\phi)$ *has strictly smaller order than that of G if one of the following happens:*

(1) *G contains a finite index Abelian subgroup and F is not Abelian;*

(2) *G contains a finite index 2-step nilpotent subgroup and F is not a 2-step nilpotent group.*

We conclude this section by pointing out that the problem of geometrical finiteness is very different in complex dimension two. Namely, it is a well known fact that any finitely generated discrete subgroup of $PU(1,1)$ is geometrically finite. This and Goldman's local rigidity theorem for uniform lattices $G \subset U(1,1) \subset PU(2,1)$ (see [GM]) suggest the following intrigue question:

Problem 3.14. *Are all finitely generated discrete groups $G \subset PU(2,1)$ with non-empty discontinuity set $\Omega(G) \subset \partial\mathbb{H}^2_{\mathbb{C}}$ geometrically finite?*

To solve this problem, one can try to decompose a given finitely generated discrete group $G \subset PU(2,1)$ with non-empty discontinuity set $\Omega(G) \subset \partial\mathbb{H}^2_{\mathbb{C}}$ into simple blocks by using a finite process of free amalgamated product and HNN-extensions. More arguments for an affirmative solution of this problem are given by the following two facts. First, due to Chen–Greenberg [CG], all pure loxodromic subgroups $G \subset PU(2,1)$ are discrete. The second fact is due to the trace classification of projective transformations [Go, VI.2]. Namely, in contrast to Kleinian groups on the plane, the subset of groups in a deformation space of a pure loxodromic group $G \subset PU(2,1)$ which have accidental parabolic elements may have real codimension 1.

4. The boundary at infinity of negatively curved manifolds

Another set of problems is related to geometry and topology of Carnot–Carathéodory manifolds which are the boundaries at infinity of negatively curved non-compact manifolds.

In a sharp contrast to the real hyperbolic case, for a compact complex manifold $M(G) = (\mathbb{H}^n_{\mathbb{C}} \cup \Omega(G))/G$, an application of Kohn–Rossi analytic extension theorem shows that the boundary of this manifold $M(G)$ is connected, and the limit set $\Lambda(G)$ is in some sense small (see [EMM] and, for quaternionic and Cayley hyperbolic manifolds, [C1]). Moreover, according to a recent result of D. Burns based on an uniformization theorem [BuM] for isolated ends of complex analytic spaces (which can be also independently obtained as a corollary of Napier–Ramachandran Theorem 4.4 in [NR]), the same claim about connectedness of the boundary $\partial M(G)$ still holds if only a boundary component is compact. Burns's theorem even claims that in dimensions $n \geq 3$, such compactness condition implies even geometrical finiteness of the whole complex n-manifold (see also [R]).

However, if $\partial M(G)$ has no compact components, and there is no finiteness condition on the holonomy group of $M(G)$, our algorithmical construction in [AX1] shows that the situation is completely different:

Theorem 4.1. *For any integers k, k_0, $k \geq k_0 \geq 0$, and $n \geq 2$, there exists a complex hyperbolic n-manifold $M = \mathbb{H}^n_{\mathbb{C}}/G$, $G \subset PU(n,1)$, whose boundary at infinity splits up into k connected $(n-1)$-manifolds, $\partial_\infty M = N_1 \cup \cdots \cup N_k$. Moreover, for each boundary component N_j, $j \leq k_0$, the inclusion $i_j : N_j \subset M(G)$ induces a homotopy equivalence of N_j to $M(G)$.*

The construction in the proof of this theorem firstly provides discrete groups $G \subset PU(n,1)$ with two connected components of the discontinuity set $\Omega(G) \subset \partial\mathbb{H}^n_{\mathbb{C}}$. Here we essentially use properties of Heisenberg inversions (3.6) in hyperchains in $\mathbb{H}^n_{\mathbb{C}}$. Then we apply an idea due to A.Tetenov [Te, KAG] in order to construct groups G with any given number k of topological G-invariant balls $\Omega_i \subset \Omega(G)$ with common boundary $\partial\Omega_i = \Lambda(G)$. The groups we construct in the proof are all however infinitely generated. The finitely generated case has a close relation to Problem 3.14:

Problem 4.2. *How many boundary components are there in a complex manifold $M(G)$ with finitely generated fundamental group $G \subset PU(n,1)$?*

Toward this problem, we can show that the situation described in Theorem 4.1 is impossible if the complex hyperbolic manifold M is geometrically finite. Namely, if the manifold $M(G)$ has non-compact boundary $\partial M = \Omega(G)/G$ with a component $N_0 \subset \partial M$ homotopy equivalent to $M(G)$, then there exists a compact homology cobordism $M_c \subset M(G)$ homotopy equivalent to $M(G)$, and $M(G)$ can be easily reconstructed from M_c by gluing up a finite number of standard open "Heisenberg collars", see [AX1]:

Theorem 4.3. *Let $G \subset PU(n,1)$ be a geometrically finite non-elementary torsion free discrete group whose Kleinian manifold $M(G)$ has non-compact boundary $\partial M = \Omega(G)/G$ with a component $N_0 \subset \partial M$ homotopy equivalent to $M(G)$. Then there exists a compact homology cobordism $M_c \subset M(G)$ such that $M(G)$ can be reconstructed from M_c by gluing up a finite number of open collars $M_i \times [0,\infty)$ where each M_i is finitely covered by the product $E_k \times B^{2n-1-k}$ of a closed $(2n-1-k)$-ball and a closed k-manifold E_k which is either flat or a nil-manifold (with 2-step nilpotent fundamental group).*

We refer the reader to [AX1] for more precise formulation and proof of this cobordism theorem.

It is still unknown what are suitable cuts of 4-manifolds, which (conjecturally) split them into geometric blocks (alike Jaco–Shalen–Johannson decomposition of 3-manifolds in Thurston's geometrization program; for a classification of 4-dimensional geometries, see [F, Wa]). However, the above homology cobordism Theorem allows one to study complex surfaces and Cauchy–Riemannian 3-manifolds at their infinity by using decomposition of such 4-manifolds along integer homology 3-spheres and "almost flat" 3-manifolds (with virtually nilpotent fundamental groups), and applying the gauge theory together with homology (cobordism) invariants. We mention that, due to Milnor [Ml], all Seifert homology 3-spheres can be seen as the boundaries at infinity of (geometrically finite) complex hyperbolic 2-orbifolds. Along this line, one can also investigate the following question (which is also related to Problem 5.4):

Problem 4.4. *Are there spherical Cauchy–Riemannian structures on integer homology 3-spheres of plumbing type or on real hyperbolic homology 3-spheres?*

One can be especially interested in this question for integer homology spheres obtained by splicing of two Seifert homology spheres along their singular fibers, see [Sa4]. We remark that, due to Chern and Hamilton [CH], every compact orientable 3-manifold admits a nondegenerate CR-structure. It is however unknown whether such a CR-structure may be spherical. Another interesting fact (due to Livingston–Myers construction [My]) is that any integer homology 3-sphere is homology cobordant to a hyperbolic one. On the other hand, as it was shown by C. T. C. Wall [Wa], the assignment of the appropriate geometry (when available) gives a detailed insight into the intrinsic structure of a complex surface. We mention here Yau's uniformization theorem [Y] which implies that every smooth complex projective 2-surface M with positive canonical bundle and satisfying the topological condition that $\chi(M) = 3\,\mathrm{Signature}(M)$, is a complex hyperbolic manifold. The necessity of homology sphere decomposition in dimension four is due to M. Freedman and L. Taylor result ([FT]):

Let M be a simply connected 4-manifold with intersection form q_M which decomposes as a direct sum $q_M = q_{M_1} \oplus q_{M_2}$, where M_1, M_2 are smooth manifolds. Then the manifold M can be represented as a connected sum $M = M_1 \#_\Sigma M_2$ along a homology sphere Σ.

As an example of such a splitting of a simply connected complex surface along $\mathbb{Z}$-homology 3-spheres, one can consider the Kummer surface $K3$. There are four disjointly embedded (Seifert fibered) $\mathbb{Z}$-homology 3-spheres in $K3$, which split it into five blocks with intersection forms $Q_{X_j} = E_8$ and $Q_{Y_i} = H$:

$$K3 = \{[z_0, z_1, z_2, z_3] \in \mathbb{CP}^3 : z_0^4 + z_1^4 + z_2^4 + z_3^4 = 0\} = X_1 \cup_\Sigma Y_1 \cup_{\Sigma'} Y_2 \cup_{-\Sigma'} Y_3 \cup_{-\Sigma} X_2 ,$$

Here the $\mathbb{Z}$-homology spheres Σ and Σ' are correspondingly the Poincaré homology sphere $\Sigma(2, 3, 5)$ and Seifert fibered homology sphere $\Sigma(2, 3, 7)$; the minus sign means the change of orientation. Their embeddings in $K3$ are described in [Lo] and [FS1].

However we note that it is unknown whether the obtained blocks may support some homogeneous 4-geometries. This raises a question whether homogeneous geometries or splitting along homology spheres (important from the topological point of view) are relevant for a geometrization of smooth 4-dimensional manifolds. For example, due to [EMM], neither of Y_i blocks in the above splitting (with the intersection form H) can support a complex hyperbolic structure (which is a natural geometric candidate since Σ has a spherical CR-structure) because each of them has two compact boundary components.

We refer to [Sa1-Sa4] and [Mat] for recent advances on Floer homology of homology 3-spheres and a new Saveliev's (presumably, homology cobordism) invariant based on Floer homology.

5. Homeomorphisms induced by group isomorphisms

The main problem we are concerned in this section is about geometric realizations of isomorphisms of geometrically finite discrete groups $G, H \subset \operatorname{Isom} X$ of isometries of a negatively curved space X:

Problem 5.1. *Given an isomorphism $\varphi : G \to H$ of geometrically finite discrete groups $G, H \subset \operatorname{Isom} X$, find subsets $X_G, X_H \subset \overline{X}$ invariant for the action of groups G and H, respectively, and an equivariant homeomorphism $f_\varphi : X_G \to X_H$ which induces the isomorphism φ. Determine metric properties of f_φ, in particular, whether it is either quasisymmetric or quasiconformal with respect to the given negatively curved metric d in X (or the induced sub-Riemannian structure on the Carnot–Carathéodory space at infinity $Y = \overline{X}\backslash\{\infty\}$).*

Such type problems were studied by several authors. In the case of lattices G and H in rank 1 symmetric spaces X, G. Mostow [Mo1] proved in his celebrated rigidity theorem that such isomorphisms $\varphi : G \to H$ can be extended to inner isomorphisms of X, provided that there is no analytic homomorphism of X onto $PSL(2, \mathbb{R})$. For that proof, it was essential to prove that φ can be induced by a quasiconformal homeomorphism of the sphere at infinity $\overline{X}$ which is a one point compactification of a Carnot group N. Quasiconformal mappings between general Carnot groups have been studied by P. Pansu [P]. For the case of complex hyperbolic spaces $X = \mathbb{H}^n_{\mathbb{C}}$ and the Heisenberg groups at their infinity, foundations of the theory of quasiconformal mappings has been made by A. Koranyi and M. Reimann [KR1, KR2].

The essential component of such rigidity results is the fact that quasiconformal mappings of Carnot–Carathéodory spaces are almost everywhere differentiable and preserve the contact structures of these spaces (horizontal vector fields). This fact is due to Pansu [P] in the case of graded nilpotent groups. In the case of general Carnot–Carathéodory spaces determined by "horizontal" subbundles of their tangent bundles, this result has been recently proven by G. Margulis and G. Mostow [MM]. The second important ingredient for such rigidity is the *ACL*-property on lines for quasiconformal mappings, with respect to the Carnot–Carathéodory metric at infinity $\overline{X}\backslash\{\infty\}$, see [Mo3, KR2, Ca, CT].

If the groups $G, H \subset \operatorname{Isom} X$ are neither lattices nor trivial and have parabolic elements, the only results on geometric realization of their isomorphisms are known for real hyperbolic spaces X (of constant negative curvature). In dimension $\dim X \geq 3$ they are due to P. Tukia [Tu] and use a natural "type-preserving" condition on the isomorphism φ, that parabolic elements are carried out only to parabolic ones. In the case of variable negative curvature, those methods cannot be used due to lack of control over convex hulls (where the convex hull of three points may be 4-dimensional), especially nearby cusps. Another (dynamical) approach due to C. Yue [Yu2, Cor.B] (and the Anosov–Smale stability theorem for hyperbolic flows) can be used only for convex cocompact groups G and H, see [Yu3]. However, as a first step in solving the above Problem 5.1, we can analyze the metric of X and the induced metrics on horospheres in X. Such analysis, a generalization of W. Floyd construction of the group completion [Fl], and the described above analysis of geometrical finiteness in X are crucial components of our isomorphism theorem [A12] constructing canonical homeomorphisms of subsets in the sphere at infinity $\overline{X}$, that induce type-preserving isomorphisms $\varphi : G \to H$ of geometrically finite groups G and H (not necessarily lattices):

Theorem 5.2. *Let $\varphi : G \to H$ be a type preserving isomorphism of two non-elementary geometrically finite discrete subgroups $G, H \subset PU(n, 1)$. Then there exists a unique equivariant homeomorphism $f_\varphi : \Lambda(G) \to \Lambda(H)$ of their limit sets that induces the isomorphism ϕ.*

Upon existence of such equivariant homeomorphism f_φ, the above problem would be reduced to the question whether f_φ is quasisymmetric with respect to the Carnot–Carathéodory metric. In the case of an affirmative answer, it may be possible to find a global G-equivariant homeomorphism (of the sphere at infinity ∂X or even the whole space $\overline{X}$ inducing the isomorphism φ. However, in contrast to the real hyperbolic case, here we have an interesting fenomenon related to possible noncompactness of the boundary $\partial M(G) = \Omega(G)/G$. Namely, even for the simplest case of cyclic groups $G \cong H \subset PU(n, 1)$, the homeomorphic CR-manifolds $\partial M(G) = \mathcal{H}^n/G$ and $\partial M(H) = \mathcal{H}^n/H$ may be not quasiconformally equivalent, see [Mn]. So it may be possible to (affirmatively) answer the following problem:

Problem 5.3. *When are there quasisymmetric homeomorphisms in a Carnot-Carathéodory space $\partial X \setminus \{\infty\}$ compatible with the action of discrete geometrically finite groups $G, H \subset \operatorname{Isom} X$ but quasisymmetrically non-extendable to the whole space?*

We note that, besides the metrical (quasisymmetric) part of this problem, some topological obstructions for extensions of equivariant homeomorphisms f_φ of the limit sets, $f_\varphi : \Lambda(G) \to \Lambda(H)$, may exist. It follows from the following example.

Example 5.4. Let $G \subset PU(1, 1) \subset PU(2, 1)$ and $H \subset PO(2, 1) \subset PU(2, 1)$ be two geometrically finite (loxodromic) groups isomorphic to the fundamental group $\pi_1(S_g)$ of a compact oriented surface S_g of genus $g > 1$. Then the equivariant homeomorphism $f_\varphi : \Lambda(G) \to \Lambda(H)$ cannot be homeomorphically extended to the whole sphere $\partial\mathbb{H}^2_{\mathbb{C}} \approx S^3$.

The obstruction in this example is topological and is due to the fact that the quotient manifolds $\mathbb{H}^2_{\mathbb{C}}/G$ and $\mathbb{H}^2_{\mathbb{C}}/H$ are not homeomorphic. Namely, these complex surfaces are disk bundles over the surface S_g and have different Toledo invariants: $\tau(\mathbb{H}^2_{\mathbb{C}}/G) = 2g-2$ and $\tau(\mathbb{H}^2_{\mathbb{C}}/H) = 0$, see [To].

We remark that some of Carnot–Carathéodory spaces are very rigid. In fact, due to P. Pansu [Pa], any quasiconformal map on the boundary ∂X of a quaternionic or octanionic hyperbolic space X (which are symmetric spaces of rank 1) is necessarily an extension of an isometry in X. This shows that any non-trivial homeomorphism $f_\varphi : \Lambda(G) \to \Lambda(H)$ (non-isometry) is non-extendable to a quasiconformal map of some open subset of ∂X.

In order to attack the above problem, in particular to construct such non-trivial geometric isomorphisms φ in both rigid (like quaternionic and octanionic spaces) and more flexible (like $\mathbb{H}^n_{\mathbb{C}}$) spaces, one probably may to generalize our block-building method [A1, A5] (whose usefulness in conformal category has been also demonstrated in [A6, A11]). Another aspect of such constructions should be

Problem 5.5. *Develop geometric combination theorems generalizing well known Maskit combination theorems for Kleinian groups, see* [Mas, A1].

So far, there are the only simplest versions (free products) of such combination theorems, see [FG]. The general case of free amalgamated product is still unknown even in the complex hyperbolic case, where one can however use a week Maskit combination, see [Be].

As the first steps in investigating Problem 5.1, it appears to be promising to study G-equivariant homeomorphisms in Carnot groups $Y = \partial X\setminus\{\infty\}$, with Carnot–Carathéodory metric ρ, from the following two classes of embeddings $f : A \hookrightarrow Y$, $A \subset Y$.

The first one consists of well known quasisymmetric embeddings f, see [TV].

The second class consists of embeddings $f : A \hookrightarrow Y$, $A \subset Y$, which generalize the so-called quasi-Möbius embeddings introduced in conformal category by V. Aseev [As, AsT] and J. Väisälä [V]. They have bounded distortion of the cross-ratio (or, equivalently, of the complex cross-ratio in the sense of Koranyi–Reimann [CR4]),

$$CR(q) = \rho(x_1, x_2)\rho(x_3, x_4)\rho(x_1, x_3)^{-1}\rho(x_2, x_4)^{-1}, \tag{5.6}$$

of quadruples $q = \{x_1, x_2, x_3, x_4\} \in (Y)^4$ in the following sense.

Let $\omega : \mathbb{R}_+ \to \mathbb{R}_+$ be a given homeomorphism, $\omega(0) = 0$. Then, for any quadruple $q \in A^4$ and its f-image $f(q) = \{f(x_1, \dots, f(x_4\} \in Y^4$, we require that

$$CR(f(q)) \leq \omega(CR(q)), \tag{5.7}$$

and call such embeddings f *quasi-CR embeddings*. By continuity, the cross ration can trivially be extended to the one-point compactification $Y \cup \{\infty\}$, so we can consider embeddings which are not necessarily fixing ∞.

In particular, for homeomorphism $\omega(t) = M \cdot \eta_\alpha(t)$ with given constants $M > 0$ and $\alpha \geq 1$ where

$$\eta_\alpha(t) = \begin{cases} t^\alpha, & \text{for } t \geq 1 \\ t^{1/\alpha}, & \text{for } 0 \leq t < 1, \end{cases}$$

we call such embeddings f satisfying (5.7) as (M, α)-CR-embeddings.

In particular, this approach can be used to study the geometric realization Problem 5.1 in the case of free geometrically finite groups $G, H \subset \operatorname{Isom} X$ whose limit sets are discontinua. Here we have the following

Conjecture 5.8. An embedding $f : A \hookrightarrow \overline{Y}$, $A \subset \overline{Y}$, in a Carnot–Carathéodory space Y is quasi-CR if and only if f preserves the notion of bounded *μ-density* of subsets $\Sigma \subset A$.

Here a set Σ is called *μ-dense* for some $\mu > 1$ if any two points $a, b \in \Sigma$ can be connected by a *μ-chain* $\{x_i\}$ of points $x_i \in \Sigma$ where $\lim_{i\to-\infty} x_i = a$, $\lim_{i\to+\infty} x_i = b$ and $\ln(CR(a, x_i, x_{i+1}, b)) \leq \ln \mu$. Naturally, the μ-density property is a characteristic property of the limit discontinua for geometrically finite groups. We also suspect that that property distinguishes parabolic subgroups of the groups G and H which cannot be quasiconformally conjugate (we expect that they should have non-maximal rank). On the other hand, we expect that an embedding $f : A \hookrightarrow \overline{Y}$ of a μ-dense set A in a Carnot–Carathéodory space Y is quasi-CR if and only if it is an (M, α)-CR embedding with some finite M and α. One can expect an affirmative solution of the above problems in the case of geometrically finite groups whose limit discontinua consist only of conical points (for the Heisenberg group Y, compare [Mn]). For study of these problems, one can use methods developed by Koranyi and Reimann [KR1-4], Capogna [Ca, CT] as well as a generalization of the technique known as Sullivan's microscope (see the next section).

6. Deformations of discrete groups and Sullivan's stability

Here we concern with problems related to deformation spaces of a geometrically finite discrete group $G \subset PU(n, 1)$ acting by isometries in the complex hyperbolic space $\mathbb{H}^n_{\mathbb{C}}$ and to the stability theorem of D. Sullivan [Su3] (which has been originally proved for planar Kleinian groups, see also [A11, MaG]). We restrict our attention to the complex hyperbolic case because other rank 1 spaces with negative variable curvature such as quaternionic and octanionic hyperbolic spaces are more rigid. In particular, due to Corlette's rigidity theorem for harmonic maps [C2], finite volume manifolds locally modeled on these spaces are super-rigid, completely analogous to Margulis super-rigidity in higher rank, see [Mg1]. Moreover, due to Gromov and Schoen [GS], all such finite volume manifolds are arithmetic.

Let $G \subset PU(n, 1)$ be a geometrically finite discrete group. Then we have a fundamental problem concerning the deformation space of the inclusion (or of the representation of the group G obtained by restricting a natural inclusion $PU(n, 1) \to PU(n+1, 1)$ to G). The corresponding problem for deformation spaces of a hyperbolically rigid real hyperbolic lattice $G \subset O(n, 1)$, $n \geq 3$, has its roots in the first construction by the author [A2] of non- trivial curves in the Teichmüller space $T(G)$ of conjugacy classes of representations of G in $O(n+1, 1)$. Then the situation was greatly clarified by Thurston's "Mickey Mouse" example [Th] showing that such deformations of a hyperbolic surface are in fact bendings along geodesics. Since that time, several authors studied such deformation spaces, see for example [A4, A6, A8, A9-A11, JM, Ko].

In the case of deformations of a complex hyperbolic group $G \subset PU(n, 1)$, we would like to assume that G is not a uniform lattice, that is the quotient $\mathbb{H}^n_{\mathbb{C}}/G$ is noncompact. Otherwise, due to a fundamental result on local rigidity of deformations [GoM], the set of "Fuchsian" representations of G, $\{\mathrm{Ad}\, h : G \to hGh^{-1},\ h \in PU(n+1, 1)\}$, is a connected component $\mathcal{R}_0$ of the variety of representations $\mathrm{Hom}(G, PU(n+1, 1))$.

On the other hand, we may deform an isomorphic uniform lattice $G' \subset PO(2, 1)$, as well as another isomorphic convex co-compact groups $G' \subset PU(2, 1)$, $G' \cong G$, if such groups have non-elementary subgroups preserving totally geodesic real planes [A13]. Such quasiconformal deformations bend the corresponding complex surfaces along any simple closed geodesic in its totally real geodesic 2-dim subsurface. For another deformations of groups $G' \subset PO(2, 1)$, see also [GP2]. We can also deform the above groups $G' \subset PO(2, 1)$ in $PO(3, 1) \subset PU(3, 1)$, so it makes sense to consider such deformations in higher dimensions, as it has been done by the author in the real hyperbolic case, compare [A9-A11].

The main problem we deal with is as follows.

Conjecture 6.1. *Let $G \subset PU(n, 1)$ be a geometrically finite group which is either convex cocompact or having parabolic subgroups of rank at least 3. Then every its representation $\rho : G \to PU(n+k, 1)$, $k \geq 0$, that close enough to a natural inclusion, is in fact discrete and faithful and, furthermore, is a quasiconformal conjugation. That means that there is an equivariant quasiconformal self-homeomorphism f of the extended Heisenberg group $\overline{\mathcal{H}^{n+k}}$ such that $\rho G = f_* G = f G f^{-1}$.*

This conjecture generalizes a remarkable structural stability theorem of D. Sullivan [Su3] for Kleinian subgroups G in $PSL(2, \mathbb{C})$, which shows that an algebraic structural stability for holomorphic perturbations implies a hyperbolicity property for the action of G on its limit set. In conformal category, we refer to [A11] for our proof of its high dimensional analog. The complex case provides a possibility to generalize a crucial Sullivan's argument, the so-called λ-Lemma proved by Mañe, Sad and Sullivan

[MSS]. Here topological stability is an intermediate open problem. We note also that the negative curvature property is crucial for algebraic stability of convex cocompact subgroups, due to G. Martin [MaG]:

Theorem 6.2. *Let* $\{G_t : t \in \mathbb{R}\}$ *be a continuous deformation of a torsion free convex cocompact group* $G \subset \mathrm{Isom}\, X$ *in a space* X *of pinched negative curvature, with convex cocompact* G_t. *Then* $\{G_t : t \in \mathbb{R}\}$, *and its closure in the topology of algebraic convergence in* Isom X, *consists entirely of isomorphic groups.*

For geometrically finite groups without parabolics, it is possible to use a generalization of the so-called Sullivan's microscope (see [Su2, A11]) to construct probable quasiconformal conjugation of the group actions on their limit sets. It is possible because one can find expanding Heisenberg coverings of the limit set of such a group, which defines a good Cauchy–Riemannian dynamics there. As another result important for a generalization of Sullivan's arguments, we mention an uniformization theorem for Cauchy–Riemannian structures which is due to Falbel and Gusevskii [FG].

Finally, we would like to mention another problems which are linked with Conjecture 6.1 and, in dimension $n = 2$, with Problem 3.14 about geometrical finiteness. These problems are related to the boundaries of the deformation spaces $\mathcal{R}(G, k) = \mathrm{Hom}(G, PU(n+k, 1))$ and $\mathcal{T}(G, k) = \mathcal{R}(G, k)/PU(n+k, 1)$, $k \geq 0$, to the number of their connected components, discreteness and geometrical finiteness of boundary representations. In particular:

Problem 6.3. *Is any representation* ρ *obtained as the limit of geometrically finite representations* $\rho_i \in \mathcal{R}(G, k)$ *discrete and faithful, that is* $\rho(G) \subset PU(n+k)$ *is a discrete group isomorphic to a given geometrically finite group* $G \subset PU(n, 1)$?

We note that, due to G. Martin [MaG], the subset of discrete faithful representations is closed in $\mathcal{R}(G, k)$.

Another question is about the boundary of the Teichmüller space $\mathcal{T}(G, k)$ of a convex cocompact group $G \subset PU(n, 1)$, $n \geq 2$, $k \geq 0$.

Problem 6.4. *Are there convex cocompact faithful representations* $\rho_t \in \mathcal{R}(G, k)$, $t \in \mathbb{R}$, *of a convex cocompact group* $G \subset PU(n, 1)$ *which converge to a boundary discrete representation* ρ *whose image* $\rho(G) \subset PU(n+k)$ *has accidental parabolic elements?*

We remark that I. Belegradek [Be] constructed a discrete faithful representation ρ of the fundamental group G of a compact hyperbolic surface into $PU(2, 1)$ such that $\rho(G)$ has parabolic elements. This construction uses a Maskit combination theorem, and it is unclear whether $\rho(G)$ lies in the boundary $\partial\mathcal{R}_0(G)$ of a component of the variety of convex co-compact representations $G \to PU(2, 1)$. However the Teichmüller space

$\mathcal{T}(G) = \mathcal{T}(G, 0)$ does have "cusps" due to a recent result by Apanasov and Gusevskii [AG2]:

Theorem 6.5. *Let $G \subset PO(2,1) \subset PU(2,1)$ be a uniform lattice isomorphic to the fundamental group of a closed surface S_g of genus $p \geq 2$. Then, for any simple closed geodesic $\alpha \subset S_p = H^2_{\mathbb{R}}/G$, there is a continuous deformation $\rho_t = f_t^*$ induced by G-equivariant quasiconformal homeomorphisms $f_t : \mathbb{H}^2_{\mathbb{C}} \to \mathbb{H}^2_{\mathbb{C}}$ whose limit representation ρ_∞ corresponds to a boundary cusp point of the Teichmüller space $\mathcal{T}(G)$. In other words, the boundary group $\rho_\infty(G)$ has an accidental parabolic element $\rho_\infty(g_\alpha)$ where $g_\alpha \in G$ represents the simple closed geodesic $\alpha \subset S_p$.*

We note that, due to our construction of such continuous quasiconformal deformations in [AG2], such deformations are independent if the corresponding simple closed geodesics $\alpha_i \subset S_p$ are disjoint. It implies the existence of a boundary group in $\partial\mathcal{T}(G)$ with "maximal" number of non-conjugate accidental parabolic subgroups:

Corollary 6.6. *Let $G \subset PO(2,1) \subset PU(2,1)$ be as in Theorem 6.5. Then there is a continuous deformation $R : \mathbb{R}^{2p-2} \to \mathcal{T}(G)$ whose boundary group $G_\infty = R(\infty)(G)$ has $3p-3$ non-conjugate accidental parabolic subgroups.*

Finally, we mention another aspect of the intrigue Problem 3.14:

Problem 6.7. *Construct a geometrically infinite (finitely generated) group $G \subset PU(n,1)$, $n \geq 2$, whose limit set is the whole sphere at infinity, $\Lambda(G) = \partial\mathbb{H}^n_{\mathbb{C}} = \overline{\mathcal{H}^n}$, and which is the limit of convex cocompact groups $G_i \subset PU(n,1)$ from the Teichmüller space $\mathcal{T}(G)$ of a convex cocompact group $G \subset PU(n,1)$. Is that possible for a Schottky group G?*

Naturally, one can expect such groups on the boundary of Teichmüller space $\mathcal{T}(G)$. Due to the Goldman–Millson [GM] local rigidity of deformations of embeddings $G \subset PU(n-1,1) \subset PU(n,1)$, $n > 2$, one may expect that some components of the space $\mathcal{T}(G)$ degenerate. Another components of $\mathcal{T}(G)$ are however big enough. The latter non-rigidity results are in particular due to the the author construction [A13] of bending deformations of complex hyperbolic surfaces along simple closed geodesics and another results on deformations of Stein manifolds [ACG, AG].

References

[Ah] Lars V. Ahlfors, Fundamental polyhedra and limit point sets of Kleinian groups, Proc. Nat. Acad. Sci. USA, 55 (1966), 251–254.

[As] V. V. Aseev, Quasi-symmetric embeddings and mappings with bounded distortion of moduli, Sib. Matem. Zh./VINITI, Dep 7190-84, 1984 (Russian).

[AsT] V. V. Aseev and D. A. Trotsenko, Quasi-symmetric embeddings, quadruples of points, and moduli distortion, Sibirsk. Mat. Zh. 28:4 (1987), 32–38 (Russian); Engl. Transl.: Siberian Math. J. 28 (1987).

[Au] L. Auslander, Bieberbach's theorem on space groups and discrete uniform subgroups of Lie groups, II, Amer. J. Math. 83 (1961), 276–280.

[A1] Boris Apanasov, Discrete Groups in Space and Uniformization Problems, Math. and Appl. 40, Kluwer Academic Publishers, Dordrecht, 1991.

[A2] ——, Nontriviality of Teichmüller space for Kleinian group in space, Riemann Surfaces and Related Topics: Proc. 1978 Stony Brook Conference (I. Kra and B. Maskit, Eds), Ann. of Math. Studies 97, Princeton Univ. Press, 1981, 21–31.

[A3] ——, Geometrically finite hyperbolic structures on manifolds, Ann. of Glob. Analysis and Geom. 1:3 (1983), 1–22.

[A4] ——, Thurston's bends and geometric deformations of conformal structures, Complex Analysis and Applications'85, Publ. Bulgarian Acad. Sci., Sofia, 1986, 14–28.

[A5] ——, Quasisymmetric embeddings of a closed ball inextensible in neighborhoods of any boundary point, Ann. Acad. Sci. Fenn. Ser. A I Math. 14 (1989), 243–255.

[A6] ——, Nonstandard uniformized conformal structures on hyperbolic manifolds, Invent. Math. 105 (1991), 137–152.

[A7] ——, Kobayashi conformal metric on manifolds, Chern–Simons and η-invariants, Internat. J. Math. 2 (1991), 361–382.

[A8] ——, Deformations of conformal structures on hyperbolic manifolds, J. Differential Geom. 35 (1992), 1–20.

[A9] ——, Quasisymmetric knots and Teichmüller spaces, Russian Acad. Sci. Dokl. Math. 354 (1997), to appear.

[A10] ——, Varieties of discrete representations of hyperbolic 3-lattices, in: Algebra (Yu. L. Ershov et al., eds.), Walter de Gruyter, Berlin–New York 1996, 7–19.

[A11] ——, Conformal geometry of discrete groups and manifolds, de Gruyter Exp. Math., Walter de Gruyter, Berlin–New York, to appear.

[A12] ——, Canonical homeomorphisms in Heisenberg groups induced by isomorphisms of discrete subgroups of $PU(n, 1)$, Russian Acad. Sci. Dokl. Math. 356 (1997), to appear.

[A13] ——, Bending deformations of complex hyperbolic surfaces, J. Reine Angew. Math., to appear;
http://www.msri.org/MSRI-preprints/online/1996-062.html.

[A14] ——, Geometry and topology of complex hyperbolic CR-manifolds, Russian Math. Surveys 52:5 (1997), to appear; preprint 1997-007, Math. Sci. Res. Inst., Berkeley, 1997;
http://www.msri.org/MSRI-preprints/online/1997-007.html.

[ACG] Boris Apanasov, Mario Carneiro and Nikolay Gusevskii, Some deformations of complex hyperbolic surfaces, in preparation.

[AG] Boris Apanasov and Nikolay Gusevskii, The boundary of Teichmüller space of complex hyperbolic surfaces, in preparation.

[AX1] B. Apanasov and X. Xie, Geometrically finite complex hyperbolic manifolds, Internat. J. Math. 8:7 (1997), to appear.

[AX2] ——, Manifolds of negative curvature and nilpotent groups, in: Complex Analysis, Differemtial Geometry and Mathematical Physics, Proc. of the third Internat. Workshop at Varna (K. Sekigawa and S. Dimiev, eds.), World Scientific, Singapore, 1997, to appear.

[AX3] ——, Discrete isometry groups of nilpotent Lie groups, preprint, Univ. of Oklahoma.

[AF] C. S. Aravinda and F. T. Farrell, Rank 1 aspherical manifolds which do not support any nonpositively curved metric, Comm. Anal. Geom. 2 (1994), 65–78.

[BGS] W. Ballman, M. Gromov and V. Schroeder, Manifolds of nonpositive curvature, Birkhäuser, Basel 1985.

[BM] A. Beardon and B. Maskit, Limit points of Kleinian groups and finite sided fundamental polyhedra, Acta Math. 132 (1974), 1–12.

[Be] I. Belegradek, Discrete surface groups actions with accidental parabolics on complex hyperbolic plane, preprint, 1995.

[Bo] B. Bowditch, Geometrical finiteness with variable negative curvature, Duke J. Math. 77 (1995), 229–274.

[BuM] D. Burns and R. Mazzeo, On geometry of cusps for $SU(n, 1)$, preprint, Univ. of Michigan, 1994.

[BS] D. Burns and S. Shnider, Spherical hypersurfaces in complex manifolds, Invent. Math. 33 (1976), 223–246.

[BK] P. Buser and H. Karcher, Gromov's almost flat manifolds, Astérisque 81 (1981), 1–148.

[Ca] L. Capogna, Optimal regularity of solutions to quasilinear equations in nilpotent Lie groups and applications, preprint, Courant Inst., 1977.

[CT] L. Capogna and P. Tang, Uniform domains and quasiconformal mappings on the Heisenberg group, Manuscripta Math. 86 (1995), 267–281.

[C1] K. Corlette, Hausdorff dimensions of limit sets, Invent. Math. 102 (1990), 521–542.

[C2] ——, Archimedian superrigidity and hyperbolic geometry, Ann. of Math. 135 (1992), 165–182.

[CG] S. Chen and L. Greenberg, Hyperbolic spaces, Contributions to Analysis, Academic Press, New York, 1974, 49–87.

[CH] S. S. Chern and R. S. Hamilton, On Riemannian metrics adapted to three-dimensional contact manifolds, Workshop Bonn 1984, Lecture Notes in Math. 1111 (1985), 279–308.

[DS] S. K. Donaldson and D. Sullivan, Quasiconformal 4-manifolds, Acta Math. 163 (1989), 181–252.

[EHS] P. Eberlein, U. Hamenstädt and V. Schroeder, Manifolds of nonpositive curvature, Proc. Symp. Pure Math. 54:3, 1993, 179–227.

[EMM] C. Epstein, R. Melrose and G. Mendoza, Resolvent of the Laplacian on strictly pseudoconvex domains, Acta Math. 167 (1991), 1–106.

[FG] E. Falbel and N. Gusevskii, Spherical CR-manifolds of dimension 3, Bol. Soc. Bras. Mat. 25 (1994), 31–56.

[FS1] Ronald Fintushel and Ronald Stern, Homotopy $K3$ surfaces containing $\Sigma(2, 3, 7)$, J. Differential Geom. 34 (1991), 255–265.

[FS2] ——, Surgery in cusp neighborhoods and the geography of irreducible 4-manifolds, Invent. Math. 117 (1994), 455–523.

[FT] M. H. Freedman and L. Taylor, Λ-splitting 4-manifolds, Topology 16 (1977), 181–184.

[GM] F. W. Gehring and G. J. Martin, Discrete quasiconformal groups, Proc. London Math. Soc. 55 (1987), 331–358.

[Go] W. Goldman, Complex hyperbolic geometry, Oxford Univ. Press, to appear.

[GoM] W. Goldman and J. Millson, Local rigidity of discrete groups acting on complex hyperbolic space, Invent. Math. 88 (1987), 495–520.

[GP1] W. Goldman and J. Parker, Dirichlet polyhedron for dihedral groups acting on complex hyperbolic space, J. of Geom. Analysis 2:6 (1992), 517–554.

[GP2] ——, Complex hyperbolic ideal triangle groups, J. Reine Angew. Math. 425 (1992), 71–86.

[Gr1] M. Gromov, Almost flat manifolds, J. Differential Geom. 13 (1978), 231–141.

[Gr2] ——, Carnot–Carathéodory spaces seen from within, preprint IHES, Bures-sur-Yvette, 1994.

[JM] D. Johnson and J. Millson, Deformation spaces associated to compact hyperbolic manifolds, in: Discrete Groups in Geometry and Analysis, Birkhäuser, Boston 1987, 48–106.

[KR1] A. Koranyi and M. Reimann, Quasiconformal mappings on the Heisenberg group, Invent. Math. 80 (1985), 309–338.

[KR2] ——, Foundations for the theory of quasiconformal mappings on the Heisenberg group, Adv. Math. 111 (1995), 1–87.

[KR3] ——, Contact transformations as limits of symplectomorphisms, C. R. Acad. Sci. Paris 318 (1994), 1119–1124.

[KR4] ——, The complex cross ratio on the Heisenberg group, Enseign. Math. 33 (1987), 291–300.

[Ko] C. Kourouniotis, Deformations of hyperbolic structures, Math. Proc. Cambridge Philos. Soc. 98 (1985), 247–261.

[LB] Claude LeBrun, Einstein metrics on complex surfaces, Geometry and Physics (J. E. Andersen a.o., eds.), Lecture Notes in Pure and Appl. Math. 184, Dekker, New York 1996.

[Lo] Eduard Looijenga, The smoothing components of a triangle singularity, Proc. Symp. Pure Math. 40 (1983), 173–184.

[MSS] R. Mañe, P. Sad and D. Sullivan, On the dynamics of rational maps, Ann. Sci. École Norm. Sup. 16 (1983), 193–217.

[Md] A. Marden, The geometry of finitely generated Kleinian groups, Ann. of Math. 99 (1974), 383–462.

[Mg1] G. A. Margulis, Discrete groups of motions of manifolds of nonpositive curvature, Amer. Math. Soc. Transl. 109 (1977), 33–45.

[Mg2] ——, Free properly discontinuous groups of affine transformations, Dokl. Acad. Sci. USSR 272 (1983), 937–940.

[MM] G. A. Margulis and G. D. Mostow, The differential of a quasiconformal mapping of a Carnot–Carathéodory space, Geom. Funct. Anal. 5 (1995), 402–433.

[MaG] G. Martin, On discrete isometry groups of negative curvature, Pacif. J. Math. 160 (1993), 109–127.

[Mas] B. Maskit, Kleinian groups, Springer-Verlag, 1987.

[Mat] R. Matveyev, A decomposition of smooth simply-connected h-cobordant 4-manifolds, preprint, Michigan State Univ., 1995.

[Ml] J. Milnor, On the 3-dimensional Brieskorn manifolds $M(p, q, r)$, Knots, groups and 3-manifolds, Ann. of Math. Studies 84, Princeton Univ. Press, 1975, 175–225.

[Mn] R. Miner, Quasiconformal equivalence of spherical CR manifolds, Ann. Acad. Sci. Fenn. Ser. A I Math. 19 (1994), 83–93.

[Mo1] George D. Mostow, Strong rigidity of locally symmetric spaces, Princeton Univ. Press, 1973.

[Mo2] ——, On a remarkable class of polyhedra in complex hyperbolic space, Pacific J. Math. 86 (1980), 171–276.

[Mo3] ——, A remark on quasiconformal mappings on Carnot groups, Mich. Math. J. 41 (1994), 31–37.

[My] R. Myers, Homology cobordisms, link concordances, and hyperbolic 3-manifolds, Trans. Amer. Math. Soc. 278 (1983), 271–288.

[NR] T. Napier and M. Ramachandran, Structure theorems for complete Kähler manifolds and applications to Lefschetz type theorems, Geom. Funct. Anal. 5 (1995), 807–851.

[Pa] P. Pansu, Métriques de Carnot–Carathéodory et quasiisométries des espaces symmétries de rang un, Ann. Math. 129 (1989), 1–60.

[Ph] M. B. Phillips, Dirichlet polyhedra for cyclic groups in complex hyperbolic space, Proc. Amer. Math. Soc. 115 (1992), 221–228.

[R] Mohan Ramachandran, Bochner–Hartogs phenomenon on complex hyperbolic manifolds, Invited talk, Geometry Conference, May 10–12, Norman/OK, 1996.

[Sa1] Nikolai Saveliev, Floer homology and 3-manifold invariants, Ph.D. Thesis, Univ. of Oklahoma, 1995.

[Sa2] ——, Floer homology and invariants of homology cobordism, preprint, Univ. of Oklahoma.

[Sa3] ——, Dehn surgery along torus knots, Topology Appl., to appear.

[Sa4] ——, Floer homology of Brieskorn homology spheres: solution to Atiyah's problem, preprint 1997-006, Math. Sci. Res. Inst., Berkeley, 1997; http://www.msri.org/MSRI-preprints/online/1997-006.html.

[St] A. N. Starkov, Parabolic fixed points of Kleinian groups and the horospherical foliation on hyperbolic manifolds, Internat. J. Math. 8 (1997), 289–299.

[Su1] D. Sullivan, Hyperbolic geometry and homeomorphisms, Geometric topology, Acad. Press, 1979.

[Su2] ——, Seminar on conformal and hyperbolic geometry, preprint IHES, 1982.

[Su3] ——, Quasiconformal homeomorphisms and dynamics, II: Structural stability implies hyperbolicity for Kleinian groups, Acta Math. 155 (1985), 243–260.

[Te] A. V. Tetenov, The discontinuity set for a Kleinian group and topology of its Kleinian manifold, Internat. J. Math. 4 (1993), 167–177.

[Th] W. P. Thurston, The geometry and topology of three-manifolds, Lect. Notes, Princeton Univ., 1979.

[To] D. Toledo, Representations of surface groups on complex hyperbolic space, J. Differential Geom. 29 (1989), 125–133.

[Tu] P. Tukia, On isomorphisms of geometrically finite Kleinian groups, Publ. Math. IHES 61 (1985), 171–214.

[TV] P. Tukia and J. Väisälä, Quasisymmetric embeddings of metric spaces, Ann. Acad. Sci. Fenn. Ser. A I Math. 5 (1980), 97–114.

[V] J. Väisälä, Quasimöbius maps, J. Analyse Math. 50 (1984/85), 218–234.

[Wa] C. T. C. Wall, Geometric structures on compact complex analytic surfaces, Topology 25 (1986), 119–153.

[Wo] J. Wolf, Spaces of constant curvature, Publ. or Perish, Berkeley, 1977.

[Ya] S.T.Yau, Calabi's conjecture and some new results in algebraic geometry, Proc. Nat. Acad. Sci 74 (1977), 1798–1799.

[Yu1] Chengbo Yue, Dimension and rigidity of quasi-Fuchsian representations, Ann. of Math. 143 (1996)331–355.

[Yu2] ——, Mostow rigidity of rank 1 discrete groups with ergodic Bowen–Margulis measure, Invent. Math. 125 (1996), 75–102.

[Yu3] ——, Private communication, Norman/OK, November 1996.

Special submanifolds of S^6 with its G_2 geometry

John Bolton and Lyndon M. Woodward

Abstract. Cayley multiplication may be used to define a vector cross product on $\mathbb{R}^7$ which in turn enables an orthogonal almost complex structure to be defined on every oriented hypersurface of $\mathbb{R}^7$. In the case of the unit sphere S^6 in $\mathbb{R}^7$, this structure is not integrable but it is nearly Kähler. We discuss the work of several authors concerning submanifolds of S^6 which are either almost complex or totally real. We also describe a characterisation of tubes around almost complex curves in terms of Hopf hypersurfaces of S^6.

1991 Mathematics Subject Classification: 53B25, 53C40, 53A10, 53C42.

1. Introduction

It is well known that the multiplicative structure on the Cayley numbers $\mathcal{O}$ may be used to define an orthogonal almost complex structure J on the unit sphere S^6 in $\mathbb{R}^7$. This almost complex structure is not integrable but is nearly Kähler and is, in a sense made precise in [16], the "best" almost complex structure on S^6.

In this expository article we will describe the work of several authors on submanifolds of S^6 which are in some way adapted to the nearly Kähler structure. The two basic types of submanifolds we discuss are the *almost complex* ones (whose tangent bundle is J-invariant) and the *totally real* ones (where J maps the tangent bundle to the normal bundle). Almost complex submanifolds must be even dimensional, but Gray [27] has shown that S^6 admits no 4-dimensional almost complex submanifolds. Thus the interesting almost complex submanifolds of S^6 are the almost complex curves. These have been studied by several authors and, after discussing the nearly Kähler S^6 in Section 2, we describe in Sections 3 and 4 the work in this area by Bryant and by Bolton, Vrancken and Woodward.

Totally real submanifolds of S^6 have dimension less than or equal to 3. Since points and curves are trivially totally real, the interesting examples are in dimensions 2 and 3. We describe two methods of constructing minimal 2-dimensional examples in a remark in Section 4, and in Section 5 we briefly describe the work of several authors on the 3-dimensional case.

In the final section we discuss a characterization, due to Berndt, Bolton and Woodward, of *Hopf hypersurfaces* of S^6 as either geodesic hypersurfaces or tubes around almost complex curves.

It is a great pleasure to acknowledge the contributions in this area made by Luc Vrancken and Jürgen Berndt. Luc first introduced us to the delights of the nearly Kähler 6-sphere and all our work on special surfaces in S^6 is joint work with him. Similarly, Jürgen told us about Hopf hypersurfaces and this led to the joint work discussed in Section 6.

2. The nearly Kähler 6-sphere

We begin by recalling the vector cross product $\times$ obtained by identifying $\mathbb{R}^7$ with the imaginary part of the Cayley numbers $\mathcal{O}$ and defining

$$u \times v = \mathcal{I}m(uv), \qquad u, v \in \mathbb{R}^7 = \mathcal{I}m\,\mathcal{O}.$$

The exceptional simple Lie group G_2 of algebra automorphisms of $\mathcal{O}$ may be naturally identified with the group of isometries of $\mathbb{R}^7$ preserving the vector cross-product. In this way, G_2 may be regarded as a compact connected subgroup of $SO(7)$ of dimension 14.

Let W be an oriented hypersurface of $\mathbb{R}^7$ with orientation determined by a choice of a unit normal vector field N and let $\langle\ ,\ \rangle$, $\bar{\nabla}$ denote the induced metric and Levi-Civita connection respectively. We recall [13] that W admits an almost complex structure J defined by

$$JX = N(p) \times X, \qquad p \in W,\ X \in T_pW.$$

Then J is an orthogonal tensor field, that is

$$\langle JX, JY\rangle = \langle X, Y\rangle$$

for all $X, Y \in T_pW$ and all $p \in W$.

In the particular case where W is the unit sphere S^6 with its standard orientation, we identify the unit normal $N(p)$ at $p \in S^6$ with the vector p in $\mathbb{R}^7$ so that

$$JX = p \times X, \qquad p \in S^6,\ X \in T_pS^6. \tag{2.1}$$

It now follows that $(\bar{\nabla}_X J)Y$ is the component of $X \times Y$ tangential to S^6, so that J is a *nearly Kähler* structure on S^6 in the sense that $(\bar{\nabla}_X J)X = 0$ for all vectors X tangential to S^6.

Finally, we note that G_2 is the group of automorphisms of the nearly Kähler structure J so that, in particular, elements of G_2 map almost complex (*resp.* totally real) submanifolds to almost complex (*resp.* totally real) submanifolds.

3. Almost complex curves

An *almost complex* curve in S^6 is a non-constant smooth map $f : S \to S^6$, from a Riemann surface S, whose differential is complex linear. Such a map is necessarily weakly conformal. Moreover, if $z = x + iy$ is a local complex coordinate on S then, since $f \times f_x = f_y$ it follows that

$$f \times (f_{xx} + f_{yy}) = 0.$$

Thus $f_{xx} + f_{yy}$ is orthogonal to S^6 and hence f is a weakly conformal harmonic map or, equivalently, a weakly conformal branched minimal immersion.

According to [5] there are four types of almost complex curves in S^6, namely

(I) linearly full and superminimal in S^6,

(II) linearly full in S^6 but not superminimal,

(III) linearly full in some totally geodesic S^5 in S^6,

(IV) totally geodesic.

For the above, we recall that a map $f : S \to S^n$ is *linearly full* if $f(S)$ is not contained in a totally geodesic S^{n-1} in S^n. Also, a weakly conformal harmonic map $f : S \to S^{2m}$ is *superminimal* if it can be constructed from holomorphic data in the sense that either f or its antipode $-f$ is the composite ρh of the Riemannian submersion $\rho : SO(2m+1)/U(m) \to SO(2m+1)/SO(2m) = S^{2m}$ and a horizontal holomorphic curve $h : S \to SO(2m+1)/U(m)$. Here we endow $SO(2m+1)/U(m)$ with the structure of a Hermitian symmetric space via its identification with $SO(2m+2)/U(m+1)$ induced by the inclusion of $SO(2m+1)$ in $SO(2m+2)$. (Compare [14], [15], [1], [11], and also [24], [7] for alternative descriptions). We recall the result of Calabi [14] that all minimal 2-spheres in S^{2m} are superminimal.

4. Constructions for almost complex curves

We first describe the work of Bryant [10], who used a twistor construction to obtain all almost complex curves in S^6 of type (I). This involves consideration of the twistor bundle $\pi : Q^5 \to S^6$ where Q^5 is the quadric in $\mathbb{C}P^6$ given by

$$Q^5 = \{[a + ib] \in \mathbb{C}P^6 : a, b \in \mathbb{R}^7,\ |a| = |b| \text{ and } a \perp b\}$$

and $\pi : Q^5 \to S^6$ is the map

$$\pi([a + ib]) = -\frac{a \times b}{|a \times b|}.$$

An alternative description of π may be given in terms of a projection between quotient spaces of G_2. Specifically, the standard action of G_2 on $\mathbb{R}^7$ induces transitive actions on S^6 and Q^5. If $\{e_1, \dots, e_7\}$ is the standard basis of $\mathbb{R}^7$ then the stabiliser of $e_4 \in S^6$ may be identified with $SU(3)$ and that of $[e_1+ie_5] \in Q^5$ with $U(2)$. Moreover, using these identifications, if $h \in U(2)$ then $he_4 = -h(e_1 \times e_5) = -e_1 \times e_5 = e_4$, so that $U(2) \subset SU(3)$. It is now easy to check that the above projection π may also be described as the standard projection map

$$\pi : Q^5 = G_2/U(2) \to G_2/SU(3) = S^6.$$

A holomorphic map $g : S \to Q^5$ is said to be *superhorizontal* if $g \times g_z = 0$. Such maps are horizontal in the sense that $g(S)$ intersects the fibres orthogonally, but superhorizontality is a rather stronger condition. (The reader is referred to [12] for the general theory of superhorizontal maps.) Although the map π above is not a Riemannian submersion, if $g : S \to Q^5$ is holomorphic and superhorizontal then the metrics induced on S by g and πg are equal.

Theorem 1 (Bryant 1982). *There is a* 1-1 *correspondence between linearly full superhorizontal maps* $g : S \to Q^5$ *and almost complex curves* $f : S \to S^6$ *of type* (I).

In fact, in the above theorem, g corresponds to the map $f = \pi g$, while f corresponds to the map $g = [a + ib]$, where $\{a, b\}$ is an ordered orthonormal basis of $\{f, f_x, f_y, f_{xx}, f_{xy}\}^\perp$ chosen so that $\{f, f_x, f_y, f_{xx}, f_{xy}, a, b\}$ determine the standard orientation on $\mathbb{R}^7$. The map g is called the *twistor lift* of f.

Bryant then obtains a "Weierstrass represention" for linearly full holomorphic superhorizontal maps $g : S \to Q^5$ and, using this result, proves the following:

Theorem 2 (Bryant 1982). *Every compact Riemann surface admits an infinite number of almost complex maps of type* (I) *into* S^6.

There is an alternative (but related) way of constructing (locally at least) almost complex curves of type (I) using the 2-dimensional open G_2-Toda equations. In fact, using methods similar to those employed in [4], it may be shown that there is a 1-1 correspondence between almost complex curves of type (I) and holomorphic superhorizontal maps [12] into G_2/T, where T is the maximal torus of G_2. This arises from the fact that the higher osculating spaces of an almost complex curve f of type (I) determine a decomposition of f^*TS^6 into an orthogonal direct sum of oriented 2-plane bundles which in turn determines an element of G_2/T. It may then be shown that, locally at least, holomorphic superhorizontal maps into G_2/T essentially correspond to solutions of the 2-dimensional open G_2-Toda equations. The relation with the construction of Bryant comes from the fact that, as may be seen from the construction of g above, if $f : S \to S^6$ is an almost complex curve of type (I) with twistor lift $g : S \to Q^5$ and corresponding superhorizontal map $\phi : S \to G_2/T$, then $g = \pi_1\phi$ where $\pi_1 : G_2/T \to G_2/U(2) = Q^5$ is the standard projection.

It is shown in [4] that almost complex curves of type (II) all arise from solutions of the affine 2-dimensional G_2-Toda equations, and at the present time this is the only way we know of constructing almost complex curves of type (II).

Almost complex curves of type (III) are intimately related to totally real, non totally geodesic, weakly conformal harmonic maps of S into $\mathbb{C}P^2$. If $f : S \to S^6$ is of type (III) we may assume, by applying an element of G_2 if necessary, that $f(S) \subset S^5 = S^6 \cap P$ where P is the hyperplane of $\mathbb{R}^7$ given by

$$P = \{(x_1, \ldots, x_7) : x_4 = 0\}.$$

Then the map $\pi : S^5 \to \mathbb{C}P^2$ given by

$$\pi(x_1, x_2, x_3, 0, x_5, x_6, x_7) = [x_1 + ix_5, x_2 + ix_6, x_3 + ix_7],$$

is the Hopf fibration and it is shown in [5] that f is horizontal and $\pi f : S \to \mathbb{C}P^2$ is a totally real, non totally geodesic, weakly conformal harmonic map of S into $\mathbb{C}P^2$. Conversely, if S is simply connected and if $\psi : S \to \mathbb{C}P^2$ is such a map then among the horizontal lifts of ψ there are exactly three almost complex curves (which are all G_2-congruent). In this way the study of almost complex curves of type (III) may be undertaken using the 2-dimensional affine $SU(3)$-Toda equations with a side condition (to ensure total reality), also called the Bullough–Dodd equation.

Remark. In fact all horizontal lifts of a totally real, non totally geodesic, weakly conformal harmonic map of S into $\mathbb{C}P^2$ give minimal surfaces in $S^5 \subset S^6$ all of which have constant Kähler angle and have ellipse of curvature a circle. Six of these lifts are totally real (that is to say they have constant Kähler angle equal to $\pi/2$), appearing as three pairs of antipodal maps. It is shown in [5] that (up to G_2-congruence) all minimal surfaces in S^6 with constant Kähler angle not equal to 0 or π and ellipse of curvature a circle arise in this way and, in particular, they are not linearly full. Totally real minimal surfaces in S^6 with noncircular ellipse of curvature are discussed in [6] where it is shown that any such surface is either contained in a totally geodesic, totally real S^3 or is contained in a totally geodesic S^5 as the orbit under a certain 1-parameter subgroup of G_2 of a curve lying in an almost complex totally geodesic S^2 of S^6. In particular, there are no totally real, linearly full minimal surfaces in S^6.

We complete the discussion of the various types of almost complex curves by noting that almost complex curves of type (IV) are just the intersections with S^6 of associative 3-planes of $\mathbb{R}^7$, that is to say 3-planes of $\mathbb{R}^7$ which are closed under the vector cross product.

The above methods of describing almost complex curves are used in [8] to obtain a complete description (originally due to Sekigawa [32]) of all almost complex curves of constant curvature in S^6. In fact, up to G_2-congruence, there are just three such curves. The simplest are those of type (IV), but more interesting ones are given by the

Boruvka 2-sphere [9], [19] in S^6 (which has constant curvature 1/6), and the almost complex torus of type (III) obtained by a suitable lifting to S^6 of the totally real flat harmonic torus in $\mathbb{C}P^2$ found by Ludden, Okumura and Yano [30].

5. Submanifolds of dimension 3

As remarked earlier, almost complex submanifolds must be even dimensional, so the condition we consider in this section is that of total reality. In this case, if M is a three-dimensional totally real submanifold of S^6 then the restriction to M of the tangent bundle TS^6 of the ambient manifold S^6 is the orthogonal direct sum $TM \oplus J(TM)$, and the term *Lagrangian* is also used to describe such manifolds.

We have already seen that almost complex curves are minimal, and Ejiri [25] has proved that every totally real 3-dimensional submanifold of S^6 is also minimal. Using the theory of calibrations, Palmer [31] has given estimates for the nullity and Morse index of such manifolds.

Lagrangian submanifolds of S^6 have also been studied by Deszcz, Dillen, Verstraelen and Vrancken. Although these submanifolds have not been classified, four types have been identified.

A. The total space of the restriction to a holomorphic curve in $\mathbb{C}P^2$ of the Hopf fibration $\pi : S^5 \to \mathbb{C}P^2$. This is a 3-dimensional submanifold of S^5 which is then included totally geodesically in S^6 in an appropriate manner.

B. The image of the map ψ defined on the unit tangent bundle of a non-totally geodesic almost complex curve f, where ψ is given by

$$\psi(v) = v \times \frac{h(v, v)}{|h(v, v)|},$$

and h is the second fundamental form of f.

C. The image of the map ψ_2 defined on the unit tangent bundle of a superminimal non-totally geodesic almost complex curve f, where the map ψ_2 acting on a unit tangent vector v at the point p is given by

$$\psi_2(v) = \frac{\sqrt{5}}{3} f(p) + \frac{2}{3}\psi(v).$$

Here, the map ψ is as in (B).

D. The image of the map ψ_3 defined on the unit tangent bundle of a superminimal non-totally geodesic almost complex curve f, where ψ_3 is given by

$$\psi_3(v) = \frac{h(v, v)}{|h(v, v)|}.$$

These types have the following properties.

1. Type (A) are the only ones which are not linearly full [23].
2. Types (A) and (B) are the only ones which satisfy Chen's equality [23], [18].
3. Apart from one exceptional example, types (A), (B) and (C) are the only ones for which the Ricci tensor has an eigenvalue with multiplicity at least 2 [20].
4. Types (A), (B) (when the almost complex curve is superminimal) and (D) are the only totally real submanifolds of S^6 which admit a unit length Killing vector field whose integral curves are great circles in S^6 [33].

The curvature properties of Lagrangian submanifolds have also been investigated. Ejiri has shown [25] that there are just two types of Lagrangian submanifolds of S^6 of constant curvature, in that the non totally geodesic ones are locally G_2-congruent to a certain Lagrangian 3-sphere of constant curvature 1/16. Subsequently, all compact Lagrangian submanifolds of S^6 whose curvature is bounded below by 1/16 were found by Dillen, Verstraelen and Vrancken [21], [22], who used this to show that any compact Lagrangian submanifold of S^6 whose curvature is pinched between 1 and 1/16 must have constant curvature.

6. Submanifolds of dimension 5

In this section we obtain a characterization of tubes around almost complex curves in terms of Hopf hypersurfaces of S^6.

Suppose M is a connected hypersurface of an almost complex metric manifold $(W, \bar{g}, J, \bar{\nabla})$. There is a natural one-dimensional foliation of M, the *Hopf foliation*, which consists of the integral curves of the distribution which is obtained by applying J to the normal bundle of M. If this foliation is totally geodesic then M is called a *Hopf hypersurface* of W. For example, the standard sphere S^{2n+1} is a Hopf hypersurface of $\mathbb{C}^{n+1}$, and the induced foliation of S^{2n+1} is precisely the well-known Hopf foliation by great circles.

Hopf hypersurfaces of simply connected complex space forms have also been considered. Those of $\mathbb{C}P^n$ have been studied by Cecil and Ryan [17], and those of $\mathbb{C}^n$ and $\mathbb{C}H^n$ by Berndt [2]. In each case, characterizations of certain types of Hopf hypersurfaces of these spaces are obtained as tubes around complex submanifolds. Similar methods (which we briefly describe below) give the following characterization of all Hopf hypersurfaces of S^6 [3].

Theorem 3. *Let M be a Hopf hypersurface of S^6. Then M is an open part of either a geodesic hypersphere of S^6 or a tube around an almost complex curve in S^6.*

To prove that all tubes around almost complex curves are Hopf hypersurfaces, it is first proved that the Hopf foliation of a hypersurface M of S^6 is a geodesic foliation if and only if it is a foliation by principal curvature directions, in which case the corresponding principal curvature function α is constant. Standard methods for the calculation of the second fundamental form of tubes [28] then show that all tubes around almost complex curves are Hopf hypersurfaces.

To prove the converse, it is assumed that M is a Hopf hypersurface of S^6 with unit normal field N, and that the principal curvature function α (which is necessarily constant) defined by the Hopf foliation is given by $\alpha = -\cot r$, $r \in (0, \pi/2]$. If $F_r : M \to S^6$ is then defined by

$$F_r(p) = \lambda_p(r), \qquad p \in M,$$

where λ_p is the geodesic of S^6 with initial vector $N(p)$, it may be proved that

(a) $dF_r(U) = 0$, where $U = JN$ generates the Hopf foliation,

(b) the dimension of the image of dF_r is constant,

(c) the image of dF_r is J-invariant.

It then follows from (b) that the image of F_r is a manifold which, from (c) is either a point or almost complex. The proof is completed by noting that the construction implies that M is (part of) the tube of radius r about the image of F_r.

In the case of Hopf hypersurfaces of $\mathbb{C}^n$, it is not yet known whether the corresponding principal curvature function α is necessarily constant, but Hopf hypersurfaces with α a non-zero constant may be characterized as tubes of radius $1/|\alpha|$ about complex submanifolds.

Similarly, step (b) has not yet been carried out for Hopf hypersurfaces of $\mathbb{C}P^n$, so this property must be assumed in a characterization in terms of tubes of radius $\operatorname{arccot}(|\alpha|)$ about complex submanifolds. A similar remark also applies to $\mathbb{C}H^n$, but in addition we need $|\alpha| > 1$ and then the corresponding radius of the tube is $\operatorname{arccoth}(|\alpha|)$. The characterization does not extend to Hopf hypersurfaces of $\mathbb{C}H^n$ with $|\alpha| \leq 1$, since in this case the corresponding map F_r will not have property (a) above for any real number r (so that the Hopf vector field does not give rise to any focal points of M). For instance, horospheres are Hopf hypersurfaces of $\mathbb{C}H^n$ with $|\alpha| = 1$.

References

[1] J. L. Barbosa, On minimal immersions of S^2 into S^{2m}, Trans. Amer. Math. Soc. 210 (1975), 75–106.

[2] J. Berndt, Über Untermannigfaltigkeiten von komplexen Raumformen, Thesis, University of Köln, 1989.

[3] J. Berndt, J. Bolton, L. M. Woodward, Almost complex curves and Hopf hypersurfaces in the nearly Kähler 6-sphere, Geometriae Dedicata 56 (1995), 237–247.

[4] J. Bolton, F. Pedit, L. M. Woodward, Minimal surfaces and the affine Toda field model, J. Reine Angew. Math. 459 (1995), 119–150.

[5] J. Bolton, L. Vrancken, L. M. Woodward, On almost complex curves in the nearly Kähler 6-sphere, Quart. J. Math. Oxford (2) 45 (1994), 407–427.

[6] J. Bolton, L. Vrancken, L. M. Woodward, Totally real minimal surfaces with non-circular ellipse of curvature in the nearly Kähler 6-sphere, to appear in J. London Math. Soc.

[7] J. Bolton, L. M. Woodward, Congruence theorems for harmonic maps of surfaces in $\mathbb{C}P^n$ and S^n, J. London Math. Soc. 45 (1992), 363–376.

[8] J. Bolton, L. M. Woodward, Almost complex curves of constant curvature in the nearly Kähler 6-sphere, in: Geometry and Topology of Submanifolds, World Scientific, Singapore 1993, 54–67.

[9] O. Boruvka, Sur les surfaces représentées par les fonctions sphériques de première espèce, J. Math. Pures Appl. 12 (1933), 337–383.

[10] R. L. Bryant, Submanifolds and special structures on the octonians, J. Diff. Geom. 17 (1982), 185–232.

[11] R. L. Bryant, Conformal and minimal immersions of compact surfaces in the 4-sphere, J. Diff. Geom. 17 (1982), 455–473.

[12] F. E. Burstall, J. H. Rawnsley, Twistor theory for Riemannian symmetric spaces, Lecture Notes in Math. 1424, Springer-Verlag, Berlin–Heidelberg–New York 1990.

[13] E. Calabi, Construction and properties of some 6-dimensional almost complex manifolds, Trans. Amer. Math. Soc. 87 (1958), 407–438.

[14] E. Calabi, Minimal immersions of surfaces in Euclidean spheres, J. Diff. Geom. 1 (1967), 111–125.

[15] E. Calabi, Quelques applications de l'analyse complexe aux surfaces d'aire minima, in: Topics in complex manifolds, University of Montreal, 1968, 59–81.

[16] E. Calabi, H. Gluck, What are the best almost-complex structures on the 6-sphere? Differential Geometry: geometry in mathematical physics and related topics (Los Angeles, CA, 1990), Proc. Sympos. Pure Math., 54, Part 2, Amer. Math. Soc., Providence, RI, 1993, 99–106.

[17] T. E. Cecil, P. J. Ryan, Focal sets and real hypersurfaces in complex projective space, Trans. Amer. Math. Soc. 269 (1982), 481–499.

[18] B.-Y. Chen, Some pinching and classification theorems for minimal submanifolds, Arch. Math. (Basel) 60 (1993), 568–578.

[19] S. S. Chern, M. do Carmo, S. Kobayashi, Minimal submanifolds of a sphere with second fundamental form of constant length, in: Functional Analysis and Related Fields, Springer-Verlag, 1970, 59–75.

[20] R. Deszcz, F. Dillen, L. Verstraelen, L. Vrancken, Quasi-Einstein totally real submanifolds of $S^6(1)$, preprint.

[21] F. Dillen, L. Verstraelen, L. Vrancken, On problems of U. Simon concerning minimal submanifolds of the nearly Kaehler 6-sphere, Bull. Amer. Math. Soc. 19 (1988), 433–438.

[22] F. Dillen, L. Verstraelen, L. Vrancken, Classification of totally real 3-dimensional submanifolds of $S^6(1)$ with $K \geq 1/16$, J. Math. Soc. Japan 42 (1990), 565–584.

[23] F. Dillen, L. Vrancken, Totally real submanifolds in S^6 satisfying Chen's equality, Trans. Amer. Math. Soc. 348 (1996), 1633–1646.

[24] J. Eells, J. C. Wood, Harmonic maps from surfaces to complex projective spaces, Adv. Math. 49 (1983), 217–263.

[25] N. Ejiri, Totally real submanifolds in a 6-sphere, Proc. Amer. Math. Soc. 83 (1981), 759–763.

[26] N. Ejiri, Equivariant minimal immersions of S^2 into $S^{2m}(1)$, Trans. Amer. Math. Soc. 297 (1986), 105–124.

[27] A. Gray, Almost complex submanifolds of the six sphere, Proc. Amer. Math. Soc. 20 (1969), 277–279.

[28] A. Gray, Tubes, Addison-Wesley, Redwood City 1990.

[29] R. Harvey, H. B. Lawson, Calibrated geometries, Acta Math. 148 (1982), 47–157.

[30] G. D. Ludden, M. Okumura, K. Yano, A totally real surface in $\mathbb{C}P^2$ that is not totally geodesic, Proc. Amer. Math. Soc. 53 (1975), 186–190.

[31] B. Palmer, Calibrations and Lagrangian submanifolds, preprint.

[32] K. Sekigawa, Almost complex submanifolds of a 6-dimensional sphere, Kodai Math. J. 6 (1983), 174–185.

[33] L. Vrancken, Killing vector fields and Lagrangian submanifolds of $S^6(1)$, preprint.

Some geometrical aspects of the 2-dimensional Toda equations

J. Bolton and L. M. Woodward

Abstract. The Toda equations are a particularly interesting example of a completely integrable Hamiltonian system and were initially studied in the 1-dimensional version in a dynamical context. Two particular forms of Toda equations, the "open" and "affine" equations, can be formulated for any simple Lie algebra and in the 2-dimensional version solutions to both forms arise naturally in a geometrical context from special types of harmonic maps of surfaces into certain homogeneous spaces associated to the Lie algebra. Generally speaking, the open case is rather easier to deal with since the solutions correspond to holomorphic curves, but the affine case is in some ways more interesting since it has certain special features not enjoyed by the open case. This article deals with some of the geometric aspects of the 2-dimensional version of these equations.

1. Introduction

In this article we consider from a differential geometric point of view the two forms of Toda equations, namely the "open" and "affine" equations, associated to a simple Lie algebra $\mathbf{g}$. We confine our attention to the elliptic abelian versions of these equations. We also remark that the "open" case is sometimes referred to as the "conformal" case. For the case $\mathbf{g} = \mathbf{su}(2)$ these equations are the Liouville equation and sinh-Gordon equation respectively and, as is well known, they have a natural geometrical interpretation as conditions on the induced metric of an oriented surface immersed in $\mathbb{R}^3$ with constant mean curvature. This can be rephrased using the Ruh-Vilms Theorem in terms of conditions on the Gauss map, namely that the Gauss map be antiholomorphic or harmonic respectively.

The purpose of this article is threefold. Firstly, to explain the geometrical interpretation of the Toda equations for the Lie algebra $\mathbf{g}$ in terms of certain harmonic maps of a Riemann surface into the flag manifold G/T, where G is a compact simple Lie group with Lie algebra $\mathbf{g}$ and maximal torus T; secondly, to describe geometrically the simplest possible solutions of the equations; and finally to indicate in this context the analogue of the classical Plücker formulae for holomorphic curves.

The layout of the paper is as follows. In Section 2 we describe the two forms of Toda equations associated to a simple Lie algebra **g** and summarise the geometrical interpretation of the case $\mathbf{g} = \mathbf{su}(2)$. Then in Section 3 we consider the G-invariant geometry of G/T including its m-symmetric space structure which is crucial for what follows. In Section 4 maps of a Riemann surface into G/T which have a simple symmetry property in terms of the m-symmetric space structure on G/T are introduced and it is noted that they have strong harmonicity properties. Then in Section 5 we show that these harmonic maps into G/T give a natural geometric interpretation of the solutions of the **g**-Toda equations. In Section 6 the maps corresponding to the simplest possible solutions are described and then in Section 7 the analogue of the Plücker formulae is discussed.

2. The Toda equations

We begin by recalling briefly some facts about simple Lie algebras and refer the reader to [17], [26] for a more detailed exposition.

Let **g** be a simple Lie algebra of rank ℓ. The root space decomposition of the complexification $\mathbf{g}^{\mathbb{C}}$ of **g** is given by

$$\mathbf{g}^{\mathbb{C}} = \mathbf{t}^{\mathbb{C}} \oplus \bigoplus_{\alpha \in \Delta} \mathbf{g}^{\alpha}, \tag{2.1}$$

where $\mathbf{t}^{\mathbb{C}}$ is a Cartan subalgebra and Δ is the set of roots. Let $\{\alpha_1, \dots, \alpha_\ell\}$ be a set of simple roots. Then among all positive roots there is a highest one

$$\theta = m_1\alpha_1 + \cdots + m_\ell\alpha_\ell, \qquad \sum_{i=1}^{\ell} m_i \text{ maximal},$$

so that if we set $\alpha_0 = -\theta$ and $m_0 = 1$ then

$$m_0\alpha_0 + \cdots + m_\ell\alpha_\ell = 0.$$

The 2-dimensional *open* **g**-*Toda equations* are the nonlinear elliptic system of partial differential equations given by

$$2\Delta\Omega + \sum_{k=1}^{\ell} m_k e^{2\alpha_k(\Omega)}\alpha_k^{\#} = 0, \tag{2.2}$$

where

$$\Delta = \frac{\partial^2}{\partial x^2} + \frac{\partial^2}{\partial y^2}, \tag{2.3}$$

$\Omega : U \to i\mathbf{t}$ is a smooth map of an open subset U of $\mathbb{R}^2$ into the purely imaginary part of $\mathbf{t}^{\mathbb{C}}$, and $(\ )^{\#} : (\mathbf{t}^{\mathbb{C}})^* \to \mathbf{t}^{\mathbb{C}}$ is the map defined using the Killing form B by

$$B(\lambda^{\#}, H) = \lambda(H), \quad \forall\, H \in \mathbf{t}^{\mathbb{C}}.$$

In the same way, the 2-dimensional *affine* $\mathbf{g}$-*Toda equations* are given by

$$2\Delta\Omega + \sum_{k=0}^{\ell} m_k e^{2\alpha_k(\Omega)} \alpha_k^{\#} = 0. \tag{2.4}$$

We will frequently just write

$$2\Delta\Omega + \sum_{k} m_k e^{2\alpha_k(\Omega)} \alpha_k^{\#} = 0, \tag{2.5}$$

where the summation is for $k = 1, \dots, \ell$ in the open case and $k = 0, \dots, \ell$ in the affine case and refer to the equations as the $\mathbf{g}$-Toda equations.

This is a slightly different formulation of the Toda equations from that found, for instance, in [20] which uses the extended Cartan matrix but (2.5) turns out to be more appropriate for geometrical purposes and has the advantage that $\Omega = 0$ is a solution of the affine Toda equations. In the case where $\mathbf{g} = \mathbf{su}(n+1)$ there is no difference between the two formulations and the equations can be written as

$$\Delta\rho_i + \sum_j K_{ij} e^{\rho_j} = 0, \tag{2.6}$$

where $\rho_i = 2\alpha_i(\Omega)$, (K_{ij}) is the extended Cartan matrix and $i, j = 1, \dots n$ in the open case and $i, j = 0, \dots n$ in the affine case.

In the 1-dimensional version the Laplacian Δ is replaced by d^2/dt^2 and the equations are then of the form $d^2y/dt^2 = F(y)$ with $F(y)$ a linear combination of exponentials as in (2.6). These ordinary differential equations arise as the equations of motion of a Hamiltonian dynamical system of masses with exponential interactions and it turns out that the system is completely integrable. See [1], [18], [27], [15] for details.

In the simplest case, namely $\mathbf{g} = \mathbf{su}(2)$, the equations (2.6) reduce to the Liouville equation

$$\Delta\rho + 2e^{\rho} = 0 \tag{2.7}$$

in the open case and the sinh-Gordon equation

$$\Delta\rho + 4\sinh\rho = 0 \tag{2.8}$$

in the affine case.

There is a natural geometric interpretation of (2.7) and (2.8) in terms of CMC surfaces in $\mathbb{R}^3$. For suppose that S is an immersed surface in $\mathbb{R}^3$ with constant mean curvature c. Let x, y be local isothermal coordinates in an open subset U of S with induced metric given by $ds^2 = e^{2\omega}(dx^2 + dy^2)$. Assuming that there are no umbilics in U there exists a function $\rho(x, y)$ such that the principal curvatures are given by

$$\kappa_1 = c + e^{\rho}, \qquad \kappa_2 = c - e^{\rho}. \tag{2.9}$$

Using the Codazzi–Mainardi equations one may show that the surface is of constant mean curvature on U if and only if the Hopf differential

$$II\left(\frac{\partial}{\partial z}, \frac{\partial}{\partial z}\right) dz^2 = \frac{1}{4}(L - N - 2iM)\, dz^2, \qquad z = x + iy, \tag{2.10}$$

is holomorphic. Thus, assuming that U is simply connected and contains no umbilics, the isothermal coordinates may be chosen uniquely in U up to multiplication by ± 1 so that the 2-form (2.10) is of the form

$$II\left(\frac{\partial}{\partial z}, \frac{\partial}{\partial z}\right) dz^2 = \frac{1}{2}\, dz^2. \tag{2.11}$$

Since $L = \kappa_1 e^{2\omega}$, $M = 0$, $N = \kappa_2 e^{2\omega}$ it follows that

$$\rho + 2\omega = 0. \tag{2.12}$$

There are two cases, namely the minimal case $c = 0$ and the case $c \neq 0$ where, without loss of generality, we may assume that $c = 1$. Then an elementary argument using (2.12) shows that (2.7) and (2.8) are the Gauss equation for the cases $c = 0$, $c = 1$ respectively. Note that in each case the unknown in the equation is essentially the metric.

In both cases there is a corresponding adapted $SO(3)$-frame F given by the unit tangent vectors e_1, e_2 to the coordinate curves and the unit normal N and the Gauss and Codazzi–Mainardi equations are precisely the Maurer–Cartan equations for F and correspond to the integrability condition $F_{z\bar{z}} = F_{\bar{z}z}$. Moreover, in terms of the special choice of coordinates, the Codazzi–Mainardi equations are given by $(\rho + 2\omega)_x = 0$ and $(\rho + 2\omega)_y = 0$, so the Maurer–Cartan equations reduce to just the Gauss equation. It then follows that, conversely, given a solution ρ to either (2.7) or (2.8) on a simply connected domain U in $\mathbb{R}^2$ with metric $ds^2 = e^{-\rho}(dx^2 + dy^2)$ there is an isometric immersion of U into $\mathbb{R}^3$ with constant mean curvature 0 or 1 respectively. The study of the 2-dimensional Toda equations for $\mathbf{g} = \mathbf{su}(2)$ is thus essentially equivalent to the local theory of CMC surfaces in $\mathbb{R}^3$ without umbilical points. See [22] for further details.

For any oriented surface S immersed in $\mathbb{R}^3$ the Gauss map is a globally defined map of S into G/T with $G = SO(3)$ and T its maximal torus. By the Ruh–Vilms Theorem the Gauss map is harmonic if and only if the surface has constant mean curvature [25]. In this sense the study of the 2-dimensional Toda equations may be interpreted in terms of harmonic maps of a surface into S^2 [9].

3. G-invariant structures on G/T

As above, let G be a compact simple Lie group with maximal torus T and let $\mathbf{g}$ and $\mathbf{t}$ denote their Lie algebras. Then G/T may identified with an $\mathrm{Ad}(G)$-orbit in $\mathbf{g}$ of a regular element H_0 of $\mathbf{t}$. Thus we identify $gT \in G/T$ with $\mathrm{Ad}(g)(H_0) \in \mathbf{g}$. For example, the embedding of $S^2 = SU(2)/T$ in $\mathbf{su}(2) = \mathbb{R}^3$ as the adjoint orbit of the matrix

$$\begin{pmatrix} i & 0 \\ 0 & -i \end{pmatrix}$$

is the standard embedding of S^2 in $\mathbb{R}^3$. We remark that the tangent space at $X \in G/T \subset \mathbf{g}$ is given by

$$T_X G/T = [\mathbf{g}, X]. \tag{3.1}$$

A crucial fact is that both G/T and its geometry are determined by $\mathbf{g}$. In particular, if $\rho : \tilde{G} \to G$ is the universal covering of G with $\tilde{T}$ a maximal torus of $\tilde{G}$ which maps onto T under ρ then ρ induces an identification of $\tilde{G}/\tilde{T}$ with G/T. Also G-invariant structures on G/T come from T-invariant structures on $\mathbf{g}/\mathbf{t}$ where T acts on $\mathbf{g}/\mathbf{t}$ via the adjoint action on $\mathbf{g}$. There are three particular structures on G/T which are important in the present context.

(i) **Complex structure.** We first identify the tangent bundle of $\mathbf{g}$ with the trivial bundle $\mathbf{g} \times \mathbf{g}$ so that we may identify the complexified tangent bundle $T^{\mathbb{C}}(G/T)$ of G/T with a subbundle of $G/T \times \mathbf{g}^{\mathbb{C}}$. Next define the complex line subbundle $[\mathbf{g}^\alpha]$ of $T^{\mathbb{C}}(G/T)$ corresponding to the root α by defining the total space to be the orbit in $\mathbf{g} \times \mathbf{g}^{\mathbb{C}}$ of $H_0 \times \mathbf{g}^\alpha$ under the diagonal adjoint action. Thus

$$[\mathbf{g}^\alpha]_{gT} = \{gT\} \times \mathrm{Ad}(g)(\mathbf{g}^\alpha). \tag{3.2}$$

Then $T^{\mathbb{C}}(G/T) = \bigoplus_{\alpha \in \Delta} [\mathbf{g}^\alpha]$ and we define an almost complex structure on G/T by setting

$$T^{1,0}G/T = \bigoplus_{\alpha \in \Delta_+} [\mathbf{g}^\alpha], \qquad T^{0,1}G/T = \bigoplus_{\alpha \in \Delta_-} [\mathbf{g}^\alpha], \tag{3.3}$$

where Δ_+, Δ_- denote the positive and negative roots respectively of $\mathbf{g}$. This almost complex structure is integrable and G-invariant [12].

(ii) ***G*-invariant metric.** There are many choices of G-invariant metric on G/T including ones which, together with the complex structure above, are Kähler. It suffices to choose an $\mathrm{Ad}(T)$-invariant metric $\langle\ ,\ \rangle$ on $\mathbf{g}$. This gives rise to a metric on G/T defined by

$$\langle Y, Z \rangle_X = \langle \mathrm{Ad}(g^{-1})Y, \mathrm{Ad}(g^{-1})Z \rangle, \qquad Y, Z \in T_X G/T, \tag{3.4}$$

where $X = \mathrm{Ad}(g)(H_0)$. This is well-defined and G-invariant and any G-invariant metric may be so obtained. We remark that, except in the case where $\mathbf{g} = \mathbf{su}(2)$, the metric arising from the Killing form is not Kähler but in what follows it will not matter which G-invariant metric we choose. For a full discussion of G-invariant metrics and which ones are Kähler see [2].

(iii) ***m*-symmetric space structure.** There is a unique element $Y \in \mathbf{t}$ called the **canonical element** which, amongst other properties, has $\mathbf{t}$ as its centraliser and the eigenvalues of $\mathrm{ad}(iY)$ lying in $\mathbb{Z}$. (See pp. 41–42 of [12].) If $h = \exp(2\pi Y/m)$, where $m = \sum_{j=0}^{\ell} m_j$, then the inner automorphism $\tau : G \to G$ given by $\tau(g) = hgh^{-1}$ is called the *Coxeter automorphism* and this has order m. (cf. [21] p. 289.) Then τ induces, for each $x \in G/T$, a map $\tau_x : G/T \to G/T$ with the property that, for all $x \in G/T$, $\tau_x : G/T \to G/T$ is of order m with x an isolated fixed point. If G/T is given a G-invariant metric then each τ_x is an isometry and thus G/T is given the structure of an m-symmetric space. For example, in the case of $G = SU(n+1)$, where G/T is the flag manifold consisting of $(n+1)$-tuples of mutually orthonormal complex lines in $\mathbb{C}^{n+1}$, the Coxeter automorphism of $\mathbf{su}(n+1)$ is given by conjugation by $\mathrm{diag}(1, \zeta, \dots, \zeta^n))$, where $\zeta = e^{2\pi i/n+1}$.

The derivative $\mathrm{Ad}(h)$ of the Coxeter automorphism also has order m so that the corresponding eigenspace decomposition of $\mathbf{g}^{\mathbb{C}}$ may be written as

$$\mathbf{g}^{\mathbb{C}} = \mathcal{M}_0 \oplus \cdots \oplus \mathcal{M}_{m-1}, \tag{3.5}$$

where $\mathcal{M}_k$ is the ζ^k-eigenspace, $\zeta = e^{2\pi i/m}$. Moreover,

$$\mathcal{M}_0 = \mathbf{t}^{\mathbb{C}} \text{ and } \mathcal{M}_1 = \bigoplus_{j=0}^{\ell} \mathbf{g}^{\alpha_j}. \tag{3.6}$$

Extending the notation used above we write $[\mathcal{M}_k]$ to denote the vector bundle over G/T determined by $\mathcal{M}_k$ so that, for example,

$$T^{\mathbb{C}}(G/T) = \bigoplus_{k=1}^{m-1} [\mathcal{M}_k], \tag{3.7}$$

and

$$[\mathcal{M}_1] = \bigoplus_{j=0}^{\ell} [\mathbf{g}^{\alpha_j}]. \tag{3.8}$$

Then, for each $x \in G/T$, $[\mathcal{M}_k]$ is the ζ^k-eigenspace of the derivative of τ_x.

4. Maps into G/T

Suppose that $\psi : S \to G/T$ is an immersion of an oriented surface S into G/T. Here, and for the rest of the article, we assume that rank $G > 1$. This is to avoid some vacuous statements and trivial counterexamples involving maps into S^2. Then ψ is τ-*adapted* if, for all $p \in S$, the differential of $\tau_{\psi(p)}$ maps $d\psi(T_pS)$ into itself by rotation through $2\pi/m$. More generally a smooth map $\psi : S \to G/T$ of a Riemann surface S is said to be τ-*adapted* if it is weakly conformal and the above condition holds. As we will see, this simple symmetry condition has some remarkable consequences. The following characterisation is easy to check.

Proposition 4.1. *Let S be a Riemann surface and $\psi : S \to G/T$ a smooth map. Then ψ is τ-adapted if and only if $d\psi(T^{1,0}S) \subseteq [\mathcal{M}_1]$.*

We recall from [12] that a map $\psi : S \to G/T$ with $d\psi(T^{\mathbb{C}}S) \subseteq [\mathcal{M}_1] \oplus [\mathcal{M}_{-1}]$ is said to be *superhorizontal*. Thus a τ-adapted map is superhorizontal.

Suppose that $\psi : S \to G/T$ is τ-adapted and let $F : U \to G$ be a local framing of ψ. Since $d\psi(T^{1,0}S) \subseteq [\mathcal{M}_1]$ it follows that

$$F^{-1}\frac{\partial F}{\partial z} = A_0 + A_1, \tag{4.1}$$

for some $A_0 \in \mathcal{M}_0$, $A_1 \in \mathcal{M}_1$. Moreover, since F is real we also have

$$F^{-1}\frac{\partial F}{\partial \bar{z}} = \bar{A}_0 + \bar{A}_1, \tag{4.2}$$

with $\bar{A}_0 \in \mathcal{M}_0$, $\bar{A}_1 \in \mathcal{M}_{m-1}$. Recall that the Maurer–Cartan equations, which are the integrability conditions for equations (4.1) and (4.2), are given by

$$\frac{\partial^2 F}{\partial z \partial \bar{z}} = \frac{\partial^2 F}{\partial \bar{z} \partial z}. \tag{4.3}$$

As we shall see in the next section for a suitable choice of framing the condition (4.3) is equivalent to the Toda equations.

The τ-adapted maps have strong harmonicity properties. Suppose that K is a closed subgroup of G containing T. A smooth map $\psi : S \to G/K$ is said to be *equiharmonic* [3] if it is harmonic with respect to any G-invariant metric on G/K. Recall (see [2], [12]) that any such metric is obtained from an $\mathrm{Ad}(K)$-invariant metric on $\mathbf{g}$. The following proposition is due to Black [3].

Proposition 4.2. *Suppose that $\psi : S \to G/T$ is τ-adapted and that K is a closed subgroup of G containing T. If $\pi : G/T \to G/K$ is the natural projection then $\pi\psi : S \to G/K$ is equiharmonic. In particular ψ itself is equiharmonic.*

5. The Toda equations and τ adapted maps

We now outline the ideas behind the central basic result that solutions of the Toda equations on a simply connected domain U of $\mathbb{R}^2$ correspond to τ-adapted maps satisfying a non-singularity condition.

We begin by considering singularities. Suppose that S is a Riemann surface and let $\psi : S \to G/T$ be a τ-adapted map. For each $j = 0, \dots, \ell$, let

$$\partial_j : T^{1,0}S \to \psi^*[\mathbf{g}^{\alpha_j}] \tag{5.1}$$

be the homomorphism of complex line bundles corresponding to be the composite of $d\psi : T^{1,0}S \to [\mathcal{M}_1]$ with projection onto $[\mathbf{g}^{\alpha_j}]$. Since $\psi^*[\mathbf{g}^{\alpha_j}]$ is a subbundle of the trivial bundle $S \times \mathbf{g}^{\mathbb{C}}$ it inherits a connection from the flat connection on $S \times \mathbf{g}^{\mathbb{C}}$ and hence, by the Koszul–Malgrange theorem, admits the structure of a holomorphic line bundle.

Proposition 5.1. *The map $\partial_j : T^{1,0}S \to \psi^*[\mathbf{g}^{\alpha_j}]$ is a holomorphic homomorphism of holomorphic line bundles and hence is either identically zero or has isolated zeros.*

Observe that the case where ∂_0 is identically zero is precisely the case where the map $\psi : S \to G/T$ is holomorphic and τ-adapted. If none of the ∂_j are identically zero ψ is said to be *τ-primitive.* (See [5], [11]).

We will return to the nature of the zeros in Section 7. For the moment in this section we restrict our attention to the case where there are no isolated zeros and relate this to solutions of the Toda equations.

Suppose that U is an open subset of a Riemann surface. A smooth map $F : U \to G$ is a *Toda frame* if there exists a complex coordinate z on U, and a smooth map $\Omega : U \to i\mathbf{t}$ such that

$$F^{-1}\frac{\partial F}{\partial z} = \frac{\partial \Omega}{\partial z} + \mathrm{Ad}(\exp \Omega)(B), \tag{5.2}$$

and

$$F^{-1}\frac{\partial F}{\partial \bar{z}} = -\frac{\partial \Omega}{\partial \bar{z}} + \mathrm{Ad}(\exp(-\Omega))(\bar{B}), \tag{5.3}$$

where $B \in \bigoplus_k g^{\alpha_k}$ and $\bar{B} \in \bigoplus_k g^{-\alpha_k}$ are given in terms of a Cartan-Weyl basis $\{X_\alpha \mid \alpha \in \Delta\}$ for $\mathbf{g}^{\mathbb{C}}$ modulo $\mathbf{t}^{\mathbb{C}}$ (see [26], [24]) by

$$B = \sum_k \sqrt{m_k} X_{\alpha_k}, \qquad \bar{B} = \sum_k \sqrt{m_k} X_{-\alpha_k}. \tag{5.4}$$

Here $k = 1, \dots, \ell$ in the open case, in which case the frame is called an *open Toda frame*, and $k = 0, \dots, \ell$ in the affine case, in which case the frame is called an *affine*

Toda frame. Thus

$$F^{-1}\frac{\partial F}{\partial z} = \frac{\partial \Omega}{\partial z} + \sum_k \sqrt{m_k} e^{\alpha_k(\Omega)} X_{\alpha_k}, \tag{5.5}$$

and

$$F^{-1}\frac{\partial F}{\partial \bar{z}} = -\frac{\partial \Omega}{\partial \bar{z}} + \sum_k \sqrt{m_k} e^{\alpha_k(\Omega)} X_{-\alpha_k}. \tag{5.6}$$

Note that these are a special form of equations (4.1) and (4.2). Furthermore, it is clear that if F is a Toda frame then the corresponding map $\psi : U \to G/T$ is τ-adapted and that none of the ∂_j has a zero with the exception of ∂_0 which in the open case is identically zero. Thus a Toda frame determines a τ-adapted map. The converse is also true.

Proposition 5.2. *Suppose that $\psi : U \to G/T$ is a τ-adapted map of a simply connected domain U in $\mathbb{C}$ and that the corresponding maps ∂_j have one of the following properties:*

(a) *∂_0 is identically zero but ∂_j has no zero for $j = 1, \dots, \ell$;*

(b) *∂_j has no zero for $j = 0, \dots, \ell$.*

Then ψ has a lift $F : U \to G$, where F is an open Toda frame in case (a), *an affine Toda frame in case (b). In each case the Toda frame is uniquely defined up to multiplication by elements of the centre $Z(G)$ of G and in case* (b) *the coordinate z is uniquely defined up to a translation and a rotation by a d-th root of unity where d is an integer determined by* $\mathbf{g}$.

The proof of this result in the affine case is given in [5]. The proof in the open case is similar but easier. We remark that in case (a) the map $\psi : U \to G/T$ is holomorphic and τ-adapted. Geometrical interpretations for the cases where $G = SU(n+1)$, $SO(2n+1)$ and the exceptional Lie group G_2 are discussed in [5] and [9], and for the other classical Lie groups in [10], [13] and [14].

The link with the Toda equations is provided by the following result which is proved by observing that, since $[X_\alpha, X_{-\beta}] = \delta_{\alpha\beta}\alpha^\#$, the Toda equations are the integrability conditions for equations (5.2) and (5.3).

Proposition 5.3. *The Toda equations are the Maurer–Cartan equations for a Toda frame $F : U \to G$.*

6. The simplest possible solutions

We now consider certain elementary solutions of the Toda equations and explain their geometrical significance. We begin with the affine case since that is the simplest.

The affine case. We recall that $\Omega = 0$ is a solution of the affine Toda equations. The equations satisfied by the corresponding Toda frame can then be written as

$$F^{-1}\frac{\partial F}{\partial z} = B, \qquad F^{-1}\frac{\partial F}{\partial \bar{z}} = \bar{B} \tag{6.1}$$

and hence, as $[B, \bar{B}] = 0$, it follows that the unique solution with $F(0) = 0$ is given by

$$F(z) = \exp(zB + \bar{z}\bar{B}).$$

This defines a homomorphism $F : \mathbb{R}^2 \to G$ and it is easy to see from (6.1) that the corresponding map $\psi : \mathbb{R}^2 \to G/T$ is a flat immersion which up to a scale factor is isometric. In certain cases, for example $G = SU(N)$, $N = 3, 4, 6$ the image is a torus (see [6]), but this is not true in general.

The open case. In this case the simplest solutions are of the form

$$2\alpha_j(\Omega) = \ln\left(\frac{a_j}{(1 + z\bar{z})^2}\right) \tag{6.2}$$

for certain real $a_j > 0$, $j = 1, \dots, \ell$ which turn out to be uniquely determined. This solution corresponds to a map $\psi : S^2 \to G/T$, (where $S^2 = SU(2)/T$), determined by a specific homomorphism $\phi : SU(2) \to G$ which in turn corresponds to a principal $\mathbf{sl}(2; \mathbb{C})$-subalgebra of $\mathbf{g}^{\mathbb{C}}$. For example, in the case where $G = SU(n+1)$, ϕ is the n-th symmetric power representation $S^n\rho : SU(2) \to SU(n+1)$ of the standard representation of $SU(2)$ on $\mathbb{C}^2$.

These are geometrically interesting examples since their projections into G/K with $T \subseteq K$ are also of constant curvature. Moreover, in the cases where G/K has a G-invariant Kähler metric, such as the case where $G = SU(n+1)$, $K = S(U(k) \times U(n-k+1))$ and G/K is the Grassmannian of k-planes in $\mathbb{C}^{n+1}$ with its usual Kähler metric, the projection into G/K also has constant Kähler angle. For a discussion of these surfaces when G/K is $\mathbb{C}P^n$ or S^n see [6], [7], [8], [19].

7. The Plücker formulae

Suppose that $\psi : S \to G/T$ is a τ-adapted map of a compact Riemann surface S into G/T. Recall from Proposition 5.1 that the bundle maps $\partial_j : T^{1,0}S \to \psi^*[\mathbf{g}^{\alpha_j}]$ are holomorphic and hence, as S is compact, are either identically zero or have only finitely many zeros. Let $r(\partial_j)$ denote the sum of the orders of the zeros of ∂_j.

Next we recall [12], [17] that we may write G/T as the quotient $G^{\mathbb{C}}/B$ of the complexification $G^{\mathbb{C}}$ of G by the Borel subgroup B. If V_j denotes the fundamental representation of $\mathbf{g}^{\mathbb{C}}$ determined by the lowest weight vector with P_j the corresponding parabolic subgroup (see [17]) then $G^{\mathbb{C}}/P_j \subseteq \mathbb{P}(V_j)$ and there is a projection π_j :

$G^{\mathbb{C}}/B \to G^{\mathbb{C}}/P_j$. Let $\tilde{\psi}_j$ denote the composite of $\psi : S \to G^{\mathbb{C}}/B$ with π_j, and let δ_j the degree of $\tilde{\psi}_j$.

Theorem 7.1. *Let S be a compact Riemann surface and let* (K_{jk}) *be the Cartan matrix of a simple Lie algebra* $\mathbf{g}$. *Then*

$$r(\partial_j) + \chi(S) = \sum_{k=1}^{\ell} K_{jk}\delta_k, \qquad j = 1, \dots, \ell, \tag{7.1}$$

where $\chi(S)$ *is the Euler characteristic of S.*

As we now briefly explain, this result is an extension of the classical Plücker formulae for holomorphic curves further details of which may be found in [16].

Suppose that $\phi = [f] : S \to \mathbb{C}P^n$ is a linearly full holomorphic curve and let $f_0, \dots, f_n$ be obtained from the derivatives $f, f^{(1)}, \dots, f^{(n)}$ by Gram–Schmidt orthogonalisation. Then the maps $\phi_0, \dots, \phi_n$, where $\phi_k = [f_k] : S \to \mathbb{C}P^n$, form the *Frenet frame* of ϕ and define a map $\psi : S \to SU(n+1)/T^n$ of S into the flag manifold $SU(n+1)/T^n$. This map is holomorphic and τ-adapted and locally corresponds to a solution of the open $\mathbf{su}(n+1)$-Toda equations. A more detailed discussion of this case is given in [9].

The *k-th associated curve* or *k-th osculating curve* of ϕ is the holomorphic curve $\tilde{\psi}_k : S \to G_{k+1}(\mathbb{C}^{n+1})$ defined by $\tilde{\psi}_k = [f \wedge f^{(1)} \wedge \dots \wedge f^{(k)}]$. This agrees with the definition of $\tilde{\psi}_k$ given above in the general case. Let $\delta_k = \deg\{\tilde{\psi}_k\}$ and, if $\beta_k(z_0)$ is the k-th ramification index at z_0, define $\beta_k = \sum_{z_0} \beta_k(z_0)$. Then the classical Plücker formulae are the equations

$$\beta_k + \chi(S) = -\delta_{k-1} + 2\delta_k - \delta_{k+1}. \tag{7.2}$$

We observe that the coefficients appearing here are precisely the coefficients in the Cartan matrix of $\mathbf{su}(n+1)$ so that (7.1) reduces to (7.2) in the open $\mathbf{su}(n+1)$ case.

Details of the proof of Theorem 1 will appear elsewhere. The idea of the proof is to establish "unintegrated Plücker formulae" which are relations between the Chern forms of the line bundles $\psi^*[\mathbf{g}^{\alpha_j}]$ and those of the pull-backs $\tilde{\psi}_j^*\mathcal{L}_j$ of the tautological line bundles $\mathcal{L}_j$ over $\mathbb{P}(V_j)$. Other proofs of Theorem 1 in the case of the open Toda equations have appeared in [23], [24] and in the case of the affine Toda equations when $\mathbf{g} = \mathbf{su}(n+1)$ in [4].

The ideas of the proof also establish the following:

(1) In the case of the open Toda equations, if there are no singularities (that is to say $r(\partial_j) = 0$ for $j = 1, \dots, \ell$) then $\chi(S) = 2$.

(2) In the case of the affine Toda equations, if $\chi(S) < 0$ then at least one $r(\partial_j)$ is non-zero.

(3) In the case of the affine Toda equations, if $\chi(S) = 0$ then $r(\partial_j) = 0$ for $j = 0, \dots, \ell$ and consequently every $\tilde{\psi}_j$ is an immersion.

References

[1] M. Adler, On a trace functional for pseudo-differential operators and the symplectic structure of Korteweg–deVries type equations, Invent. Math. 50 (1979), 219–248.

[2] D. V. Alekseevski and A. M. Perelomov, Invariant Kähler-Einstein metrics on compact homogeneous spaces, Functional Anal. Appl. 20 (1986), 171–182.

[3] M. Black, Harmonic Maps into Homogeneous Spaces, Longman, 1991.

[4] J. Bolton, G. Jensen, M. Rigoli and L. M. Woodward, On conformal minimal immersions of S^2 into $\mathbb{C}P^n$, Math. Ann. 279 (1988), 599–620.

[5] J. Bolton, F. Pedit and L. M. Woodward, Minimal surfaces and the affine Toda field model, J. Reine Angew. Math. 459 (1995), 119–150.

[6] J. Bolton and L. M. Woodward, Congruence theorems for harmonic maps of surfaces into $\mathbb{C}P^n$ and S^n, J. Lond. Math. Soc. 45 (1992), 365–376.

[7] J. Bolton and L. M. Woodward, Minimal surfaces in $\mathbb{C}P^n$ with constant curvature and Kähler angle, J. Lond. Math. Soc. 45 (1992), 365–376.

[8] J. Bolton and L. M. Woodward, Almost complex curves of constant curvature in the nearly Kähler 6-sphere, in: Geometry and Topology of Submanifolds, World Scientific, Singapore 1993, 54–67.

[9] J. Bolton and L. M. Woodward, The affine Toda equations and minimal surfaces, in: Harmonic Maps and Integrable Systems (A. P. Fordy and J. C. Wood, eds.), Vieweg 1993, 59–82.

[10] J. Bolton and L. M. Woodward, Minimal surfaces and the Toda equations for the classical groups, in: Geometry and Topology of Submanifolds VIII, World Scientific, Singapore 1996, 22–30.

[11] F. E. Burstall and F. Pedit, Harmonic maps via Adler–Kostant–Symes theory, in: Harmonic Maps and Integrable Systems (A. P. Fordy and J. C. Wood, eds.), Vieweg 1993, 221–272.

[12] F. E. Burstall and J. H. Rawnsley, Twistor Theory for Riemannian Symmetric Spaces with Applications to Harmonic Maps of Riemann Surfaces, Lecture Notes in Math. 1424, Springer-Verlag, Berlin 1990.

[13] A. Doliwa, Holomorphic curves and Toda systems, preprint, Institute of Theoretical Physics, Warsaw University.

[14] A. Doliwa, Harmonic maps and periodic Toda systems, preprint, Institute of Theoretical Physics, Warsaw University.

[15] M. Guest, Integrable Systems, Loop Groups, and Harmonic Maps, London Math. Soc. Lecture Note Ser., Cambridge University Press, 1997.

[16] P. A. Griffiths and J. Harris, Principles of Algebraic Geometry, Wiley, 1978.

[17] W. Fulton and J. Harris, Representation Theory, Springer-Verlag, 1991.

[18] B. Kostant, The solution to a generalized Toda lattice and representation theory, Adv. Math. 34 (1979), 195–338.

[19] Y. Ohnita, Minimal surfaces with constant curvature and Kähler angle in complex space forms, Tsukuba J. Math 13 (1989), 191–207.

[20] D. Olive and N. Turok, The symmetries of the Dynkin diagram and the reduction of Toda field equations, Nucl. Phys.B 215 (1983), 470–494.

[21] A. L. Onishchik and E. B. Vinberg, Lie Groups and Algebraic Groups, Springer-Verlag, Berlin–Heidelberg–New York 1990.

[22] U. Pinkall and I. Sterling, On the classification of constant mean curvature tori, Ann. Math. 130 (1989), 407–451.

[23] L. E. Positsel'skii, Local Plücker formulas for a semisimple Lie group, Functional Anal. Appl. 25 (1991), 291–292.

[24] A. V. Razumov and M. V. Saveliev, Differential geometry of Toda systems, Comm. Anal. Geom. 461 (1994).

[25] E. A. Ruh and J. Vilms, The tension field of the Gauss map, Trans. Amer. Math. Soc. 149 (1970), 569–573.

[26] H. Samelson, Notes on Lie Algebras, Springer-Verlag, 1990.

[27] W. W. Symes, Systems of Toda type, inverse spectral problems and representation theory, Invent. Math. 59 (1982), 13–51.

A new look at the vortex equations and dimensional reduction

Steven Bradlow[1], *James Glazebrook*[2] *and Franz Kamber*[3]

Abstract. In order to use the technique of dimensional reduction, it is usually necessary for there to be a symmetry coming from a group action. In this paper we consider a situation in which there is no such symmetry, but in which a type of dimensional reduction is nevertheless possible. We obtain a relation between the Coupled Vortex equations on a closed Kähler manifold, X, and the Hermitian–Einstein equations on certain $\mathbb{P}^1$-bundles over X. Our results thus generalize the dimensional reduction results of Garcia-Prada, which apply when the Hermitian–Einstein equations are on $X \times \mathbb{P}^1$.

1991 Mathematics Subject Classification: Primary 58C25; Secondary 58A30, 53C12, 53C21, 53C55, 83C05.

Key words: Vortex equation, projective bundles, Hermitian–Einstein equation.

§1. Introduction

Dimensional reduction is a technique for studying special solutions to partial differential equations. The technique is applicable when there is a *symmetry*, i.e. a group action, in which case it makes sense to look for *invariant* solutions. The term 'dimensional reduction' refers to the fact that the invariant solutions to the original equation can be interpreted as ordinary solutions to a related set of equations on the (lower dimensional) orbit space of the group action.

In this paper we will describe a situation in which an effective dimensional reduction of equations is possible, even though there is no global group action. The equations in question are the Hermitian–Einstein (HE) equations for special metrics

---

1 Supported in part by the NSF grant DMS-9303545.

2 Supported in part by the NSF grant DMS-9208182.

3 Supported in part by the NSF grants DMS-9208182, DMS-9504084 and by an A. O. Beckman Research Award from the University of Illinois.

on holomorphic bundles. As a result of our dimensional reduction procedure, we relate special solutions of the HE equations to solutions of the so-called Coupled Vortex equations. These latter equations are also for special metrics on holomorphic bundles, but involve more data than HE equations. The extra data is in the form of prescribed holomorphic sections.

A dimensional reduction from the Hermitian–Einstein to the Vortex equations is not new. Indeed in [GP], Garcia-Prada described just such a relation between the equations. In the situation considered by Garcia-Prada the vortex equations are defined on bundles over a closed Kähler manifold, say X, the Hermitian–Einstein equations are on bundles over $X \times \mathbb{P}^1$, and the reduction is with respect to the standard $SU(2)$-action on $\mathbb{P}^1$.

The main difference between Garcia-Prada's results and those described in this paper is that we have replaced $X \times \mathbb{P}^1$ by certain *non-trivial* $\mathbb{P}^1$-bundles over X. One way to view such a situation is as one in which there is no group action but in which the (group) orbits are still evident. In fact, the group orbits are now the leaves of a foliation structure. Instead of reducing to an orbit space of a group action, we thus reduce to the leaf space of a foliation. Such a reduction with respect to a foliation has been investigated in the context of foliations on Riemannian manifolds in [GKPS]. Interestingly, despite the suggestiveness of the foliation aspects of our construction, the techniques we use actually rely more on the holomorphic aspects, and are thus much closer to those of Garcia-Prada than to those of [GKPS].

Indeed, the results we describe in this paper show that the Garcia-Prada techniques readily extend to the case where $X \times \mathbb{P}^1$ is replaced by a *projectively flat* $\mathbb{P}^1$-bundle over X. More precisely, we assume that the $\mathbb{P}^1$-bundle has a flat $PU(2)$ structure. This assumption allows us to extend any $PU(2)$-invariant structure on $\mathbb{P}^1$ to the total space, say M, of the $\mathbb{P}^1$-bundle over X. In particular, the Kähler form of the Fubini–Study metric can be extended to become a global form on M, and the canonical bundle on $\mathbb{P}^1$ extends to become the relative canonical bundle $K_{M/X}$. By carefully exploiting these constructions we are able to identify an interesting class of holomorphic bundles over M on which the Garcia-Prada techniques can be made applicable.

In Section 2 we give some background material on the Hermitian–Einstein equations and the Coupled Vortex equations, and in Section 3 we review the Garcia-Prada dimensional reduction techniques. In Section 4 we introduce our Projective Flatness Condition and discuss its implications. Our main result is given in Section 5, where it appears as Theorem 5.1. Section 6 contains some computations for the special case that the projective bundle is the projectivization of a vector bundle. Finally, in Section 7 we indulge in some discussion of our results, with particular emphasis on possible interpretations from the point of view of foliation theory.

Acknowledgements. The authors are pleased to acknowledge useful conversations with, and informative comments from Tony Pantev, Robert Friedman, Domingo Toledo and Andrei Tyurin. The first author is a member of the VBAC group of Europroj. The third named author is grateful for the hospitality and support provided by the

Erwin Schrödinger International Institute for Mathematical Physics in Vienna and the Mathematics Institute of Aarhus University.

§2. The equations

The Hermitian–Einstein equations determine special metrics on holomorphic bundles over Kähler manifolds (cf. [Ko] or [LT] for a complete review). Indeed, if (M, ω_M) is any closed Kähler manifold, and $\mathcal{E}$ is a holomorphic bundle over M, then the Hermitian–Einstein equations for a Hermitian metric H are

$$i\Lambda_M F_H = \lambda \mathbf{I}_E \, . \tag{2.1}$$

Here E denotes the smooth complex bundle underlying $\mathcal{E}$, F_H is the curvature of the metric connection on $\mathcal{E}$ determined by H, $\Lambda_M F_H$ is the contraction of F_H with the Kähler form ω_M, $\mathbf{I}_E \in \Omega^0(\mathrm{End}\, E)$ is the identity and λ is a constant determined by ω_M, the rank of $\mathcal{E}$, and its degree. Recall that the degree of a complex bundle over a Kähler manifold is defined by

$$\deg(E) = \int_M c_1(E) \wedge \omega_M^{m-1} = \int_M \Lambda_M c_1(E) \frac{\omega_M^m}{m} \, , \tag{2.2}$$

where m is the dimension of M and $c_1(E)$ is the first Chern class of E. It follows (by the Chern–Weil formula for $c_1(E)$) that the λ in (2.1) is given by

$$\lambda = \frac{2\pi}{(m-1)!} \frac{1}{\mathrm{Vol}(M)} \frac{\deg(E)}{\mathrm{rank}(E)} \, .$$

We remind the reader that the $\deg(E)$ is used to define the *slope*, $\mu(E)$, of the bundle by

$$\mu(E) = \frac{\deg(E)}{\mathrm{rank}(E)} \, .$$

For a holomorphic bundle $\mathcal{E}$, these definitions extend to coherent subsheaves, and are used to define the algebro-geometric notion of *stability*. We record, for the sake of completeness, that a bundle $\mathcal{E}$ is called stable if $\mu(\mathcal{E}') < \mu(\mathcal{E})$ for all coherent subsheaves $\mathcal{E}' \subset \mathcal{E}$.

Remark 2.1. The Hermitian–Einstein equations have many interesting features, among which are the following:

(1) They are the Kähler versions of the anti-self-dual (ASD) equations. What we mean by this is that, like the ASD equations, the HE equations can be obtained as the minimizing condition for the Yang–Mills functional on the space of unitary connections on a complex vector bundle. Furthermore the ASD and HE equations are equivalent in the case where they are both defined, i.e. on unitary bundles with vanishing first Chern class over closed Kähler surfaces.

(2) The existence of solutions is closely related to the stability of a holomorphic bundle. This is known as the Hitchin–Kobayashi correspondence.

The Coupled Vortex Equations also determine special metrics, but on a *holomorphic triple* over a Kähler manifold (cf. [GP], [BGP]). If (X, ω_X) is a closed Kähler manifold of dimension n, then by definition a holomorphic triple on X is a triple $(\mathcal{E}_1, \mathcal{E}_2, \Phi)$ consisting of two holomorphic vector bundles $\mathcal{E}_1$ and $\mathcal{E}_2$ on X together with a homomorphism $\Phi : \mathcal{E}_2 \longrightarrow \mathcal{E}_1$, i.e. an element $\Phi \in H^0(\mathrm{Hom}(\mathcal{E}_2, \mathcal{E}_1))$. The coupled vortex equations on $(\mathcal{E}_1, \mathcal{E}_2, \Phi)$ are defined to be

$$\begin{aligned} i\Lambda_X F_{H_1} + \Phi\Phi^* &= \tau_1 \mathbf{I}_{E_1} \\ i\Lambda_X F_{H_2} - \Phi^*\Phi &= \tau_2 \mathbf{I}_{E_2} \,. \end{aligned} \tag{2.3}$$

In these equations, Φ^* is the adjoint of Φ with respect to the metrics on $\mathcal{E}_1$ and $\mathcal{E}_2$, and τ_1 and τ_2 are real parameters. If the ranks of the bundles are r_1 and r_2 respectively, and we denote their degrees by d_1 and d_2, then the parameters τ_1 and τ_2 satisfy the constraint

$$r_1\tau_1 + r_2\tau_2 = \frac{2\pi}{(n-1)!}\frac{1}{\mathrm{Vol}(X)}(\deg E_1 + \deg E_2) \,. \tag{2.4}$$

Remark 2.2. The Coupled Vortex equations may be viewed as a natural generalization of the abelian vortex equations (cf. [JT]). Originally introduced in the so-called Ginsburg-Landau theory of superconductivity, these equations were later generalized to describe special Hermitian metrics on holomorphic bundles with prescribed holomorphic sections. In that setting the equations play as important a role as do the Hermitian–Einstein equations in the study of stable holomorphic bundles. Most recently, the abelian vortex equations have appeared in Seiberg–Witten theory, where they emerge as the Kähler version of the Seiberg–Witten equations (cf. [W]).

§3. Garcia-Prada dimensional reduction

Suppose now that $M = X \times \mathbb{P}^1$. Following Garcia-Prada in [GP], we can consider the situation in which the product $X \times \mathbb{P}^1$ has an $SU(2)$-action in which $SU(2)$ acts trivially on X and via the identification with the homogeneous space $SU(2)/U(1)$ on $\mathbb{P}^1$. We can thus consider $SU(2)$-equivariant bundles on M, where a smooth bundle $V \longrightarrow M$ is called $SU(2)$-equivariant if there is an action of $SU(2)$ on V covering the action on $X \times \mathbb{P}^1$. In particular, we can consider smooth bundles over $M = X \times \mathbb{P}^1$ of the form

$$V = \pi_1^* E_1 \oplus \pi_1^* E_2 \otimes \pi_2^* H^{\otimes 2} \,. \tag{3.1}$$

Here π_1 and π_2 are the projections from $X \times \mathbb{P}^1$ to the first and second factors, E_i is a smooth vector bundle over X, and H is the smooth line bundle over $\mathbb{P}^1$ with

Chern class 1. The $SU(2)$-action on V is trivial on the $\pi_1^* E_i$ factors and standard on $H^{\otimes 2}$.

Not all the $SU(2)$-equivariant bundles are of the form in (3.1); in general they can be direct sums where each summand is of the form $\pi_1^* E \otimes \pi_2^* H^{\otimes k}$. The special case represented by (3.1) is chosen with a view towards the dimensional reduction from the Hermitian–Einstein equations to the Coupled Vortex equations. For our purposes, the key feature of such bundles, given in the next Proposition, has to do with the $SU(2)$-equivariant *holomorphic* structures that they admit. Recall that a holomorphic bundle can be viewed as an underlying smooth bundle with a holomorphic structure, and that a holomorphic bundle $\mathcal{V}$ is $SU(2)$-equivariant if it is $SU(2)$-equivariant as a smooth bundle and in addition the action of $SU(2)$ on $\mathcal{V}$ is holomorphic.

Proposition 3.1 ([Prop. 3.9 in [GP]). *There is a one-to-one correspondence between $SU(2)$-equivariant holomorphic vector bundles $\mathcal{E}$ with underlying $SU(2)$-equivariant C^∞ structure given by* (3.1), *and holomorphic extensions of the form*

$$0 \longrightarrow \pi_1^*\mathcal{E}_1 \longrightarrow \mathcal{E} \longrightarrow \pi_1^*\mathcal{E}_2 \otimes \pi_2^*\mathcal{O}(2) \longrightarrow 0\,, \tag{3.2}$$

where $\mathcal{E}_1$ and $\mathcal{E}_2$ are the bundles in (3.1), *equipped with holomorphic structures. Moreover, every such extension is defined by an element $\Phi \in \mathrm{Hom}(\mathcal{E}_2, \mathcal{E}_1)$.*

The characterization of the extension classes in the above Proposition depends on the fact that extensions over $X \times \mathbb{P}^1$ of the form (3.2) are parametrized by $H^1(X \times \mathbb{P}^1, \pi_1^*(\mathcal{E}_1 \otimes \mathcal{E}_2^*) \otimes \pi_2^*\mathcal{O}(-2))$, and that by the Kunneth formula this is isomorphic to $H^0(X, \mathcal{E}_1 \otimes \mathcal{E}_2^*) \otimes H^1(\mathbb{P}^1, \mathcal{O}(-2))$. The result thus follows from the fact that $H^1(\mathbb{P}^1, \mathcal{O}(-2)) \cong \mathbb{C}$.

It follows immediately from Proposition 3.1 that there is a one-to-one correspondence between extensions over $X \times \mathbb{P}^1$ of the type given in (3.2), and holomorphic triples $(\mathcal{E}_1, \mathcal{E}_2, \Phi)$ on X. Thus, at the level of holomorphic objects,

> *a holomorphic triple on X can be viewed as the dimensional reduction of an $SU(2)$-equivariant holomorphic bundle on $X \times \mathbb{P}^1$.*

We now turn to the dimensional reduction at the level of equations, i.e. we examine the relation between the Hermitian–Einstein equations on an extension as in (3.2) and the Coupled Vortex equations on the corresponding triple.

In order to define the equations, we need to fix Kähler structures. We let ω_X be the (fixed) Kähler form on X, and let $\omega_{\mathbb{P}^1}$ denote the Kähler form of the Fubini–Study metric on $\mathbb{P}^1$ normalized so that

$$\mathrm{Vol}(\mathbb{P}^1) = \int_{\mathbb{P}^1} \omega_{\mathbb{P}^1} = 1\,.$$

On $X \times \mathbb{P}^1$ there is a natural 1-parameter family of $SU(2)$-equivariant Kähler metrics, with Kähler forms

$$\omega_\sigma = \pi_1^*\omega_X + \sigma\pi_2^*\omega_{\mathbb{P}^1}\,. \tag{3.3}$$

Using these Kähler structures, one finds:

Proposition 3.2 (Proposition 3.11 [GP]). *Let $T = (\mathcal{E}_1, \mathcal{E}_2, \Phi)$ be a holomorphic triple. Let $\mathcal{E}$ be the $SU(2)$-equivariant holomorphic bundle over $X \times \mathbb{P}^1$ associated to T, that is, let $\mathcal{E}$ be given as an extension*

$$0 \longrightarrow \pi_1^*\mathcal{E}_1 \longrightarrow \mathcal{E} \longrightarrow \pi_1^*\mathcal{E}_2 \otimes \pi_2^*\mathcal{O}(2) \longrightarrow 0\,.$$

Suppose that τ_1 and τ_2 are related by (2.4) *and let σ be chosen to satisfy*

$$\frac{1}{\sigma} = \frac{(r_1 + r_2)\frac{\tau_1 \mathrm{Vol}(X)}{2\pi} - \frac{(\deg E_1 + \deg E_2)}{(n-1)!}}{2r_2 \mathrm{Vol}(X)} = \frac{\tau_1 - \tau_2}{4\pi}\,. \tag{3.4}$$

Fix the Kähler metric on $X \times \mathbb{P}^1$ such that the Kähler form is ω_σ, as given by (3.3).

Then $\mathcal{E}_1$ and $\mathcal{E}_2$ admit metrics satisfying the Coupled Vortex equations (i.e. equations (2.3)*) if and only if $\mathcal{E}$ admits an $SU(2)$-invariant metric which satisfies the Hermitian–Einstein equation (i.e. equation* (2.1)*) with respect to ω_σ.*

Remark 3.3. Now by the Hitchin–Kobayashi correspondence, the existence of a solution to the Hermitian–Einstein equation is equivalent (cf. [UY], [Do], [NS])to the polystability of the holomorphic bundle $\mathcal{E}$. There is a similar correspondence for holomorphic triples, which relates the existence of solutions to the coupled vortex equations to a notion of stability for a holomorphic triple (cf. [BGP]). As one might expect, Proposition 3.2 thus has an analog which relates *stable* holomorphic extensions on $X \times \mathbb{P}^1$ to *stable* triples on X (cf. Theorem 4.1 in [BGP]). It should be noted that, with fixed Kähler structures on X and $\mathbb{P}^1$, the notion of stability for the holomorphic triple on X depends explicitly on a parameter τ, while the notion of stability for the holomorphic bundles on $X \times \mathbb{P}^1$ depends on the choice of Kähler structure on $X \times \mathbb{P}^1$. If the Kähler form on $X \times \mathbb{P}^1$ is taken to be ω_σ, then the theorem relates the stability of $\mathcal{E}$ to the τ-stability of the corresponding triple on X, where τ is the same as τ_1 in Proposition 3.2. This analog to Proposition 3.2 can be be thought of as a dimensional reduction result for *stable holomorphic objects* (rather than for *equations*).

§4. Non-trivial $\mathbb{P}^1$-bundles over X

§4.1. Dimensional reduction of bundles

It is natural to view $X \times \mathbb{P}^1$ as a special case of a $\mathbb{P}^1$-fibration over X. One is then lead to consider whether any of the results of the previous section carry over to this more general situation. Henceforth we thus replace $X \times \mathbb{P}^1$ by

$$\mathbb{P}^1 \hookrightarrow M \longrightarrow X,$$

a holomorphic $\mathbb{P}^1$-bundle over X.

In its most general form this means that M has a description as a $\mathbb{P}^1$-bundle associated to a principal $PGL(2,\mathbb{C})$-bundle. Denoting the principal $PGL(2,\mathbb{C})$-bundle by $\tilde{P}$, we can thus write

$$M = \tilde{P} \times_{PGL(2,\mathbb{C})} \mathbb{P}^1 .$$

Remark. If X is a Riemann surface, so that M is a ruled surface, then the $PGL(2,\mathbb{C})$ structure always lifts to $GL(2,\mathbb{C})$ (cf. [GH] or [H]) and M can be described as $M = \mathbb{P}(E)$, where $E \longrightarrow X$ is a rank two holomorphic bundle. If $\dim_{\mathbb{C}} X > 1$, then this is not always possible. In general there is an obstruction to such a lift, with the obstruction located in $H^3(X,\mathbb{Z})$.

Our initial goal is to find a class of holomorphic bundles on M whose structure is determined by *data on X*. In general there is no longer an $SU(2)$-action on M, so it does not make sense to study $SU(2)$-equivariant bundles over M. Instead, it turns out that the appropriate replacements for the extensions described in (3.2) are extensions of the form

$$0 \longrightarrow \pi^*\mathcal{E}_1 \longrightarrow \mathcal{E} \longrightarrow \pi^*\mathcal{E}_2 \otimes \mathcal{K}^*_{M/X} \longrightarrow 0 , \tag{4.1}$$

where $\pi : M \longrightarrow X$ is the projection, $\mathcal{E}_i$ are holomorphic bundles over X, and $\mathcal{K}^*_{M/X}$ is the dual of the relative canonical bundle on M.

Notice that if $M = X \times \mathbb{P}^1$, then $\mathcal{K}^*_{M/X} = \pi_2^*\mathcal{O}(2)$ and we recover the extensions of the form (3.2). Furthermore, an important part of Proposition 3.1 remains true for extensions of this sort, namely that every such extension is defined by an element $\Phi \in \mathrm{Hom}(\mathcal{E}_2, \mathcal{E}_1)$. More precisely,

Proposition 4.1. *There is a natural isomorphism*

$$H^{0,1}(M, \pi^*(\mathcal{E}_1 \otimes \mathcal{E}_2^*) \otimes \mathcal{K}_{M/X}) \cong H^0(X, \mathcal{E}_1 \otimes \mathcal{E}_2^*) . \tag{4.2}$$

Proof. This follows by the Leray spectral sequence, plus the fact that the direct image sheaves of the relative canonical bundle satisfy (cf. [H])

$$\begin{aligned} R^q\pi_*\mathcal{K}_{M/X} &\cong 0 \text{ if } q \neq 1 , \\ R^1\pi_*\mathcal{K}_{M/X} &\cong \mathcal{O}_X . \end{aligned} \tag{4.3}$$

□

What this says is:

Corollary 4.2. *Let M be any holomorphic $\mathbb{P}^1$-bundle over X. There is a one to one correspondence between holomorphic triples $(\mathcal{E}_1, \mathcal{E}_2, \Phi)$ on X and extensions*

$$0 \longrightarrow \pi^*\mathcal{E}_1 \longrightarrow \mathcal{E} \longrightarrow \pi^*\mathcal{E}_2 \otimes \mathcal{K}^*_{M/X} \longrightarrow 0 ,$$

over M.

§4.2. Projective flatness condition

To proceed further with the dimensional reduction, i.e. to relate the Hermitian–Einstein and Coupled Vortex equations, we need to consider the Kähler structures on M. In general, it seems to be a difficult matter to find a Kähler metric on M such that the resulting Hermitian–Einstein equations on an extension (4.1) will 'dimensionally reduce' to the Coupled Vortex equations on the corresponding triple over X. We have, however, discovered a sufficient condition on M in order for this to be possible. It remains an interesting open question whether our condition is also necessary.

The condition we impose on M is the

Projective Flatness Condition. *We assume that* $\mathbb{P}^1 \hookrightarrow M \longrightarrow X$ *is a flat* $PU(2)$*-bundle.*

The real significance of the projective flatness condition lies in the fact, that under this assumption, we are able to construct the following:

(1) a class of bundles which correspond to the $SU(2)$-equivariant extensions in the Garcia-Prada construction,

(2) a suitable family of Kähler metrics on M, generalizing the family corresponding to the Kähler forms given in (3.3), and

(3) explicit descriptions of the forms representing the extension classes in (4.1).

In short, we can construct bundles which are are "locally equivariant" and retain enough symmetry so that the key constructions required for dimensional reduction still apply.

Before describing these constructions, it is useful to elaborate somewhat on the projective flatness condition. There are three equivalent ways of viewing this condition:

(1) Using the definition of a flat bundle in terms of locally constant transition functions, the projective flatness condition means that we can choose a cover $\{U_i\}$ for X, and local trivializations $M|_{U_i} \cong U_i \times \mathbb{P}^1$, such that

$$M = (\coprod_i U_i \times \mathbb{P}^1)/g_{ij}\,, \tag{M1}$$

where the transition functions

$$g_{ij} : U_i \cap U_j \longrightarrow PU(2)\,,$$

are locally constant maps. In particular, the coordinate transformations from $U_j \times \mathbb{P}^1$ to $U_i \times \mathbb{P}^1$, given by $(x, \lambda) \mapsto (x, g_{ji}\lambda)$, are holomorphic maps. The description (M1) thus describes M as a complex manifold.

(2) Using the cover $\{U_i\}$ and transition functions g_{ij}, we can construct a flat principal $PU(2)$-bundle. This is the bundle $\tilde{P}$ in the description given at the beginning

of §4.1. Hence we see that a second way to formulate the projective flatness condition, is to say that

$$M = \tilde{P} \times_{PU(2)} \mathbb{P}^1 , \tag{M2}$$

where $\tilde{P}$ is a *flat* principal $PU(2)$-bundle.

(3) Finally, we can use the fact that a flat bundle $\tilde{P}$ can be associated to a representation

$$\rho : \pi_1(X) \longrightarrow PU(2) .$$

This gives a description of a projectively flat M as

$$M = \tilde{X} \times_\rho \mathbb{P}^1 , \tag{M3}$$

where $\tilde{X}$ is the universal cover of X, acted on by $\pi_1(X)$ via deck transformations.

One way that such flat $PU(2)$-bundles arise, is as the projectivization of rank two holomorphic vector bundles. If $E \longrightarrow X$ is a vector bundle, then the projective bundle $\mathbb{P}(E)$ is defined as follows. Let E be defined by local trivializations $E|_{U_i} \cong U_i \times \mathbb{C}^2$, where $\{U_i\}$ is an open cover of X, and transition functions

$$\hat{g}_{ij} : U_i \cap U_j \longrightarrow GL(n+1, \mathbb{C}) ,$$

i.e.

$$E = (\coprod_i U_i \times \mathbb{C}^{n+1})/\hat{g}_{ij} .$$

Then $\mathbb{P}(E)$ is defined as

$$\mathbb{P}(E) = (\coprod_i U_i \times \mathbb{P}^n)/g_{ij} ,$$

where the transition functions $g_{ij} : U_i \cap U_j \longrightarrow PGL(n, \mathbb{C})$ are obtained from the $\hat{g}_{ij}$ by the projection from $GL(n+1, \mathbb{C})$ to $PGL(n, \mathbb{C})$. The proof of the following well known fact can be found in [Ko].

Lemma 4.3. *Let E be a rank $(n+1)$ holomorphic vector bundle, and let H be a Hermitian metric on E. If the corresponding metric connection is projectively flat, i.e. if its curvature F_H is given by*

$$F_H = \alpha \mathbf{I} ,$$

where α is a 2-form on X and $\mathbf{I} \in \Omega^0(\mathrm{End}\, E)$ is the identity map, then $\mathbb{P}(E)$ has a flat $PU(n+1)$-structure.

If X is a Riemann surface, i.e. $\dim_{\mathbb{C}} X = 1$, then the flat $PU(n)$-bundles which arise in this way are characterized by the following:

Proposition 4.4. *Let $E \longrightarrow X$ be a rank $(n+1)$ holomorphic vector bundle over a closed Riemann surface. Then E admits a projectively flat Hermitian metric if and* *only if E is polystable, i.e. if and only if*

$$E = \bigoplus_i E_i \ ,$$

where each summand E_i is a stable holomorphic bundle with

$$\frac{\deg(E_i)}{\operatorname{rank}(E_i)} = \frac{\deg(E)}{\operatorname{rank}(E)} \ .$$

In particular, if $n = 1$, then $\mathbb{P}(E)$ has a flat $PU(2)$-structure if and only if either

(1) *E is a stable rank 2 holomorphic bundle, or*

(2) *$E = L_1 \oplus L_2$, where L_1 and L_2 are holomorphic line bundles with*

$$\deg(L_1) = \deg(L_2) = \frac{1}{2}\deg(E) \ .$$

Proof. On a Riemann surface, the condition $F_H = \alpha\mathbf{I}$ is equivalent (cf. [Ko]) to the Hermitian–Einstein condition $\Lambda F_H = \lambda\mathbf{I}$. The result thus follows from the Hitchin–Kobayashi correspondence (or, in this case, the theorem of Narasimhan and Seshadri). □

If $\dim_{\mathbb{C}} X > 1$, then the projective flatness condition is in general stronger than the Hermitian–Einstein condition. However, if X has a Kähler metric with Kähler form ω, and if the Chern classes of E satisfy the Bogomolov relation

$$\int_X ((r-1)c_1(E)^2 - 2rc_2(E)) \wedge \omega^{n-2} = 0 \ ,$$

(where $r = \operatorname{rank}(E)$) then the Hermitian–Einstein condition is again equivalent to the projective flatness of the full curvature tensor (cf. [LT], also [Ko], p. 114). We thus get:

Proposition 4.5. *Let (X, ω) be a closed Kähler manifold, and let $E \longrightarrow X$ be a rank two holomorphic bundle over X. Then $\mathbb{P}(E)$ has a flat $PU(2)$-structure if and only if either*

(1) *E is stable (with respect to ω) and*

$$\int_X (c_1(E)^2 - 4c_2(E)) \wedge \omega^{n-2} = 0 \ ,$$

or

(2) *$E = L_1 \oplus L_2$, where L_1 and L_2 are holomorphic line bundles with*

$$c_1(L_1) = c_1(L_2) \ .$$

Proof. Part (1) follows immediately from the Bogomolov relation. For part (2), we use the fact that the first Chern classes are represented by forms of type $(1, 1)$. They can thus be expressed as $c_1(L_i) = \alpha_i\omega + \beta_i$, where (for $i = 1, 2$) $\alpha_i \in \mathbb{R}$, and the β_i are primitive forms of type $(1, 1)$, i.e. $\int_X (\beta_i \wedge \omega^{n-1}) = 0$. Furthermore, since E is polystable, it follows that $\alpha_1 = \alpha_2$. The Bogomolov relation thus reduces, in this case, to

$$\int_X (\beta_1 - \beta_2) \wedge (\beta_1 - \beta_2) \wedge \omega^{n-2} = 0 .$$

Using the Hodge-Riemann bilinear relations for real primitive (1,1) forms (cf. [GH], p.207), we see that the left hand side of this expression is strictly negative unless $(\beta_1 - \beta_2) = 0$. □

Remark. It should be noted that not all flat $PU(2)$-bundles arise in this way. As pointed out in the Remark at the beginning of §4.1, there is an obstruction to realizing a $\mathbb{P}^1$-bundle as $\mathbb{P}(E)$, where E is a vector bundle.

§4.3. Equivariant bundles

Given a $PU(2)$-equivariant holomorphic bundle $\mathcal{V} \longrightarrow \mathbb{P}^1$, we can use the description of M as $M = \tilde{P} \times_{PU(2)} \mathbb{P}^1$ to construct a holomorphic bundle

$$\tilde{\mathcal{V}} = \tilde{P} \times_{PU(2)} \mathcal{V} \longrightarrow M = \tilde{P} \times_{PU(2)} \mathbb{P}^1 . \tag{4.4}$$

We call $\tilde{\mathcal{V}}$ the *extension of* $\mathcal{V}$ *to* M . In fact this construction applies equally well if $PU(2)$ is replaced by $PGL(2, \mathbb{C})$, and does not require $\tilde{P}$ to have a flat structure. If however, $\tilde{P}$ does have a flat structure, so that descriptions (M1-3) apply, then we get:

Lemma 4.6. *Let $\tilde{P}$ be a flat $PU(2)$-bundle, described as above by either the representation $\rho : \pi_1(X) \longrightarrow PU(2)$ or by the locally constant transition functions $g_{ij} : U_i \cap U_j \longrightarrow PU(2)$. Let $\mathcal{V} \longrightarrow \mathbb{P}^1$ be a $PU(2)$-equivariant holomorphic bundle.*

Then the extension of $\mathcal{V}$ to M can be described in the following two equivalent ways:

(1)
$$\tilde{\mathcal{V}} = \tilde{X} \times_\rho \mathcal{V} , \tag{4.5a}$$

(2)
$$\tilde{\mathcal{V}} = (\textstyle\coprod_i U_i \times \mathcal{V})/g_{ij} . \tag{4.5b}$$

Remark 4.7. Using the above construction, we can extend any invariant structure on $\mathcal{V}$ to $\tilde{\mathcal{V}}$. In particular, an invariant Hermitian bundle metric on $\mathcal{V}$ extends to define a Hermitian metric on $\tilde{\mathcal{V}}$.

When $\mathcal{V}$ is a line bundle, we can describe the $PU(2)$-equivariant bundles over $\mathbb{P}^1$. They are precisely the $SU(2)$-equivariant bundles on which the element $-I \in SU(2)$

acts as the identity. The $PU(2)$-action is then obtained by lifting to an $SU(2)$-action. We conclude therefore that the $PU(2)$-equivariant holomorphic line bundles on $\mathbb{P}^1$ are precisely the even powers of the tautological bundle $\mathcal{O}_{\mathbb{P}^1}(-1)$, i.e. are the bundles $\mathcal{O}_{\mathbb{P}^1}(2k)$, for any integer k. We denote the corresponding extensions to M by the notation $\tilde{\mathcal{O}}(2k)$. For our purposes, the most important special case is given by $\tilde{\mathcal{O}}(-2)$.

Lemma 4.8. *Let M be a projectively flat $PU(2)$-bundle, as above. Let $K_{M/X}$ be the relative canonical bundle. Then we can identify*

$$K_{M/X} = \tilde{\mathcal{O}}(-2) .$$

§4.4. Equivariant metrics

If M is projectively flat (and thus has descriptions as in (M1-3)), then we can construct a 1-parameter family of Kähler metrics which generalizes the construction used in [GP]. This is made possible by the fact that $\mathbb{P}^1$ has an $SU(2)$-equivariant (and thus a $PU(2)$-equivariant) Kähler metric, viz. the Fubini–Study metric. The corresponding Kähler form, which we denote by $\omega_{\mathbb{P}^1}$, is thus a $PU(2)$-invariant form of holomorphic type $(1, 1)$ on $\mathbb{P}^1$. It therefore extends to a form of type $(1, 1)$ on M. This form, which we denote by $\tilde{\omega}_{\mathbb{P}^1}$ can be described in two ways (corresponding to the two descriptions (M3) and (M1)):

(1) Let $\pi_2^*\omega_{\mathbb{P}^1}$ be the pull-back of $\omega_{\mathbb{P}^1}$ to $\tilde{X} \times \mathbb{P}^1$, and let

$$q : \tilde{X} \times \mathbb{P}^1 \longrightarrow \tilde{X} \times_\rho \mathbb{P}^1 ,$$

be the quotient map. Then $\tilde{\omega}_{\mathbb{P}^1}$ is the form on M such that $q^*\tilde{\omega}_{\mathbb{P}^1} = \pi_2^*\omega_{\mathbb{P}^1}$.

(2) Let $\omega_{\mathbb{P}^1}^{(i)}$ be the pull-back of $\omega_{\mathbb{P}^1}$ to $U_i \times \mathbb{P}^1$, where $\{U_i\}$ is the cover of X used in the description (M1). Because of the $PU(2)$-invariance of $\omega_{\mathbb{P}^1}$, and the fact that the transition functions g_{ij} are locally constant, the $\omega_{\mathbb{P}^1}^{(i)}$ patch together to define a globally defined form. The form they define is $\tilde{\omega}_{\mathbb{P}^1}$.

Notice that $\tilde{\omega}_{\mathbb{P}^1}$ is closed and restricts to $\omega_{\mathbb{P}^1}$ on the $\mathbb{P}^1$-fibers of M. If $\pi : M \longrightarrow X$ is the projection map, and ω_X denotes the Kähler form on X, then we can define the 1-parameter family

$$\omega_\sigma = \pi^*\omega_X + \sigma\tilde{\omega}_{\mathbb{P}^1} . \tag{4.6}$$

This is clearly a family of closed, non-degenerate, positive forms of type $(1, 1)$. In fact ω_σ is the Kähler form for the Hermitian metric on M which (using description (M1)) pulls back to the (weighted) product metric on each neighborhood $U_i \times \mathbb{P}^1$. This can also be described as the metric which descends from the weighted product metric on $\tilde{X} \times \mathbb{P}^1$.

In the case where $M = X \times \mathbb{P}^1$, this is exactly the family of Kähler forms described in [GP]. If M is not a product, but is projectively flat with a metric of the above sort, then M is 'close enough' to the product case for the following to be true.

Lemma 4.9. *Let $f : X \longrightarrow R$ be a smooth function on X. Then*

$$\int_M \pi^*(f) \mathrm{dvol}_\sigma = \sigma \mathrm{Vol}(\mathbb{P}^1) \int_X f \, \mathrm{dvol}_X \,, \tag{4.7a}$$

where $\mathrm{dvol}_\sigma = \frac{\omega_\sigma^{n+1}}{(n+1)!}$ is the volume element on M, and $\mathrm{dvol}_X = \frac{\omega_X^n}{n!}$ is the volume element on X. In particular,

(1)
$$\mathrm{Vol}_\sigma(M) = \sigma \mathrm{Vol}(X) \mathrm{Vol}(\mathbb{P}^1) \,, \tag{4.7b}$$

(2) if $V \longrightarrow X$ is a complex bundle on X, then

$$\deg_\sigma(\pi^* V) = n\sigma \deg(V) \,, \tag{4.7c}$$

where for any bundle, say W, on M, $\deg_\sigma(W) = \int_M c_1(W) \wedge \omega_\sigma^n$ is the degree with respect to the Kähler form in (4.6).

Proof. Equation (4.7a) is an immediate consequence of the fact that the metric is locally a product metric on the open sets homeomorphic to $U_i \times \mathbb{P}^1$, and the fact that

$$\frac{\omega_\sigma^{n+1}}{(n+1)!} = \pi^*(\frac{\omega_X^n}{n!}) \wedge \sigma \tilde{\omega}_{\mathbb{P}^1} \,.$$

For (4.7b), take $f = 1$. For (4.7c), we use the fact that $c_1(\pi^* V) = \pi^* c_1(V)$, and also the identity

$$n\alpha \wedge \omega_X^{n-1} = (\Lambda_X \alpha) \omega_X^n \,,$$

where α is any complex 2-form on X. Using equation (4.6) we then compute

$$\begin{aligned} \deg_\sigma \pi^* V &= n\sigma \int_M \pi^*(c_1(V) \wedge \omega_X^{n-1}) \wedge \tilde{\omega}_{\mathbb{P}^1} \\ &= \sigma \int_M \pi^*(n! \Lambda_X c_1(V)) \mathrm{dvol}_\sigma \,. \end{aligned}$$

Applying (4.7a) with $f = n! \Lambda_X c_1(V)$ then gives the result. □

§4.5. Explicit description of extension classes

In order to adapt the dimensional reduction procedure of Garcia-Prada, we need to have an explicit description of the extension class in (4.1), i.e. we need explicit representatives for the classes in $H^{0,1}(M, \pi^*(\mathcal{E}_1 \otimes \mathcal{E}_2^*) \otimes \mathcal{K}_{M/X})$. Using the Projective

Flatness condition, we now show that these can be taken to be of the form $\pi^*\Phi \otimes \tilde{\eta}$, where $\Phi \in \Omega^0(X, \mathcal{E}_1 \otimes \mathcal{E}_2^*)$ and $\tilde{\eta}$ is the extension to $\mathcal{K}_{M/X}$ of a uniquely determined invariant element $\eta \in \Omega^{(0,1)}(\mathbb{P}^1, \mathcal{O}(-2))$.

We begin with a description of η. Up to a scale factor, this is uniquely determined by the requirements that

(1) η is $SU(2)$- (and thus also $PU(2)$-) invariant,

(2) η represents a generator of $H^{(0,1)}(\mathbb{P}^1, \mathcal{O}(-2)) \cong \mathbb{C}$.

We can give an explicit description of η if we use the identification of $\mathcal{O}(-2)$ with $K_{\mathbb{P}^1}$ i.e. with the holomorphic cotangent bundle. Let z be a local coordinate on $\mathbb{P}^1$, and use $\{dz\}$ as a local frame for $K_{\mathbb{P}^1}$. With respect to this frame, and with respect to the local coordinate z, we get

$$\eta(z) = \eta_0 d\bar{z} \otimes \frac{dz}{(1+|z|^2)^2}\,, \tag{4.8}$$

where η_0 is the undetermined scale factor. Now if we interpret the Fubini–Study metric as a bundle metric on the holomorphic tangent bundle $T^{1,0}\mathbb{P}^1$, then with respect to the local coordinate z, and the local frame $\frac{\partial}{\partial z}$, this metric is given by

$$k\left(\frac{\partial}{\partial z}, \frac{\partial}{\partial z}\right) = \frac{1}{(1+|z|^2)^2}\,. \tag{4.9}$$

Thus taking duals with respect to this metric, we can write

$$\eta(z) = \eta_0 d\bar{z} \otimes \left(\frac{\partial}{\partial z}\right)^*.$$

Or, if η^* denotes the 'conjugate adjoint', i.e. the section of $\Omega^{(1,0)}(\mathbb{P}^1, \mathcal{O}(2))$ obtained from η by complex conjugation on the form part of η and taking the adjoint (with respect to the metric induced by k) of the section of $\mathcal{O}(-2)$, then $\eta^* = \overline{\eta_0} dz \otimes \frac{\partial}{\partial z}$. We thus get that

$$\eta \wedge \eta^* = i|\eta_0|^2 \omega_{\mathbb{P}^1}\,. \tag{4.10}$$

With a view towards the next section, we henceforth fix η_0 such that

$$|\eta_0|^2 = \frac{1}{\sigma}\,, \tag{4.11}$$

where σ is the weighting factor in the Kähler metric on M, i.e. in (4.6).

Definition 4.10. We define

$$\tilde{\eta} \in \Omega^{(0,1)}(M, \tilde{\mathcal{O}}(-2)) \cong \Omega^{(0,1)}(M, \mathcal{K}_{M/X})\,,$$

to be the extension of η (defined by (4.8) and (4.11)).

A local calculation confirms that this is a $\bar{\partial}$-closed form. Furthermore, the cohomology class which it represents in $H^{(0,1)}(\mathbb{P}^1, \mathcal{O}(-2)) \cong \mathbb{C}$ is non-trivial since, for example, it restricts to the generator of $H^{(0,1)}(\mathbb{P}^1, \mathcal{O}(-2))$ on each $\mathbb{P}^1$ fiber of M. We thus get

Proposition 4.11. *The isomorphism*

$$H^0(X, \mathcal{E}_1 \otimes \mathcal{E}_2^*) \cong H^{0,1}(M, \pi^*(\mathcal{E}_1 \otimes \mathcal{E}_2^*) \otimes \mathcal{K}_{M/X}),$$

is realized by the map

$$\Phi \longrightarrow \pi^*\Phi \otimes \tilde{\eta} \in \Omega^{0,1}(M, \pi^*(\mathcal{E}_1 \otimes \mathcal{E}_2^*) \otimes \mathcal{K}_{M/X}). \tag{4.12}$$

Proof. We need to check that $\pi^*\Phi \otimes \tilde{\eta}$ is $\overline{\partial}$-closed (so that the map is well defined at the level of the cohomology groups), and that the map is injective (and thus an isomorphism). The first issue is clear, since Φ is assumed to be a holomorphic section of $\mathcal{E}_1 \otimes \mathcal{E}_2^*$ and $\tilde{\eta}$ is also $\overline{\partial}$-closed. The injectivity of the map follows from the above comment about the non-triviality of the cohomology class of $\tilde{\eta}$. □

The following property of $\tilde{\eta}$ will be used in the next section:

Lemma 4.12. *Let $\tilde{k}$ be the extension to $\tilde{\mathcal{O}}(2) \cong \mathcal{K}_{M/X}^*$ of the metric k on $\mathcal{O}(2)$. Using this metric, and the metric induced by it on $\tilde{\mathcal{O}}(-2) \cong \mathcal{K}_{M/X}$, we get*

$$\tilde{\eta} \wedge \tilde{\eta}^* = \frac{i}{\sigma} \tilde{\omega}_{\mathbb{P}^1}. \tag{4.13}$$

Proof. This is a pointwise property. But in local coordinates on a neighborhood biholomorphic to $U_i \times \mathbb{P}^1$, this corresponds precisely to the fact that $\eta \wedge \eta^* = \frac{i}{\sigma}\omega_{\mathbb{P}^1}$. □

§5. Dimensional reduction of the Hermitian–Einstein equations

We are now ready to examine the Hermitian–Einstein equations for an extension of the form in (4.1) over a projectively flat $PU(2)$-bundle M, with respect to a Kähler metric as above. The setup we consider is the following:

Let M be a projectively flat $PU(2)$-bundle over X, with descriptions as in (M1), (M2), and (M3). Fix a Kähler metric on M with Kähler form ω_σ as in equation (4.6). Let $\mathrm{Vol}_\sigma(M)$ denote the volume with respect to this metric (cf. Lemma 4.9), and for any vector bundle $V \longrightarrow M$, set

$$\deg_\sigma(V) = \int_M c_1(V) \wedge \omega_\sigma^n.$$

Consider a holomorphic extension over M of the form

$$0 \longrightarrow \pi^*\mathcal{E}_1 \longrightarrow \mathcal{E} \longrightarrow \pi^*\mathcal{E}_2 \otimes \mathcal{K}^*_{M/X} \longrightarrow 0\,,$$

as in (4.1). The extension class is represented by a holomorphic section in

$$\Omega^{(0,1)}(M, \mathrm{Hom}(\pi^*\mathcal{E}_2 \otimes \mathcal{K}^*_{M/X}, \pi^*\mathcal{E}_1)) \cong \Omega^{(0,1)}(M, \pi^*(\mathcal{E}_1 \otimes \mathcal{E}_2^*) \otimes \mathcal{K}_{M/X})\,.$$

Using the isomorphism given in Proposition 4.11, we can take this to be

$$\beta = \pi^*\Phi \otimes \tilde{\eta}\,. \tag{5.1}$$

Theorem 5.1. *Let $\mathcal{E}$ be a holomorphic extension as above, with extension class β as in* (5.1). *Let* $\mathbf{h}$ *be a Hermitian metric on $\mathcal{E}$ given by*

$$\mathbf{h} = \pi^*h_1 \oplus \pi^*h_2 \otimes \tilde{k}\,,$$

where, for $i = 1, 2$, h_i is a Hermitian metric on $\mathcal{E}_i$ and $\tilde{k}$ is the Hermitian metric on $\mathcal{K}^*_{M/X} = \tilde{\mathcal{O}}(2)$ described in Lemma 4.12.

Fix $\sigma > 0$, and set

$$\lambda = \frac{2\pi}{n!}\frac{1}{\mathrm{Vol}_\sigma(M)}\frac{\deg_\sigma(\mathcal{E})}{\mathrm{rank}\,\mathcal{E}}\,. \tag{5.2}$$

Let parameters τ_1 and τ_2 be given by

$$\begin{aligned} \tau_1 &= \lambda \\ \tau_2 &= \lambda - \frac{2\pi}{n!\mathrm{Vol}_\sigma(M)}\deg_\sigma(\mathcal{K}^*_{M/X})\,. \end{aligned} \tag{5.3}$$

Then the following are equivalent:

(1) The metric $\mathbf{h}$ satisfies the Hermitian–Einstein equation

$$i\Lambda_\sigma F_{\mathbf{h}} = \lambda \mathbf{I}_{\mathcal{E}}\,,$$

where Λ_σ denotes contraction against the Kähler form

$$\omega_\sigma = \pi^*\omega_X + \sigma\tilde{\omega}_{\mathbb{P}^1}\,.$$

(2) The metrics h_1 and h_2 satisfy the coupled vortex equations:

$$\begin{aligned} i\Lambda_X F_{h_1} + \Phi\Phi^* &= \tau_1 \mathbf{I}_{\mathcal{E}_1} \\ i\Lambda_X F_{h_2} - \Phi^*\Phi &= \tau_2 \mathbf{I}_{\mathcal{E}_2}\,, \end{aligned}$$

where the adjoint in Φ^* is with respect to the metrics h_1 and h_2, and Λ_X denotes contraction against the Kähler form ω_X on X.

Proof. Because of the projectively flat structure on M and our choice of Kähler structure, the geometry of our situation is *locally* indistinguishable from the case of

$M = X \times \mathbb{P}^1$. The proof of this theorem is thus essentially the same as that of the corresponding result in [GP]. We proceed as follows.

Given a Hermitian metric $\mathbf{h}$ on $\mathcal{E}$, we can analyse the Hermitian–Einstein tensor $i\Lambda_\sigma F_{\mathbf{h}}$. With respect to a smooth orthogonal splitting $\mathcal{E} = \pi^*\mathcal{E}_1 \oplus \pi^*\mathcal{E}_2 \otimes \mathcal{K}^*_{M/X}$, we can write

$$i\Lambda_\sigma F_{\mathbf{h}} = i\Lambda_\sigma \begin{pmatrix} F_{\mathbf{h}_1} - \beta\wedge\beta^* & D'\beta \\ -D''\beta^* & F_{\mathbf{h}_2} - \beta^*\wedge\beta \end{pmatrix} . \tag{5.4}$$

Here $\mathbf{h}_1$ and $\mathbf{h}_2$ are the metrics induced on the sub- and quotient bundles, the $F_{\mathbf{h}_i}$ are the curvatures of the corresponding metric connections, and

$$D = D' + D'' : \Omega^1(\pi^*(\mathcal{E}_1 \otimes \mathcal{E}_2^*) \otimes \mathcal{K}_{M/X}) \to \Omega^2(\pi^*(\mathcal{E}_1 \otimes \mathcal{E}_2^*) \otimes \mathcal{K}_{M/X})$$

is constructed in the standard way from the metrics on $\pi^*(\mathcal{E}_1)$ and $\pi^*(\mathcal{E}_2^*) \otimes \mathcal{K}_{M/X}$. The summands D' and D'' are the $(1,0)$ and $(0,1)$ parts respectively. The "$*$" in β^* denotes the adjoint on sections (with respect to the bundle metrics) and conjugation on forms, so that $\beta^* \in \Omega^{(1,0)}(M, \pi^*(\mathcal{E}_2 \otimes \mathcal{E}_1^*) \otimes \mathcal{K}^*_{M/X})$. We can write

$$\begin{aligned} F_{\mathbf{h}_1} &= \pi^* F_{h_1} \\ F_{\mathbf{h}_2} &= \pi^* F_{h_2} + F_{\tilde{k}} \otimes \mathbf{I}_2 . \end{aligned} \tag{5.5}$$

Now the $SU(2)$-invariant metric k on $\mathcal{O}_{\mathbb{P}^1}(2)$ (described in (4.9)) is Hermitian–Einstein with respect to the Fubini–Study metric on $\mathbb{P}^1$. Because of our choice of Kähler structure on M it thus follows that $\tilde{k}$ is a Hermitian–Einstein metric on $\mathcal{K}^*_{M/X}$, so that we have

$$i\Lambda_\sigma F_{\tilde{k}} = c = \frac{2\pi}{n!\mathrm{Vol}_\sigma(M)} \deg_\sigma(\tilde{\mathcal{O}}(2)) . \tag{5.6}$$

Also, with β as in (5.1), we get

$$i\Lambda_\sigma(\beta\wedge\beta^*) = \pi^*\Phi\Phi^* \otimes i\Lambda_\sigma(\tilde{\eta}\wedge\tilde{\eta}^*) , \tag{5.7}$$

where the adjoint in Φ^* is with respect to the metrics on $\mathcal{E}_1$ and $\mathcal{E}_2$, and $\tilde{\eta}^*$ is determined by the metric $\tilde{k}$. The proof of the theorem thus depends on the following lemma:

Lemma 5.2. *With $\tilde{\eta}$, β, and Λ_σ as above, we have*

(1) $i\Lambda_\sigma(\tilde{\eta}\wedge\tilde{\eta}^*) = -1$,

(2) $\Lambda_\sigma D'\beta = 0$, *and*

(3) $\Lambda_\sigma D''\beta^* = 0$.

Furthermore, for $i = 1, 2$, we get

(4) $\Lambda_\sigma \pi^* F_{h_i} = \pi^*\Lambda_X F_{h_i}$.

Proof of Lemma 5.2. Part (1) follows immediately from Lemma 4.12.

For part (2), write

$$D'\beta = D'(\pi^*\Phi)\otimes\tilde{\eta} + \pi^*\Phi\otimes D'\tilde{\eta} .$$

We are abusing notation here, since the covariant derivatives denoted by D' are not the same in the three terms in this expression. We can be more precise if we restrict to local coordinates on a neighborhood biholomorphic to $U_i \times \mathbb{P}^1$. The formula then becomes

$$D'\beta = \pi_1^* D'_{\mathcal{E}_1 \otimes \mathcal{E}_2^*} \Phi \otimes \pi_2^* \eta + \pi_1^* \Phi \otimes \pi_2^* D'_{\mathcal{O}(-2)} \eta\,, \tag{5.8}$$

where the π_i are the projections onto the first and second factors of $U_i \times \mathbb{P}^1$. The covariant derivatives are now those corresponding to the metric connections on $\mathcal{E}_1 \otimes \mathcal{E}_2^*$ and on $\mathcal{O}_{\mathbb{P}^1}(-2)$. Furthermore, in these local coordinates,

$$\omega_\sigma = \pi_1^* \omega_X + \sigma \pi_2^* \omega_{\mathbb{P}^1}\,.$$

The rest of the proof is essentially the same as in [GP], with the key points being:

(i) The term $\Lambda_\sigma \pi_1^* D'_{\mathcal{E}_1 \otimes \mathcal{E}_2^*} \Phi \otimes \pi_2^* \eta$ vanishes because the Kähler form has no contribution from 'mixed' terms, i.e. terms in $\Omega^{(1,0)}(U_i) \wedge \Omega^{(0,1)}(\mathbb{P}^1)$, and

(ii) the term $\pi_1^* \Phi \otimes \Lambda_\sigma \pi_2^* D'_{\mathcal{O}(-2)} \eta$ vanishes because

$$\Lambda_\sigma \pi_2^* D'_{\mathcal{O}(-2)} \eta = \pi_2^* (\sigma \Lambda D'_{\mathcal{O}(-2)} \eta) = 0\,.$$

The proof of part (3) is similar to (2) (alternatively, one can simply observe that $\Lambda_\sigma D'' \beta^* = -(\Lambda_\sigma D' \beta)^*$).

For part(4), we note that with $\omega_\sigma = \pi^* \omega_X + \sigma \tilde{\omega}_{\mathbb{P}^1}$, we get

$$\Lambda_\sigma \pi^* F_{h_i} = \pi^* \Lambda_X F_{h_i} + \sigma (\pi^* F_{h_i}, \tilde{\omega}_{\mathbb{P}^1})\,, \tag{5.9}$$

where the inner product in the second term on the right is on $\Omega^2(M, \mathbb{C})$. A computation in local coordinates verifies that for any $\alpha \in \Omega^2(X, \mathbb{C})$, we get $(\pi^* \alpha, \tilde{\omega}_{\mathbb{P}^1}) = 0$. □

Remark 5.3. The proof of Lemma 5.2 uses the fact that M is projectively flat and the special properties of the Kähler metrics that we fix on M. In a more general situation, it is not so clear which (if any) Kähler structures are suitable for the carrying out of a dimensional reduction of the Hermitian–Einstein equations. It is tempting to speculate that the properties listed in the lemma should serve as the *definition* of suitable Kähler structures.

Completion of Proof of Theorem 5.1. Using (5.2), (5.6), and Lemma 5.2, we thus find that the Hermitian–Einstein condition on $\mathbf{h}$ is equivalent to

$$\begin{pmatrix} \pi^*(i\Lambda_X F_{h_1} + \Phi\Phi^*) & 0 \\ 0 & \pi^*(i\Lambda_X F_{h_2} - \Phi^*\Phi) + c \otimes \mathbf{I}_2 \end{pmatrix} = \lambda \begin{pmatrix} \mathbf{I}_1 & 0 \\ 0 & \mathbf{I}_2 \end{pmatrix} \tag{5.10}$$

where c is as in (5.6). The results follows from this. □

Remark 5.4. By using (5.3), (5.4), and (4.7c) we can compute

$$\begin{aligned} r_1\tau_1 + r_2\tau_2 &= \frac{2\pi}{n!\mathrm{Vol}_\sigma(M)}(\deg_\sigma(\mathcal{E}) - \deg_\sigma(\mathcal{K}^*_{M/X})) \\ &= \frac{2\pi}{n!\mathrm{Vol}_\sigma(M)}(\deg_\sigma(\pi^*(\mathcal{E}_1 \oplus \mathcal{E}_2)) \\ &= \frac{2\pi}{(n-1)!\mathrm{Vol}(X)}\deg(\mathcal{E}_1 \oplus \mathcal{E}_2)\,, \end{aligned}$$

thus verifying that the relation given by (2.4) holds. Also, by (5.4) and (4.7b), we have

$$\tau_1 - \tau_2 = \frac{2\pi}{\sigma}\frac{\deg_\sigma(\mathcal{K}^*_{M/X})}{n!\mathrm{Vol}(X)}\,. \tag{5.11}$$

We can compute $\deg_\sigma(\mathcal{K}^*_{M/X})$, using the fact that $\mathcal{K}^*_{M/X} = \tilde{\mathcal{O}}(2)$:

Lemma 5.5. *Let $\tilde{\mathcal{O}}(2k)$ be the extension to M of the bundle $\mathcal{O}_{\mathbb{P}^1}(2k)$. Then the first Chern class $c_1(\tilde{\mathcal{O}}(2k))$ can be represented by the $(1,1)$-form $2k\tilde{\omega}_{\mathbb{P}^1}$, where $\tilde{\omega}_{\mathbb{P}^1}$ is the extension of the Fubini Study Kähler form (as in §4.4). Hence (assuming that the Fubini–Study metric is normalized so that* $\mathrm{Vol}(\mathbb{P}^1) = 1$*) we get*

$$\deg_\sigma(\tilde{\mathcal{O}}(2k)) = 2kn!\mathrm{Vol}(X)\,.$$

Proof. The first Chern class of $\mathcal{O}_{\mathbb{P}^1}(2k)$ can be represented by the (1,1)-form $2k\omega_{\mathbb{P}^1}$. This in turn can be obtained as $\frac{i}{2\pi}\partial\bar{\partial}log(h)$, where h is an ($SU(2)$-invariant) hermitian metric on $\mathcal{O}_{\mathbb{P}^1}(2k)$. Since h is $PU(2)$-invariant, it extends to a metric $\tilde{h}$ on $\tilde{\mathcal{O}}(2k)$. One can compute that

$$\frac{i}{2\pi}\partial_M\bar{\partial}_M log(\tilde{h}) = 2k\tilde{\omega}_{\mathbb{P}^1}\,,$$

where ∂_M and $\bar{\partial}_M$ are the holomorphic and anti-holomorphic parts of the exterior derivative on M. It follows immediately from this that $c_1(\tilde{\mathcal{O}}(2k))$ can be represented by $2k\tilde{\omega}_{\mathbb{P}^1}$. The formula for $\deg_\sigma(\tilde{\mathcal{O}}(2k))$ then follows from the form of ω_σ (cf. (4.6)), and by the computation for $\mathrm{Vol}_\sigma(M)$ (cf. (4.7b)). □

Applying Lemma 5.5. to $\mathcal{K}^*_{M/X} = \tilde{\mathcal{O}}(2)$, we get that $\deg_\sigma(\mathcal{K}^*_{M/X}) = 2n!\mathrm{Vol}(X)$. Equation (5.11), which applies in the case where M is a flat $PU(2)$-bundle, is therefore the same as the relation given in Proposition 3.1, which refers to the case where $M = X \times \mathbb{P}^1$.

§6. Special case where $M = \mathbb{P}(E)$

As an interesting special case, we consider the situation in which M comes from a holomorphic vector bundle with a projectively flat Hermitian structure. In this case, the parameter computations of the previous section can then be carried out quite explicitly. We thus assume that:

(1) $M = \mathbb{P}(E)$, where $E \longrightarrow X$ is a rank 2 holomorphic bundle, and

(2) there is a Hermitian metric h on E such that

$$i\Lambda_X F_h = \text{const.}\,\mathbf{I} = \frac{\pi \deg(E)}{(n-1)!\mathrm{Vol}(X)}\,\mathbf{I}.$$

Using the description $M = \mathbb{P}(E)$, we get a canonically defined line bundle on M, namely the tautological line bundle on $\mathbb{P}(E)$. This bundle, which we denote by $\mathcal{O}_M(-1)$, is a subbundle of $\mathbb{P}(E) \times E$ and restricts to the tautological line bundle on each $\mathbb{P}^1$-fiber of $\mathbb{P}(E)$. The main result of this section is :

Proposition 6.1.

$$\deg_\sigma(\mathcal{O}_M(-1)) = \frac{n\sigma \deg(E)}{2} - n!\mathrm{Vol}(X) \tag{6.1a}$$

$$= \frac{\deg_\sigma(\pi^*(\det E))}{2} - n!\mathrm{Vol}(X)\,. \tag{6.1b}$$

Proof. The proof is a computation, using Equation (2.2) and the Chern–Weil formula for $c_1(E)$ in terms of curvature. Since $\mathcal{O}_M(-1)$ is a subbundle of $\mathbb{P}(E) \times \pi^*E$, the metric h on E induces a fiber metric, say H, on $\mathcal{O}_M(-1)$. Let F_H be the curvature of its corresponding metric connection. We thus need to compute $\frac{i}{2\pi}\Lambda_\sigma F_H$ at a point of $\mathbb{P}(E)$.

Any point in $\mathbb{P}(E)$ can be represented by a unit vector ξ_0 in a fiber E_{x_0} of E. As in [Ko] (p. 89)], we choose a normal holomorphic frame $s = (s_1, s_2)$ for E with $s_2(x_0) = \xi_0$. This gives local coordinates $(z^1, \ldots, z^n, \xi^1, \xi^2)$ on E, where $(z^1, \ldots, z^n)$ are local coordinates near x_0 on X, and (ξ^1, ξ^2) are the coordinates with respect to the frame s of a point in the fiber. If we denote the points in E by $\xi(z)$, and the corresponding points in $\mathbb{P}(E)$ by $[\xi(z)]$. Then

$$[\xi(z)] \mapsto (z^1, \ldots, z^n, [\xi^1, \xi^2])\,,$$

gives a local trivialization $\mathbb{P}(E)|_U \cong U \times \mathbb{P}^1$, and thus defines a local holomorphic frame for $\mathcal{O}_M(-1)$. Using these local coordinates, the computation in [Ko2 (p. 90 and I.4)] shows that

$$\begin{aligned}(iF_H)_{[\xi_0]} &= i\sum R_{i\bar{k}\alpha\bar{\beta}}\,\xi_0^i\bar{\xi}_0^k\,dz^\alpha \wedge d\bar{z}^\beta \;-\; (\omega_{\mathbb{P}^1})_{[0,1]} \\ &= ih(R(\xi_0), \xi_0) \;-\; (\omega_{\mathbb{P}^1})_{[0,1]}\,,\end{aligned} \tag{6.2}$$

where R is the curvature of the metric connection for h on E. But , since h satisfies the Hermitian–Einstein condition, we get that

$$in\, h(R(\xi_0), \xi_0) \wedge \omega_X^{n-1} = \frac{\pi \deg(E)}{\mathrm{Vol}(X)(n-1)!} |\xi_0|^2\, \omega_X^n . \tag{6.3}$$

Also, using the local coordinate system on $\mathbb{P}(E)$ near $[\xi_0]$, the Kähler form on $M = \mathbb{P}(E)$ is given by $\omega_\sigma = \omega_X + \sigma\omega_{\mathbb{P}^1}$. Thus, using $|\xi_0|^2 = 1$, we get

$$\begin{aligned} \frac{(iF_H)}{2\pi} \wedge \omega_\sigma^n &= \frac{(in\sigma)}{2\pi} h(R(\xi_0), \xi_0) \wedge \omega_X^{n-1} \wedge \omega_{\mathbb{P}^1} - \omega_X^n \wedge \omega_{\mathbb{P}^1} \\ &= \Big(\frac{n\sigma \deg(E)}{2\mathrm{Vol}(X)} - n!\Big)\Big(\frac{\omega_X^n}{n!} \wedge \omega_{\mathbb{P}^1}\Big) \\ &= \Big(\frac{n\sigma \deg(E)}{2\mathrm{Vol}(X)} - n!\Big)\Big(\frac{\omega_\sigma^{n+1}}{(n+1)!}\Big) . \end{aligned} \tag{6.4}$$

The first result follows from this, Lemma 4.9, and the fact that $\int_{\mathbb{P}^1} \omega_{\mathbb{P}^1} = 1$. The second result then follows by (4.7c) in Lemma 4.9, applied to $\det E$. □

We can define $\mathcal{O}_M(1)$ to be the dual bundle to $\mathcal{O}_M(-1)$, and set

$$\mathcal{O}_M(\pm k) = \mathcal{O}_M(\pm 1)^{\otimes k} .$$

Proposition 6.2 (cf. [Ko]).

$$\mathcal{K}_{M/X}^* \equiv \tilde{\mathcal{O}}(2) = \mathcal{O}_M(2) \otimes \pi^*(\det E) ,$$

where $\det E$ *is the determinant bundle of E, and* $\pi : M \longrightarrow X$ *is as in the previous section.*

Combining Propositions 6.1 and 6.2 we thus get

Proposition 6.3.

(1) *For* $k \in \mathbb{Z}$,

$$\begin{aligned} \deg_\sigma \mathcal{O}_M(k) &= k\Big(n!\mathrm{Vol}(X) - \frac{\deg_\sigma \pi^*(\det E)}{2}\Big) \\ &= k\Big(n!\mathrm{Vol}(X) - \frac{n\sigma \deg(E)}{2}\Big) , \end{aligned} \tag{6.5}$$

(2)
$$\deg_\sigma(\mathcal{K}_{M/X}^*) = 2n!\mathrm{Vol}(X) . \tag{6.6}$$

Remark 6.4. Notice that as a result of (6.6) and (5.11) we see that when $M = \mathbb{P}(E)$ the parameters τ_1, τ_2, and σ (in Theorem 5.1) are related in precisely the same way as in the case where $M = X \times \mathbb{P}^1$, i.e. as in Proposition 3.2.

§7. Discussion of results

It is clear from the results described above that dimensional reduction should be thought of in more general terms than was previously expected. In particular, the role of a global symmetry in the form of a group action is not as central as we assumed. On the other hand, what makes the reduction possible is certainly some kind of geometric order, even if cannot quite be called a global symmetry. In this section we explore what kind of geometric structure is evident in our situation, and look at how it compares to the Garcia-Prada case.

In the case that $M = X \times \mathbb{P}^1$, there is an action of $SU(2)$ on M as described in [GP]. The action is trivial on X and via the identification $\mathbb{P}^1 = SU(2)/U(1)$ on $\mathbb{P}^1$. The group orbits are thus of the form $x \times \mathbb{P}^1$, and the orbit space is X.

In order to exploit this symmetry (i.e. the group action), one must consider *invariant* solutions for whatever equations are to be dimensionally reduced. In order for such solutions to exist, one needs to be in an *equivariant* setting, and it is for this reason that the bundles considered in [GP] are $SU(2)$-equivariant bundles.

As soon as the structure of M as a $\mathbb{P}^1$-bundle over X becomes non-trivial, the $SU(2)$-action is destroyed . Notice however that the group orbits survive in the form of fibers of the bundle. As we shall see, it is perhaps more pertinent to describe the fiber $\mathbb{P}^1$'s as the *leaves of a foliation*. Indeed, given a foliation on M, there is a natural replacement for the notion of an equivariant bundle, namely that of a *foliated bundle* (cf. [KT]).

Roughly speaking, a foliated bundle has a flat structure along each leaf of the foliation. The relation between such bundles and the equivariant bundles (which can exist when the foliation comes from a global group action) is made even clearer if we introduce the holonomy groupoid, $\mathcal{G}_{\mathcal{F}}$, associated to the foliation $\mathcal{F}$. The foliated bundles we consider can be described as $\mathcal{G}_{\mathcal{F}}$-equivariant bundles, where an $\mathcal{G}_{\mathcal{F}}$-equivariant bundle is one which supports an action of $\mathcal{G}_{\mathcal{F}}$ on its fibers. The action in question is in fact via holonomy transport along the leaves of the foliation.

This language of foliations provides a convenient framework for understanding the dimensional reduction we have carried out in §5. It is illuminating even in the case where $M = X \times \mathbb{P}^1$. In that case the $SU(2)$-equivariant bundles over $X \times \mathbb{P}^1$ can indeed be considered as foliated bundles - but in way that is slightly unexpected. The novelty is due to the fact that $X \times \mathbb{P}^1$ is foliated in *two* ways; in one foliation the leaves are the copies of $\mathbb{P}^1$ in $X \times \mathbb{P}^1$, and in the other way the leaves are the copies of X. We will refer to these as the *fiber foliation* and the *base foliation* respectively.

Lemma 7.1. *The bundles over $X \times \mathbb{P}^1$ that are foliated with respect to the fiber foliation are of the form $\pi_1^* E$, where $E \longrightarrow X$ is any smooth vector bundle over X and π_1 denotes the projection onto the first factor of $X \times \mathbb{P}^1$.*

The line bundles over $X \times \mathbb{P}^1$ that are foliated with respect to the base foliation are of the form $\pi_2^* H^{\otimes k}$, where $H \longrightarrow \mathbb{P}^1$ is the line bundle of degree 1 over $\mathbb{P}^1$ and π_2 denotes the projection onto the second factor of $X \times \mathbb{P}^1$.

Thus the $SU(2)$-equivariant bundles that appear in the dimensional reduction construction , i.e. the bundles of the form $V = \pi_1^* E \otimes \pi_2^* H^{\otimes k}$, can be thought of as *doubly foliated bundles*. The next Proposition shows that it is this doubly foliated property that is retained when the trivial fibration $X \times \mathbb{P}^1$ is replaced by a flat projective fibration $M \longrightarrow X$.

Proposition 7.2. *Let $M \longrightarrow X$ be a flat projective $PU(2)$-bundle over X, as in (§4). Then M has two natural foliations; one in which the leaves are the $\mathbb{P}^1$-fibers of $M \longrightarrow X$, and one in which the leaves are copies of $\tilde{X}$, the universal cover of X. The leaves of the first foliation intersect the leaves of the second transversally.*

Proof. This follows immediately form the descriptions (M1-3) of M given in §4. □

We will denote the foliation by the $\mathbb{P}^1$-fibers by $\mathcal{F}_\pi$, and the other foliation by $\mathcal{F}_\alpha$. The next result shows that the bundles with the structure of extensions as in (4.1) are indeed the analogues of the bundles considered in [GP].

Proposition 7.3. *Let $M \longrightarrow X$ be a flat projective $PU(2)$-bundle over X, as in (§4).*

(1) *There is a one to one correspondence between vector bundles W over X and $\mathcal{G}_{\mathcal{F}_\pi}$-equivariant bundles on M, given by $W \longrightarrow \pi^* W$.*

(2) *There is a one to one correspondence between α-equivariant vector bundles V over $\mathbb{P}^1$ and $\mathcal{G}_{\mathcal{F}_\alpha}$-equivariant bundles on M, given by $V \longrightarrow \tilde{V}$, where $\tilde{V}$ is the extension as defined in §4.2*

Finally, we remark that $\mathbb{P}^1$-bundles are almost certainly not the only objects on which our techniques will work. In particular, there is good reason to expect that $\mathbb{P}^1 = SU(2)/U(1)$ can be replaced by any Kählerian homogeneous space G/H. We will return to this question in a later publication.

References

[BGP] Bradlow, S. B., and García-Prada, O., Stable triples, equivariant bundles and dimensional reduction, Math. Ann. 304 (1996), 225–252.

[Do] Donaldson, S. K., Anti-self-dual Yang–Mills connections on a complex algebraic surface and stable vector bundles, Proc. Lond. Math. Soc. 3 (1985), 1–26.

[GP] García-Prada, O., Dimensional reduction of stable bundles, vortices and stable pairs, Int. J. Math. 5 (1994), 1–55.

[GKPS] Glazebrook, J. F., Kamber, F. W., Pedersen, H., and Swann, A., Foliation reduction and self duality, Geometric Study of Foliations, ed. Mizutani et.al. World Scientific, Singapore 1994, 219–249.

[GH] Griffiths, P., and Harris, J., Principles of Algebraic Geometry, J. Wiley and Sons, 1978.

[H] Hartshorne, R., Algebraic Geometry, Springer-Verlag, 1977.

[KT] Kamber F., and Tondeur, P., Foliated bundles and characteristic classes, Lecture Notes in Math. 493, Springer-Verlag, Berlin–Heidelberg–New York 1975.

[Ko] Kobayashi, S., Differential Geometry of Complex Vector Bundles, Princeton University Press, 1987.

[LT] Lübke, M., and Teleman, A., The Kobayashi–Hitchin Correspondence, World Scientific, Singapore 1996.

[JT] Jaffe, A., and Taubes, C., Vortices and Monopoles, Progress in Physics 2, Birkhäuser, Basel 1980.

[NS] Narasimhan, M. S., and Seshadri, C. S, Stable and unitary vector bundles over a Riemann surface, Ann. of Math. 82 (1965), 540–567.

[UY] Uhlenbeck, K. K., and Yau, S. T., On the existence of Hermitian–Yang–Mills connections in stable vector bundles, Comm. Pure. Appl. Math. 39 (1986), 5257–5293.

[W] Witten, E., Monopoles and Four Manifolds, Math. Res. Lett. 1 (1994), 769–796.

The universal cover of an affine three-manifold with holonomy of discompactedness two

*Suhyoung Choi**

Abstract. An affine manifold is a manifold with an affine structure. We give background to the claim that every closed affine manifold with holonomy group of discompactedness less than or equal to d is d-convex. We explain using the Morse theory that the universal cover of a closed affine three-manifold with holonomy group of discompactedness less than or equal to two is diffeomorphic to $\mathbb{R}^3$. In particular, the universal cover of a closed affine three-manifold with parallel volume form is diffeomorphic to $\mathbb{R}^3$.

1991 Mathematics Subject Classification: 57M50, 53A20, 53C15.

1. Introduction

We begin by recalling what an affine structure on a manifold is. Affine geometry is determined by a pair of the vector space $\mathbb{R}^n$, $n \geq 1$, and the group of affine transformations $\mathrm{Aff}(\mathbb{R}^n)$. An affine structure on a smooth n-manifold is given by an atlas of charts to $\mathbb{R}^n$ where transition functions of the charts are restrictions of affine transformations on $\mathbb{R}^n$. That is, an affine structure gives a way to lift affine geometry on a manifold locally and consistently. (An affine structure can also be given by a torsion-free flat affine connection.) An affine manifold is a manifold with an affine structure. Associated with an affine structure on an n-manifold M, there is a development pair $(\boldsymbol{dev}, h)$ where $\boldsymbol{dev} : \tilde{M} \to \mathbb{R}^n$ is an immersion and $h : \pi_1(M) \to \mathrm{Aff}(\mathbb{R}^n)$ is a homomorphism such that $h(\vartheta) \circ \boldsymbol{dev} = \boldsymbol{dev} \circ \vartheta$ for $\vartheta \in \pi_1(M)$. Every such pair $(\boldsymbol{dev}', h')$ for the given affine structure on M satisfies

$$(\boldsymbol{dev}', h'(\cdot)) = (\varphi \circ \boldsymbol{dev}, \varphi \circ h(\cdot) \circ \varphi^{-1}) \text{ for } \varphi \in \mathrm{Aff}(\mathbb{R}^n).$$

$\boldsymbol{dev}$ is said to be a *developing map* and h the *holonomy homomorphism* associated with $\boldsymbol{dev}$. A *holonomy group* of an affine manifold is the image of h. The holonomy group is well-defind up to conjugation.

*This research was partially supported by GARC-KOSEF

Benzécri proved that only closed surfaces admitting affine structures are homeomorphic to tori or Klein bottle (see [2]). Recently, Carrière [3] proved a remarkable result that a closed affine manifold M with holonomy in the Lorentzian isometry group of $\mathbb{R}^n$ is complete. (Note Lorentzian isometry is an affine transformation.) A *complete* affine manifold is one affinely diffeomorphic to a quotient of $\mathbb{R}^n$ by a properly discontinuous and free action of a discrete subgroup of $\mathrm{Aff}(\mathbb{R}^n)$.

Carrière's work is a partial and important answer to the conjecture of L. Markus [13] that a closed affine manifold is complete if and only if it admits a parallel volume form. Markus's conjecture is true if Γ is Abelian ([16]), Γ is nilpotent ([8]), Γ is solvable of rank n ([9]), or $L(\Gamma)$ is distal ([7]). A hope is that the following weakened version of the Markus's conjecture may be proved: a closed affine manifold has a universal cover diffeomorphic to $\mathbb{R}^n$ if it admits a parallel volume. (This was conjectured by Carrière around 1988.) We will sketch in our paper the proof of this for affine 3-manifolds.

The affine space $\mathbb{R}^n$ has a Euclidean metric μ and the distance metric $\boldsymbol{d}$ induced from it. $\boldsymbol{dev}$ induces a Riemannian metric μ on $\tilde{M}$ from the Euclidean metric of $\mathbb{R}^n$. (An arc is *straight* if it is a geodesic for μ.) μ induces a distance metric $\boldsymbol{d}$ on $\tilde{M}$. The completion $\check{M}$ of M is obtained by Cauchy completing $\tilde{M}$ with respect to $\boldsymbol{d}$. Let $\tilde{M}_\infty = \check{M} - \tilde{M}$ be the frontier set. $\check{M}$ and $\tilde{M}_\infty$ are topologically independent of the choice of $\boldsymbol{dev}$. The developing map $\boldsymbol{dev}$ and every deck transformation ϑ extend uniquely on $\check{M}$ to a distance-nonincreasing map and a self-homeomorphism respectively. For convenience, the extended maps are denoted by same symbols $\boldsymbol{dev}$ and ϑ respectively (see [12] and [5]).

An *m-simplex* in $\check{M}$ is a compact m-ball A imbedded in $\check{M}$ such that $A^o \subset \tilde{M}$ and $\boldsymbol{dev}|A$ is a (topological) imbedding onto a closed nondegenerate affine m-simplex in $\mathbb{R}^n$. We introduce the definition given by Carrière [3].

Definition 1. We say that M is *m-convex*, $0 < m < n$, if the following holds. If $T \subset \check{M}$ be an $(m+1)$-simplex with faces $F_1, F_2, \dots, F_{m+2}$ such that $F_2, \dots, F_{m+2}$ do not meet $\tilde{M}_\infty$, then F_1 do not meet $\tilde{M}_\infty$ and $T \subset \tilde{M}$.

Obviously, for a complete affine manifold M, since M is affinely isomorphic to the quotient of $\mathbb{R}^n$ by an action of a group of affine transformations, we may identify $\tilde{M}$ with $\mathbb{R}^n$ and hence $\tilde{M}_\infty = \emptyset$. Thus, M is m-convex for all $m, 0 < m < n$.

Remove the origin O from $\mathbb{R}^n$, $n \geq 3$, and let ϑ be the dilatation by factor 2. Then $M = \mathbb{R}^n - \{O\}/ < \vartheta >$ has a natural affine structure induced from $\mathbb{R}^n$. We can identify $\tilde{M}$ with $\mathbb{R}^n - \{O\}$ and $\check{M}$ with $\mathbb{R}^n$. M is not $(n-1)$-convex since there exists an n-simplex with face in the plane given by $x_n = 0$ meeting O in its interior. It is easy to see that i-convexity implies j-convexity whenever $i \leq j \leq n$.

Theorem 1. *Let M be a closed n-dimensional affine manifold. If the holonomy group of M has discompactedness less than or equal to d, then M is d-convex.*

The above theorem almost answers the first conjecture of Carrière [3, Section 3.3]. The proof is an essentially straightforward generalization of Carrière's argument [3] with some modification involving the definition of discompactedness. (However, we feel that our definition is just as useful in application.)

Theorem 2. *Let M be a closed two-convex affine three-manifold. Then $\tilde{M}$ is diffeomorphic to $\mathbb{R}^3$.*

Consequently, we get the second conjecture of Carrière [3, Section 3.3] for affine three-manifolds.

Corollary 1. *The universal cover of a closed affine three-manifold with holonomy group of discompactedness less than or equal to 2 is diffeomorphic to $\mathbb{R}^3$.*

Unfortunately, we don't see a way to generalize our argument to arbitrary dimensions.

The holonomy group of an affine three-manifold M with parallel volume form is of discompactedness less than or equal to 2 (see equation 2.2). Hence, $\tilde{M}$ is diffeomorphic to $\mathbb{R}^3$. In particular, M is aspherical.

2. Discompactedness

In this section, we give a definition of discompactedness, which is stronger than that of Carrière [3], but probably just as useful.

Recall $\boldsymbol{d}$ the standard Euclidean metric on $\mathbb{R}^n$. Let $\boldsymbol{d}'$ be the pull-back metric $f^*\boldsymbol{d}$ for a linear automorphism $f : \mathbb{R}^n \to \mathbb{R}^n$; let Γ be a subgroup of $\mathrm{GL}(n, R)$ acting on $\mathbb{R}^n$ in a standard manner. For $\epsilon > 0$, let E_ϵ be the standard n-$\boldsymbol{d}'$-ball of radius ϵ given by $\{v \in \mathbb{R}^n \mid \boldsymbol{d}'(O, v) \leq \epsilon\}$ where O is the origin. For $i \geq 1$, an *i-$\boldsymbol{d}'$-ball E^i_ϵ of radius* ϵ is the intersection of E_ϵ with a vector subspace of dimension i of $\mathbb{R}^n$. We say that Γ has *discompactedness* less than or equal to d for $0 \leq d \leq n-1$, or say $\operatorname{disc}\Gamma \leq d$, if given ϵ, $\epsilon > 0$, there exists a positive constant δ depending only on ϵ such that for every $\vartheta \in \Gamma$ there exists an $(n-d)$-$\boldsymbol{d}'$-ball $E_{\epsilon,\vartheta}$ of radius ϵ such that

$$\boldsymbol{d}'(\delta\vartheta(E_{\epsilon,\vartheta}), O) \geq \delta. \tag{2.1}$$

(See Figure 1.) Since ϑ is linear, the condition is independent of ϵ. That is, if the condition is satisfied for a certain ϵ, then the condition is satisfied for every positive ϵ. Moreover, the condition is independent of which Euclidean metric we use: that is, if $\boldsymbol{d}'$ is given by the pull-back metric $g^*\boldsymbol{d}$ for another linear automorphism $g : \mathbb{R}^n \to \mathbb{R}^n$, then the above condition is true for a different choice of δ. This follows since $\boldsymbol{d}'$ and $\boldsymbol{d}$ are uniformly compatible metrics on $\mathbb{R}^n$. Thus, we need to verify equation 2.1 only for a certain metric.

We give some examples: Let G be a connected reductive subgroup of $\mathrm{GL}(n, \mathbb{R})$. Then G admit an internal Cartan decomposition

$$G = KTK$$

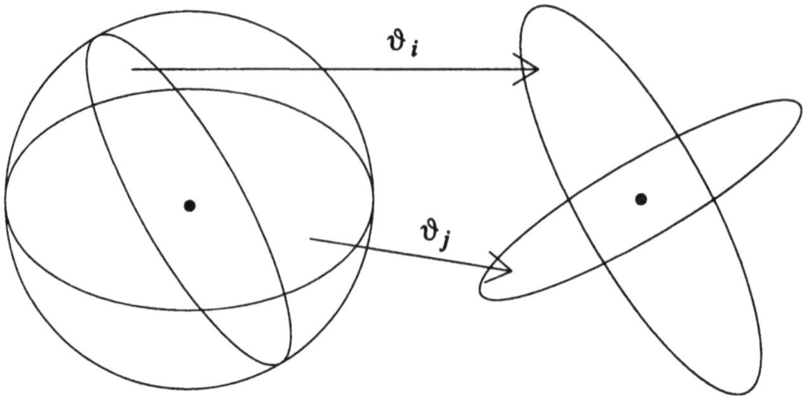

Figure 1: A diagram for discompactedness one.

where $K \subset G$ is a maximal compact subgroup and T a real maximal torus (see [10, p. 249]). Suppose that we can conjugate simultaneously so that $K \subset O(n)$, and $T \subset D(n)$, where $O(n)$ is the orthogonal subgroup and $D(n)$ the group of diagonal matrices. Thus, with respect to a fixed basis $\mathbf{f}$ of $\mathbb{R}^n$ the matrix $M(g)$ of every element g of G can be written $M(g) = R_1(g)D(g)R_2(g)$ where $R_1(g), R_2(g) \in O(n)$ and $D(g)$ is a diagonal matrix with positive diagonal element. We may further assume that the diagonal entries of $D(g)$ can be written in the decreasing order, i.e., $a_{11} \geq a_{22} \geq \cdots \geq a_{nn}$. For each element g of G, let $n(g)$ be the number of diagonal elements of $D(g)$ less than 1.

If $n(g) \leq d$ for every $g \in G$, then we claim that the discompactedness of G is less than or equal to d: Let $\boldsymbol{d}'$ be the metric pulled-back from $\boldsymbol{d}$ by the linear map sending the basis $\mathbf{f}$ to the standard basis of $\mathbb{R}^n$. We consider the $(n-d)$-$\boldsymbol{d}'$-ball E_ϵ^{n-d} of radius ϵ that is the intersection of E_ϵ with the vector subspace given by $x'_{n-d+1} = 0, \ldots, x'_n = 0$ where $x'_1, \ldots, x'_n$ are coordinates associated with $\mathbf{f}$. For each $g \in G$, let $E_{\epsilon,g}$ be $R_2(g)^{-1}(E_\epsilon^{n-d})$, an $(n-d)$-$\boldsymbol{d}'$-ball of radius ϵ. Then $g(E_{\epsilon,g}) = R_1(g)D(g)(E_\epsilon^{n-d})$. Since $a_{11}, \ldots, a_{n-d,n-d} \geq 1$, it follows that

$$\boldsymbol{d}'(O, \delta R_1(g)D(g)(E_\epsilon^{n-d})) \geq \epsilon.$$

Let $\mathrm{SL}(n, \mathbb{R})$ be the group of $n \times n$-matrices of determinant 1, and $\mathrm{Sp}(2n, \mathbb{R})$ the group of symplectic linear maps $\mathbb{R}^{2n} \to \mathbb{R}^{2n}$ with respect to the standard symplectic form. Let $O(p, q)$ denote the subgroup of $\mathrm{GL}(n, \mathbb{R})$ of linear maps preserving the quadratic form given by

$$x_1^2 + \cdots + x_p^2 - x_{p+1}^2 - \cdots - x_{p+q}^2, \; p + q = n.$$

It is easy to see that the discompactedness of a Lie group with finitely many components is less than or equal to the dimension of its real maximal torus. From the dimension f the real maximal torus and their diagnalizability (see [3] and [14, p. 310]), we can educe

$$\operatorname{disc} \mathrm{SL}(n, \mathbb{R}) \leq n-1, \operatorname{disc} O(2, q) \leq 2 (2 \leq q), \operatorname{disc} \mathrm{Sp}(2n, \mathbb{R}) \leq n. \quad (2.2)$$

There is a homomorphism L from the group $\mathrm{Aff}(\mathbb{R}^n)$ of affine transformations of $\mathbb{R}^n$ to $\mathrm{GL}(n, \mathbb{R})$ given by sending g to its linear part $L(g)$. A subgroup G of $\mathrm{Aff}(\mathbb{R}^n)$ of $\mathbb{R}^n$ has *discompactedness less than or equal to d* if its linear part $L(G)$ has discompactedness less than or equal to d.

For the proof of Theorem 1, see Choi [6].

3. Properties of completions

Let M be a simply-connected affine n-manifold without boundary. We will gather the needed facts on the completion $\check{M}$ here. A *convex subset* of M is a subset A such that every two points of A can be connected by a segment in A. Obviously, if A is a convex subset of M, then $\boldsymbol{dev}|A$ is an isometry with respect to the metrics denoted by $\boldsymbol{d}$ on M and $\mathbb{R}^n$. Hence, $\boldsymbol{dev}|A$ is an imbedding onto a convex subset of $\mathbb{R}^n$.

Lemma 1. *Let A be a convex subset of M. Then for the closure* $\mathrm{Cl}(A)$ *of A in $\check{M}$, $\boldsymbol{dev}|$* $\mathrm{Cl}(A)$ *is an imbedding onto a closed convex set* $\mathrm{Cl}(\boldsymbol{dev}(A))$. *Moreover, if A is $\boldsymbol{d}$-bounded, then* $\mathrm{Cl}(A)$ *is a compact subset of $\check{M}$.*

A *tame* subset of $\check{M}$ is a compact ball A imbedded $\check{M}$ such that $\boldsymbol{dev}|A$ is an imbedding onto a convex subset of $\mathbb{R}^n$ and $A^o \subset M$. For example, a simplex is tame.

Lemma 2. *The closure of a $\boldsymbol{d}$-bounded convex subset of M is a tame subset of $\check{M}$. Conversely, the interior of a tame subset is a convex subset of M.*

Lemma 3. *Let A and B be tame k-balls in $\check{M}$, and $\boldsymbol{dev}|A$ and $\boldsymbol{dev}|B$ are maps into an affine subspace P^k of dimension k $(k \leq n)$. Suppose that $A^o \cap B \neq \emptyset$, and $\boldsymbol{dev}(A) \cap \boldsymbol{dev}(B)$ is a path-connected compact k-manifold in $\mathbb{R}^n$ with interior equal to $\boldsymbol{dev}(A)^o \cap \boldsymbol{dev}(B)^o$. Then $\boldsymbol{dev}|A \cup B$ is a homeomorphism onto $\boldsymbol{dev}(A) \cup \boldsymbol{dev}(B)$.*

Lemma 4. *Let $\{K_i\}$ be a sequence of uniformly $\boldsymbol{d}$-bounded tame subsets of $\check{M}$ where $K_i \subset K_{i+1}$ for each i, $i = 1, 2, \ldots$. Then the closure K' of $\bigcup_{i=1}^{\infty} K_i$ is a tame subset of $\check{M}$, and $\boldsymbol{dev}(K')$ equals the closure of $\bigcup_{i=1}^{\infty} \boldsymbol{dev}(K_i)$.*

4. Tetrahedral convexity

Let M be a two-convex open affine simply-connected three-manifold. The purpose of this section is to show that two-convexity implies tetrahedral two-convexity, Lemma

5, and further cylindrical two-convexity, Theorem 3.

A point x of a convex subset A of $\mathbb{R}^n$ is called *exposed* if there exists a supporting affine hyperplane H at x such that $H \cap A = \{x\}$ (see Berger [1, p. 361]).

Lemma 5. *Let T be a convex compact three-simplex in $\check{M}$, F_1 a face of T such that $T \cap \tilde{M}_\infty = F_1 \cap \tilde{M}_\infty$, and P the affine two-space including $\mathbf{dev}(F_1)$. Then the convex hull K of $\mathbf{dev}(F_1 \cap \tilde{M}_\infty)$ in P has no exposed point belonging to $\mathbf{dev}(F_1^o)$.*

Let $(\mathbf{dev}, h)$ be a development pair of M. We have affine coordinate functions $x_1, x_2, x_3 : \mathbb{R}^3 \to \mathbb{R}$. For convenience, we let x_i denote the map $x_i \circ \mathbf{dev}$ defined on M for each i, $i = 1, 2, 3$. (We do not extend this map over to $\check{M}$.) Let M_t denote $x_1^{-1}(t)$, which is a closed subset of M.

Lemma 6. *For each t, $t \in \mathbb{R}$, each component of M_t is a properly imbedded open surface. In particular, the union of a collection of disjoint simple closed curves in M_t forms the boundary of at most one subsurface of M_t.*

Proof. Since M is without boundary, M_t is also without boundary. Let U be a component of M_t. Since $\mathbf{dev}|U$ is an immersion onto an open subset of an affine hyperplane of $\mathbb{R}^3$, U is not compact. Hence, U is an open surface. The conclusion follows. □

We now define cylindrical maps: Given an embedded compact surface Ω in the plane $\mathbb{R}^2$ with smooth boundary and the interval $[a, b]$ of $\mathbb{R}$ for $a, b \in \mathbb{R}$, $a < b$, think of $\Omega \times [a, b]$ as sitting in the Euclidean space $\mathbb{R}^2 \times \mathbb{R}$. A map $f : \Omega \times [a, b] \to \check{M}$ is called a *cylindrical map* if f satisfies the following conditions:

(i) $f|\Omega \times (a, b]$ is an injective map into M.

(ii) $\mathbf{dev} \circ f|\Omega \times [a, b]$ is a C^1-immersion.

(iii) $x_1 \circ \mathbf{dev} \circ f|\Omega \times \{t\}$ is a constant map with value t for each t, $t \in [a, b]$.

(iv) $f|\delta\Omega \times [a, b]$ is an embedding into M.

In particular, $f|\Omega \times \{t\}$ is an embedding into M_t for each t, $t \in (a, b]$, and $f(\Omega \times \{a\})$ may or may not intersect M_∞. We will often call M_t and $\Omega \times \{t\}$ *t-level sets*. Thus, f preserves levels.

Lemma 7. *Let $f : \Omega \times [a, b] \to \check{M}$ be a cylindrical map. For every point x of $\Omega^o \times \{a\}$, there exists a compact neighborhood V such that $f|V$ is an imbedding onto a three-simplex T in $\check{M}$ with a face F_1 such that $T - F_1 \subset M$, $f(x) \in F_1^o$, and $x_1 \circ \mathbf{dev}(F_1) = \{a\}$.*

Theorem 3. *Let f be a cylindrical map. Then f is a map into M.*

Proof. By condition (i), $f|\Omega \times \{a\}$ is a map into $\check{M}$. Define $K = (f|\Omega \times \{a\})^{-1}(\tilde{M}_\infty)$. By condition (iv), K is a compact subset of $\Omega^o \times \{a\}$. Suppose that K is not empty.

Let r be the function on $\mathbb{R}^3$ defined by $r = x_2^2 + x_3^2$. Let $r' = r \circ \boldsymbol{dev} \circ f$, and choose a maximum point x of r' restricted to K. By above lemma, $\check{M}$ includes a convex three-simplex T with a face F_1 such that $f(x) \in F_1^o$ and $T - F_1 \subset M$.

Since $F_1 \subset f(\Omega \times \{a\})$, we have $\tilde{M}_\infty \cap F_1 \subset f(K)$; hence, $\tilde{M}_\infty \cap F_1 = f(K) \cap F_1$. Then $\boldsymbol{dev} \circ f(x)$ is an exposed point of the convex hull of $\boldsymbol{dev}(F_1 \cap f(K)) = \boldsymbol{dev}(F_1 \cap \tilde{M}_\infty)$ since x is a maximum point of r' in K. This contradicts Lemma 5. Hence, K is empty. □

We now state the strongest form of two convexity. Let $\delta\Omega$ be the boundary of Ω, which is a union of disjoint simple closed curves.

Theorem 4. *Let $f : \delta\Omega \times [a, b] \to M$ be a differentiable map satisfying the following conditions:*

(i) *$f | \delta\Omega \times \{t\}$ is an embedding into a union of disjoint simple closed curves in M_t for each t.*

(ii) *$f | \delta\Omega \times \{b\}$ extends to an embedding $f' : \Omega \times \{b\} \to M_b$.*

Then f and f' extend to a common cylindrical embedding $\mathcal{F} : \Omega \times [a, b] \to M$.

5. Filling a disk with three critical points

Let Ω be a compact surface. Suppose that $F : \Omega \to M$ is a smooth embedding with the property that $F(\delta\Omega)$ is a subset of M_t for some $t \in \mathbb{R}$, $x_1 \circ F$ is a Morse function with k critical values, and t not a critical value. Then F is said to be an *amenable embedding* of level t with k critical values. We say that the image of F is an *amenable surface*, and if $x_1 \circ F$ increases in the inner direction on $\delta\Omega$, then F is a *positive* amenable embedding; otherwise, F is a *negative* amenable embedding.

The level sets of $x_1 \circ F$ form a singular foliation on Ω with components of $\delta\Omega$ as leaves. Let us call the foliation the *level-set foliation* on Ω for F. For such foliations indices of singularities are well-defined. (See Casson-Bleiler [4, p. 71].)

Lemma 8. *The sum of indices of all singularities of the level-set foliation on Ω for F equals the Euler characteristic of Ω.*

Proof. This follows since the boundary components are leaves of the foliation. See [4, p. 72]. □

Proposition 1. *Let $\mathcal{D}$ be a compact disk with smooth boundary, and $F : \mathcal{D} \to M$ an amenable embedding of level t with one critical value. Then there exists an embedding $G : \mathcal{D} \to M_t$ homotopic to F relative to $\delta\mathcal{D}$ with $F(\mathcal{D}^o) \cap G(\mathcal{D}^o) = \emptyset$. Moreover, the union of the imbedded disks $F(\mathcal{D})$ and $G(\mathcal{D})$ is the boundary of a compact three-ball.*

Proof. Assume without loss of generality that F is positive. Let $\Sigma = F(\mathcal{D})$, and let f_1 denote $x_1|\Sigma$. Let k be the unique critical value of f_1 and z the critical point. Then z is the maximum point of f_1. Choose a convex open ball neighborhood B of z. Then there exists a real number k' near k, $k' < k$, so that the level set $f^{-1}(k')$ at k' is a simple closed curve in $\Sigma \cap B$, say α, which equals the boundary of a disk Σ_α in Σ^o. Since $M_{k'} \cap B$ is a convex open disk, α is the boundary of a disk $D^2_{k'}$ in $M_{k'}$. Now, the union of $D^2_{k'}$ and Σ_α is the boundary of a three-ball in B. Let us call the ball B_1.

By Theorem 4, it follows that there exists a disk D^2_t in the level t so that the union of $D^2_{k'}$, D^2_t, and the embedded annulus $f_1^{-1}([k', k''])$ is the boundary of a three-ball. Let us call it B_2. Then $B_1 \cup B_2$ is the three-ball with boundary $\Sigma \cup D^2_t$. The lemma follows easily now. □

Lemma 9. *Let A be a compact annulus, and $F : A \to M$ a positive amenable embedding of level t with two critical values, one of which is a local maximum and the other corresponds to a saddle point. Then one of the following holds:*

(α) *$F(\delta A)$ is the boundary of the union of two disjoint disks D_1 and D_2 in M_t, so that $D_1 \cup D_2 \cup F(A)$ is the boundary of a compact subset of M homeomorphic to a three-ball.*

(β) *There exists an embedding $G : A \to M_t$ homotopic to F relative to δA, so that $G(A) \cup F(A)$ equals the boundary of a compact subset in M, homeomorphic to a solid torus.*

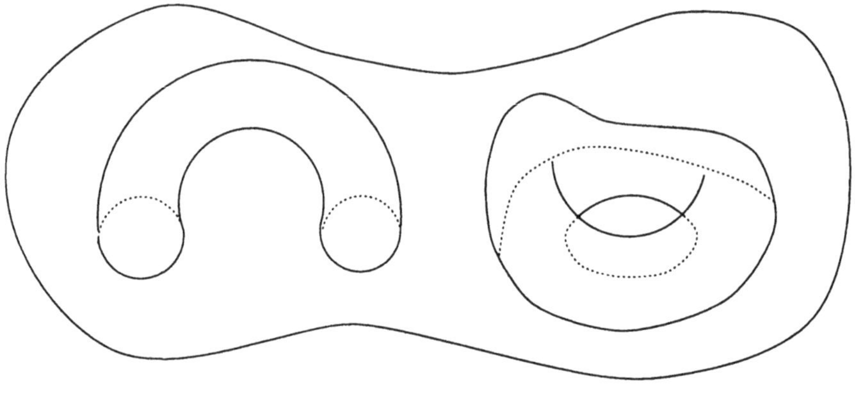

Figure 2: Diagrams of annuli corresponding to (α) and (β).

We come to the main result of this section.

Proposition 2. *Let $F : \mathcal{D} \to M$ be a positive amenable embedding of level t with three critical values. Then there exists an embedding $G : \mathcal{D} \to M_t$ homotopic to F relative to $\delta\mathcal{D}$.*

6. The incompressible level surfaces

We prove that every amenable disk with boundary in a level set is homotopic to an embedding into the level set.

By Proposition 1, the following *induction hypothesis* is satisfied for $k = 1$:

Definition 2. Let $\mathcal{D}$ be a standard disk and $F : \mathcal{D} \to M$ a positive amenable embedding of level t with k critical values. Then there exists an embedding $G : \mathcal{D} \to M_t$ homotopic to F relative to $\delta\mathcal{D}$.

Now assume that the induction hypothesis holds for $k \leq i - 1$ for an integer i, $i > 1$. We will show that the induction hypothesis holds for $k = i$. Let $F : \mathcal{D} \to M$ be a positive amenable embedding of level t with i critical values. Let $f = x_1 \circ F$, v the maximum of the critical values of index 1, and z the saddle point corresponding to v. We choose ϵ, $\epsilon > 0$, so that $[v - \epsilon, v]$ contains no other critical value. Letting $v' = v - \epsilon$, we see that the components of $f^{-1}(v')$ are simple closed curves $\alpha_1, \ldots, \alpha_n$ and the components of $f^{-1}([v', \infty))$ smooth subsurfaces $A_1, \ldots, A_m$ of $\mathcal{D}^o$ with boundaries each of whose components equals α_i for some i. We may assume without loss of generality that exactly one component A_1 contains the unique saddle point z in the interior. Lemma 8 and the Euler characteristic argument show that A_1 is diffeomorphic to (i) a disk, or (ii) an annulus. Since there is no saddle point in A_j for $j > 1$, the Euler characteristic argument shows that each A_j for $j = 2, \ldots, m$ is diffeomorphic to a compact disk and contains a unique critical point, which is a local maximum point.

(i) Every A_i is diffeomorphic to a disk. Since each A_i has a unique boundary component, we have $m = n$ and may assume without loss of generality that α_i is the boundary of A_i in $\mathcal{D}$ for each i, $i = 1, 2, \ldots, n$. In this case, A_1 has three critical points, one of which is a saddle point and the other two are local maximum or minimum points. Since by Propositions 1 and 2, each $F(\alpha_i)$ for $i = 1, \ldots, n$ is the boundary of a compact disk in $M_{v'}$, the closure of precisely one component of $M_{v'} - F(\alpha_i)$ is homeomorphic to a disk by Lemma 6. Therefore, we can define an innermost closed curve among $F(\alpha_1), \ldots, F(\alpha_n)$ as follows; $F(\alpha_j)$ is an *innermost closed curve* (with respect to $M_{v'}$) if the disk with boundary $F(\alpha_j)$ in $M_{v'}$ includes no $F(\alpha_i)$ for $i \neq j$. We consider an innermost $F(\alpha_j)$, $j \neq 1$, included in the disk in $M_{v'}$ with boundary $F(\alpha_1)$. By smoothing, we obtain an amenable embedding $F' : \mathcal{D} \to M$ homotopic to F relative to $\mathcal{D} - N(A_j)^o$ for a compact-disk neighborhood $N(A_j)$ of A_j with equal number of critical points, and the number of components of $f'^{-1}(v')$ where $f' = x_1 \circ F'$ is less than that of $f^{-1}(v')$ by one. Hence, by an induction, we get a positive amenable embedding $F'' : \mathcal{D} \to M$ homotopic to F relative to

$$\mathcal{D} - \bigcup_{i=2}^{n} N(A_j)^o$$

so that $F''(\alpha_1)$ is now an innermost component and the number of critical points of

F'' is less than or equal to that of F. (We may assume that $N(A_j)$ is disjoint from A_1 and $N(A_k)$ for $k \neq j$, $k = 2, \ldots, n$.) Then by Proposition 2, $F|A_1$ is homotopic to an embedding onto a disk in $M_{v'}$ relative to δA_1. By smoothing, we get an amenable embedding $F''' : \mathcal{D} \to M$ homotopic to F'' relative to $\mathcal{D} - A_1^o$ and $x_1 \circ F'''$ has two less number of critical points than $x_1 \circ F''$. Then by the induction hypothesis, F is homotopic to an embedding $G : \mathcal{D} \to M_t$ relative to $\delta\mathcal{D}$.

(ii) We have $m = n - 1$, and may assume without loss of generality that α_1 and α_n are the boundary components of A_1 and that α_i is the boundary component of a compact disk A_i for each i, $i = 2, \ldots, n-1$. We may further assume without loss of generality that α_1 is the boundary of a disk $\mathcal{D}_1$ in $\mathcal{D}^o$ such that $\alpha_n \subset \mathcal{D}_1^o$. Let $\mathcal{D}_2$ be the disk with boundary α_n in $\mathcal{D}$; we have $\mathcal{D}_2 \subset \mathcal{D}_1^o$. Then $F|\mathcal{D}_1$ is a positive amenable embedding of level v'. The map $F|\mathcal{D}_2$ is a negative amenable embedding of level v' with the number of critical points less than that of F by two or more. By reversing the direction of x_1 and applying the induction hypothesis, we are able to show that $F|\mathcal{D}_2$ is homotopic to an embedding $H_2 : \mathcal{D}_2 \to M_{v'}$ relative to $\delta\mathcal{D}_2$. By smoothing maps, we get a positive amenable embedding $H_1 : \mathcal{D}_1 \to M$ homotopic to $F|\mathcal{D}_1$ relative to $\mathcal{D}_1 - \mathcal{D}_2^o$ so that $x_1 \circ H_1$ has only three critical points, which are a local maximum point, a saddle point, and a local minimum point respectively.

By Proposition 2, we get an embedding $H : \mathcal{D}_1 \to M_{v'}$ homotopic to $F|\mathcal{D}_1$ relative to $\delta\mathcal{D}_1$. Let W be the image disk $H(\mathcal{D}_1)$, where $\delta W = F(\alpha_1)$. Let us define a map $F' : \mathcal{D} \to M$ defined by $F'(x) = F(x)$ if $x \in \mathcal{D} - \mathcal{D}_1^o$ and $F'(x) = H(x)$ if $x \in \mathcal{D}_1$. Now F' may not be an imbedding but is homotopic to F relative to $\mathcal{D} - \mathcal{D}_1^o$. For each i, $i = 2, \ldots, n-1$, A_i either is a subset of $\mathcal{D}_2^o$ or is disjoint from $\mathcal{D}_1$, and $F(\alpha_i)$ either is included in W^o or disjoint from W. Let

$$J = \{j \mid F(\alpha_j) \subset W^o, A_j \not\subset \mathcal{D}_2^o, 2 \leq j \leq n-1\}.$$

Since α_j, $j \in J$, is the boundary of a disk A_j, Proposition 1 shows that $F'|A_j$ is homotopic relative to δA_j to an embedding $A_j \to W^o$. Suppose that $F(\alpha_j)$, $j \in J$, is an innermost one among $F(\alpha_k)$ for $k \in J$. We may assume that F' is homotopic relative to $\mathcal{D} - N(A_j)^o$ to a map $F'' : \mathcal{D} \to M$ so that $x_1 \circ F''(N(A_j)) < v'$. By an induction eliminating such innermost α_j, $j \in J$, we obtain an imbedding $L : \mathcal{D} \to M$ so that $L(x) \in W$ if and only if $x \in \mathcal{D}_1$ and $x_1 \circ L(N(A_j)) < v'$ for $j \in J$ hold, L is homotopic to F relative to

$$\mathcal{D} - \mathcal{D}_1^o - \bigcup_{j \in J} N(A_j)^o,$$

where $N(A_j)$ is obtained as above for each j, and L is smooth except on $\delta\mathcal{D}_1$. (The neighborhoods $N(A_j)$, $j = 2, \ldots, n-1$, are mutually disjoint.)

By smoothing, we get an amenable embedding $L' : \mathcal{D} \to M$ homotopic to L relative to $\mathcal{D} - \mathcal{D}_1^o$, and have a unique critical point in $\mathcal{D}_1^o$, which is a local maximum point. Thus we reduced the number of critical points by two or more. The induction hypothesis implies that there exists an embedding $G : D \to M_t$ homotopic to L' relative to $\delta\mathcal{D}$ and, hence, to F. This completes the induction argument.

We proved:

Theorem 5. *Let $F : D \to M$ be a positive amenable embedding of level t. Then there exists an embedding $G : D \to M_t$ homotopic to F relative to δD.*

We follow Chapter 6 of [11]. (A number of topologists told me that the following lemma is true.) Let M be an orientable smooth open three-manifold. A *surface* is a properly imbedded connected two-manifold. We say that a surface F is incompressible in M if none of the following condition is satisfied.

- F is a two-sphere that is the boundary of a homotopy three-cell in M, or
- there is a two-cell D in M such that $D \cap F = \delta D$, and δD not contractible in F.

Lemma 10. *Let S be a two-sided surface in M, and let $i_* : \pi_1(S) \to \pi_1(M)$ be the homomorphism induced by the inclusion map. Then if the kernel of i_* is not trivial, then there is an embedded two-cell D in M, transversal to S, such that $D \cap S = \delta D$, and δD is not contractible in S.*

Corollary 2. *If S is a two-sided incompressible surface in M, then i_* is injective.*

Now, let M be a simply connected open two-convex affine three-manifold. Recall that x_1 is a function on M defined by $x_1 = x_1 \circ \boldsymbol{dev}$ for a coordinate function x_1 on $\mathbb{R}^3$. We defined M_t to be the inverse image $x_1^{-1}(t)$ in M for each real number t. Let F be a component of M_t, which is a properly imbedded submanifold of M. By Lemma 6, F is not homeomorphic to a sphere. We claim that F is incompressible. Suppose that there is an embedded two-cell D in M, transversal to F, such that $D \cap F = \delta D$. It is easy to see that D can be modified so that $x_1|D$ is a Morse function. Theorem 5 shows that δD is contractible in F.

Since M is simply connected, F is simply connected by Corollary 2. Hence F is diffeomorphic to $\mathbb{R}^2$. We conclude that M is foliated by leaves that are diffeomorphic to $\mathbb{R}^2$. (The leaves are closed in M since $x^{-1}(t)$ is closed in M.) By Corollary 3 of Palmeira [15], M is diffeomorphic to $\mathbb{R}^3$. This completes the proof of Theorem 2.

A detailed version of this paper will appear elsewhere.

References

[1] Berger, M., Geometry I, Universitext, Springer-Verlag, Berlin 1987.

[2] Benzécri, J. P., Variétés localement affines, Sem. Topologie et Géom. Diff., Ch. Ehresmann (1958–60) No. 7 (Mai 1959).

[3] Carrière, Y., Autour de la conjecture de L. Markus sur les variétés affines, Invent. Math. 95 (1989), 615–628.

[4] Casson, A., Bleiler, S., Automorphisms of surfaces after Nielsen and Thurston, London Math. Soc. Stud. Texts 9, Cambridge University Press, Cambridge 1988.

[5] Choi, S., Convex decomposition of real projective surfaces. I: π-annuli and convexity, J. Differential Geom. 40 (1994), 165–208.

[6] Choi, S., The universal cover of an affine three-manifold with holonomy of discompactedness two, submitted.

[7] Fried, D., Distality, completeness, and affine structures, J. Differential Geom. 24 (1986), 265–273.

[8] Fried, D., Goldman, W., Hirsch, M., Affine manifolds with nilpotent holonomy, Comment. Math. Helv. 56 (1981), 487–496.

[9] Goldman, W., Hirsch, M., Affine manifolds and orbits of algebraic groups, Trans. Amer. Math. Soc. 295 (1986), 175–190.

[10] Helgason, S., Differential geometry, Lie groups, and symmetric spaces, Pure Appl. Math. 80, Academic Press, New York 1978.

[11] Hempel, J., 3-manifolds, Ann. of Math. Stud. 86, Princeton University Press, Princeton 1976.

[12] Kamishima, Y., Tan, S., Deformation spaces on geometric structures, in: Aspects of Low Dimensional Manifolds (Y. Matsumoto, S. Morita, eds.), Adv. Stud. Pure Math. 20, Kinokuniya, Tokyo 1992.

[13] Markus, L., Cosmological models in differential geometry, mimeographed notes, University of Minnesota 1962.

[14] Onishchick, A. L., Vinberg, E. B., Lie groups and algebraic groups, Springer-Verlag, Berlin–Heidelberg–New York 1990.

[15] Palmeira, C., Open manifolds foliated by planes, Ann. of Math. 107 (1978), 109–131.

[16] Smillie, J., Affinely flat manifolds, Doctoral Dissertation, University of Chicago 1977.

Geometric quantization for the moduli space of vector bundles with parabolic structure

Georgios D. Daskalopoulos and Richard A. Wentworth

Abstract. We initiate a study of the geometric quantization of Chern–Simons gauge theory on Riemann surfaces with punctures. We construct a moduli space of flat connections using weighted Sobolev spaces, and then by analogy with the compact case, we construct a line bundle over this moduli space. The line bundle exists only for certain holonomies which correspond in a one-to-one way with representations. When the bundle does exist, its holomorphic sections reproduce the space of states defined by Segal.

1991 Mathematics Subject Classification: 32G13; 14D20, 14H60, 32G07.

1. Introduction

This paper was first written in June 1991 and revised in February 1992. The motivation derived from the link, first elucidated by Witten [Wi1], between Chern–Simons theory, conformal field theory, and the moduli spaces of flat connections on Riemann surfaces. This point of view was further developed in the Oxford notes [Ox] by Atiyah, Hitchin, and Segal. Our work was an attempt to understand and make rigorous some of the ideas presented in these references.

The main mathematical question was to prove a factorization theorem for certain non-abelian versions of level k theta functions. This goal was achieved in [D-W2, D-W3], where a proof of the Verlinde formula for $SU(2)$ theta functions can also be found. This paper has remained unpublished; however, since some of the results are quoted without proof in the above references, we have decided that it might be worth having this initial work appear in the literature, even with the substantial delay.

We now outline the organization of the paper and the main results: in Section 2, we review some standard results concerning spaces of $\mathbf{G}$-representations of the fundamental group of a Riemann surface with boundaries. These spaces admit natural complex structures via the theorem of Mehta and Seshadri which identifies them with moduli spaces of vector bundles with *parabolic structure*. We review this work in some detail. Moreover, we show how these spaces may be used to form a correspondence between moduli spaces of vector bundles of different degrees. This is essentially the notion of a *Hecke correspondence* (cf. [N-Rn]). For simplicity, we shall throughout the paper restrict ourselves to the case of a single boundary component. All of our results can be generalized to the case of many components, with certain important modifications. Details will be presented elsewhere (see [D-W3]).

In Section 3, we give the gauge theoretic description of these representation spaces. We achieve this by pushing the boundary to infinity and requiring exponential decay both on connections and gauge transformations. This technique was first used by Taubes [T1] in the setting of four dimensional topology.

To be more precise, let ∇_0 be a base flat connection on a trivial **G**-bundle over the Riemann surface Σ obtained from $\overline{\Sigma}$ by removing a point (the "puncture"). Assume that at infinity (i.e. a disk about the point) ∇_0 has the form $A_0 = i\alpha d\theta$ with holonomy $\exp(2\pi i\alpha)$. Let $\mathcal{A}_{F,\delta}$ denote the space of flat connections $\nabla_0 + A$ such that both A and $\nabla_0 A$ have exponential decay, and let $\mathcal{G}_\delta$ denote the bundle automorphisms g with $\nabla_0 g$, $\nabla_0^2 g$ exponentially decaying. The main theorem of Section 3 is the following (see Theorem 3.8):

Theorem A. *The space of equivalence classes of flat connections $\mathcal{A}_{F,\delta}/\mathcal{G}_\delta$ is homeomorphic to the space of equivalence classes of* **G**-*representations of* $\pi_1(\Sigma)$ *with value at infinity conjugate to* $\exp(2\pi i\alpha)$.

Having identified the spaces, we investigate their topological structure in Section 4. Since we are interested in constructing line bundles, we focus on the low dimensional homotopy and cohomology groups. The results are obtained by means of transversality arguments of the type used in [D-U]. The analytic description of the moduli spaces allows us to define a candidate Ω for the curvature two-form of our line bundle. The form does not, however, always define an integral class, and in Theorem 4.13, we show that for $k \in \mathbb{Z}$, $k\Omega/2\pi$ is integral if and only if $ik\alpha$ is in the coroot lattice of $\mathfrak{g} = \mathrm{Lie}\,\mathbf{G}$.

By contrast, on the quotient $\mathcal{F}_{\alpha,\delta} = \mathcal{A}_{F,\delta}/\mathcal{G}_{0,\delta}$, where $\mathcal{G}_{0,\delta}$ denotes the subgroup of automorphisms which are the identity at infinity, the form $k\Omega/2\pi$ is always integral. In Section 5, we show how to use the cocycle (see (5.1)) to construct a line bundle $\tilde{L}_\alpha^{\otimes k}$ with connection $\tilde{\nabla}$ and curvature $-ik\Omega$ over $\mathcal{F}_{\alpha,\delta}$. The question of integrality of Ω on the quotient $\mathcal{A}_{F,\delta}/\mathcal{G}_\delta$ may now be recast as follows: the residual gauge group $\mathcal{G}_\delta/\mathcal{G}_{0,\delta}$ is, for generic holonomies, a maximal torus $\mathbf{T} \subset \mathbf{G}$. The infinitesimal action of **T** is given via $\tilde{\nabla}$. The line bundle and connection push down to a bundle $L_\alpha^{\otimes k}$ precisely when this action may be exponentiated. This gives the following correspondence between holonomies and representations (see Theorem 6.1):

Theorem B. *Fix a maximal torus* $\mathbf{T} \subset \mathbf{G}$. *There is a one-to-one correspondence between characters* $\lambda : \mathbf{T} \to U(1)$ *invariant under the center* $\mathbb{Z}_n$ *of* **G** *and elements* $i\alpha \in \mathrm{Lie}\,\mathbf{T}$ *such that* $k\Omega/2\pi$ *represents an integral class on* $\mathcal{F}_{\alpha,\delta}/\mathbf{T}$.

Invariance under the center is required for a free action on the base (recall that we are restricting to the case of one puncture).

Finally, in Section 6, we show how the line bundles constructed in the above manner solve a problem in geometric quantization. Namely, consider the compact surface $\overline{\Sigma}$. Let $\overline{\mathcal{A}}_s$ denote the smooth stable connections on a trivial $\mathbf{G}$-bundle over $\overline{\Sigma}$, and let $\overline{\mathcal{G}}^{\mathbb{C}}$ denote the group of smooth complex automorphisms. The determinant line bundle $\Delta \to \overline{\mathcal{A}}_s$ descends via the action of $\overline{\mathcal{G}}^{\mathbb{C}}$ to generate the Picard group of the moduli space $\overline{\mathcal{A}}_s/\overline{\mathcal{G}}^{\mathbb{C}}$. $\Delta^{\otimes k}$ may be regarded as a "quantum" line bundle at level k over the symplectic manifold $\overline{\mathcal{A}}_s$, and the quotient is the corresponding quantum bundle over the moduli space.

Instead of taking the $\overline{\mathcal{G}}^{\mathbb{C}}$ invariant sections, one may choose a highest weight λ and consider the multiplicity space (cf. [G-S1])

$$\mathcal{V}_\lambda = \mathrm{Hom}_{\overline{\mathcal{G}}^{\mathbb{C}}}\left(V_\lambda^*, H^0(\overline{\mathcal{A}}_s, \Delta^{\otimes k})\right) ,$$

where V_λ^* is the (dual) representation space of λ. The homomorphisms are required to intertwine the action of $\overline{\mathcal{G}}^{\mathbb{C}}$ on H^0 with the action on V_λ obtained by evaluation at the puncture. This is essentially Segal's definition of the space of states in [Ox]. We prove the following (see Theorem 6.6):

Theorem C. *Assume* $\mathbf{G} = \mathrm{SU}(2)$ *and* $g > 3$. *Fix a dominant weight* $\lambda : \mathbf{T} \to U(1)$ *invariant under the center, and let* α *correspond to* λ *as in Theorem B above. Then* $\mathcal{V}_\lambda \simeq H^0\left(\mathcal{F}_{\alpha,\delta}/\mathbf{T}, L_\alpha^{\otimes k}\right)$.

We shall show in [D-W1] that $\mathcal{V}_\lambda$ is zero for all but finitely many λ.

In addition to the papers which motivated this study back in 1991, there has since been a plethora of work on topics related to parabolic bundles, geometric quantization, and the Verlinde formula. We cannot hope to compile an exhaustive collection of references, and so we shall content ourselves instead with briefly mentioning some of those articles having a more or less close connection with the present work and its sequels [DW1-3]: For proofs of the Mehta–Seshadri theorem [Me-Se] in the spirit of Donaldson and Uhlenbeck–Yau, see [Biq], [Po], and [Si1-2]. For more on the moduli space of parabolic bundles on Riemann surfaces, see [Ba], [Bd], [Bd-Hu], [Bs-Rm], [Fu-St], [Ni], [Ns-St]. Geometric quantization of the moduli space of vector bundles, first studied in [Ax] and [Ax-DP-W] (see also [A], [Ox], [Wi1]), has since been considered by many authors. For some of these ideas, see [Bis], [Ch], [J-W1], [R-S-W], and [V-V]. The Verlinde dimension formula [V] was the motivation for this work and many, many others. For a description of the formula in the context of stable bundles, see [B]. Various aspects, partial proofs, and proofs of the formula can be found in [Bv-L], [Be], [Be-Sz], [Don2], [DW2-3], [Fa], [G-P], [J-K], [J-W2], [K-N-Rm], [N-R], [Sz1-2], [Te], [Th1-2], [T-U-Y], [Wi2], [Z]. Finally, we would like to mention the article [Bv] where the reader will find a beautiful exposition of some of the recent developments in this subject.

Acknowledgements. We have benefited from conversations with Scott Axelrod, Aaron Bertram, Dan Freed, Bob Friedman, Lisa Jeffrey, Gordana Matic, Roberto Silvotti, and Karen Uhlenbeck. Our thanks go to them and to the MSRI, where the majority of this work was completed. We would also like to thank Jonathan Poritz for pointing out a mistake in the earlier version of the proof of Proposition 3.5. G.D. is supported in part by NSF grant DMS-9504297; R.W. is supported in part by NSF grant DMS-9503635 and a Sloan Fellowship.

2. Moduli of G-bundles over punctured Riemann surfaces

In this section we introduce the basic geometric object we shall be concerned with throughout this paper. This is the moduli space of vector bundles on a Riemann surface with punctures. For bundles on a compact surface, one has to restrict to an appropriate subspace of the space of vector bundles, namely the *stable* bundles, in order to obtain a "good" moduli space. The famous theorem of Narasimhan and Seshadri states that when stability is taken into account, the moduli space is just the equivalence classes of irreducible unitary representations of the fundamental group.

For vector bundles on Riemann surfaces with punctures, stability must be replaced with Seshadri's notion of *parabolic stability*. That this is the correct generalization follows from the theorem of Mehta and Seshadri which again states that the moduli space, now of parabolic stable bundles, is isomorphic to the space of irreducible representations of the fundamental group of the non-compact surface.

2.1. Representations of the fundamental group

Let $\overline{\Sigma}$ be a compact Riemann surface of genus $g > 0$ with fixed coordinate disk $(\tilde{D}, z)$, i.e. $z : \tilde{D} \to D$ is a complex analytic isomorphism with the unit disk $D \subset \mathbb{C}$. Let $p = z^{-1}(0)$, and denote by Σ the non-compact surface $\overline{\Sigma} \setminus \{p\}$. We may define new coordinates on $\tilde{D}^* = \tilde{D} \setminus \{p\}$: set $w = -\log z$. Then

$$w : \tilde{D}^* \longrightarrow \mathbb{C}/(\tau, \theta) \sim (\tau, \theta + 2\pi)$$

maps $\tilde{D}^*$ analytically to the semi-infinite cylinder

$$C = \{(\tau, \theta) : \tau \geq 0, 0 \leq \theta \leq 2\pi\}/(\tau, 0) \sim (\tau, 2\pi) .$$

If we use polar coordinates (r, ϕ) on D^*, then the map $w \circ z^{-1} : D^* \to \mathbb{C}$ is simply $\tau = -\log r$, $\phi = \theta$. We also make a choice of base point $x_0 \in \tilde{D}^* \subset \Sigma$ and set $(\tau_0, \theta_0) = w(x_0)$. Finally, we denote by Σ_0 the Riemann surface with boundary: $\Sigma_0 = \Sigma \setminus \{x : \tau(x) > \tau_0\}$.

Let $\mathbf{G} = \mathrm{SU}(n)$, and let P be the trivial principal $\mathbf{G}$-bundle on Σ, E the vector bundle associated to P via the standard representation. Let $\mathcal{A}$ denote the space of all

C^∞ **G**-connections on P. We identify $\mathcal{A} \simeq \Omega^1(\Sigma, \mathfrak{g})$ and endow it with the obvious Fréchet topology on $\Omega^1(\Sigma, \mathfrak{g})$. Let $\mathcal{G}$ denote the group of C^∞ gauge transformations on P, endowed with the obvious Fréchet topology induced by the inclusion $\mathrm{Map}(\Sigma, \mathbf{G}) \subset \mathrm{Map}(\Sigma, \mathfrak{gl}(n, \mathbb{C}))$. Let $\mathcal{G}_0 \subset \mathcal{G}$ be the subgroup of $\mathcal{G}$ preserving the base point x_0. The groups $\mathcal{G}_0, \mathcal{G}$ act smoothly on $\mathcal{A}_F$, the flat connections in $\mathcal{A}$. Let $\mathcal{M}_0 = \mathcal{A}_F/\mathcal{G}_0$, $\mathcal{M} = \mathcal{A}_F/\mathcal{G}$ be the quotient spaces.

Define the evaluation map $e_0 : \mathcal{G} \to \mathbf{G}$ which takes $g \mapsto g(x_0)$. In view of the exact sequence of groups $1 \longrightarrow \mathcal{G}_0 \longrightarrow \mathcal{G} \xrightarrow{e_0} \mathbf{G} \longrightarrow 1$, we can identify $\mathbf{G} \simeq \mathcal{G}/\mathcal{G}_0$. Thus the group $\mathbf{G}$ acts on $\mathcal{M}_0$, and we have the identifications $\mathcal{A}_F/\mathcal{G} \simeq \mathcal{M}_0/\mathbf{G} \simeq \mathcal{M}$. The spaces $\mathcal{M}, \mathcal{M}_0$ are called the *moduli spaces of flat connections* and *based flat connections* on P, respectively.

We choose generators $\{a_1, \ldots, a_g, b_1, \ldots, b_g\}$ for $\pi_1(\Sigma, x_0)$ whose intersection numbers in the quotient $H_1(\Sigma, \mathbb{Z})$ satisfy $\#a_i \cdot b_j = \delta_{ij}$, $\#a_i \cdot a_j = \#b_i \cdot b_j = 0$. Given $\nabla \in \mathcal{A}_F$, let $\rho = \rho(\nabla)$ be the corresponding holonomy homomorphism. This defines a map $\mathcal{A}_F \to \mathrm{Hom}\,(\pi_1(\Sigma, x_0), \mathbf{G}) \simeq \mathbf{G}^{2g}$ which is $\mathcal{G}_0$-invariant, where the last identification is from $\{a_i, b_i\} \to \{A_i, B_i\}$. It is well known that this induces a homeomorphism of $\mathcal{M}_0 \simeq \mathcal{A}_F/\mathcal{G}_0$ with $\mathrm{Hom}\,(\pi_1(\Sigma, x_0), \mathbf{G}) \simeq \mathbf{G}^{2g}$. This proves

Lemma 2.1. *$\mathcal{M}_0$ is a smooth manifold diffeomorphic to $\mathbf{G}^{2g}$.*

Let c denote the loop $\prod_{i=1}^g [a_i, b_i]$ in $\pi_1(\Sigma, x_0)$ and let $q : \mathcal{M}_0 \to \mathbf{G}$ be the map measuring the holonomy around c. We clearly have

Lemma 2.2. *Under the identification $\mathcal{M}_0 \simeq \mathbf{G}^{2g}$ the map q corresponds to the map $\{A_1, \ldots, A_g, B_1, \ldots, B_g\} \longrightarrow \prod_{i=1}^g [A_i, B_i]$.*

The next proposition summarizes some important properties of the map q in the case $\mathbf{G} = \mathrm{SU}(2)$ which we shall need later on (see Theorem 4.1):

Proposition 2.3. *For $\mathbf{G} = \mathrm{SU}(2)$, the map q has the following properties:*

(i) *q is surjective;*

(ii) *{critical points of q}={reducible connections};*

(iii) *{reducible connections} $\subset q^{-1}(I)$;*

(iv) *$q^{-1}(pt.) \setminus$ {reducible connections} is a smooth manifold of dimension $6g - 3$;*

(v) *$q : \mathcal{M}_0 \setminus q^{-1}(I) \to \mathbf{G} \setminus \{I\}$ is a trivial fiber bundle.*

Proof. For (i)–(iv) see [Ak-M], Proposition 2.1 and Corollary 2.2. The result (v) follows from (i)–(iii), since q is clearly proper and the base is contractible. □

We return to the general case $\mathbf{G} = \mathrm{SU}(n)$. Let $\mathbf{G} \cdot a$ denote the adjoint orbit of a in $\mathbf{G}$, and let $\mathcal{M}_{0,a}$ be the subspace of irreducible connections in $q^{-1}(\mathbf{G} \cdot a)$. Clearly,

$\mathcal{M}_{0,a}$ is preserved by the action of **G**, so we set $\mathcal{M}_a = \mathcal{M}_{0,a}/\mathbf{G}$. The next lemma follows immediately from the definition of the map q:

Lemma 2.4. (i) $\mathcal{M}_{0,a}$ *may be identified with the space of irreducible representations* $\rho : \pi_1(\Sigma, x_0) \to \mathbf{G}$ *satisfying* $\rho(c) \in \mathbf{G} \cdot a$.

(ii) $\mathcal{M}_a$ *may be identified with the quotient of the space in* (i) *by the adjoint action of* **G**.

Lemma 2.5. *For every* $a \in \mathbf{G}$, $\mathrm{PU}(n) \to \mathcal{M}_{0,a} \to \mathcal{M}_a$ *is a principal* $\mathrm{PU}(n)$ *bundle.*

Proof. The group $\mathrm{PU}(n) = \mathrm{SU}(n)/\mathbb{Z}_n$ acts freely on $\mathcal{M}_{0,a}$. The lemma follows from a general theorem on group actions (cf. [V], Lemma 2.9.11). □

Let $\mathcal{F}_a$ denote the subspace of irreducible flat connections in $q^{-1}(a)$ over $a \in \mathbf{G}$. Then $\mathcal{F}_a$ is a smooth manifold of dimension $(2g-1)(n^2-1)$. Let $\mathbf{P}_a$ denote the normalizer of a in **G**. For example, for $\mathbf{G} = \mathrm{SU}(2)$, $\mathbf{P}_a$ is a maximal torus for $a \neq \pm I$, and $\mathbf{P}_{\pm I} = \mathbf{G}$. Then $\mathbf{P}_a/\mathbb{Z}_n$, the normalizer modulo the center of **G**, acts freely on $\mathcal{F}_a$ and we have an isomorphism $\mathcal{M}_a \simeq \mathcal{F}_a/\mathbf{P}_a$. Moreover, for every $a \in \mathbf{G}$, the principal bundle of Lemma 2.5 restricts to a principal bundle

$$\mathbf{P}_a/\mathbb{Z}_n \longrightarrow \mathcal{F}_a \xrightarrow{\pi} \mathcal{M}_a \tag{2.6}$$

Fibration (2.6) will be of fundamental importance in the subsequent sections. We finish this section with the following fairly standard

Lemma 2.7. *Let* ∇ *be a flat connection of holonomy* $a \in \mathbf{G}$ *around* p. *Fix* $i\alpha \in \mathfrak{g}$ *with* $a = \exp(2\pi i\alpha)$. *Then we can find* $g \in \mathcal{G}_0$ *such that over* $\tilde{D}^*$ *we can write (in the coordinates* w*)* $\nabla^g = d + i\alpha d\theta$.

A similar result holds for the group U(n). We shall refer to the form of the connection $d + i\alpha d\theta$ as *temporal gauge.*

Proof. As a first step, we show that for flat connections we can find a smooth gauge transformation g such that for $\tau \geq \tau_0$, $\nabla^g = d + A(\theta)d\theta$. Indeed, write

$$\nabla = d + a(\tau, \theta)d\theta + b(\tau, \theta)d\tau$$

and let g be a solution to the ODE: $dg/d\tau + b(\tau, \theta)g = 0$, $g(\tau_0) = I$. Such a solution exists for all $\tau \geq \tau_0$ and is invertible, hence also $g^{-1}dg/d\tau + g^{-1}bg = 0$. Since b is periodic in θ we extend the solution to the semi-infinite cylinder C. Then we extend

g to all of Σ as an element of $\mathcal{G}_0$ by a cut-off function. We have

$$\begin{aligned}\nabla^g &= d + g^{-1}dg + g^{-1}agd\theta + g^{-1}bgd\tau \\ &= d + \left(g^{-1}\frac{\partial g}{\partial\theta} + g^{-1}ag\right)d\theta + \left(g^{-1}\frac{\partial g}{\partial\tau} + g^{-1}bg\right)d\tau \\ &= d + A(\tau,\theta)d\theta\ , \quad \text{for } \tau \geq \tau_0\ .\end{aligned}$$

By assumption, the curvature $F_\nabla = 0$ so $\partial A/\partial\tau = 0$, and this completes the proof of the first step. In order to make A independent of θ, we notice that on the circle any connection is gauge equivalent to constant one. We extend the gauge transformation to Σ, constant on $\tau \geq \tau_0$. □

2.2. The theorem of Mehta and Seshadri

In this section we review some fundamental results involving complex structures on the spaces $\mathcal{M}_a$ from our point of view. The standard references are [Me-Se], [Se], [Si1], [Si2]. We begin by stating the fundamental theorem:

Theorem 2.8 ([Me-Se], Theorem 5.3). *Let $\mathcal{M}_a$ denote the space of isomorphism classes of irreducible flat connections on P with holonomy around the point p conjugate to $a \in \mathbf{G}$. If all the eigenvalues of a are of the form* $\exp(2\pi i\alpha_i)$*, where the α_i are rational numbers, then $\mathcal{M}_a$ has the structure of a complex manifold of dimension* $(n^2-1)(g-1) + \dim \mathbf{G}/\mathbf{P}_a$*, where $\mathbf{P}_a$ is the normalizer of a in* $\mathbf{G}$.

Let us briefly review this theorem from our point of view. First, recall (cf. Lemma 2.7) that given $[\nabla]$ an equivalence class of flat unitary connections on E of holonomy $a \in \mathrm{U}(n)$ around p we can choose a unitary frame $e_1, \ldots, e_n$ over the punctured disk $\tilde{D}^*$ with respect to which $\nabla = d + i\hat{\alpha}d\theta$ where $\hat{\alpha}$ is the matrix $\mathrm{diag}(\hat{\alpha}_1, \ldots, \hat{\alpha}_n)$ and $0 \leq \hat{\alpha}_1 \leq \cdots \leq \hat{\alpha}_n < 1$. Let $\bar{\partial}_\nabla$ be the corresponding $\bar{\partial}$-operator. Notice that if we define a new basis $\{f_i\}$ by $f_i = |z|^{\hat{\alpha}_i}e_i$, then the f_i are holomorphic with respect to $\bar{\partial}_\nabla$. Let $D \times \mathbb{C}^n \to D$ be the trivial bundle over the disk with the trivial $\bar{\partial}$-operator. Let $\bar{E} \to \bar{\Sigma}$ denote the bundle obtained by gluing E with $D \times \mathbb{C}^n$ via the f_i's, and let $\bar{\partial}_{\bar{E}}$ denote the resulting $\bar{\partial}$-operator on $\bar{E}$. It is easy to check that $\bar{E}$ is a holomorphic bundle of Chern class $l = -\sum_{i=1}^n \hat{\alpha}_i$. (Note that the sum of the $\hat{\alpha}_i$'s is always an integer.)

The f_i's determine a complete flag of the fiber $\bar{E}_p \simeq \mathbb{C}^n$. In general, given a holomorphic bundle $\bar{E}$ of degree l, a choice of flag $Fl\,(\overline{E}) \in \mathbf{G}/\mathbf{P}_a$ of the fiber $\overline{E}_p \simeq \mathbb{C}^n$ is called a *quasi-parabolic structure*. Furthermore, the pair $(Fl\,(\overline{E}); \hat{\alpha}_1, \ldots, \hat{\alpha}_n)$, where $\hat{\alpha}_1, \ldots, \hat{\alpha}_n$ are weights associated to $Fl\,(\overline{E})$, is called a *parabolic structure* on $\overline{E}$, and $\overline{E}$ is called a *parabolic bundle* (see also [Se]).

Let $\mathcal{B}$ denote the space of holomorphic structures on $\overline{E}$, $\mathcal{K}$ the group of complex automorphisms of $\overline{E}$, and $\mathcal{K}_p$ its subgroup of *parabolic automorphisms*, i.e. the elements of $\mathcal{K}$ preserving the flag $Fl\,(\overline{E})$. Given a holomorphic structure on $\overline{E}$, let us denote also by $\overline{E}$ the resulting holomorphic bundle. Let $\overline{F} \subset \overline{E}$ be a holomorphic subbundle of rank m. The flag $Fl\,(\overline{E})$ induces by restriction a flag $Fl\,(\overline{F})$ on $\overline{F}_p$ with associated weights $0 \leq \hat{\alpha}'_1 \leq \cdots \leq \hat{\alpha}'_m < 1$. The next definition is essential:

Definition 2.9 ([Me-Se], 1.11 and 1.13). Given a subbundle $\overline{F}$ of $\overline{E}$ we define the *parabolic degree* of $\overline{F}$ by $\text{pardeg}(\overline{F}) = \deg(\overline{F}) + \sum_{i=1}^{m} \hat{\alpha}'_i$. The bundle $\overline{E}$ is called *parabolic stable* if for any subbundle $\overline{F}$ of $\overline{E}$, $\text{pardeg}(\overline{F}) < 0$.

Henceforth, we shall denote by $\mathcal{B}_{p.s.}$ the subspace of all parabolic stable holomorphic structures. The space $\mathcal{B}_{p.s.}$ is an open submanifold of $\mathcal{B}$, and it is thus naturally (via the fixed complex structure on $\overline{\Sigma}$) an infinite dimensional complex manifold. The group $\mathcal{K}_p$ is an infinite dimensional complex Lie group acting on $\mathcal{B}_{p.s.}$ in a holomorphic way. Therefore, the quotient $\mathcal{B}_{p.s.}/\mathcal{K}_p$ is naturally a complex manifold whose tangent space at $[\bar{\partial}_{\bar{E}}] \in \mathcal{B}_{p.s.}/\mathcal{K}_p$ can be identified with $H^1\left(\bar{\Sigma}, End^+\,\bar{E}\right)$, where $End^+\,\bar{E}$ is the sheaf of holomorphic sections of $\text{End}\,\bar{E}$ preserving the flag at p. It is useful, however, to have the following slightly different description of $\mathcal{B}_{p.s.}/\mathcal{K}_p$. First, note that given $g \in \mathcal{K}$, the value g_p at the point p acts on the space of flags $\mathbf{G}/\mathbf{T}_a \simeq \mathbf{G}^{\mathbb{C}}/\mathbf{B}_a^+$. Let $\mathcal{P}$ denote the subspace of $\mathcal{B} \times \mathbf{G}/\mathbf{T}_a$ consisting of pairs $(\bar{\partial}_{\bar{E}}, f)$, where $\bar{\partial}_{\bar{E}}$ is parabolic stable with respect to the image of the flag f under the projection $\mathbf{G}/\mathbf{T}_a \to \mathbf{G}/\mathbf{P}_a$ and the weights $\hat{\alpha}_i$. The action of the group $\mathcal{K}$ preserves $\mathcal{P}$, and we denote by $\mathcal{P}/\mathcal{K}$ the quotient space. Then we have the following

Proposition 2.10. *The space $\mathcal{P}/\mathcal{K}$ is homeomorphic to a $\mathbf{P}_a/\mathbf{T}_a$ bundle over $\mathcal{B}_{p.s.}/\mathcal{K}_p$. In particular, if all of the weights are distinct, then $\mathcal{P}/\mathcal{K}$ is homeomorphic to $\mathcal{B}_{p.s.}/\mathcal{K}_p$.*

Returning to our equivalence class $[\nabla]$, the basis $\{f_i\}$ along with the weights $\{\hat{\alpha}_i\}$ determine a flag $f \in \mathbf{G}/\mathbf{P}_a$ as in [M-S], p. 211, and ∇ and $\{f_i\}$ determine a holomorphic structure $\bar{\partial}_{\bar{E}}$. The theorem of Mehta and Seshadri states that if ∇ is irreducible then $\bar{\partial}_{\bar{E}}$ is parabolic stable with respect to f. In light of Proposition 2.10, we have a map

$$\mu : [\nabla] \longmapsto [\bar{\partial}_{\bar{E}}] \tag{2.11}$$

which identifies $\mathcal{B}_{p.s.}/\mathcal{K}_p$ with the space of flat irreducible unitary connections on E of holonomy conjugate to a. Therefore, $\mathcal{M}_a$ is a complex manifold provided we can fix the determinant (recall that we are taking $\mathbf{G} = \text{SU}(n)$ rather than $\text{U}(n)$). Indeed, let $J_l(\overline{\Sigma})$ be the degree l component of the Picard group of $\overline{\Sigma}$, and let

$$\det : \mathcal{B}_{p.s.}/\mathcal{K}_p \longrightarrow J_l(\overline{\Sigma}) \tag{2.12}$$

denote the determinant map associating to a given vector bundle $\overline{E} \in \mathcal{B}_{p.s.}/\mathcal{K}_p$ its determinant line bundle $\det \overline{E} = \bigwedge^n \overline{E} \in J_l(\overline{\Sigma})$. The map det is a holomorphic map of full rank whose fiber may be identified with the space $\mathcal{M}_a$. This defines a complex structure on $\mathcal{M}_a$. The details behind the identification (2.11) are, of course, quite intricate, and we refer to [Me-Se], [Si1], and [Si2] for elaboration. In Section 4, we shall make use of this description of $\mathcal{M}_a$ to extract some useful topological information.

2.3. Correspondence varieties

In this subsection, we would like to show how the spaces $\mathcal{M}_a$ form a correspondence between different moduli spaces of rank 2 bundles over $\overline{\Sigma}$. This is an elaboration of the concluding remarks in [Me-Se] (see also [Be]); this description, however, will be crucial for us in Section 6, so we explain it in some detail. Throughout this section, we set $\mathbf{G} = SU(2)$.

Let $\mathcal{M}(2, -1)$ denote the moduli space of semistable vector bundles of rank 2 and fixed determinant of degree -1 over the compact Riemann surface $\overline{\Sigma}$. In fact, we shall assume the determinant is $\mathcal{O}(-p)$, where p is the puncture. Assume $a \in \mathbf{G}$, $a \neq \pm I$, $\mathbf{T}_a$ the maximal torus in $\mathbf{G}$ containing a, and parameterize the parabolic weights $0 < \alpha < 1/2$.

Proposition 2.13. *For the choice of weights above, $\mathcal{B}_{p.s.}$ is equal to $\mathcal{B}_s$, the set of stable holomorphic structures on $\overline{E}$, and $\mathcal{M}_a \simeq \mathcal{B}_s \times_{\mathcal{K}} \mathbf{G}/\mathbf{T}_a$. Moreover, there exists a holomorphic projection $p_1 : \mathcal{M}_a \to \mathcal{M}(2, -1)$ which realizes $\mathcal{M}_a$ as a holomorphic $\mathbb{P}^1 \simeq \mathbf{G}/\mathbf{T}_a$ bundle over $\mathcal{M}(2, -1)$.*

Proof. Let $\overline{E}$ be a rank two parabolic stable bundle over $\overline{\Sigma}$ with determinant $\mathcal{O}(-p)$. If $\overline{F} \subset \overline{E}$ is a parabolic subbundle, then $\operatorname{pardeg} \overline{F} < 0$. In particular, $\deg \overline{F} < 0$, and since $\deg \overline{F}$ is an integer, $\deg \overline{F} \leq -1$. In any case, $\mu(\overline{F}) = \deg \overline{F} = -1 < -1/2 = \mu(\overline{E})$, and the first claim follows. The map p_1 is just projection onto the first factor in Proposition 2.10. □

We next construct a map onto the degree zero moduli space. Let $\mathcal{M}(2, 0)$ denote the moduli space of semistable rank two vector bundles over $\overline{\Sigma}$ with trivial determinant, and let $\mathcal{M}_s(2, 0)$ be the Zariski open subset consisting of stable bundles. Suppose $\overline{E}$ is a parabolic stable bundle. In particular, we have seen that $\overline{E}$ is stable with determinant $\mathcal{O}(-p)$. Moreover, the choice of a quasi-parabolic structure σ at p defines, via evaluation at the point p, a natural map

$$\overline{E}^* \longrightarrow \mathcal{C}_p \longrightarrow 0\,, \tag{2.14}$$

where $\mathcal{C}_p$ denotes the torsion sheaf obtained by extending $\overline{E}^*_p/\sigma^*$ by zero. Let $\overline{F}$ be the kernel of (2.14). It is easy to verify that $\overline{F}$ has trivial determinant and is semistable. We define $\overline{F} = p_0(\overline{E})$ to be its equivalence class in $\mathcal{M}(2, 0)$. All in all, we have shown

Proposition 2.15 (cf. [Me-Se]). *The spaces $\mathcal{M}_a$ for $a \neq \pm I$ form a correspondence*

$$\begin{array}{ccc} & \mathcal{M}_a & \\ {}^{p_0}\swarrow & & \searrow^{p_1} \\ \mathcal{M}(2,0) & & \mathcal{M}(2,-1) \end{array}$$

between $\mathcal{M}(2,0)$ and $\mathcal{M}(2,-1)$.

Proposition 2.16. *Assume $g > 3$. Then $\mathcal{U}_a = p_0^{-1}(\mathcal{M}_s(2,0))$ in $\mathcal{M}_a$ is a Zariski open set with complement of codimension at least 2.*

Proof. Under the natural biholomorphism $\mathcal{M}(2,-1) \simeq \mathcal{M}(2,1)$ obtained by taking duals, it suffices to show that the image of $\mathcal{U}_a$ in $\mathcal{M}(2,1)$ is Zariski open with complement of codimension at least 2. If however $\overline{E} \in \mathcal{M}_a$, then its dual $\overline{E}^*$ is stable of degree 1, and therefore $\overline{E}^*$ is (0,1)-semistable in the sense of Narasimhan and Ramanan (cf. [N-Rn] Definition 5.1). Moreover, if $\overline{E}$ does not belong to $\mathcal{U}_a$, then it is easy to see that $\overline{E}^*$ is not (0,1)-stable. Hence, $\mathcal{M}_a \setminus \mathcal{U}_a$ is mapped by duality inside the variety of (0,1)-semistable and not (0,1)-stable holomorphic bundles. By the same computation as that in the proof of Proposition 5.4 of [N-Rn], the codimension of this variety is at least $g-2$, which proves the assertion. □

Next, we will proceed by analogy with Proposition 2.10 to construct another $\mathbb{P}^1$-bundle over $\mathcal{M}(2,0)$. Let $\overline{\mathcal{A}}_s$ denote the stable holomorphic structures on a trivial rank 2 bundle on $\overline{\Sigma}$ and denote by $\overline{\mathcal{G}}^{\mathbb{C}}$ the group of complex automorphisms. Then we have

Proposition 2.17. *For $a \neq \pm I$, let $\mathcal{N}_a = \overline{\mathcal{A}}_s \times_{\overline{\mathcal{G}}^{\mathbb{C}}} \mathbf{G}/\mathbf{T}_a$, where $\overline{\mathcal{G}}^{\mathbb{C}}$ acts on $\mathbf{G}/\mathbf{T}_a$ by evaluation at p. Then the natural projection $pr_1 : \mathcal{N}_a \to \mathcal{M}_s(2,0)$ is a holomorphic $\mathbb{P}^1$ bundle.*

The next theorem will be important in Section 6.

Theorem 2.18. *There is a holomorphic embedding $\varphi : \mathcal{N}_a \to \mathcal{M}_a$ onto $\mathcal{U}_a$ with holomorphic inverse ψ which makes the following diagram commute:*

$$\begin{array}{ccc} \mathcal{N}_a & \xrightarrow{\varphi} & \mathcal{M}_a \\ \big\downarrow{\scriptstyle pr_1} & & \big\downarrow{\scriptstyle p_0} \\ \mathcal{M}_s(2,0) & \subset & \mathcal{M}(2,0) \end{array}$$

Proof. We define our map φ first. Let $(\overline{F}, f) \in \mathcal{N}_a$. This means $\overline{F}$ is a stable rank 2 bundle with trivial determinant and $f \in \mathbf{G}/\mathbf{T}_a$ can be used to define a line in $\overline{F}_p$, the fiber at p, via a fixed identification $\mathbb{P}(\overline{F}_p) \simeq \mathbf{G}/\mathbf{T}_a$. Let $\mathcal{C}_p$ denote the torsion sheaf obtained as the cokernel of the natural map $0 \to \mathcal{O}(-p) \to \mathcal{O}$. By Grothendieck duality, we can identify

$$\mathrm{Ext}^1(\mathcal{C}_p, \overline{F}) \simeq H^0\left(\overline{\Sigma}, \overline{F}^* \otimes \omega^*_{\overline{\Sigma}} \otimes \mathcal{C}_p\right)^* \simeq \overline{F}_p ,$$

and any f as above defines an extension

$$0 \longrightarrow \overline{F} \longrightarrow \overline{E}^* \longrightarrow \mathcal{C}_p \longrightarrow 0 \tag{2.19}$$

with $\overline{E}^*$ locally free. Moreover, by evaluation at the point p, extension (2.19) naturally defines a line in $\overline{E}^*_p$ which gives a quasi-parabolic structure σ on $\overline{E}$. In fact, it is clear that for a choice of weights as before, $\overline{E}$ is parabolic stable (cf. Proposition 2.13). We then set $\varphi(\overline{F}, f) = (\overline{E}, \sigma)$. One can verify that φ is indeed a morphism. Moreover, as $p_0(\overline{E}) = \overline{F}$, the diagram of Theorem 2.18 commutes and the image of φ is $\mathcal{U}_a$. Therefore, we only need construct an inverse morphism $\psi : \mathcal{U}_a \to \mathcal{N}_a$. This can be done as follows: an element of $\mathcal{U}_a$ consists of a stable bundle $\overline{E}$ of determinant $\mathcal{O}(-p)$ and a quasi-parabolic structure σ at p. Again, as in the definition of the map p_0, σ defines a projection of $\overline{E}^*$ to a torsion sheaf $\mathcal{C}_p$ supported at p and an exact sequence (2.19). But this defines a stable bundle $\overline{F}$ with a flag f, dual to the kernel of the map $\overline{E}^* \to \mathcal{C}_p \to 0$ restricted to p. We set $\psi(\overline{E}) = (\overline{F}, f)$. It is easily verified that ψ is a morphism and an inverse to φ. □

3. The L^2_δ-theory

We shall now take a different approach and describe the spaces $\mathcal{F}_a$ and $\mathcal{M}_a$ via weighted Sobolev spaces. Although much of the material in this section is quite standard (cf. [T1], [T2], [M], and [A-P-S]), the L^2_δ description of these spaces is the key to describing the quantum line bundle in Section 5 using the cocycle.

3.1. Flat connections and the gauge group

We continue to consider the punctured Riemann surface $\Sigma = \overline{\Sigma} \setminus \{p\}$ with fixed coordinate $w = (\tau, \theta) : \tilde{D}^* \to C$ where C is the semi-infinite cylinder from Section 2.1. In addition, we make a choice of metric K on Σ compatible with the complex structure of Σ and such that on $\tilde{D}^*$, $K = w^*(K_C)$, where $K_C = d\tau^2 + d\theta^2$ is the standard flat metric on C. It is important to extend τ to a C^∞-function defined on the whole surface Σ. We achieve this by setting $\tau = 0$ outside a small tubular neighborhood of $\partial\Sigma_0$ in Σ_0.

For $\delta \in \mathbb{R}$, we define the weighted L^p spaces of sections of E, denoted $L^p_\delta(E)$, as the completion of the space of compactly supported sections $C_0^\infty(E)$ in the norm

$$\|\sigma\|_{L^p_\delta} = \left\{\int_\Sigma e^{\tau\delta}|\sigma|^p\right\}^{1/p} .$$

Given also a positive integer k we define the weighted Sobolev space $L^p_{k,\delta}(E)$ as the completion of $C_0^\infty(E)$ in the norm

$$\|\sigma\|_{L^p_{k,\delta}} = \left\{\int_\Sigma e^{\tau\delta}\left(|\nabla_0^{(k)}\sigma|^p + \cdots + |\nabla_0\sigma|^p + |\sigma|^p\right)\right\}^{1/p} ,$$

where ∇_0 is a base connection in temporal gauge with holonomy $a = \exp(2\pi i\alpha)$ (see Lemma 2.7). It is not difficult to see that the spaces $L^p_{k,\delta}$ do not change if we replace ∇_0 with any other connection which agrees with ∇_0 on the cylinder C at infinity.

The weighted Sobolev spaces admit multiplication and embedding theorems similar to those for compact manifolds. Therefore, all the gauge theory developed for compact manifolds (e.g. as in [F-U]) extends quite easily to the non-compact case provided we replace the standard Sobolev spaces by the weighted ones.

Following [T1], Section 7, or [M], Section 3, we make the following definitions: let ∇_0 be a base connection in temporal gauge with holonomy a. We define our space of connections $\mathcal{A}_\delta = \nabla_0 + L^2_{1,\delta}(T^*\Sigma \otimes \mathfrak{g}_E)$. To define the appropriate gauge group we first consider the following space of sections of $\mathfrak{gl}(E)$. Define

$$\mathcal{R} = \left\{\varphi \in L^2_{2,\mathrm{loc.}}(\mathfrak{gl}(E)) : \|\nabla_0\varphi\|_{L^2_{1,\delta}} < \infty\right\} ,$$

and let $\mathcal{H}$ denote the subspace of $\mathcal{R}$ consisting of *harmonic* sections, i.e. $\mathcal{H} = \left\{\varphi \in \mathcal{R} : e^{-\tau\delta}\nabla_0^* e^{\tau\delta}\nabla_0\varphi = 0\right\}$, where $\nabla^* = *\nabla*$ is the formal L^2 adjoint.

There is another space of sections closely related to $\mathcal{H}$. Let $\nabla_{0,\partial}$ denote ∇_0 restricted to the bundle $\mathfrak{gl}(E)|_{\partial\Sigma_0}$. Let $\ker\nabla_{0,\partial}$ denote the space of parallel sections of $\mathfrak{gl}(E)|_{\partial\Sigma_0}$ with respect to $\nabla_{0,\partial}$. Since over $\partial\Sigma_0$, $\nabla_{0,\partial} = d/d\theta + i\alpha d\theta$, $a = \exp(2\pi i\alpha)$, $\ker\nabla_{0,\partial}$ can be identified with the normalizer of a in $\mathfrak{gl}(n,\mathbb{C})$. The next two propositions contain all the important properties of the gauge group we shall need.

Proposition 3.1 (cf. [M], Theorem 3.1) There is a direct sum decomposition $\mathcal{R} = L^2_{2,\delta}(\mathfrak{gl}(E)) \oplus \mathcal{H}$. *Moreover, there is a well-defined map* $r : \mathcal{R} \to \ker\nabla_{0,\partial}$ *given by* $r(\varphi)(\theta) = \lim_{\tau\to\infty}\varphi(\tau,\theta)$. *The map* r *satisfies* $r^{-1}(0) = L^2_{2,\delta}(\mathfrak{gl}(E))$ *and* $r : \mathcal{H} \to \ker\nabla_{0,\partial}$ *is an isomorphism.*

Proposition 3.2 (cf. [M], Appendix). *The norm*

$$\|\varphi\|^2_{\mathcal{R}} = \|\nabla_0\varphi\|^2_{L^2_{1,\delta}} + \int_{\partial\Sigma_0}|r(\varphi)|^2$$

gives $\mathcal{R}$ a Banach space structure in which the natural projection onto $L^2_{1,\delta}(\mathfrak{gl}(E))$, $\mathcal{H}$ are continuous. Moreover, pointwise multiplication $\mathcal{R} \times \mathcal{R} \to \mathcal{R}$ is well-defined and continuous, and $r : \mathcal{R} \to \ker \nabla_{0,\bar\partial}$ is a continuous linear map.

We now make the definitions

$$\mathcal{G}_\delta = \left\{\varphi \in \mathcal{R} : \varphi\varphi^* = I, \det\varphi = 1\right\}, \quad \mathcal{G}_{0,\delta} = \{\varphi \in \mathcal{G}_\delta : r(\varphi) = I\}$$

It is not difficult to see that $\mathcal{G}_\delta$, $\mathcal{G}_{0,\delta}$ are Banach Lie groups with Lie algebras

$$\mathrm{Lie}(\mathcal{G}_\delta) = \left\{\varphi \in \mathcal{R} : \varphi^* = -\varphi\right\}, \quad \mathrm{Lie}(\mathcal{G}_{0,\delta}) = \left\{\varphi \in L^2_{1,\delta} : \varphi^* = -\varphi\right\}.$$

Moreover, $\mathcal{G}_{0,\delta}$ is a closed, normal subgroup of $\mathcal{G}_\delta$ and $\mathcal{G}_\delta/\mathcal{G}_{0,\delta}$ may be identified with the normalizer $\mathbf{P}_a$ of a in $\mathbf{G}$. The groups $\mathcal{G}_\delta$, $\mathcal{G}_{0,\delta}$ act on the space $\mathcal{A}_\delta$ of connections by push forward, the usual way. The action is well-defined and smooth.

Given a connection $\nabla = \nabla_0 + A \in \mathcal{A}_\delta$, let $F_\nabla = \nabla_0 A + \frac{1}{2}[A, A]$ denote the curvature of ∇. It is easy to verify that the map $\nabla \mapsto F_\nabla$ is smooth from $\mathcal{A}_\delta$ to $L^2_\delta\left(\bigwedge^2 T^*\Sigma \otimes \mathfrak{g}_E\right)$. Let $\mathcal{A}_{\mathrm{F},\delta}$ denote the subspace of irreducible flat connections. The groups $\mathcal{G}_\delta$, $\mathcal{G}_{0,\delta}$ preserve $\mathcal{A}_{\mathrm{F},\delta}$. Finally, we denote by $\mathcal{F}_{\alpha,\delta} = \mathcal{A}_{\mathrm{F},\delta}/\mathcal{G}_{0,\delta}$, $\mathcal{M}_{\alpha,\delta} = \mathcal{A}_{\mathrm{F},\delta}/\mathcal{G}_\delta$, the quotient spaces. Then $\mathcal{M}_{\alpha,\delta} = \mathcal{F}_{\alpha,\delta}/\mathbf{P}_a$ under the identification $\mathbf{P}_a \simeq \mathcal{G}_\delta/\mathcal{G}_{0,\delta}$. We conclude this subsection with the following useful

Proposition 3.3. *The space $\mathcal{G}_{0,\delta}$ is connected and path connected.*

Proof. Being a Banach manifold, $\mathcal{G}_{0,\delta}$ is clearly locally path connected, hence the path components and components are identical. It suffices then to find a dense subset of $\mathcal{G}_{0,\delta}$ which is connected. Actually, we shall do this for $\mathcal{G}^{\mathbb{C}}_{0,\delta}$, the complex gauge group – this is sufficient, since the quotient $\mathcal{G}^{\mathbb{C}}_{0,\delta}/\mathcal{G}_{0,\delta}$ is clearly a contractible space.

Let $\overline{\mathcal{G}}^{\mathbb{C}}_1$ be the group of smooth maps $\overline{\Sigma} \to \mathbf{G}^{\mathbb{C}}$ which are the identity in some open neighborhood of p. Denote by $\overline{\mathcal{G}}^{\mathbb{C}}_0$ the group of smooth maps $\Sigma \to \mathbf{G}^{\mathbb{C}}$ which are the identity at p. $\overline{\mathcal{G}}^{\mathbb{C}}_0$ is nothing but the based complex gauge group for connections on a trivial $\mathbf{G}$ bundle over $\overline{\Sigma}$, and since $\mathbf{G}$ is simply connected, $\overline{\mathcal{G}}^{\mathbb{C}}_0$ is path connected. We claim that $\overline{\mathcal{G}}^{\mathbb{C}}_1$ is a dense, connected subset of $\mathcal{G}^{\mathbb{C}}_{0,\delta}$. First, density: the group of smooth compactly supported maps from Σ to $\mathfrak{gl}(n, \mathbb{C})$ is dense in $L^2_{2,\delta}$ by definition. Any $g \in \mathcal{G}^{\mathbb{C}}_{0,\delta}$ may be written as $1 + (-1 + g)$ with $-1 + g \in L^2_{2,\delta}$ (see Proposition 3.1). Hence, there exist smooth, compactly supported $\tilde{g}_j \to -1 + g$ as $j \to \infty$. For large j, $1 + \tilde{g}_j$ is invertible, therefore

$$g_j = \frac{1 + \tilde{g}_j}{\det(1 + \tilde{g}_j)^{1/n}} \longrightarrow g$$

and $\{g_j\}$ is a sequence in $\overline{\mathcal{G}}^{\mathbb{C}}_1$ (here we are using the multiplication theorems). This proves that $\overline{\mathcal{G}}^{\mathbb{C}}_1$ is dense. Second, we show that $\overline{\mathcal{G}}^{\mathbb{C}}_1$ is path connected, hence connected.

Given $g \in \overline{\mathcal{G}}_1^{\mathbb{C}}$, since $\overline{\mathcal{G}}_0^{\mathbb{C}}$ is path connected we can find a smooth path γ_t in $\overline{\mathcal{G}}_0^{\mathbb{C}}$ with $\gamma_0 = I, \gamma_1 = g$. We want to push γ_t into $\overline{\mathcal{G}}_1^{\mathbb{C}}$. But this is easily done: choose a small neighborhood U of the origin on which $\exp : \mathfrak{g}^{\mathbb{C}} \to \mathbf{G}^{\mathbb{C}}$ is a diffeomorphism. For a sufficiently small disk B about p, we may assume $\gamma_t(z) \in U$ for all $z \in B$ and $0 \leq t \leq 1$. Hence, we may write $\gamma_t(z) = \exp \Gamma_t(z)$, where Γ_t is a smooth map $B \to U, \Gamma_0 \equiv 0$. Choosing B sufficiently small, we may also assume $\Gamma_1 \equiv 0$. Now choose a smooth cut-off function $\eta : B \to [0, 1]$ which is one in a neighborhood of ∂B and zero in a neighborhood of the origin. Define $\tilde{\gamma}_t$ to be equal to γ_t on $\Sigma \setminus B$, and $\tilde{\gamma}_t(z) = \exp(\eta(z)\Gamma_t(z))$ for $z \in B$. Then $\tilde{\gamma}_t$ is clearly smooth and in $\overline{\mathcal{G}}_1^{\mathbb{C}}$. Moreover, $\tilde{\gamma}_0 \equiv I$ and $\tilde{\gamma}_1 = g$, so this gives the desired path and completes the proof. □

3.2. Application of the index theorem

Our next goal is to show that the spaces $\mathcal{F}_{\alpha,\delta}, \mathcal{M}_{\alpha,\delta}$ are finite dimensional manifolds of given dimension. To do this we will have to digress a bit to study the Fredholm properties of the $\bar{\partial}$-operator defined on the $L^2_{k,\delta}$ spaces via the index theorem of Atiyah, Patodi, and Singer.

Let ∇ be a flat connection, and let us denote also by ∇ the associated operator on $\mathfrak{g}$-valued forms. Let

$$\delta_\nabla : L^2_{1,\delta}\left(T^*\Sigma \otimes \mathfrak{g}_E\right) \longrightarrow L^2_\delta\left(\bigwedge^2 T^*\Sigma \otimes \mathfrak{g}_E\right) \oplus L^2_\delta(\mathfrak{g}_E) \tag{3.4}$$

denote the operator $\delta_\nabla = \left(\nabla, e^{-\tau\delta}\nabla^* e^{\tau\delta}\right)$. As we shall see presently, the kernel of the operator δ_∇ is closely related to the tangent spaces of $\mathcal{F}_{\alpha,\delta}, \mathcal{M}_{\alpha,\delta}$ at the point $[\nabla]$. We first prove

Proposition 3.5. *There exists a* $\delta_0 > 0$ *such that for every* $\delta \in (0, \delta_0)$, *the operator* δ_∇ *is a bounded, Fredholm operator of index* $2(g-1)(n^2-1) + \dim(\mathbf{G}/\mathbf{P}_a)$. *Moreover,* $\dim \operatorname{coker} \delta_\nabla = \dim \mathbf{P}_a$.

Proof. First, the cokernel of δ_∇ is the kernel of the operator

$$(\delta_\nabla)^* : L^2_{1,\delta}(\mathfrak{g}_E) \oplus L^2_{1,\delta}\left(\bigwedge^2 T^*\Sigma \otimes \mathfrak{g}_E\right) \longrightarrow L^2_\delta\left(T^*\Sigma \otimes \mathfrak{g}_E\right)$$

$$(\delta_\nabla)^*(\varphi, \omega) = \nabla\varphi + e^{-\tau\delta}\nabla^* e^{\tau\delta}\omega\,.$$

Since ∇ is flat, $\ker(\delta_\nabla)^*$ consists of pairs (φ, ω) such that $\nabla\varphi = 0, e^{-\tau\delta}\nabla^* e^{\tau\delta}\omega = 0$. Since φ, ω are exponentially decaying the only possibility is $\varphi = 0$, $*\omega \in e^{-\tau\delta}\mathbf{P}_a$, hence $\dim \operatorname{coker} \delta_\nabla = \dim \mathbf{P}_a$.

Next, we observe that it suffices to prove the result for the operator $\tilde{\delta}_\nabla$ defined by composing δ_∇ with the continuous inclusion

$$L^2_\delta\left(\bigwedge^2 T^*\Sigma \otimes \mathfrak{g}_E\right) \oplus L^2_\delta(\mathfrak{g}_E) \hookrightarrow L^2_{\delta/2}\left(\bigwedge^2 T^*\Sigma \otimes \mathfrak{g}_E\right) \oplus L^2_{\delta/2}(\mathfrak{g}_E)\,.$$

Indeed, the kernels of δ_∇ and $\tilde{\delta}_\nabla$ coincide and δ_∇ has closed range if $\tilde{\delta}_\nabla$ does. Moreover, the cokernels are isomorphic under the continuous injection

$$L^2_{\delta/2}\left(\bigwedge^2 T^*\Sigma \otimes \mathfrak{g}_E\right) \oplus L^2_{\delta/2}(\mathfrak{g}_E) \hookrightarrow L^2_\delta\left(\bigwedge^2 T^*\Sigma \otimes \mathfrak{g}_E\right) \oplus L^2_\delta(\mathfrak{g}_E)\,,$$

coming from multiplication by $e^{-\tau\delta/2}$. This is clear, since by the previous paragraph, if $\psi = (0, \omega) \in \operatorname{coker}\ \delta_\nabla$, then $e^{\tau\delta}\psi$ is bounded, so in particular $e^{\tau\delta/2}\psi$ is in $L^2_{\delta/2}$.

We also may assume without loss of generality that $\nabla = \nabla_0$ is a connection in temporal gauge (see Lemma 2.7). Thus, locally near the puncture $\nabla = d + i\alpha d\theta$. This is because for general ∇, $\tilde{\delta}_\nabla - \tilde{\delta}_{\nabla_0}$ is a multiplication operator by an element of $L^2_{1,\delta}$. Since the inclusion $L^2_{1,\delta} \hookrightarrow L^4_{\delta/2}$ is compact, and the multiplication $L^4_{\delta/2} \times L^4_{\delta/2} \to L^2_{\delta/2}$ is continuous, we see that $\tilde{\delta}_\nabla - \tilde{\delta}_{\nabla_0}$ is in fact a compact operator. Therefore, $\tilde{\delta}_\nabla$ is Fredholm if and only if $\tilde{\delta}_{\nabla_0}$ is Fredholm, and index δ_∇ = index $\tilde{\delta}_\nabla$ = index $\tilde{\delta}_{\nabla_0}$. Let

$$\tilde{\delta}^0_{\nabla_0} : L^2_{1,\delta}\left(T^*\Sigma \otimes \mathfrak{g}_E\right) \longrightarrow L^2_{\delta/2}\left(\bigwedge^2 T^*\Sigma \otimes \mathfrak{g}_E\right) \oplus L^2_{\delta/2}(\mathfrak{g}_E)$$

denote the operator (∇_0, ∇_0^*). Since all the derivatives of τ are in L^∞, the operator $\tilde{\delta}_{\nabla_0} - \delta^0_{\nabla_0}$ is a multiplication operator by a function in L^∞, hence by the compactness theorems for weighted Sobolev spaces $\tilde{\delta}_{\nabla_0} - \tilde{\delta}^0_{\nabla_0}$ is a compact operator. Then $\tilde{\delta}_{\nabla_0}$ is Fredholm if and only if $\tilde{\delta}^0_{\nabla_0}$ is Fredholm, and again they have the same indices.

Via the identification $T^*\Sigma \otimes \mathfrak{g}_E \simeq T^*\Sigma^{0,1} \otimes \mathfrak{g}_E^{\mathbb{C}}$, $\bigwedge^2 T^*\Sigma \otimes \mathfrak{g}_E \oplus \mathfrak{g}_E \simeq \mathfrak{g}_E^{\mathbb{C}}$, the operator $\tilde{\delta}^0_{\nabla_0}$ is nothing but the formal L^2-adjoint of the $\bar{\partial}$-operator associated to ∇_0:

$$\bar{\partial}^*_{\nabla_0} : L^2_{1,\delta}\left(T^*\Sigma^{0,1} \otimes \mathfrak{g}_E^{\mathbb{C}}\right) \longrightarrow L^2_{\delta/2}(\mathfrak{g}_E^{\mathbb{C}}) \tag{3.6}$$

Following Atiyah, Patodi, and Singer, we can define the same operator $\bar{\partial}^*_{\nabla_0}$ on the modified spaces:

$$\bar{\partial}^*_{\nabla_0} : L^2\left(T^*\Sigma^{0,1} \otimes \mathfrak{g}_E^{\mathbb{C}}\right) \longrightarrow L^2_{\text{ext.}}(\mathfrak{g}_E^{\mathbb{C}}) \tag{3.6$'$}$$

By $L^2_{\text{ext.}}(\mathfrak{g}_E^{\mathbb{C}})$ we mean the space of extended L^2-sections of $\mathfrak{g}_E^{\mathbb{C}}$ as defined in [A-P-S], Proposition 3.11.

At this point, we require δ_0 to be less than the first positive eigenvalue of $\nabla_{0,\partial}$ on $E|_{\partial\Sigma_0}$, and let $\delta \in (0, \delta_0)$. It follows in exactly the same way as in [A-P-S], Proposition 3.11, that the kernel of (3.6) is the same as the kernel of (3.6$'$) and hence is finite dimensional (see also Lemma 3.12 below). We claim also that the cokernels are the same. Indeed, the cokernel of (3.6) may be identified with

$$\left\{\varphi \in L^2_{\delta/2}(\mathfrak{g}_E^{\mathbb{C}}) : e^{-\tau\delta}\bar{\partial}_{\nabla_0} e^{2\tau\delta}\varphi = 0\right\}$$

which, under the isometry $e^{\tau\delta/4} : L^2_{\delta/2} \to L^2$, corresponds to

$$\left\{\varphi \in L^2(\mathfrak{g}_E^{\mathbb{C}}) : \bar{\partial}_{\nabla_0} e^{\tau\delta/4}\varphi = 0\right\} .$$

The cokernel of (3.6′) is the space $\left\{\psi \in L^2_{\text{ext.}}(\mathfrak{g}_E^{\mathbb{C}}) : \bar{\partial}_{\nabla_0}\psi = 0\right\}$. We assert that the map $\varphi \mapsto \psi = \varphi e^{\tau\delta/4}$ is an isomorphism. This follows as in [A-P-S] by the following argument: let $\varphi \in L^2(\mathfrak{g}_E^{\mathbb{C}})$, $\bar{\partial}_{\nabla_0}(e^{\tau\delta/4}\varphi) = 0$. Expand φ near the boundary $\varphi(\tau, \theta) = \sum_\lambda f_\lambda(\tau)\varphi_\lambda(\theta)$ where $\varphi_\lambda(\theta)$ form an orthonormal basis of eigenfunctions of $\nabla_{0,\partial}$. Since $\bar{\partial}_{\nabla_0}(e^{\tau\delta/4}\varphi) = 0$, we have

$$\frac{d}{d\tau}\left(e^{\tau\delta/4} f_\lambda(\tau)\right) + \lambda\left(e^{\tau\delta/4} f_\lambda(\tau)\right) = 0 ,$$

hence $\varphi(\tau, \theta) = \sum_\lambda e^{-(\lambda+\delta/4)\tau} f_\lambda(0)\varphi_\lambda(\theta)$. Since $\varphi \in L^2$, the summation is over $\lambda \geq 0$, hence $\varphi(\tau, \theta) = e^{-\tau\delta/4}\sum_{\lambda\geq 0} e^{-\lambda\tau} f_\lambda(0)\varphi_\lambda(\theta)$. But then the image $\psi(\tau, \theta) = e^{\tau\delta/4}\varphi(\tau, 0) = \sum_{\lambda\geq 0} e^{-\lambda\tau} f_\lambda(0)\varphi_\lambda(\theta)$ belongs to the cokernel of (3.6′). The inverse map is constructed in the same way.

By the Atiyah, Patodi, Singer Index Theorem (see [A-P-S], Theorem 3.10), $\bar{\partial}^*_{\nabla_0}$ is Fredholm, hence so is $\bar{\delta}^0_{\nabla_0}$, and index $\bar{\delta}^0_{\nabla_0} = \text{index}(\bar{\partial}^*_{\nabla_0})$. Moreover,

$$\text{index}(\bar{\partial}^*_{\nabla_0}) = -(n^2 - 1)\int_\Sigma c_1 - \frac{h}{2} , \quad h = \dim(\mathbf{P}_a \otimes \mathbb{C}) .$$

Hence, since $\int_\Sigma c_1$ is the Euler characteristic of $\Sigma = 1 - 2g$, we have

$$\text{index}(\bar{\partial}^*_{\nabla_0}) = (n^2 - 1)(2g - 1) - \dim \mathbf{P}_a = 2(g - 1)(n^2 - 1) + \dim(\mathbf{G}/\mathbf{P}_a) .$$

□

Let us denote by $H^1_{\delta,\nabla}$ the kernel of the operator δ_∇ of (3.4). According to Proposition 3.5, $H^1_{\delta,\nabla}$ is a vector space of dimension $(2g - 1)(n^2 - 1)$. Let $\mathcal{H}$ be the space of harmonic 0-forms, and let $r_0 : \mathcal{H} \to H^1_{\delta,\nabla}$ be the map $r_0(\varphi) = \nabla\varphi$. Let $H^1_{\delta+,\nabla} = H^1_{\delta,\nabla}/\operatorname{im} r_0$. The importance of the spaces $H^1_{\delta,\nabla}$, $H^1_{\delta+,\nabla}$ is illustrated in the theorem below.

Theorem 3.7. (i) *$\mathcal{F}_{\alpha,\delta}$ is a smooth manifold of dimension $(2g-1)(n^2-1)$. Moreover, for $[\nabla] \in \mathcal{F}_{\alpha,\delta}$, $T_{[\nabla]}\mathcal{F}_{\alpha,\delta} \simeq H^1_{\delta,\nabla}$.*

(ii) *$\mathcal{M}_{\alpha,\delta}$ is a smooth manifold of dimension $2(g - 1)(n^2 - 1) + \dim(\mathbf{G}/\mathbf{P}_a)$. Moreover, for $[\nabla] \in \mathcal{M}_{\alpha,\delta}$, $T_{[\nabla]}\mathcal{M}_{\alpha,\delta} \simeq H^1_{\delta+,\nabla}$.*

(iii) *$\mathcal{M}_{\alpha,\delta}$ naturally has the structure of an almost complex manifold.*

Proof. The proof of (i) and (ii) follows from standard arguments (cf. [Ko]) using our Proposition 3.5 and Theorem 3.4 of [M] (see also Sections 7 and 8 of [T1] for a very similar situation). In order to prove (iii) we proceed as follows: Given $[\nabla] \in \mathcal{M}_{\alpha,\delta}$, lift to a representative $\nabla \in \mathcal{A}_{F,\delta}$ and let $\bar{\partial}^*_\nabla$ denote the operator associated to ∇ as in (3.6′). Let $\sigma_\nabla : H^1_{\delta,+} \to \ker \bar{\partial}^*_\nabla$ be the linear map defined as follows: Given $X \in H^1_{\delta,+} \simeq H^1_\delta / \operatorname{im} r_0$, lift to

$$\alpha_X \in H^1_\delta \subset L^2_{1,\delta}\left(T^*\Sigma \otimes \mathfrak{g}_E\right) \subset L^2\left(T^*\Sigma \otimes \mathfrak{g}_E\right)$$

perpendicularly in the L^2-norm to the subspace $\operatorname{im} r_0 \subset H^1_\delta$. Let β_X be the image of α_X under the natural isomorphism $L^2\left(T^*\Sigma \otimes \mathfrak{g}_E\right) \simeq L^2\left(T^*\Sigma^{0,1} \otimes \mathfrak{g}_E^{\mathbb{C}}\right)$, and let $\gamma_X = \sigma_\nabla(X)$ be the L^2-projection of β_X in $\ker \bar{\partial}^*_\nabla$. We claim that σ_∇ is an isomorphism. Indeed, as both spaces have the same dimension, it suffices to check that σ_∇ is injective. Assume therefore that $\gamma_X = 0$. Then $\beta_X = \bar{\partial}_\nabla \psi$, for some $\psi \in L^2_{\text{ext.}}\left(\mathfrak{g}_E^{\mathbb{C}}\right)$, and hence $\alpha_X = \nabla\varphi$, $\varphi = \psi - \psi^*$. But since $0 = e^{-\tau\delta}\nabla^* e^{\tau\delta}\alpha_X = e^{-\tau\delta}\nabla^* e^{\tau\delta}\nabla\varphi$, we obtain that $\varphi \in \mathcal{H}$, and therefore $\alpha_X \in \operatorname{im} r_0$. By the assumption that α_X is perpendicular to the latter, $\alpha_X = 0$, and σ_∇ is therefore an isomorphism.

The almost complex structure on $\mathcal{M}_{\alpha,\delta}$ is now defined by pulling back the complex structure from $\ker \bar{\partial}^*_\nabla$ to $T_{[\nabla]}\mathcal{M}_{\alpha,\delta}$ via σ_∇. This completes the proof of Theorem 3.7.

□

3.3. Equivalence of the two descriptions

We shall now show that the L^2_δ description produces the same moduli spaces as the ones described in Section 2 via representations of the fundamental group. Moreover, the almost complex structure on $\mathcal{M}_{\alpha,\delta}$ defined in Theorem 3.7 coincides with that of Mehta and Seshadri (cf. Section 2.2). We start by showing

Theorem 3.8. *There is a natural* $\mathbf{P}_a \simeq \mathcal{G}_\delta/\mathcal{G}_{0,\delta}$ *diffeomorphism* $\Phi : \mathcal{F}_{\alpha,\delta} \to \mathcal{F}_a$. *Moreover,* Φ *induces a diffeomorphism* $\mathcal{M}_{\alpha,\delta} \to \mathcal{M}_a$.

Proof. The construction of Φ proceeds as follows: let $[\nabla] \in \mathcal{F}_{\alpha,\delta}$. First, choose a smooth representative $\nabla \in \mathcal{A}_{F,\delta}$ such that over the punctured disk $\nabla = d + i\alpha d\theta$. This can be achieved as in Lemma 2.7 by noticing that if in the ODE in the proof of 2.7, $b \in L^2_{1,\delta}$, then $g \in \mathcal{G}_\delta$; we then use g multiplied by the inverse of its value at ∞ (cf. Propositions 3.1 and 3.2). We set $\Phi[\nabla]$ to be equal to the class of ∇ in $\mathcal{M}_0$. Clearly, $\Phi[\nabla] \in \mathcal{F}_a$. We claim that Φ is surjective. Let $[\hat{\nabla}] \in \mathcal{F}_a$. Using Lemma 2.2, choose a representative $\hat{\nabla} = d + A_s$ such that on the disk $\tilde{D}^*$, $A_s = i\alpha d\theta$. Let $A = \hat{\nabla} - \nabla_0$, and observe that A is compactly supported. Hence, $\nabla_0 + A \in \mathcal{A}_\delta$ and $\Phi[\nabla_0 + A] = [\hat{\nabla}]$. To show that Φ is injective, let $\Phi[\nabla_1] = \Phi[\nabla_2]$. Choose representatives $\nabla_1 = \nabla_0 + A_1$, $\nabla_2 = \nabla_0 + A_2$ as before. By assumption, $\nabla_2 = \nabla_1^g$ where $g \in \mathcal{G}_0$. Hence, over $\tilde{D}^*$ and via the cylindrical coordinates (τ, θ) we obtain

$\partial g/\partial\tau = 0, \nabla_{0,\partial} g = 0$. The above equations together with the fact that g at the base point is I imply $g|_{\tilde{D}^*} \equiv I$, hence $g \in \mathcal{G}_{0,\delta}$. Thus $[\nabla_1] = [\nabla_2]$ in $\mathcal{F}_{\alpha,\delta}$, proving that Φ is injective. We have shown Φ to be a bijection – we must verify that Φ and Φ^{-1} are smooth. The smoothness of Φ follows from the fact that the holonomy map is smooth, plus the commutativity of the diagram:

$$\begin{array}{ccccc} & & \mathcal{A}_{\mathrm{F},\delta} & \xrightarrow{\text{hol.}} & \mathrm{Hom}\,(\pi_1(\Sigma), \mathbf{G}) \\ & & \downarrow P & & \downarrow \\ \mathcal{A}_{\mathrm{F},\delta}/\mathcal{G}_{0,\delta} & = & \mathcal{F}_{\alpha,\delta} & \xrightarrow{\Phi} & \mathcal{F}_a \subset \mathcal{M}_0 \end{array}$$

The smoothness of Φ^{-1} is obtained as follows: let $\{\rho_\xi\}$ be any smooth family in $\mathrm{Hom}\,(\pi_1(\Sigma), \mathbf{G})$, and let $\{A_\xi\}$ be any smooth family in $\mathcal{A}_{\mathrm{F},\delta}$ such that holonomy $A_\xi = \rho_\xi$. Then $\Phi^{-1}(\rho_\xi) = P(A_\xi)$, and hence is smooth in ξ (recall that the C^∞-structure on $\mathcal{M}_0$, and hence also that on $\mathcal{F}_a$, was defined via the identification $\mathcal{M}_0 \simeq \mathrm{Hom}\,(\pi_1(\Sigma), \mathbf{G})$. In order to complete the proof of Theorem 3.8, we need to check the equivariance of Φ. Let $\phi \in \mathbf{P}_a \simeq \mathcal{G}_\delta/\mathcal{G}_{0,\delta}$. Let g_ϕ be the unique harmonic gauge transformation satisfying $r(g_\phi) = \phi$. Then

$$\Phi(\phi \cdot [\nabla]) = \Phi\left[\nabla^{g_\phi}\right] = \Phi\left[\nabla^{(g_\phi \cdot \phi^{-1}) \cdot \phi}\right] = \Phi[\nabla]^\phi = \phi \cdot \Phi[\nabla]\,.$$

The diffeomorphism $\mathcal{M}_{\alpha,\delta} \simeq \mathcal{M}_a$ follows. □

To prepare for the proof of the equivalence of almost complex structures, recall from Section 2.2 that a flat unitary connection ∇ gives rise to a flag f. Let us denote by $\mathfrak{b}^+$ (resp. $\mathfrak{b}^-$) the endomorphisms of the fiber $\bar{E}_p$ which are upper triangular (resp. lower triangular) with respect to the flag, and by $\mathfrak{u}^+$ and $\mathfrak{u}^-$ we mean the strictly upper and lower triangular endomorphisms, respectively.

Let $\mathcal{S}$ denote the sheaf of meromorphic sections of $\mathit{End}\,\bar{E} \otimes \Omega^1$ with at most a simple pole at the parabolic point p. We denote also by $\mathcal{S}_0$ the subsheaf of traceless endomorphisms, and by $\mathcal{S}^-$ and $\mathcal{S}_0^-$ those with residue strictly lower triangular at p with respect to the fixed flag f. We have the following

Lemma 3.9. *Hermitian conjugation gives a complex anti-linear injection* $\ker \bar{\partial}_\nabla^* \hookrightarrow H^0\left(\bar{\Sigma}, \mathcal{S}_0^-\right)$.

Proof. The hermitian conjugate of an element of $\ker \bar{\partial}_\nabla^*$ is an L^2 holomorphic (1,0) form ω with values in End E. We must show ω extends as a meromorphic form on $\bar{\Sigma}$ with residue strictly lower triangular. Near the point p we write $\omega(z) = \omega_{ij}(z)dz \otimes f_i \otimes f_j^*$, where ω_{ij} are holomorphic functions on the punctured disk $\tilde{D}^*$ and $\{f_i\}$ is the basis of holomorphic sections introduced in Section 2.2. Writing in terms of the orthonormal basis $\{e_i\}$, $\omega(z) = |z|^{\hat{\alpha}_i - \hat{\alpha}_j}\omega_{ij}(z)dz \otimes e_i \otimes e_j^*$, hence

$$\|\omega\|_{L^2}^2 \sim \int |dz|^2 |\omega_{ij}|^2 |z|^{2(\hat{\alpha}_i - \hat{\alpha}_j)} < +\infty\,.$$

Since $|\hat{\alpha}_i - \hat{\alpha}_j| < 1$, the ω_{ij} may have at most a simple pole at $z = 0$. If $\hat{\alpha}_i - \hat{\alpha}_j > 0$, the residue may be arbitrary, but if $\hat{\alpha}_i - \hat{\alpha}_j \leq 0$, the residue must vanish. This proves the lemma. □

Lemma 3.10. *Serre duality gives a complex linear isomorphism* $H^0\left(\bar{\Sigma}, \mathcal{S}^-\right) \xrightarrow{\sim} H^1\left(\bar{\Sigma}, End^-\,\bar{E}\right)^*$.

Proof. Since $End\,\bar{E}$ is locally free we have the exact sequence

$$0 \longrightarrow End\,\bar{E} \otimes \mathcal{O}(-p) \longrightarrow \mathrm{End}\bar{E} \longrightarrow \mathfrak{gl}^{\mathbb{C}} \longrightarrow 0\,.$$

Restricting to lower triangular endomorphisms, we have via evaluation $End^-\,\bar{E} \to \mathfrak{b}^- \to 0$. Let $\mathcal{Q}$ denote the kernel of the sequence above. Then we have the following commutative diagram of exact sequences:

$$\begin{array}{ccccccccc}
 & & & & 0 & & 0 & & \\
 & & & & \downarrow & & \downarrow & & \\
0 & \longrightarrow & \mathcal{Q} & \longrightarrow & End^-\,\bar{E} & \longrightarrow & \mathfrak{b}^- & \longrightarrow & 0 \\
 & & \| & & \downarrow & & \downarrow & & \\
0 & \longrightarrow & End\,\bar{E} \otimes \mathcal{O}(-p) & \longrightarrow & End\,\bar{E} & \longrightarrow & \mathfrak{gl}^{\mathbb{C}} & \longrightarrow & 0 \\
 & & & & \downarrow & & \downarrow & & \\
 & & & & \mathfrak{u}^+ & = & \mathfrak{u}^+ & & \\
 & & & & \downarrow & & \downarrow & & \\
 & & & & 0 & & 0 & &
\end{array}$$

where the isomorphism $\mathcal{Q} \simeq End\,\bar{E} \otimes \mathcal{O}(-p)$ is easily proven. Looking at the exact (dual) sequence in cohomology, we have

$$\begin{array}{ccccccc}
0 & & & & & & \\
\uparrow & & & & & & \\
\mathfrak{b}^+ & \overset{j}{\longleftarrow} & H^1(\bar{\Sigma}, \mathcal{Q})^* & \longleftarrow & H^1(\bar{\Sigma}, End^-\,\bar{E})^* & \longleftarrow & 0 \\
\uparrow & & \| & & & & \\
\mathfrak{gl}^{\mathbb{C}} & \overset{res}{\longleftarrow} & H^1\left(\bar{\Sigma}, End\,\bar{E} \otimes \mathcal{O}(-p)\right)^* & & & & \\
\uparrow & & & & & & \\
\mathfrak{u}^- & & & & & & \\
\uparrow & & & & & & \\
0 & & & & & &
\end{array}$$

By Serre duality, $H^1\left(\bar{\Sigma}, End\,\bar{E} \otimes \mathcal{O}(-p)\right)^*$ may be identified with the holomorphic sections of $End\,\bar{E} \otimes \mathcal{O}(p) \otimes \Omega^1$, and the map to $\mathfrak{gl}^{\mathbb{C}}$ is just the residue. Requiring the residue to be strictly lower triangular identifies $H^0(\bar{\Sigma}, \mathcal{S}^-)$ with $\ker(\jmath) \simeq H^1(\bar{\Sigma}, End^-\,\bar{E})^*$. This proves the lemma. □

Combining Lemmas 3.9 and 3.10 with the identification (cf. Section 2.2)

$$T_{[\bar{\partial}_{\bar{E}}]}\left(\mathcal{B}_{p.s.}/\mathcal{K}_p\right) \simeq H^1\left(\bar{\Sigma}, End^+\,\bar{E}\right) \xrightarrow{\ *\ } H^1\left(\bar{\Sigma}, End^-\,\bar{E}\right)^*$$

and the holomorphic inclusion $\mathcal{S}_0^- \hookrightarrow \mathcal{S}^-$, we obtain

Proposition 3.11. *The injection* $\ker \bar{\partial}_\nabla^* \hookrightarrow T_{[\bar{\partial}_{\bar{E}}]}\left(\mathcal{B}_{p.s.}/\mathcal{K}_p\right)$ *is complex linear.*

Now let $\Psi = \Phi^{-1}$ in Theorem 3.8. If $\mathcal{A}_F^a$ denotes the space of smooth traceless hermitian flat connections on E with holonomy around p conjugate to a, then we clearly can lift Ψ to a map $\tilde{\Psi} : \mathcal{A}_F^a \to \mathcal{M}_{\alpha,\delta}$. Given $\nabla \in \mathcal{A}_F^a$, the map $a \to a^{0,1}$ induces a linear isomorphism between $\ker \bar{\partial}_\nabla^*$ in (3.6′) and space

$$\mathcal{W} = \left\{a \in L^2\left(T^*\Sigma \otimes \mathfrak{g}_E\right) : \nabla a = 0\ ,\ \nabla * a = 0\right\}\ .$$

Recall the map σ_∇ from Theorem 3.7. We have the following

Lemma 3.12. *Under the identification* $\mathcal{W} \simeq \ker \bar{\partial}_\nabla^*$*, the restriction of* Ψ_* *to* $\mathcal{W}$ *corresponds to* σ_∇^{-1}.

Proof. Let $a \in \mathcal{W}$. By our choice of δ and the equality $e^{\tau\delta} = |z|^{-\delta}$, it is evident from the proof of Lemma 3.9 above that $a^{0,1} \in L^2_{1,\delta}\left(T^*\Sigma^{0,1} \otimes \mathfrak{g}_E^{\mathbb{C}}\right)$, hence $a \in L^2_{1,\delta}\left(T^*\Sigma \otimes \mathfrak{g}_E\right)$ (see also, [A-P-S], Proposition 3.11). But $\sigma_\nabla^{-1}(a)$ is nothing but the class of a in $H^1_{\delta,+}$, which by definition of $\tilde{\Psi}$ equals $\Psi_*(a)$. □

Finally, note that the map μ from (2.11) lifts to a map $\tilde{\mu} : \mathcal{A}_F^a \to \mathcal{B}_{p.s.}/\mathcal{K}_p$. It is clear from Lemmas 3.9, 3.10, and 3.12 that the following diagram commutes:

$$\begin{array}{ccccccc}
 & \sigma_\nabla^{-1} & \ker \bar{\partial}_\nabla^* & \xrightarrow{\ *\ } & H^0\left(\bar{\Sigma}, \mathcal{S}_0^-\right) & \longrightarrow & H^0\left(\bar{\Sigma}, \mathcal{S}^-\right) \\
 & \swarrow & \| & & & & \| \\
T_{[\nabla]}\mathcal{M}_{\alpha,\delta} & & \mathcal{W} & & & & H^1\left(\bar{\Sigma}, End^-\,\bar{E}\right)^* \\
\| & \tilde{\Psi}_* & \cap & & \tilde{\mu}_* & & \|* \\
T_{[\nabla]}\mathcal{M}_{\alpha,\delta} & \longleftarrow & T_\nabla\mathcal{A}_F^a & \text{———} & \text{———} & \longrightarrow & T_{[\bar{\partial}_{\bar{E}}]}\mathcal{B}_{p.s.}/\mathcal{K}_p \\
\| & & \downarrow & & & & \| \\
T_{[\nabla]}\mathcal{M}_{\alpha,\delta} & \longleftrightarrow & T_{[\nabla]}\mathcal{M}_a & \longleftarrow & \text{———} & \longrightarrow & T_{[\bar{\partial}_{\bar{E}}]}\mathcal{B}_{p.s.}/\mathcal{K}_p
\end{array}$$

By Proposition 3.11 we now have

Theorem 3.13. *With respect to the almost complex structure defined on $\mathcal{M}_{\alpha,\delta}$ in Theorem 3.7 and the complex structure on $\mathcal{M}_a$ defined via the Mehta–Seshadri theorem, the map $\Phi : \mathcal{M}_{\alpha,\delta} \to \mathcal{M}_a$ is biholomorphic.*

4. Topological results

We now examine the topological properties of the spaces we have defined. The key idea is the interplay between the different descriptions. The analytic description will allow us to define a closed, integral two-form on the spaces $\mathcal{F}_{\alpha,\delta}$ which will serve as the curvature for the pre-quantum line bundle. Topological restrictions will prevent the line bundle descending, except in special cases.

4.1. The topology of the moduli spaces

In this section we analyze the topology of the spaces $\mathcal{M}_a \simeq \mathcal{F}_a/\mathbf{P}_a$ for $a \in \mathbf{G}$. The case $\mathbf{G} = \mathrm{SU}(2)$ is rather simple and essentially follows via Proposition 2.3 from the results of Atiyah and Bott [A-B]. For higher rank, there is not such a simple description, and instead we will make use of certain transversality results developed in [D-U]. In this general case, we have to make the additional assumption that the genus of $\overline{\Sigma}$ is strictly greater than two.

To begin, let us assume that $\mathbf{G} = \mathrm{SU}(2)$. Recall from Proposition 2.3 that the map $q : \mathcal{M}_0 \to \mathbf{G}$ is a fibration away from the identity. Therefore, the fiber $\mathcal{F}_a$ of q over $a \neq I$ is homotopy equivalent to $\mathcal{F}_{-I}$. By the results of Atiyah and Bott [A-B], $\pi_1(\mathcal{F}_{-I}) = 0$, $\pi_2(\mathcal{F}_{-I}) = \mathbb{Z}$, and therefore the same is true for $\mathcal{F}_a$, $a \neq I$. Looking at the long exact sequence in homotopy for the fibration (2.6) for $a \neq \pm I$, we obtain

Theorem 4.1. *Suppose $\mathbf{G} = \mathrm{SU}(2)$, $a \in \mathbf{G}$, and $a \neq \pm I$. Then $\mathcal{M}_a$ is connected and simply connected, and $\pi_2(\mathcal{M}_a)$ is free abelian of rank 2.*

We wish to define a pair of generators for $\pi_2(\mathcal{M}_a) \simeq H_2(\mathcal{M}_a, \mathbb{Z})$, $a \neq \pm I$, explicitly. Let β be the image of a generator of $\pi_2(\mathcal{F}_a) \simeq \mathbb{Z}$ in $\pi_2(\mathcal{M}_a)$ under the map π_*. Let γ be an element of $\pi_2(\mathcal{M}_a)$ whose image under the boundary homomorphism ∂ is a generator of $\pi_1(\mathbf{T}_a/\mathbb{Z}_2) \simeq \mathbb{Z}$ and which is perpendicular to β with respect to the dual pairing. This defines β and γ up to sign. We fix the signs to be compatible with the complex structure on $\mathcal{M}_a$. Recall the map $\varphi : \mathcal{N}_a \to \mathcal{U}_a \subset \mathcal{M}_a$ of principal $\mathbb{P}^1$ bundles over $\mathcal{M}_s(2,0)$ (see Theorem 2.18). Let $\overline{\beta}, \overline{\gamma}$ denote the generators of $\pi_2(\mathcal{N}_a)$ defined via projection onto the first and second factors, respectively. Note that $(pr_1)_*\overline{\beta}$ generates the free part of $\pi_2(\mathcal{M}(2,0)) \simeq \mathbb{Z} \oplus \mathbb{Z}_2$. The torsion in $\pi_2(\mathcal{M}(2,0))$ implies another important fact: if $[\mathbb{P}^1]$ denotes the fundamental class of the fiber in the fibration $\mathbb{P}^1 \xrightarrow{\iota} \mathcal{N}_a \xrightarrow{pr_1} \mathcal{M}(2,0)$, then in $H_2(\mathcal{N}_a, \mathbb{Z})$, $\iota_*[\mathbb{P}^1] = 2\overline{\gamma}$. We now prove

Proposition 4.2. *Assume $a \neq \pm I$, and $g > 3$. Then the map φ induces an isomorphism $\varphi_* : \pi_2(\mathcal{N}_a) \to \pi_2(\mathcal{M}_a)$ given by $\overline{\beta} \to \beta$, $\overline{\gamma} \to \gamma$.*

Proof. By Theorem 2.18 and Proposition 2.16, φ clearly gives rise to an isomorphism $\varphi_* : \pi_2(\mathcal{N}_a) \simeq H_2(\mathcal{N}_a, \mathbb{Z}) \to \pi_2(\mathcal{M}_a) \simeq H_2(\mathcal{M}_a, \mathbb{Z})$. In addition, we have the homotopy equivalences $\mathcal{N}_a \sim B(\overline{\mathcal{G}}_p^{\mathbb{C}}/\mathbb{Z}_2)$ and $\mathcal{M}_a \sim B(\mathcal{K}_p/\mathbb{Z}_2)$ (see Propositions 2.10 and 2.17). Therefore, tensoring with the rationals, there is an isomorphism $\varphi_* : H_2(B\overline{\mathcal{G}}_p^{\mathbb{C}}, \mathbb{Q}) \to H_2(B\mathcal{K}_p, \mathbb{Q})$. According to [A-B], Section 2 and [Ni], Proposition 3.2, the $\mathbb{P}^1$ fibrations $B\overline{\mathcal{G}}_p^{\mathbb{C}}$ and $B\mathcal{K}_p$ over $B\overline{\mathcal{G}}^{\mathbb{C}}$ and $B\mathcal{K}$, respectively, are trivial in rational homology. Moreover, it easy to see that $H_2(B\overline{\mathcal{G}}^{\mathbb{C}}, \mathbb{Q}) \simeq H_2(B\overline{\mathcal{G}}_0^{\mathbb{C}}, \mathbb{Q})$, and similarly for $B\mathcal{K}$ and $B\mathcal{K}_0$. Hence, φ_* defines a map

$$\varphi_* : H_2(B\overline{\mathcal{G}}_0^{\mathbb{C}}, \mathbb{Q}) \oplus H_2(\mathbf{G}/\mathbf{T}_a, \mathbb{Q}) \longrightarrow H_2(B\mathcal{K}_0, \mathbb{Q}) \oplus H_2(\mathbf{G}/\mathbf{T}_a, \mathbb{Q}) .$$

By the definition of φ, this map preserves the summands. Now from [A-B], Section 2, $\overline{\beta}$ is half the generator of $\pi_2(B\overline{\mathcal{G}}_0^{\mathbb{C}})$, and by the Mehta–Seshadri theorem (cf. [Si1] and [Si2]), $B\mathcal{K}_0$ is homotopy equivalent to $B\mathcal{G}_{0,\delta} \simeq \mathcal{F}_{\alpha,\delta}$. The assertion that $\overline{\beta} \to \beta$ follows from Theorem 3.8, our definition of β, and the fact that φ_* is an isomorphism on integral homology. The Proposition then follows from the commutativity of the diagram

$$\begin{array}{ccccccccc}
0 & \to & \pi_2(\mathcal{F}_a) & \to & \pi_2(\mathcal{F}_a/\mathbf{T}_a) & \to & \pi_1(\mathbf{T}_a/\mathbb{Z}_2) & \to & 0 \\
 & & \| & & \| & & & & \\
0 & \leftarrow & \pi_2(B\mathcal{K}_o) & \leftarrow & \pi_2(B\mathcal{K}_p) & \leftarrow & \pi_2(\mathbf{G}/\mathbf{T}_a) & \leftarrow & 0 .
\end{array}$$

□

We return now to the general case $\mathbf{G} = \mathrm{SU}(n)$, $g \geq 3$. We shall analyze the topology of $\mathcal{M}_a$ via the techniques developed in [D-U]. We first recall from Section 2.2 that $\mathcal{M}_a$ can be identified with the fiber of the determinant map det $: \mathcal{B}_{p.s.}/\mathcal{K}_p \to J_l(\overline{\Sigma})$, where $\mathcal{B}_{p.s.}$ denotes the subspace of parabolic stable holomorphic structures on $\overline{E}$ and $\mathcal{K}_p$ is the group of parabolic automorphisms of $\overline{E}$. The next proposition parallels [D-U], Theorem 2.4.

Proposition 4.3. *The inclusion $\mathcal{B}_{p.s.}$ in $\mathcal{B}$ is an homotopy equivalence up to dimension $2(n-1)(g-2)+1$.*

Proof. The proof is similar to the one in [D-U] with the only difference being the existence of a parabolic structure. From Theorem 2.4 in [D-U], the inclusion of $\mathcal{B}_{p.s.}$ is $\mathcal{B}$ is an homotopy equivalence up to an integer d where

$$d + 2\{(1-g)m(n-m) + ml - nc\} < 0 .$$

Here l is the degree of $\overline{E}$ and c is the degree of a destabilizing subbundle of rank m. The above inequality has to be satisfied for every possible such subbundle of $\overline{E}$, i.e. for every integer m between 1 and $n-1$ and every integer c such that $c+\sum_{i=1}^{m}\alpha_i' \geq 0$, where $\{\alpha_i'\}$ is a subcollection of the weights $\{\alpha_i\}$. In other words, the inclusion of $\mathcal{B}_{p.s.}$ in $\mathcal{B}$ is an homotopy equivalence if

$$d < 2\left\{(g-1)m(n-m)+mn\left(\frac{c}{m}-\frac{l}{n}\right)\right\}$$

for all m, c as above. But since $\operatorname{pardeg} E = l+\sum_{i=1}^{n}\alpha_i = 0$, we obtain $-l/n = -\sum_{i=1}^{n}\alpha_i/n$. On the other hand, $\operatorname{pardeg} F = c+\sum_{i=1}^{m}\alpha_i' \geq 0$, hence $c/m \geq -\sum\alpha_i'/m$. Therefore,

$$\begin{aligned}
&\left\{(g-1)m(n-m)+mn\left(\frac{c}{m}-\frac{l}{n}\right)\right\} \\
&\qquad \geq \left\{(g-1)m(n-m)+mn\left(-\frac{1}{m}\sum_{i=1}^{m}\alpha_i'+\frac{1}{n}\sum_{i=1}^{n}\alpha_i\right)\right\} \\
&\qquad = \left\{(g-1)m(n-m)+mn\left(\frac{1}{m}\sum_{i=1}^{n-m}\alpha_i''-\frac{n-m}{mn}\sum_{i=1}^{m}\alpha_i'\right)\right\} \\
&\qquad > \{(g-1)m(n-m)-m(n-m)\} = (g-2)m(n-m)\,.
\end{aligned}$$

Since both quantities are integral,

$$(g-1)m(n-m)+mn\left(\frac{c}{m}-\frac{l}{m}\right) \geq (g-2)m(n-m)+1\,.$$

The right hand side is minimized at $m=1$, and the proposition follows. □

Therefore, for $g \geq 3, n \geq 3$, the inclusion $\mathcal{B}_{p.s.} \hookrightarrow \mathcal{B}$ is an homotopy equivalence, at least up to dimension 5. This is sufficient to compute the low dimensional homotopy and cohomology groups of the moduli space $\mathcal{M}_a$ that we are interested in.

From the Proposition 4.3, we obtain, as in the ordinary case [D-U],

$$\pi_i\left(\mathcal{B}_{p.s.}/\mathcal{K}_p\right) \simeq \pi_i\left(B(\mathcal{K}_p/\mathbb{C}^*)\right)\ ;\ i=0,1,2.$$

We are going first to compute $\pi_i(B\mathcal{K})$; $i=0,1,2$. Let $\mathcal{K}$ denote as before the full complex gauge group of $\overline{E}$. Then the relation between $B\mathcal{K}_p$ and $B\mathcal{K}$ is given in the following fibration (cf. [Ni], eq. (3.2))

$$\mathbf{G}/\mathbf{P}_a \longrightarrow B\mathcal{K}_p \longrightarrow B\mathcal{K} \tag{4.4}$$

(observe that if $\mathbf{N}_a$ denotes the normalizer of a in $\mathrm{U}(n)$, and $\mathbf{P}_a$ the normalizer in $\mathbf{G}=\mathrm{SU}(n)$, then $\mathrm{U}(n)/\mathbf{N}_a \simeq \mathbf{G}/\mathbf{P}_a$). Since $\pi_1(\mathbf{G})=\pi_2(\mathbf{G})=0$, we obtain

$$\pi_1(\mathbf{G}/\mathbf{P}_a) \simeq \pi_0(\mathbf{G}/\mathbf{P}_a) \simeq 0\,, \quad \pi_2(\mathbf{G}/\mathbf{P}_a) \simeq \pi_1(\mathbf{P}_a)\,.$$

Thus, via the long exact sequence in homotopy associated to (4.4) we obtain for $i = 0, 1, \pi_i(B\mathcal{K}_p) \simeq \pi_i(B\mathcal{K})$, whereas for $i = 2$ we have the short exact sequence

$$0 \longrightarrow \pi_1(\mathbf{P}_a) \longrightarrow \pi_2(B\mathcal{K}_p) \longrightarrow \pi_2(B\mathcal{K}) \longrightarrow 0 \tag{4.5}$$

Finally, in order to compute $\pi_i(B\overline{\mathcal{K}}_p)$; $i = 0, 1, 2$, we consider the fibrations

$$B\mathbb{C}^* \longrightarrow B\mathcal{K}_p \longrightarrow B\left(\mathcal{K}_p/\mathbb{C}^*\right) \tag{4.6}$$

induced from $B\mathbb{C}^* \to B\mathcal{K} \to B\left(\mathcal{K}/\mathbb{C}^*\right)$. Via the long exact sequence in homotopy associated to (4.6) and using (4.4) we obtain the short exact sequence

$$0 \longrightarrow \pi_1(\mathbf{P}_a) \longrightarrow \pi_2(B\mathcal{K}_p) \longrightarrow \pi_2\left(B(\mathcal{K}_p/\mathbb{C}^*)\right) \longrightarrow 0\,.$$

In particular, since the rank of $\pi_2\left(B(\mathcal{K}/\mathbb{C}^*)\right)$ is one (cf. [A-B], Section 2), we obtain $\operatorname{rank}\pi_2\left(B(\mathcal{K}_p/\mathbb{C}^*)\right) = \operatorname{rank}\pi_1(\mathbf{P}_a) + 1$. Summarizing, we have shown that

$$\begin{aligned} \pi_0\left(\mathcal{B}_{p.s.}/\mathcal{K}_p\right) &= 0\,, \\ \pi_1\left(\mathcal{B}_{p.s.}/\mathcal{K}_p\right) &\simeq H_1\left(\overline{\Sigma}, \mathbb{Z}\right)\,, \\ \operatorname{rank}\pi_2\left(\mathcal{B}_{p.s.}/\mathcal{K}_p\right) &= \operatorname{rank}\pi_1(\mathbf{P}_a) + 1\,. \end{aligned}$$

In order to compute $\pi_i(\mathcal{M}_a)$ for $i = 0, 1, 2$, we consider the long exact sequence in homotopy associated to the fibration det. Since $J_l(\overline{\Sigma})$ is a torus and det induces an isomorphism $\det_* : \pi_1\left(\mathcal{B}_{p.s.}/\mathcal{K}_p\right) \simeq \pi_1(B\mathcal{K}) \to \pi_1(J_l\left(\overline{\Sigma}\right))$ (cf. [A-B], Section 9), we obtain $\pi_0(\mathcal{M}_a) = \pi_1(\mathcal{M}_a) = 0$, and $\operatorname{rank}\pi_2(\mathcal{M}_a) = \operatorname{rank}\pi_1(\mathbf{P}_a) + 1$. Combining this with Theorem 4.1, we can now state

Theorem 4.7. *Assume that either* $\mathbf{G} = \mathrm{SU}(2)$, $g \geq 2$, *or* $\mathbf{G} = \mathrm{SU}(n)$, $n \geq 3$, $g \geq 3$. *Then the moduli space* $\mathcal{M}_a$ *is connected, simply connected, and* $\operatorname{rank}\pi_2(\mathcal{M}_a) = r+1$, *where* $r = \operatorname{rank}\pi_1(\mathbf{P}_a)$.

Corollary 4.8. *The second cohomology of* $\mathcal{M}_a$ *is free of rank* $r + 1$.

Proof. Since $\mathcal{M}_a$ is simply connected, $H^2(\mathcal{M}_a, \mathbb{Z})$ is torsion free. The rest follows from Theorem 4.7 and the Hurewicz isomorphism. □

As in the case of SU(2), we would like to understand explicitly the generators of $H^2(\mathcal{M}_a, \mathbb{Z})$; hence, we need to understand the topology of the space $\mathcal{F}_a$. By the long exact sequence in homotopy associated to (2.6) we obtain

$$0 \longrightarrow \pi_2(\mathcal{F}_a) \xrightarrow{\pi_*} \pi_2(\mathcal{M}_a) \xrightarrow{\partial} \pi_1(\mathbf{P}_a/\mathbb{Z}_n) \longrightarrow \pi_1(\mathcal{F}_a) \longrightarrow \pi_1(\mathcal{M}_a) = 0\,. \tag{4.9}$$

Since $\operatorname{rank}\pi_2(\mathcal{M}_a) = \operatorname{rank}(\mathbf{P}_a) + 1$ and $\operatorname{rank}\pi_1(\mathbf{P}_a/\mathbb{Z}_n) = \operatorname{rank}\pi_1(\mathbf{P}_a)$, we clearly have $\operatorname{rank}\pi_2(\mathcal{F}_a) = 1$ and $\operatorname{rank}\pi_1(\mathcal{F}_a) = 0$. Let β be the image of a generator of $\operatorname{Free}\pi_2(\mathcal{F}_a)$ in $\pi_2(\mathcal{M}_a) \simeq H_2(\mathcal{M}_a, \mathbb{Z})$ under the map π_*. Let $\gamma_1, \ldots, \gamma_r$ be the elements of $\operatorname{Free}\pi_2(\mathcal{M}_a) \subset \pi_2(\mathcal{M}_a) \simeq H_2(\mathcal{M}_a, \mathbb{Z})$ such that their image under ∂ form a set of generators for $\operatorname{Free}\pi_1(\mathbf{P}_a)$. Then we have shown

Corollary 4.10. *Assume* $\mathbf{G} = \mathrm{SU}(n)$, $n \geq 3$, $g \geq 3$. *The free part of the group* $H_2(\mathcal{M}_a, \mathbb{Z})$ *is generated by the elements* $\beta, \gamma_1, \ldots, \gamma_r$ *as above, where* $r =$ rank $\pi_1(\mathbf{P}_a)$.

Remark 4.11. Analogous results to 4.7, 4.8, and 4.10 hold also for the spaces $\mathcal{F}_a/\mathbf{T}_a$, where $\mathbf{T}_a$ is a maximal torus in $\mathbf{P}_a$.

4.2. Integrality of the symplectic form

In this subsection, we shall consider the class of a certain closed two-form Ω on the spaces $\mathcal{F}_{\alpha,\delta}$, $\mathcal{M}_{\alpha,\delta}$. This is in preparation for the next section, where we shall construct a line bundle over $\mathcal{F}_{\alpha,\delta}$ with a connection whose curvature is $-i\Omega$. Whether the connection pushes down to $\mathcal{M}_{\alpha,\delta}$ amounts to the question of integrality of the class Ω on $\mathcal{M}_{\alpha,\delta}$.

On the space $\mathcal{A}_{\mathrm{F},\delta}$, defined as in Section 3.1 with respect to some base connection ∇_0 which at infinity is in temporal gauge, we define the two-form

$$\Omega_\nabla(\beta_1, \beta_2) = \frac{1}{2\pi} \int_\Sigma \mathrm{Tr}\,(\beta_1 \wedge \beta_2) \ .$$

This is well-defined, since by the inclusion $L^2_{1,\delta} \hookrightarrow L^2$ the integral is convergent. Ω is a closed two-form, and integration by parts shows that it is degenerate in the $\mathcal{G}_{0,\delta}$ directions. Hence, Ω descends to a closed form on $\mathcal{F}_{\alpha,\delta}$, and the normalization has been chosen such that $\Omega/2\pi$ is a representative of an integral class in $H^2(\mathcal{F}_{\alpha,\delta}, \mathbb{Z})$.

Recall that for $a \in \mathbf{G}$, $a = \exp(2\pi i\alpha)$, we denote by $\mathbf{P}_a$ the normalizer of a and by $\mathbf{T}_a$ some choice of maximal torus in $\mathbf{P}_a$. Suppose β is an arbitrary tangent vector to $\mathcal{F}_{\alpha,\delta}$ and $\phi_1, \phi_2 \in \mathrm{Lie}\,\mathbf{T}_a$. Then letting $\tilde{\phi}_1, \tilde{\phi}_2$ denote the unique harmonic extensions satisfying $r(\tilde{\phi}_i) = \phi_i$, $i = 1, 2$, we have

$$\begin{aligned}
\Omega_\nabla(\beta + \nabla\tilde{\phi}_1, \nabla\tilde{\phi}_2) &= \frac{1}{2\pi} \int_\Sigma \mathrm{Tr}\left((\beta + \nabla\tilde{\phi}_1) \wedge \nabla\tilde{\phi}_2\right) \\
&= -\frac{1}{2\pi} \int_\Sigma \mathrm{Tr}\left(\nabla(\beta + \nabla\tilde{\phi}_1) \wedge \tilde{\phi}_2\right) + \frac{1}{2\pi} \int_{\partial\Sigma} \mathrm{Tr}\left(\nabla\tilde{\phi}_1 \cdot \tilde{\phi}_2\right) \\
&= \frac{1}{2\pi} \int_{\partial\Sigma} \mathrm{Tr}\,([A_0, \phi_1]\phi_2) = \frac{1}{2\pi} \int_{\partial\Sigma} \mathrm{Tr}\,(A_0[\phi_1, \phi_2]) = 0 \ ,
\end{aligned}$$

since ϕ_1 and ϕ_2 commute. In the above computation, we have written $\nabla_0 = d + A_0$, and by the integral over $\partial\Sigma$ we mean the usual limiting procedure $\tau \to \infty$.

We conclude from the above that Ω descends as a closed two-form on $\mathcal{F}_{\alpha,\delta}/\mathbf{T}_a$. We now have the following

Theorem 4.12. *For* $k \in \mathbb{Z}$, *the form* $k\Omega/2\pi$ *on* $\mathcal{F}_{\alpha,\delta}/\mathbf{T}_a$. *defines an integral class if and only if* $ik\alpha$ *is in the coroot lattice. Thus, a must have the form* $a =$ $\mathrm{diag}(e^{2\pi i\lambda_1/k}, \ldots, e^{2\pi i\lambda_n/k})$, *where* $\lambda_1 + \cdots + \lambda_n = 0$ *and all the* λ_i*'s are integers.*

We should point out that this is the form of the holonomy matrix which is important in the geometric quantization of Chern–Simons theory (cf. [A]). We shall have more to say about the consequences of this in Section 6 and in [D-W1].

Proof. By the simple connectivity of $\mathcal{F}_{\alpha,\delta}/\mathbf{T}_a$ and the Hurewicz isomorphism, it suffices to evaluate $k\Omega$ on the generators $\beta, \gamma_1, \ldots, \gamma_r$ from Proposition 4.10 (cf. also Remark 4.11 and note that in this case, $r = n-1$). On β, $k\Omega/2\pi$ is integral by the choice of normalization. For the γ_i's, we use the following explicit realization: fix a winding number $\phi = \operatorname{diag}(\phi_1, \ldots, \phi_n)$, where ϕ_i's are all integers, and $i\phi$ is in the integer lattice of Lie $\mathbf{T}_a$. Choose m, p integers, $1 \leq p \leq n$, and let ϵ_p denote the $n \times n$ matrix with a single entry 1 in the p-th row, p-th column. Set

$$f(t) = \exp\left\{2\pi i t\phi + 2\pi i t m\left(\frac{1}{n}\cdot I - \epsilon_p\right)\right\}.$$

This defines a map $[0,1] \to \mathbf{T}_a$ which descends to a closed loop on $\mathbf{T}_a/\mathbb{Z}_n$, and every such loop can be obtained from loops of this form. Fix a point $[\nabla] \in \mathcal{F}_{\alpha,\delta}$, with representative $\nabla_0 + A$. We write $\nabla_0 = d + A_0$ and take $A \in L^2_{1,\delta}(\nabla_0)$. Let $A_s = A_0 + A$. Then $A_s^f = f^{-1}A_s f$ defines a closed loop in $\mathcal{F}_{\alpha,\delta}$ which projects to a the point $[\nabla] \in \mathcal{M}_a$. By definition of the boundary homomorphism in (4.9) we may find a disk D_{A_s} with $\partial D_{A_s} = \operatorname{im}\left(A_s^f\right)$. The projection of D_{A_s} in $\mathcal{M}_a$ is then a sphere and is the realization of the element of $\pi_2(\mathcal{M}_a)$ corresponding to ϕ. Let $\tilde{d}$ denote the trivial connection on the unit disk $D \subset \mathbb{C}$ and A the map $D \to D_{A_s}$. Then

$$\begin{aligned}
\frac{k}{2\pi}\int_{D_{A_s}} \Omega &= \frac{k}{4\pi^2}\int_D\int_\Sigma \operatorname{Tr}\left(\tilde{d}A \wedge \tilde{d}A\right) = \frac{k}{8\pi^2}\int_\Sigma\int_{\partial D} \operatorname{Tr}\left(A\tilde{d}A\right) \\
&= \frac{k}{8\pi^2}\int_\Sigma\int_0^1 dt\, \operatorname{Tr}\left(f^{-1}A_s f[f^{-1}A_s f, f^{-1}f']\right) \\
&= \frac{k}{8\pi^2}\int_\Sigma\int_0^1 dt\, \operatorname{Tr}\left(f' f^{-1}[A_s, A_s]\right)
\end{aligned}$$

Since A_s is flat, $[A_s, A_s] = -2d(A + A_0)$. Integrating by parts, and using the fact that $A_0 = i\alpha d\theta = i\operatorname{diag}(\alpha_1, \ldots, \alpha_n)d\theta$ at infinity,

$$\frac{k}{2\pi}\int_{D_{A_s}} \Omega = \frac{-k}{4\pi^2}\int_0^1 dt \int_{\partial\Sigma} \operatorname{Tr}\left(f' f^{-1} A_0\right) = k\left\{\sum_{i=1}^n \alpha_i\phi_i + \frac{m}{n}\sum_{i=1}^n \alpha_i - m\alpha_p\right\}$$

The middle term vanishes, since $i\alpha \in \mathfrak{g}$. By choosing different ϕ_i's, m, and p, we see that the expression above is an integer for all such choices only when $k\alpha_p \in \mathbb{Z}$ for $p = 1, \ldots, n$. This completes the proof of the theorem. □

5. Construction of the line bundle

We now proceed to define a line bundle with a connection 1-form over the spaces $\mathcal{F}_{\alpha,\delta}$ via a cocycle defined on the space of connections. The topological restrictions of Section 4.2 will prevent this line bundle from descending to $\mathcal{F}_{\alpha,\delta}/\mathbf{T}_a$, $\mathbf{T}_a \subset \mathbf{P}_a$, except in certain cases. We make this precise in Theorem 5.8, which is the main result of this section.

5.1. The cocycle and connection

Fix a base connection $\nabla_0 = d + A_0$ where A_0 is in temporal gauge with associated holonomy matrix $i\alpha$. Let $\mathcal{G}_{0,\delta}$ be the gauge group defined as in Section 3.1. We define $\tilde{\mathcal{G}}_{0,\delta}$ to be the space of smooth paths γ in $\mathcal{G}_{0,\delta}$ which are the identity at an endpoint; hence we take $\gamma : [0, 1] \to \mathcal{G}_{0,\delta}$ with $\gamma(0) = I$. The evaluation map $e_1 : \tilde{\mathcal{G}}_{0,\delta} \to \mathcal{G}_{0,\delta}$ which takes γ to $\gamma(1)$ is smooth, and by Proposition 3.3 it is surjective. We define a map

$$C_{\nabla_0} : L^2_{1,\delta}(\mathfrak{g}_E) \times \tilde{\mathcal{G}}_{0,\delta} \longrightarrow \mathrm{U}(1) \tag{5.1}$$

$$C_{\nabla_0}(A, \tilde{g}) = \exp\left\{\frac{i}{4\pi}\int_\Sigma \mathrm{Tr}\left((A + A_0)dgg^{-1}\right) - \frac{i}{12\pi}\int_{\Sigma\times[0,1]} \mathrm{Tr}\left(\tilde{d}\tilde{g}\tilde{g}^{-1}\right)^3\right\}.$$

where $g = e_1(\tilde{g})$ and $\tilde{d} = d + d/dt$. The terms in the exponential are well-defined: the first integral is convergent because A, dgg^{-1} are in L^1 and L^2, and A_0 is in L^∞; the second integral is convergent, since for each $t \in [0, 1]$, $d\tilde{g}\tilde{g}^{-1}$ is in L^2, and $(d\tilde{g}/dt)\tilde{g}^{-1}$ is in L^∞.

The sum in the exponential of (5.1) is formally the integral of the Chern–Simons form over $\Sigma \times [0, 1]$. As was shown in [R-S-W] for the case of SU(2) (see also [Do-Sa] and [Mick]), this expression gives rise to a cocycle in the case of smooth connections on the compact surface $\overline{\Sigma}$ defining a line bundle with a natural connection which pushes down to a holomorphic line bundle over the moduli space. The point of our construction, as will be shown, is that the same technique may be used for the moduli space of vector bundles with parabolic structure. We have the following

Lemma 5.2. *The $U(1)$-valued function defined in* (5.1) *depends only on $e_1(\tilde{g})$ and therefore descends to a smooth map $\mathcal{A}_\delta \times \mathcal{G}_{0,\delta} \to U(1)$.*

Proof. Fix $\tilde{g}$, $e_1(\tilde{g}) = g$. Any other extension can be gotten from $\tilde{g}$ by multiplying by $h \in \tilde{\mathcal{G}}_{0,\delta}$ satisfying $e_1(h) = I$. Then

$$C_{\nabla_0}(A, \tilde{g}h) = C_{\nabla_0}(A, \tilde{g})\exp\left\{\frac{i}{12}\pi\int_{\Sigma\times[0,1]} \mathrm{Tr}\left(\tilde{d}(\tilde{g}h)h^{-1}\tilde{g}^{-1}\right)^3 - \frac{i}{12\pi}\int_{\Sigma\times[0,1]} \mathrm{Tr}\left(\tilde{d}\tilde{g}\tilde{g}^{-1}\right)^3\right\} =$$

$$= C_{\nabla_0}(A,\tilde{g})\exp\left\{\frac{i}{12\pi}\int_{\Sigma\times[0,1]}\operatorname{Tr}\left(\tilde{d}hh^{-1}\right)^3 + \frac{i}{4\pi}\int_{\Sigma\times[0,1]}\tilde{d}\operatorname{Tr}\left(\tilde{d}\tilde{g}\tilde{g}^{-1}\tilde{d}hh^{-1}\right)\right\}$$

The second term in the exponential clearly vanishes upon integration by parts. By the embedding theorems, we know that h extends as a continuous map $\overline{\Sigma}\times S^1 \to \mathbf{G}$. The first integral is $2\pi i\times$ the integral of the generator of $H^3(\mathbf{G},\mathbb{Z})$ over the image $h(\overline{\Sigma}\times S^1)$, and is therefore an integer (cf. [Mick]). The exponential is thus always one, and the lemma follows. □

By a straightforward computation along the lines indicated in the above lemma, we also have

Lemma 5.3. *The map $C_{\nabla_0}(A,g)$ satisfies the cocycle condition, i.e. for any $g,h\in\mathcal{G}_{0,\delta}$, $C_{\nabla_0}(A,gh) = C_{\nabla_0}(A^g,h)C_{\nabla_0}(A,g)$.*

Hence, we may use C_{∇_0} to define a twisted principal U(1) bundle $\mathcal{U} = \mathcal{A}_{\mathrm{F},\delta}\times_{\mathcal{G}_{0,\delta}} \mathrm{U}(1)$ over $\mathcal{F}_{\alpha,\delta} = \mathcal{A}_{\mathrm{F},\delta}/\mathcal{G}_{0,\delta}$, where the map $\mathcal{U}\to\mathcal{F}_{\alpha,\delta}$ is just projection onto the first factor. We define the line bundle $\tilde{L}_\alpha\to\mathcal{F}_{\alpha,\delta}$ to be the one associated to $\mathcal{U}$ via the standard representation.

We now define a connection on $\tilde{L}_\alpha$. Let ω denote the one-form on $\mathcal{A}_\delta$ given by the expression $\omega_{\nabla_0+A}(\beta) = \frac{1}{4\pi}\int_\Sigma \operatorname{Tr}((A+A_0)\wedge\beta)$. The integral makes sense, as before, by the inclusions $L^2_{1,\delta}\hookrightarrow L^1$, $L^2_{1,\delta}\hookrightarrow L^2$ and the fact that $A_0\in L^\infty$. By adding the Maurer-Cartan form, ω trivially induces a connection $\hat{\omega}$ on $\mathcal{A}_{\mathrm{F},\delta}\times\mathrm{U}(1)$.

Lemma 5.4. *$\hat{\omega}$ descends to a connection on $\mathcal{U}$.*

Proof. We must show that $\hat{\omega}$ is invariant under the twisting by $\mathcal{G}_{0,\delta}$. It suffices to show that for vectors V vertical to the $\mathcal{G}_{0,\delta}$ action, $\hat{\omega}(V) = L_V\hat{\omega} = 0$. The vertical tangent space is generated by vectors of the form $(\nabla\phi, dC_{\nabla_0}(\nabla\phi,\phi))$, $\phi\in L^2_{1,\delta}$. A simple computation shows that $\hat{\omega}$ on the vector V associated to ϕ is

$$\begin{aligned}\hat{\omega}(V) &= \omega(V) + d\log C_{\nabla_0}(\nabla\phi,\phi)\\ &= \frac{i}{4\pi}\int_\Sigma \operatorname{Tr}((A+A_0)\wedge d_{A+A_0}\phi) + \frac{i}{4\pi}\int_\Sigma \operatorname{Tr}((A+A_0)\wedge d\phi)\\ &= \frac{i}{2\pi}\int_\Sigma \operatorname{Tr}((A+A_0)\wedge d\phi) + \frac{i}{4\pi}\int_\Sigma \operatorname{Tr}(\phi[A+A_0,A+A_0])\end{aligned}$$

Since $0 = d(A+A_0) + \frac{1}{2}[A+A_0,A+A_0]$ and ϕ is exponentially decaying, the above is

$$= \frac{i}{2\pi}\int_\Sigma \operatorname{Tr}(A+A_0\wedge d\phi) - \frac{i}{2\pi}\int_\Sigma \operatorname{Tr}(\phi d(A+A_0)) = 0\,.$$

upon integration by parts. From this, it follows that $L_V\hat{\omega} = i_V d\hat{\omega} = i_V\Omega = 0$ by that result of Section 4.2. This completes the proof of the lemma. □

We summarize the results of Lemmas 5.2 to 5.4 with the following

Proposition 5.5. *The trivial line bundle $\mathcal{A}_{F,\delta} \times \mathbb{C}$ with connection $\hat{\omega}$ pushes down to an hermitian line bundle $\tilde{L}_\alpha \to \mathcal{F}_{\alpha,\delta}$ and connection $\tilde{\nabla}$ with curvature $-i\Omega$.*

Actually, in what follows, we shall always be interested in an arbitrary power of $\tilde{L}_\alpha$, $\tilde{L}_\alpha^{\otimes k}$, where $k \in \mathbb{Z}$.

5.2. Pushing down the line bundle

In this subsection, we wish to go one step further; namely, we would like to obtain a line bundle with connection $L_\alpha^{\otimes k} \to \mathcal{F}_{\alpha,\delta}/\mathbf{T}_a$ by pushing down our construction of the previous section via the action of the torus $\mathbf{T}_a$. Clearly, there is an obstruction to this, for if the bundle $\tilde{L}_\alpha^{\otimes k}$ with connection $\tilde{\nabla}$ pushed down, the curvature $-ik\Omega$ would have to be $2\pi i\times$ an integral class. By Theorem 4.12, we know that this occurs precisely when $ik\alpha$ is in the coroot lattice.

There is a natural action of $\mathbf{T}_a$ on sections of $\tilde{L}_\alpha^{\otimes k}$; a section is to be regarded as a complex valued function ψ on $\mathcal{A}_{F,\delta}$ satisfying the cocycle condition

$$\psi(A^g) = C_{\nabla_0}(A, g)^k \psi(A) \tag{5.6}$$

for $g \in \mathcal{G}_{0,\delta}$. If $t \in \mathbf{T}_a$, define $t \cdot \psi(A) = \psi(A^t)$. It is easily verified that $t \cdot \psi$ still satisfies (5.6), and hence is a section of $\tilde{L}_\alpha^{\otimes k}$.

Now consider a tangent vector $V = \nabla\phi$ at $\nabla \in \mathcal{A}_{F,\delta}$, vertical with respect to the $\mathbf{T}_a$ action. A simple computation shows that the connection 1-form evaluated on this vector is $-ik\omega_\nabla(V) = -k\,\mathrm{Tr}(\phi\alpha)$. Under the assumption that $ik\alpha$ is a coroot, this exponentiates to a representation $\lambda : \mathbf{T}_a \to \mathrm{U}(1)$. Note that λ is invariant under the center $\mathbb{Z}_n$ of $\mathbf{G}$ (see Theorem 6.1), and so it descends to a representation of $\mathbf{T}_a/\mathbb{Z}_n$. The sections of $\tilde{L}_\alpha^{\otimes k}$ which are λ-covariant with respect to the $\mathbf{T}_a/\mathbb{Z}_n$ action, i.e. those which satisfy $t \cdot \psi = \lambda(t)\psi$, then automatically satisfy $\tilde{\nabla}_V\psi = 0$ for any vertical vector V. Conversely, any section satisfying this condition is λ-covariant.

We therefore define a line bundle $L_\alpha^{\otimes k} \to \mathcal{F}_{\alpha,\delta}/\mathbf{T}_a$ by declaring its sheaf of sections to be

$$\Gamma\left(\mathcal{F}_{\alpha,\delta}/\mathbf{T}_a, L_\alpha^{\otimes k}\right) = \Gamma\left(\mathcal{F}_{\alpha,\delta}, \tilde{L}_\alpha^{\otimes k}\right)^\lambda, \tag{5.7}$$

where the right hand side denotes the λ-covariant smooth sections of $\tilde{L}_\alpha^{\otimes k}$ (cf. [Ax]). The connection ∇ on $L_\alpha^{\otimes k}$ may be defined from the connection $\tilde{\nabla}$ on $\tilde{L}_\alpha^{\otimes k}$ by the equation $\nabla_X\psi = \tilde{\nabla}_{\tilde{X}}\psi$, where $\tilde{X}$ is any vector which projects to X via the map $\mathcal{F}_{\alpha,\delta} \to \mathcal{F}_{\alpha,\delta}/\mathbf{T}_a$. The λ-covariance guarantees that this is well-defined, and it is easy to see that ∇ is a connection on $L_\alpha^{\otimes k}$ with curvature $-ik\Omega$. To summarize, we have the

Theorem 5.8. *Suppose $ik\alpha$ is in the coroot lattice of* $\mathbf{G}$ *and* λ *is the associated character* $\mathbf{T}_a \to U(1)$. *There exists an hermitian line bundle* $L_\alpha^{\otimes k} \to \mathcal{F}_{\alpha,\delta}/\mathbf{T}_a$ *with connection* ∇ *and curvature* $-ik\Omega$ *whose global smooth sections are given by* (5.7), *where the superscript* λ *denotes the subspace of sections satisfying* $t \cdot \psi = \lambda(t)\psi$ *for all* $t \in \mathbf{T}_a$.

6. A problem in geometric quantization

In this final section, we would like to address the issue of holomorphic structure on the line bundles we have constructed. We shall show that in the cases where the line bundles descend to the space $\mathcal{F}_{\alpha,\delta}/\mathbf{T}_a$, the space of holomorphic sections form the multiplicity space of a problem in geometric quantization, at least in the case $\mathbf{G} = \mathrm{SU}(2)$. The multiplicities are associated to some representation λ of the group $\mathbf{G}$. The holonomy matrices $i\alpha \in \mathfrak{g}$ correspond to the λ's in a one to one way. We begin by making this relationship between holonomies and representations explicit.

Theorem 6.1. *Fix $k \in \mathbb{Z}$ and a maximal torus* $\mathbf{T} \subset \mathbf{G}$. *Then there is a one-to-one correspondence between characters* $\lambda : \mathbf{T} \to U(1)$ *which are invariant under the center and elements* $i\alpha \in \mathrm{Lie}\,\mathbf{T}$ *such that* $k\Omega/2\pi$ *defines an integral class on* $\mathcal{F}_{\alpha,\delta}/\mathbf{T}$.

Proof. By Theorem 4.12, $k\Omega/2\pi$ is an integral class on $\mathcal{F}_{\alpha,\delta}/\mathbf{T}$ if and only if $k\alpha = \lambda = \mathrm{diag}(\lambda_1, \dots, \lambda_n)$, where the λ_i's are integers. The coroot $i\lambda$ induces a character of $\mathbf{T}$ as follows: an arbitrary element $t \in \mathbf{T}$ may be written $t = \mathrm{diag}(e^{i\theta_1}, \dots, e^{i\theta_n})$, $\sum \theta_j = 0$. Hence we define $\lambda(t) = \exp\left(i \sum \theta_j \lambda_j\right)$. The θ_j's are determined only modulo 2π, but that does not change the definition of $\lambda(t)$, since the λ_j's are integers. It is also easy to verify that the condition $\sum \lambda_j = 0$ ensures that for t in the center of $\mathbf{G}$, $\lambda(t) = 1$.

Conversely, suppose λ is a character of $\mathbf{T}$. Write λ as

$$t = \mathrm{diag}(t_1, \dots, t_n) \longrightarrow \lambda(t) = t_1^{\lambda_1} \cdots t_n^{\lambda_n},$$

where the λ_i's are integers. The multi-index $(\lambda_1, \dots, \lambda_n)$ is only determined up to addition of multiples of $(1, \dots, 1)$. If we require λ to be invariant under the center, we have $\sum \lambda_j = 0 \mod n$. Therefore, the index $(\lambda_1, \dots, \lambda_n)$ is completely determined by the requirement $\sum \lambda_j = 0$. Now the matrix $i\alpha = \mathrm{diag}(i\lambda_1/k, \dots, i\lambda_n/k)$ is an element of $\mathrm{Lie}\,\mathbf{T}$, and by Theorem 4.12, $k\Omega/2\pi$ defines an integral class on $\mathcal{F}_{\alpha,\delta}/\mathbf{T}$. This completes the proof of Theorem 6.1. □

Example 6.2. For the sake of clarity, let us dwell for a moment on the case $\mathbf{G} = \mathrm{SU}(2)$. A representation $\lambda : \mathbf{T} \to \mathrm{U}(1)$ is characterized by a half-integer $j \in \frac{1}{2}\mathbb{Z}$, the *spin* of the representation, by writing $\lambda_1 - \lambda_2 = 2j$. Then as above, if $t = \mathrm{diag}(e^{i\theta}, e^{-i\theta})$, $\lambda(t) = \exp(2i(j\theta))$. Requiring λ to be invariant under the center means that j must in fact be an integer. If we take $\lambda_1 + \lambda_2 = 0$, then $\lambda_1 = j$, $\lambda_2 = -j$, and the corresponding holonomy matrix is $\alpha = \mathrm{diag}(j/k, -j/k)$. We wish to emphasize that $k\alpha$ corresponds to the spin j of the representation; the spin is required to be integral to guarantee invariance under the center.

In the case where $k\Omega/2\pi$ is an integral class in $\mathcal{F}_{\alpha,\delta}/\mathbf{T}_a$, we have constructed in Theorem 5.8 a line bundle $L_\alpha^{\otimes k} \to \mathcal{F}_{\alpha,\delta}/\mathbf{T}_a$ with connection and curvature $-ik\Omega$. By Theorems 3.7 and 3.13, it is easy to see that Ω is of type (1,1). The line bundle $L_\alpha^{\otimes k}$ therefore inherits a holomorphic structure compatible with the connection. We wish to identify the space $H^0\left(\mathcal{F}_{\alpha,\delta}/\mathbf{T}_a, L_\alpha^{\otimes k}\right)$ of global holomorphic sections of $L_\alpha^{\otimes k}$.

At this point, we restrict ourselves to the case $\mathbf{G} = \mathrm{SU}(2)$. The reason for this is that we have need of the correspondences between moduli spaces discussed in Section 2.3, and these were only defined for rank two. Therefore, *for the rest of the paper* we let $\mathbf{G} = \mathrm{SU}(2)$. In addition, so that the maps we use are defined off a set of high codimension, we assume $g > 3$.

Let us recall the construction of Section 2.3. We let $\overline{\mathcal{A}}_s$ denote the smooth stable connections on a trivial $\mathbf{G}$-bundle over $\overline{\Sigma}$ and $\overline{\mathcal{G}}^{\mathbb{C}}$ the group of complex automorphisms. Choose a non-negative integer λ. Regarding λ as the spin of a representation, we have the associated representation space V_λ of dimension $2\lambda + 1$. As in Example 6.2, we may also regard λ as a character $\lambda : \mathbf{T}_a \to \mathrm{U}(1)$. Then we have the natural holomorphic bundles, $\Delta^{\otimes k}$ and B_λ over $\overline{\mathcal{A}}_s \times \mathbf{G}/\mathbf{T}_a$; Δ is the trivial extension of the determinant bundle over $\overline{\mathcal{A}}_s$, and B_λ is the trivial extension of the Borel-Weil-Bott bundle over $\mathbf{G}/\mathbf{T}_a$ determined by the character λ. Using a computation similar to that of Theorem 4.12, we find $c_1(B_\lambda) = 2\lambda$, and V_λ may be identified with $H^0(\mathbf{G}/\mathbf{T}_a, B_\lambda)$ via the Borel-Weil-Bott theorem (cf. [G-S2]). If λ is invariant under the center $\mathbb{Z}_2$ of $\mathbf{G}$, then the center acts trivially on the total space of the line bundle $\Delta^{\otimes k} \otimes B_\lambda \to \overline{\mathcal{A}}_s \times \mathbf{G}/\mathbf{T}_a$, and we may take the quotient by $\overline{\mathcal{G}}^{\mathbb{C}}$ to obtain a holomorphic bundle $S_\lambda \to \mathcal{N}_a$. Let α be the matrix corresponding to λ as in Theorem 6.1. In the case where $a = \exp(2\pi i\alpha) \neq \pm I$, we have by Theorem 2.18 a holomorphic embedding $\varphi : \mathcal{N}_a \to \mathcal{F}_{\alpha,\delta}/\mathbf{T}_a$ with an inverse ψ on the image. This gives rise to

Proposition 6.3. *For $a \neq \pm I$, the bundles $\psi^* S_\lambda$ and $L_\alpha^{\otimes k}$ over the image of φ have the same first Chern class.*

Proof. As in Proposition 4.2, let us denote the generators of $H^2(\mathcal{N}_a, \mathbb{Z}) \simeq \pi_2(\mathcal{N}_a)$ by $\overline{\beta}, \overline{\gamma}$. The natural symplectic form $\overline{\Omega}$ on $\overline{\mathcal{A}}_s$ descends to a representative of the class $\overline{\beta}$. Using, for example, the Quillen connection on $\Delta^{\otimes k}$, one has $c_1(\Delta^{\otimes k}) = ik\overline{\Omega}/2\pi$ (cf. [Q], [Don1]). On the other hand, the Chern class of B_λ is 2λ, and because of the torsion class in $\pi_2(\mathcal{M}(2,0))$, we have $c_1(S_\lambda) = k\overline{\beta} + \lambda\overline{\gamma}$ (see the discussion preceding Proposition 4.2). The result now follows from Proposition 4.2 and (the proof of) Theorem 4.12. □

Proposition 6.4. *Let α correspond to λ as in Theorem* 6.1. *Let $\mathbf{T}_a$ be a maximal torus containing* $\exp(2\pi i\alpha)$ *and a point $a \neq \pm I$. Then there is an isomorphism* $H^0(\mathcal{N}_a, S_\lambda) \simeq H^0(\mathcal{F}_{\alpha,\delta}/\mathbf{T}_a, L_\alpha^{\otimes k})$.

Proof. In the case $\exp(2\pi i\alpha) \neq \pm I$, we may take $a = \exp(2\pi i\alpha)$. Then the manifold $\mathcal{F}_{\alpha,\delta}/\mathbf{T}_a$ is compact and simply connected, so by the exponential sequence, bundles are classified by their first Chern classes. By Theorem 2.18 and Proposition 6.3, $\psi^* S_\lambda$ and $L_\alpha^{\otimes k}$ agree outside a set of codimension ≥ 2, and so their extensions are holomorphically equivalent. Now, since ψ is a biholomorphism outside a set of codimension ≥ 2, the sections of S_λ are in one-to-one correspondence with sections of $\psi^* S_\lambda$. Notice that the above argument shows that line bundles on $\mathcal{N}_a$ are in fact classified by their first Chern classes. In the case $\exp(2\pi i\alpha) = I$, we have from Proposition 2.17 and Theorem 3.8 that $\mathcal{F}_{\alpha,\delta}/\mathbf{T}_a$ is biholomorphic to $\mathcal{N}_a$, hence the Proposition holds in this case again by a Chern class computation. In the case $\exp(2\pi i\alpha) = -I$, we argue as follows: by the theorem of Narasimhan-Seshadri, $\mathcal{M}(2, -1)$ is naturally diffeomorphic to $\mathcal{F}_{-I}/\mathrm{SO}(3)$ (cf. [A-B], Section 6). By Theorem 3.8, $\mathcal{F}_{\alpha,\delta}/\mathbf{T}_a$ is then diffeomorphic to a $\mathbb{P}^1$-bundle over $\mathcal{M}(2, -1)$ and thus inherits a natural complex structure. The construction of Section 2.3 produces a holomorphic map $\varphi : \mathcal{N}_a \to \mathcal{F}_{\alpha,\delta}/\mathbf{T}_a$ as before, and the Chern class computation above proves the Proposition in this case as well. □

We can now easily identify the space of global holomorphic sections of our bundle $L_\alpha^{\otimes k}$ in terms of the sections of $\Delta^{\otimes k}$ and the character λ. Let $i\alpha \in \mathfrak{g}$ be the holonomy matrix corresponding to λ via Theorem 6.1. We shall denote by $H^0(\overline{\mathcal{A}}_s, \Delta^{\otimes k})$ the infinite dimensional space of global holomorphic sections of $\Delta^{\otimes k}$ over $\overline{\mathcal{A}}_s$. Following Guillemin and Sternberg [G-S1], we define the *λ-covariant subspace* of $H^0(\overline{\mathcal{A}}_s, \Delta^{\otimes k})$, $\mathcal{V}_\lambda = \mathrm{Hom}_{\overline{\mathcal{G}}^{\mathbb{C}}}(V_\lambda^*, H^0(\overline{\mathcal{A}}_s, \Delta^{\otimes k}))$ as follows: first, notice that the gauge group $\overline{\mathcal{G}}^{\mathbb{C}}$ acts on V_λ by evaluating the gauge transformation at the puncture p, and on $H^0(\overline{\mathcal{A}}_s, \Delta^{\otimes k})$ by its standard action (cf. [Q], [Don1]). Then $\mathcal{V}_\lambda$ consists of homomorphisms that intertwine the $\overline{\mathcal{G}}^{\mathbb{C}}$ action. Thus, $\mathcal{V}_\lambda$ is equivalent to the $\overline{\mathcal{G}}^{\mathbb{C}}$ invariant elements of $H^0(\overline{\mathcal{A}}_s, \Delta^{\otimes k}) \otimes V_\lambda$, and so

$$\mathcal{V}_\lambda = \left\{H^0\left(\overline{\mathcal{A}}_s, \Delta^{\otimes k}\right) \otimes V_\lambda\right\}^{\overline{\mathcal{G}}^{\mathbb{C}}} = \left\{H^0\left(\overline{\mathcal{A}}_s, \Delta^{\otimes k}\right) \otimes H^0(\mathbf{G}/\mathbf{T}_a, B_\lambda)\right\}^{\overline{\mathcal{G}}^{\mathbb{C}}}$$
$$= H^0\left(\overline{\mathcal{A}}_s \times \mathbf{G}/\mathbf{T}_a, \Delta^{\otimes k} \otimes B_\lambda\right)^{\overline{\mathcal{G}}^{\mathbb{C}}} = H^0(\mathcal{N}_a, S_\lambda) \simeq H^0\left(\mathcal{F}_{\alpha,\delta}/\mathbf{T}_a, L_\alpha^{\otimes k}\right),$$

by Proposition 6.4. We can now state our main result as follows:

Theorem 6.6. *For* $\mathbf{G} = SU(2)$ *and* $g > 3$, *let* $L_\alpha^{\otimes k}$ *be the holomorphic line bundle constructed in Section* 5, λ *the character corresponding to* α *via Theorem* 6.1. *Then* $\mathcal{V}_\lambda = H^0\left(\mathcal{F}_{\alpha,\delta}/\mathbf{T}_a, L_\alpha^{\otimes k}\right)$.

We conclude with a few remarks. First, our λ-covariant subspace $\mathcal{V}_\lambda$ is essentially Segal's definition of the space of states (or conformal blocks) associated to a Riemann surface with boundary labeled by λ [Ox]. Therefore, our Theorem 6.6 exhibits a quantum line bundle on the moduli space of vector bundles with parabolic structure constructed via the Chern–Simons functional and whose space of holomorphic sections reproduces Segal's space. Notice also that the restriction that λ be invariant under the center is no restriction at all: since $-I$ acts trivially on the determinant bundle Δ, $\mathcal{V}_\lambda$ vanishes identically if λ is not invariant. Second, it should not be difficult to verify that all of the higher cohomology of $L_\alpha^{\otimes k}$ vanishes and that therefore $\dim H^0\left(\mathcal{F}_{\alpha,\delta}/\mathbf{T}_a, L_\alpha^{\otimes k}\right)$ is independent of the underlying complex structure. This would involve computing the canonical bundle and using a vanishing theorem as in [H]. Third, it would be interesting to generalize Theorem 6.6 to higher rank and low genus. For this it is necessary to construct the correct correspondence variety as in Section 2.3. Finally, we note that $H^0\left(\mathcal{F}_{\alpha,\delta}/\mathbf{T}_a, L_\alpha^{\otimes k}\right)$ may naturally be identified with holomorphic sections of a vector bundle over the moduli space $\mathcal{M}(2,0)$. This is seen by pushing forward $L_\alpha^{\otimes k}$ via the fibration

$$\begin{array}{ccc} \mathbf{G}/\mathbf{T}_a & \longrightarrow & \mathcal{F}_{\alpha,\delta}/\mathbf{T}_a \\ & & \downarrow \\ & & \mathcal{M}(2,0) \end{array}$$

from Proposition 2.15. We shall discuss the implications of this point of view in detail in [D-W1].

References

[Ak-M] Akbulut, S., McCarthy, J. D., Casson's Invariant for Oriented Homology 3-Spheres: An Exposition, Math. Notes 36, Princeton University Press 1990.

[A] Atiyah, M. F., The Mathematics and Physics of Knots, Cambridge University Press, Cambridge 1990.

[A-B] Atiyah, M. F., Bott, R. , The Yang–Mills equations over Riemann surfaces, Philos. Trans. Roy. Soc. Lond. A 308 (1982), 523–615.

[A-P-S] Atiyah, M. F., Patodi, V. K., Singer, I. M., Spectral asymmetry and Riemannian geometry I, Math. Proc. Cambridge Philos. Soc. 77 (1975), 43–69.

[Ax] Axelrod, S., Geometric Quantization of Chern–Simons Gauge Theory, Princeton thesis, 1991.

[Ax-DP-W] Axelrod, S., Della Pietra, S., Witten, E., Geometric quantization of Chern–Simons gauge theory, J. Differential Geom. 33 (1991), 787–902.

[Ba] Bauer, S., Parabolic bundles, elliptic surfaces, and SU(2) representation spaces of genus zero Fuchsian groups, Math. Ann. 290 (1991), 509–526.

[Bv] Beauville, A., Vector bundles on curves and generalized theta functions: recent results and open problems, in: Current topics in complex algebraic geometry (Berkeley, CA 1992/1993), MSRI Publ. 28, Cambridge University Press, Cambridge 1995, 17–33.

[Bv-L] Beauville, A., Laszlo, Y., Conformal blocks and generalized theta functions, Commun. Math. Phys. 164 (1994), 385–419.

[Be] Bertram, A., Generalized SU(2) theta functions, Invent. Math. 113 (1993), 351–372.

[Be-Sz] Bertram, A., Szenes, A., Hilbert polynomials of moduli spaces of rank 2 vector bundles II, Topology 32 (1993), 599–609.

[Bs-Rm] Bhosle, U., Ramanathan, A., Moduli of parabolic G-bundles on curves, Math. Z. 202 (1989), 161–180.

[Biq] Biquard, O., Fibrés paraboliques stables et connexions singulières plates, Bull. Soc. Math. France 119 (1991), 231–257.

[Bis] Biswas, I., Determinant line bundle over the universal moduli space of vector bundles over the Teichmüller space, preprint 1995.

[Bd] Boden, H., Representations of orbifold groups and parabolic bundles, Comment. Math. Helv. 66 (1991), 389–447.

[Bd-Hu] Boden, H., Hu, Y., Variations of moduli of parabolic bundles, Math. Ann. 301 (1995), 539–559.

[B] Bott, R., Stable bundles revisited, in: Surveys in Differential Geometry, 1-18, J. Diff. Geom. Supp. 1 1991.

[Ch] Chang, S., Non-abelian θ functions, geometric quantization, and a mathematical proof of Verlinde's formula, MIT preprint 1992.

[D-U] Daskalopoulos, G., Uhlenbeck, K. K., An application of transversality to the topology of moduli space of stable bundles, Topology 34 (1995), 203–215.

[D-W1] Daskalopoulos, G., Wentworth, R., The integrability criterion in Chern–Simons gauge theory, Commun. Math. Phys. 151 (1993), 437–445.

[D-W2] Daskalopoulos, G., Wentworth, R., Local degeneration of the moduli space of vector bundles and factorization of rank two theta functions I, Math. Ann. 297 (1993), 417–466.

[D-W3] Daskalopoulos, G., Wentworth, R., Factorization of rank two theta functions II, Proof of the Verlinde formula. Math. Ann. 304 (1996), 21–51.

[Don1] Donaldson, S. K., Infinite determinants, stable bundles, and curvature, Duke Math. J. 54 (1987), 231–247.

[Don2] Donaldson, S. K., Gluing techniques in the cohomology of moduli spaces, in: Topological methods in modern mathematics (Stony Brook, NY 1991), Publish or Perish, Houston, TX 1993, 137–170.

[Do-Sa] Dostoglou, S. and D. A. Salamon, Instanton homology and symplectic fixed points, London Math. Soc. Lecture Note Ser. 192, Cambridge University Press, Cambridge 1993.

[E-M-S-S] Elitzur, S., Moore, G., Schwimmer, A., Seiberg, N., Remarks on the canonical quantization of Chern–Simons–Witten theory, Nuclear Phys. B 320 (1989), 108–134.

[Fa] Faltings, G., A proof for the Verlinde formula, J. Algebraic Geom. 3 (1994), 347–374.

[F-U] Freed, D., Uhlenbeck, K. K., Instantons and Four-Manifolds, MSRI Publ. 1, Springer-Verlag, New York 1984.

[Fu-St] Furuta, M., Steer, B., Seifert fibred homology 3-spheres and the Yang–Mills equations on Riemann surfaces with marked points, Adv. Math. 96 (1992), 38–102.

[G-S1] Guillemin, V., Sternberg, S., Geometric quantization and multiplicities of group representations, Invent. Math. 67 (1982), 515–538.

[G-S2] Guillemin, V., Sternberg, S., Symplectic Techniques in Physics, Cambridge University Press, Cambridge 1984.

[H] Hitchin, N., Flat connections and geometric quantization, Commun. Math. Phys. 131 (1990), 347–380.

[J-K] Jeffrey, L., Kirwan, F., Intersection pairings in moduli spaces of holomorphic bundles on a Riemann surface, Elect. Res. Ann. AMS 1 (1995), 57–71.

[J-W1] Jeffrey, L., Weitsman, J., Torus actions, moment maps, and the symplectic geometry of the moduli space of flat connections on a two-manifold, in: Low-dimensional topology and quantum field theory, NATO Adv. Sci. Inst. Ser. B. Phys. 315, Plenum, New York 1993, 307–316.

[J-W2] Jeffrey, L., Weitsman, J., Bohr–Sommerfeld orbits in the moduli space of flat connections and the Verlinde dimension formula, Commun. Math. Phys. 150 (1992), 593–630.

[Ko] Kobayashi, S., Differential Geometry of Complex Vector Bundles, Princeton University Press 1987.

[K-N-Rm] Kumar, S., Narasimhan, M.S., Ramanathan, A., Infinite grassmannians and moduli spaces of G-bundles, Math. Ann. 300 (1994), 41–75.

[M] Matic, G., SO(3)-connections and rational homology cobordisms, J. Differential Geom. 28 (1988), 277–307.

[Me-Se] Mehta, V., Seshadri, C., Moduli of vector bundles on curves with parabolic structures, Math. Ann. 248 (1980), 205–239.

[Mick] Mickelsson, J., Current Algebras and Groups, Plenum Press, New York 1989.

[N-R] Narasimhan, M. S., Ramadas, T. R., Factorisation of generalized theta functions I, Invent. Math. 114 (1993), 565–623.

[N-Rn] Narasimhan, M. S., Ramanan, S., Geometry of Hecke cycles - I, in: C. P. Ramanujam: A Tribute, Springer-Verlag, New York 1978.

[Ns-St] Nasatyr, E. B., Steer, B., The Narasimhan–Seshadri theorem for parabolic bundles: an orbifold approach, Philos. Trans. Roy. Soc. London Ser. A. 353 (1995), 137–171.

[Ni] Nitsure, N., Cohomology of the moduli of parabolic bundles, Proc. Ind. Math. Soc. 95 (1986), 61–77.

[Ox] Oxford seminar on Jones–Witten theory, Michaelmas term 1988.

[Po] Poritz, J., Parabolic vector bundles and Hermitian–Yang–Mills connections over a Riemann surface, Internat. J. Math. 4 (1993), 467–501.

[Q] Quillen, D., Determinants of Cauchy–Riemann operators over a Riemann surface, Funct. Anal. Appl. 19 (1985), 31–34.

[R] Ramadas, T. R., Chern–Simons gauge theory and projectively flat vector bundles on $\mathcal{M}_g$, Commun. Math. Phys. 128 (1990), 421–426.

[R-S-W] Ramadas, T. R., Singer, I. M., Weitsman, J., Some comments on Chern–Simons gauge theory, Commun. Math. Phys. 126 (1989), 409–420.

[Se] Seshadri, C., Fibrés Vectoriels sur les Courbes Algébriques, Astérisque 96, 1982.

[Si1] Simpson, C. T., Constraining variations of Hodge structure using Yang–Mills theory and applications to uniformization, J. Amer. Math. Soc. 1 (1988), 867–918.

[Si2] Simpson, C. T., Harmonic bundles on noncompact curves, J. Amer. Math. Soc. 3 (1991), 713–770.

[Sz1] Szenes, A., Hilbert polynomials of moduli spaces of rank 2 vector bundles I, Topology 32 (1993), 587–597.

[Sz2] Szenes, A., The combinatorics of the Verlinde formula, in: Vector bundles in algebraic geometry (Durham, 1993), London Math. Soc. Lecture Note Ser. 208, Cambridge University Press, Cambridge 1995, 241–253.

[T1] Taubes, C., Gauge theory on asymptotically periodic 4-manifolds, J. Differential Geom. 25 (1987), 363–430.

[T2] Taubes, C., L^2 moduli spaces on 4-manifolds with cylindrical ends, International Press, Boston 1993.

[Te] Teleman, C., Verlinde factorization and Lie algebra cohomology, Invent. Math. 126 (1996), 249–263.

[Th1] Thaddeus, M., Conformal field theory and the cohomology of the moduli space of stable bundles, J. Differential Geom. 35 (1992), 131–149.

[Th2] Thaddeus, M., Stable pairs, linear systems, and the Verlinde formula, Invent. Math. 117 (1994), 317–353.

[T-U-Y] Tsuchiya, T., Ueno, K., Yamada, T., Conformal field theory on universal family of stable curves with gauge symmetries, Adv. Stud. Pure Math. 19 (1989), 459–565.

[G-P] Van Geemen, B., Previato, E., Prym varieties and the Verlinde formula, Math. Ann. 294 (1992), 741–754.

[V] Varadarajan, V. S., Lie Groups, Lie Algebras, and Their Representations, Springer-Verlag, New York 1984.

[V] Verlinde, E., Fusion rules and modular transformations in 2D conformal field theory, Nuclear Phys. B 300 (1988), 89–95.

[V-V] Verlinde, E., Verlinde, H., Conformal field theory and geometric quantization, in: Superstrings '89 (Trieste, 1989), World Scientific, River Edge, NJ 1990, 422–449.

[Wi1] Witten, E., Quantum field theory and the Jones polynomial, Commun. Math. Phys. 121 (1989), 351–399.

[Wi2] Witten, E., Two dimensional gauge theories revisited, J. Geom. Phys. 9 (1992), 303–368.

[Z] Zagier, D., On the cohomology of moduli spaces of rank 2 vector bundles over curves, in: The moduli space of curves (Texel Island, 1994), Progr. Math. 129, Birkhäuser, Boston, MA 1995, 533–563.

The eta invariant, manifolds of positive scalar curvature, and equivariant bordism

Peter B. Gilkey†

Abstract. Let M be a closed manifold with Abelian fundamental group of dimension $m \geq 5$. Assume the universal cover of M is spin. We use the eta invariant and equivariant spin bordism to study when M admits a metric of positive scalar curvature

1991 Mathematics Subject Classification: Primary 58G25, 53C21.

Key words: Gromov–Lawson conjecture, positive scalar curvature.

§0. Introduction

Let M be a closed, connected manifold of dimension m. The scalar curvature $\tau := R_{ijji}$ is the simplest local invariant of a Riemannian metric g_M on M. It has strong geometrical content. If $\nu_M(r, x)$ is the volume of the geodesic ball of radius r about a point $x \in M$, then $\nu_M(r, x) = \nu_{\mathbb{R}^m}(r, 0)(1 - \tau(x)r^2/(6m) + O(r^4))$. Thus if the metric has positive scalar curvature, the volume of small geodesic balls in M grows less rapidly than it does in flat space $\mathbb{R}^m$. If $m = 2$, we may use the Gauss–Bonnet theorem $\chi(M) = \int_M (4\pi)^{-1}\tau(x)|dvol(x)|$ to express the Euler–Poincaré characteristic $\chi(M)$ in terms of the integrated scalar curvature. If M admits a metric of positive scalar curvature, then $\chi(M) > 0$ so $M = S^2$ or $M = \mathbb{RP}^2$. Since these manifolds do admit metrics of positive scalar curvature, this gives a necessary and sufficient condition for M to admit a metric of positive scalar curvature.

In higher dimensions, the situation is not as simple. We shall assume that $m \geq 5$ to ensure that certain surgery arguments work; these arguments fail in dimensions $m = 3$ and $m = 4$ as recent work using Seifert–Witten theory has shown. The Gromov–Lawson–Rosenberg conjecture for a group π proposes that if M is a closed manifold of dimension $m \geq 5$ with fundamental group π whose universal cover is spin, then M admits a metric of positive scalar curvature $\iff$ $\alpha_\pi(M)$ vanishes where $\alpha_\pi(M)$ is a generalized index of the Dirac operator. To simplify the discussion,

† Research partially supported by the NSF (USA) and MPIM (Germany)

we shall assume here that π is a finite Abelian group; we will define α_π in §2.8 for the cases relevant to this paper and refer to [24, 25, 26] for the general definition.

In §1, we summarize our results concerning the Gromov–Lawson–Rosenberg conjecture; this is joint work with Botvinnik and Stolz. In §2, we introduce equivariant bordism groups and connective K-theory groups and reduce the question to a problem in algebraic topology. In §3, we introduce the eta invariant; the eta invariant provides the characteristic numbers of the connective K-theory groups in question. In §4, we will sketch the proof of the results in §1; we refer to [6, 8, 9, 10, 13] for further details. The Gromov–Lawson–Rosenberg conjecture deals with the existence of metrics of positive scalar curvature. There is also the question of uniqueness. That is beyond the scope of this note. We refer to [5, 7] to see that in certain circumstances, the moduli space of metrics of positive scalar curvature on M has an infinite number of arc components.

It is a pleasant task to thank the referee for helpful suggestions.

§1. Metrics of positive scalar curvature

1.1. The A-roof genus. In higher dimensions, the $\hat{A}$ genus plays a role similar to that played by the Euler-Poincaré characteristic χ in dimension 2. We assume for the moment that M admits a spin structure. Let sw_i be the Stiefel–Whitney classes. Then M is spin if $sw_1(M) = 0$ and $sw_2(M) = 0$ or equivalently if M is orientable and if the transition functions of the tangent bundle of M can be lifted from the special orthogonal group SO(m) to the spin group SPIN(m); we refer to [21] for further details. Fix a spin structure s_M on M; inequivalent spin structures are parametrized by real line bundles over M i.e. by representations of $\pi_1(M)$ to the cyclic group $\mathbb{Z}_2$ of order 2. Let $D := D(g_M, s_M)$ be the Dirac operator acting on the space of smooth spinors over M. If $m \equiv 4$ mod 8, we may decompose the spinors into positive and negative chirality and decompose $D = D_+ + D_-$. Let

$$\hat{A}(M, g_M, s_M) := \dim\ker(D_+(g_M, s_M)) - \dim\ker(D_-(g_M, s_M)) \in \mathbb{Z}$$

be the index of the spin complex; $\hat{A}(M) = \hat{A}(M, g_M, s_M)$ is independent of the particular spin structure and metric chosen. The Atiyah–Singer index theorem expresses $\hat{A}(M)$ as the integral of a certain polynomial in the Pontrjagin classes of M so this is a charactistic number. If $m \equiv 1$ mod 8, we reduce the dimension of the kernal mod 2 to define $\hat{A}(M, s_M) := \dim\ker(D(g_M, s_M)) \in \mathbb{Z}_2$; this is independent of the metric chosen but does depend on the spin structure chosen. Similarly, if $m \equiv 2$ mod 8, $\hat{A}(M, s_M) := \dim\ker(D(g_M, s_M))/2 \in \mathbb{Z}_2$ is a well defined invariant of the spin structure. We set $\hat{A} = 0$ in the remaining dimensions.

1.2. The Lichnerowicz formula [22]. Let ∇^s be the spin connection. Then

$$D^2 = -\mathrm{Tr}\nabla^s\nabla^s + \tau(g_M)/4 \text{ so } (\phi, D\phi)_{L^2} = |\nabla^s\phi|^2_{L^2} + (\phi, \tau\phi)_{L^2}.$$

In particular, if τ is positive, there are no harmonic spinors and thus $\hat{A}(M) = 0$. For example, the Kummer surface K is a complex hypersurface of real dimension 4 in $\mathbb{CP}^3$ which is defined by the homogeneous equations $z_0^4 + z_1^4 + z_2^4 + z_3^4 = 0$; K is simply connected spin manifold with $\hat{A}(K) = 2$. Thus K does not admit a metric of positive scalar curvature. Stolz [32] has solved the Gromov–Lawson–Rosenberg conjecture in the simply connected case.

1.3. Theorem. *Let M be a closed simply connected spin manifold of dimension $m \geq 5$. Then M admits a metric of positive scalar curvature* $\iff$ $\hat{A}(M) = 0$.

By a theorem of Kwasik and Schultz [19], the Gromov–Lawson–Rosenberg conjecture holds for a finite group π $\iff$ the conjecture holds for all the Sylow subgroups of π. Thus we may assume henceforth that π is a p group. All the torsion in the spin bordism ring Ω_* is 2 torsion so the prime 2 plays a distinguished role.

1.4. Cyclic groups of odd order. Let $\mathbb{Z}_n$ be the cyclic group of order $n > 1$. Let n be odd. If $\pi_1(M) = \mathbb{Z}_n$ and if M is spin, then the spin structure on M is unique and we shall ignore the role of the spin structure and simply write $\hat{A}(M)$. We refer to Kwasik and Schultz [19] for the following theorem; see also Botvinnik and Gilkey [6] for a different proof. Both approaches use the result of Stolz cited above.

1.5. Theorem. *Let M be a closed connected spin manifold of dimension $m \geq 5$ with $\pi_1(M) = \mathbb{Z}_n$ for n odd. Then M admits a metric of positive scalar curvature* $\iff$ $\hat{A}(M) = 0$.

1.6. Cyclic groups of even order. If $\pi_1(M) = \mathbb{Z}_n$ for n even and if M is spin, then there are two inequivalent spin structures on M and the A-roof genus can depend on which spin structure is chosen if $m \equiv 1, 2 \bmod 8$. We refer to Botvinnik, Gilkey, and Stolz [10] for the proof of the following result; a slightly different proof may also be found in [8].

1.7. Theorem. *Let M be a closed connected spin manifold of dimension $m \geq 5$ with $\pi_1(M) = \mathbb{Z}_n$ for n even. Then M admits a metric of positive scalar curvature* $\iff$ $\hat{A}(M, s_M) = 0$ *for both spin structures on M.*

1.8. Odd order Abelian groups of higher rank. Let $\pi(n, k) = (\mathbb{Z}_n)^k$ where $n = p^\ell$ is an odd prime power. We refer to Botvinnik and Gilkey [9] for the proof of the following theorem; the assertion for $k = 2$ and $n = p$ also follows from work of Schultz [31].

1.9. Theorem. *Let M be a closed connected spin manifold of dimension $m \geq 5$ with $\pi_1(M) = \pi(n,k)$ for $n = p^\ell$ an odd prime power and $k = 2$ or $k = 3$. If $k = 3$, assume that m is odd. If $\ell > 1$, assume that $2(p-1) \geq m$. Then M admits a metric of positive scalar curvature $\iff \hat{A}(M) = 0$.*

1.10. Orientable manifolds whose universal cover is spin. In the results cited above, we assumed M is spin. We now assume that the universal cover $\tilde{M}$ is spin, that M is orientable, and that M is not spin. Assume $\pi_1(M) = \mathbb{Z}_n$; necessarily n is even. We refer to Botvinnik and Gilkey [8] for the proof of the following result; Kwasik and Schultz [20] have also proved this result using different methods.

1.11. Theorem. *Let M be a closed connected orientable manifold of dimension $m \geq 5$ with $\pi_1(M) = \mathbb{Z}_n$. Assume the universal cover $\tilde{M}$ of M is spin but that M is not spin. Then M admits a metric of positive scalar curvature $\iff \hat{A}(\tilde{M}) = 0$.*

1.12. Non orientable manifolds whose universal cover is spin. Let $\mathrm{PIN}^-(m)$ be the pinor group; this group double covers the orthogonal group $\mathrm{O}(m)$. A manifold M admits a pin$^-$ structure if $sw_1(M)^2 + sw_2(M) = 0$ or equivalently if we can lift the transition functions from $\mathrm{O}(m)$ to $\mathrm{PIN}^-(m)$. If M admits a pin$^-$ structure, then the universal cover is spin. If $m \equiv 2 \bmod 8$, let $\hat{A}(M) := \hat{A}(\tilde{M}) \in \mathbb{Z}_2$; if $m \equiv 6 \bmod 8$, let $\hat{A}(M) = 0$. We refer to [13] for the proof of the following theorem.

1.13. Theorem. *Let M be a closed connected non orientable pin$^-$ manifold of dimension $m = 4k+2 \geq 6$. Assume that $\pi_1(M) = \mathbb{Z}_n$ for $n = 2^\ell > 2$. Then M admits a metric of positive scalar curvature $\iff \hat{A}(M) = 0$.*

1.14. Other results. In addition to the results cited above, the Gromov–Lawson–Rosenberg conjecture has also been proved for spherical space form groups [10], and for a (short) list of infinite groups including free groups, free Abelian groups, and fundamental groups of orientable surfaces [28].

1.15. Moduli space of metrics of positive scalar curvature. If we perturb a metric of positive scalar slightly, it remains a metric of positive scalar curvature. Thus the space $\mathcal{R}(M)$ of metrics of positive scalar curvature is an open subset of the space of all Riemannian metrics on M. Let $\mathcal{G}(M)$ be the group of diffeomorphisms of M. Let $\mathcal{M}(M) := \mathcal{R}(M)/\mathcal{G}(M)$ be the moduli space. The following theorem follows from a more general result of Botvinnik and Gilkey [5, 7].

1.16. Theorem. *Let M be a closed connected orientable manifold of odd dimension $m \geq 5$. Assume the fundamental group π of M is a finite Abelian group which contains an element of order at least 3. Assume that the universal cover of M is spin and that M admits a metric of positive scalar curvature. Then the moduli space $\mathcal{M}(M)$ has an infinite number of arc components.*

§2. Equivariant spin bordism and connective K-theory

As noted in §1, we can work one prime at a time in proving the Gromov–Lawson–Rosenberg conjecture so we may assume π is a p group. In this section, we will reformulate the Gromov–Lawson–Rosenberg conjecture in terms of a question about equivariant bordism if p is odd in Corollary 2.3 and about connective K-theory if $p = 2$ in Corollary 2.6.

2.1. Twisted spin structures. Let ξ be a real vector bundle over the classifying space $B\pi$ of a finite group π. We consider a triple (M, s, f) where M is a closed manifold of dimension m, where $f : M \rightarrow B\pi$ gives M a π structure, and where s is a spin structure on $T(M) \oplus f^*(\xi)$. If M is connected and if π is Abelian, then f is equivalent to a representation from $\pi_1(M)$ to π. If $\pi_1(M) = \pi$, there is a canonical π structure f so f_* is the identity map from $\pi_1(M)$ to π. The twisted bordism group $\Omega_m(B\pi, \xi)$ is defined by introducing the equivalence relation $(M_1, s_1, f_1) \equiv (M_2, s_2, f_2)$ if there exists a compact manifold N with boundary $M_1 \dot{\sqcup} -M_2$ so that the structures extend over N. Note that only the values of the first two Stiefel–Whitney classes $sw_1(\xi)$ and $sw_2(\xi)$ are relevant in defining $\Omega_m(B\pi, \xi)$. We refer to [14, 15, 30] for the proof of the first assertion and to [23, 24, 25, 26] for the proof of the second assertion in the following theorem. The restriction that the surgeries be in codimension at least 3 is what forces the requirement $m \geq 5$ in the results of §1.

2.2. Theorem.

a) *Let M be a closed manifold which need not be connected and which admits a metric of positive scalar curvature. Then any manifold which can be obtained from M by performing surgeries in codimension at least 3 also admits a metric of positive scalar curvature.*

b) *Let M be a connected closed manifold of dimension $m \geq 5$ with fundamental group π. Give M the canonical π structure f. Suppose that we have $[(M, s, f)] = [(M_1, s_1, f_1)]$ in $\Omega_m(B\pi, \xi)$. If M_1 admits a metric of positive scalar curvature, then so does M.*

Theorem 2.2 reduces the question of constructing a metric of positive scalar curvature to a question in bordism. Let $\Omega_m^+(B\pi, \xi)$ be the subgroup of $\Omega_m(B\pi, \xi)$ which is generated by manifolds which admit metrics of positive scalar curvature. Let α_π be the generalized $\hat{A}$ genus of the Gromov–Lawson–Rosenberg conjecture. It is easy to use the index theorem for manifolds with boundary of Atiyah et al [1, 2, 3] to show that α_π extends to $\Omega_*(B\pi, \xi)$ and vanishes on $\Omega_*^+(B\pi, \xi)$. The following reformulation of the Gromov–Lawson–Rosenberg conjecture is now immediate:

2.3. Corollary. *Let π be a finite group. Let $m \geq 5$. The following assertions are equivalent:*

a) *Let M be a closed connected manifold of dimension m with fundamental group π and canonical π structure f. Assume $T(M) \oplus f^*\xi$ admits a spin structure. Then M admits a metric of positive scalar curvature* $\iff \alpha_\pi(M) = 0$.

b) *We have* $\Omega_m^+(B\pi, \xi) = \ker \alpha_\pi \cap \Omega_m(B\pi, \xi)$.

2.4. Connective K-theory. All the torsion in the spin bordism ring Ω_* is 2 torsion so that prime is distinguished in this subject. As noted above, it suffices to work one prime at a time; we now assume π is an Abelian 2-group. This avoids having to localize at the prime 2. Let $\mathbb{HP}^2$ denote quaternion projective space with the usual homogeneous metric of positive scalar curvature. Let

$$\mathbb{HP}^2 \to E \to B$$

be a fiber bundle where the transition functions are the group of isometries PSp(3) of $\mathbb{HP}^2$. Since $\mathbb{HP}^2$ is simply connected, the projection $p : E \to B$ induces an isomorphism on the fundamental group; any π structure on E arises from a π structure on B. Let $T_m(B\pi, \xi)$ be the subgroup of $\Omega_m(B\pi, \xi)$ generated by manifolds E arising in this fashion. Let $ko_m(B\pi, \xi)$ be the twisted connective K-theory groups. The following result follows from a more general result of Stolz [33].

2.5. Theorem. *If π is an Abelian 2-group,* $ko_m(B\pi, \xi) = \Omega_m(B\pi, \xi)/T_m(B\pi, \xi)$.

Since the elements of $T_m(B\pi, \xi)$ all admit metrics of positive scalar curvature, α_π vanishes on these manifolds and therefore extends to the connective K-theory groups. Let $ko_m^+(B\pi, \xi)$ be the image of $\Omega_m^+(B\pi, \xi)$ under the natural projection. The following assertion is now immediate; it is an improvement over Corollary 2.3 if $p = 2$ since the connective K-theory groups are smaller than the bordism groups.

2.6. Corollary. *Let π be an Abelian 2-group. Let $m \geq 5$. The following assertions are equivalent:*

a) *Let M be a closed connected manifold of dimension m with fundamental group π and canonical π structure f. Assume $T(M) \oplus f^*\xi$ admits a spin structure. Then M admits a metric of positive scalar curvature* $\iff \alpha_\pi(M) = 0$.

b) *We have* $ko_m^+(B\pi, \xi) = \ker \alpha_\pi \cap ko_m(B\pi, \xi)$.

2.7. Twisted spin structures for cyclic groups. Let ξ_0 be the trivial real line bundle over the classifying space $B\pi$ of a finite group π. If M is a closed spin manifold with fundamental group π, then $[(M, \cdot)] \in \Omega_m(B\pi, \xi_0)$; we may identify $\Omega_m(B\pi, \xi_0) = \Omega_m(B\pi)$. Let $\pi = \mathbb{Z}_n = \{\lambda \in \mathbb{C} : \lambda^n = 1\}$ be the group of nth roots of unity. Let $\rho_s(\lambda) := \lambda^s$ for $\lambda \in \mathbb{Z}_n$. We suppose n is even. Let ξ_1 be the real 2-plane bundle defined by the complex representation ρ_1, let ξ_2 be the real line bundle defined by the real representation $\rho_{n/2}$, and let ξ_3 be the real 3-plane bundle $\xi_3 := \xi_1 \oplus \xi_2$. Let

$H^1(B\mathbb{Z}_n;\mathbb{Z}_2)=x_1\mathbb{Z}_2$ and $H^2(B\mathbb{Z}_n;\mathbb{Z}_2)=x_2\mathbb{Z}_2$;

$$sw_1(\xi_0)=0,\ sw_2(\xi_0)=0,\ sw_1(\xi_1)=0,\ sw_2(\xi_1)=x_2,$$
$$sw_1(\xi_2)=x_1,\ sw_2(\xi_2)=0,\ sw_1(\xi_3)=x_1,\ sw_2(\xi_3)=x_2.$$

Let M be a closed manifold of dimension m with fundamental group $\mathbb{Z}_n$. We assume the universal cover of M is spin. If M is spin, then $[(M,\cdot)]\in\Omega_m(B\mathbb{Z}_n,\xi_0)$. If M is orientable, but not spin, then $[(M,\cdot)]\in\Omega_m(B\mathbb{Z}_n,\xi_1)$. If M is not orientable but $sw_2(M)=0$, then $M\in\Omega_m(B\mathbb{Z}_n,\xi_2)$; we may identify the equivariant pinor bordism group $\Omega_m^{\text{pin}}(B\mathbb{Z}_n)$ with the twisted bordism group $\Omega_m(B\mathbb{Z}_n,\xi_2)$. Finally, if $sw_1(M)\neq 0$ and $sw_2(M)\neq 0$, then $M\in\Omega_m(B\mathbb{Z}_n,\xi_3)$. Thus these 4 twisted bordism groups handle the case of a closed connected manifold M with cyclic fundamental group and with spin universal cover.

2.8. The invariant α_π. We can express α_π in terms of the $\hat{A}$ genus for the cases we shall be considering. Let π be a non-trivial finite Abelian p group. Let $\tilde{M}_f$ be the principal π bundle defined by a π structure f on M; $\tilde{M}$ will be spin for the twisted bordism groups we are considering. If $m\equiv 0$ mod 4, let $\alpha_\pi(M)=\hat{A}(\tilde{M})\in\mathbb{Z}$; if M admits a spin structure, $\alpha_\pi(M)=|\pi|\hat{A}(M)$.

(1) Let p be odd. If $m\equiv 1,2$ mod 8, let $\alpha_\pi(M)=\hat{A}(M)\in\mathbb{Z}_2$. Set $\alpha_\pi(M)=0$ if $m\not\equiv 0,1,2,4$ mod 8.

(2) Let $\xi=\xi_0$ and let $\pi=\mathbb{Z}_n$ for $n=2^\ell>1$. If $m\equiv 1,2$ mod 8, let $\alpha_\pi(M)=\hat{A}(M,s_1)\oplus\hat{A}(M,s_2)\in\mathbb{Z}_2\oplus\mathbb{Z}_2$ where s_1 and s_2 are the two inequivalent spin structures on M. Set $\alpha_\pi(M)=0$ if $m\not\equiv 0,1,2,4$ mod 8.

(3) Let $\xi=\xi_1$ and let $\pi=\mathbb{Z}_n$ for $n=2^\ell>1$. Set $\alpha_\pi(M)=0$ for $m\not\equiv 0,4$ mod 8.

(4) Let $\xi=\xi_2$ and let $\pi=\mathbb{Z}_n$ for $n=2^\ell>2$. If $m\equiv 6$ mod 8, let $\alpha_\pi(M)=0$. If $m\equiv 2$ mod 8, let $\alpha_\pi(M)=\hat{A}(\tilde{M})\in\mathbb{Z}$.

§3. The eta invariant

3.1. The eta invariant with coefficients in a representation. Let P be an elliptic self-adjoint partial differential operator on a closed manifold M. Let $\eta(z,P):=\{\mathrm{Tr}_{L^2}(P(P^2)^{-(z+1)/2}\}$. This is holomorphic for $\mathrm{Re}(z)>0$ and has a meromorphic extension to $\mathbb{C}$ which is regular at $z=0$; see [12] for details. Let

$$\eta(P):=2^{-1}\{\eta(z,P)+\dim\ker(P)\}|_{z=0}\in\mathbb{R}$$

be a measure of the spectral asymmetry of P; this is the eta invariant of Atiyah et al [1, 2, 3]. Let $f:M\to B\pi$ and let ρ be a representation of π. Let P_ρ be P with coefficients in the locally flat bundle defined by ρ. The map $\rho\to\eta(P_\rho)$ extends naturally to the group representation ring $R(\pi)$.

3.2. spin^c structures. Let $\mathrm{SPIN}^c(\cdot) := \mathrm{SPIN}(\cdot) \times \mathrm{U}(1)/\mathbb{Z}_2$ be the complexification of the spinor group where we identify (g, λ) to $(-g, -\lambda)$; see Hitchin [16] for details. The map $(g, \lambda) \to \lambda^2$ defines the determinant representation $\det : \mathrm{SPIN}^c \to \mathrm{U}(1)$. Let ξ be a real vector bundle over the classifying space $B\pi$ of a finite group π. We assume ξ has a spin^c structure with flat associated determinant line bundle. This is the case, for example, for the bundles ξ_0 and ξ_1 over $B\mathbb{Z}_n$ defined in §2.7. If $f : M \to B\pi$ gives M a π structure and if s is a spin structure on $T(M) \oplus f^*(\xi)$, there is a natural spin^c structure induced on M with flat associated determinant line bundle. If m is odd, we let P be the associated Dirac operator and let $\eta(M, s, f)(\rho)$ be the associated eta invariant for any $\rho \in R(\pi)$; we shall also denote this invariant by $\eta(M)(\rho)$ when (s, f) are understood; we supress the dependence upon the Riemannian metric g. Let $R_0(\pi)$ denote the augmentation ideal of virtual representations of virtual dimension 0. The following result is an immediate consequence of the index theorem for manifolds with boundary of Atiyah et al; see [8] for details.

3.3. Theorem. *Let ξ be a real vector bundle over the classifying space of a finite group π which admits a spin^c structure with flat determinant line bundle. Let $\rho \in R(\pi)$ and let m be odd. If $m \equiv 3$ mod 4, assume* $\dim \rho = 0$. *Then the map $(M, s, f) \to \eta(M, s, f)(\rho)$ extends to a homomorphism $\eta_\rho : \Omega_m(B\pi; \xi) \to \mathbb{R}/\mathbb{Z}$.*

If π is a 2 group, a bit more can be said. It is not hard to show $\eta(M, s, f)(\rho)$ vanishes on $\mathbb{HP}^2$ fibrations so η extends to connective K-theory. Furthermore, we can lift the invariants from $\mathbb{R}/\mathbb{Z}$ to $\mathbb{R}/2\mathbb{Z}$ in certain cases using an appropriate quaternion structure. More general results of [8] imply the following theorem.

3.4. Theorem. *Let ξ be a real vector bundle over the classifying space of a finite Abelian 2 group π which admits a spin^c structure with flat determinant line bundle. Let $\rho \in R(\pi)$ and let m be odd. If $m \equiv 3$ mod 4, assume* $\dim \rho = 0$. *Then the map $(M, s, f) \to \eta(M, s, f)(\rho)$ extends to a homomorphism $\eta_\rho : ko_m(B\pi; \xi) \to \mathbb{R}/\mathbb{Z}$. If ξ admits a spin structure, if $m \equiv 3$ mod 8, and if ρ is of real type, then η_ρ is well defined with values in $\mathbb{R}/2\mathbb{Z}$.*

3.5. pin$^-$ structures. Let $n = 2^\ell > 2$. Recall we identified $\Omega_m(B\pi, \xi_2)$ with $\Omega_m^{pin}(B\pi)$. Let M be an even dimensional pin$^-$ manifold with a $\mathbb{Z}_n$ structure. We can define an operator P of Dirac type on M which is analogous to the ordinary Dirac operator. If $\rho \in R(\pi)$, let $\eta(M, P_\rho)$ be the associated eta invariant. We refer to [13] for the proof of the following assertion:

3.6. Theorem. *Let $n = 2^\ell > 2$ and let $\rho \in R(\pi)$. If m is even, then the map $M \to \eta(M, P_\rho)$ extends to a map $\eta_\rho : ko_m(B\mathbb{Z}_n; \xi_2) \to \mathbb{R}/\mathbb{Z}$. If $m \equiv 2$ mod 8 and if ρ is real, then η_ρ can be extended to take values in $\mathbb{R}/2\mathbb{Z}$.*

3.7. Computing the eta invariant. We shall need some combinatorial formulas to apply Theorems 3.3, 3.4, and 3.6. We note that Cartesian product makes $\Omega_*(B\pi, \xi)$ into an Ω_* module. The spin complex is multiplicative with respect to Cartesian products. The following is an immediate consequence of that observation, see Atiyah et al [1,2, 3].

3.8. Lemma. *Let* $[(M, s, f)] \in \Omega_m(B\pi, \xi)$ *and let* $[N] \in \Omega_{4k}$. *If* $\rho \in R(\pi)$, *then* $\eta(M \times N, s, f)(\rho) = \eta(M, s, f)\hat{A}(N)$.

3.9. Lens spaces. In §2.7, we identified $\mathbb{Z}_n$ with the group of nth roots of unity and defined linear representations $\rho_s(\lambda) = \lambda^s$. If $\vec{a}$ is a collection of indices coprime to n, let $\sigma(\vec{a}) := \rho_{a_1} \oplus \cdots \oplus \rho_{a_k}$ define a fixed point free representation of $\mathbb{Z}_n$ to the unitary group $\mathrm{U}(k)$. If n is odd, we choose the a_i even. We let

$$L^{2k-1}(n; \vec{a}) := S^{2k-1}/\sigma(\vec{a})(\mathbb{Z}_n)$$

be the resulting quotient We refer to [8] for the proof of the following result; this uses formulas of Donnelly [11]. Define:

If n is odd, let $\psi(\vec{a})(\lambda) := \lambda^{-|\vec{a}|/2}\det(I - \sigma(\vec{a}))$.

If n is even and if k is even, let $\psi(\vec{a})(\lambda) := \lambda^{-|\vec{a}|/2}\det(I - \sigma(\vec{a}))$.

If n is even and if k is odd, let $\psi(\vec{a})(\lambda) := \lambda^{-(|\vec{a}|+1)/2}\det(I - \sigma(\vec{a}))$.

3.10. Lemma. *Let* $M = L^{2k-1}(n; \vec{a})$.

a) *If n is odd, then* $[(M, \cdot)] \in \Omega_{2k-1}(B\mathbb{Z}_n, \xi_0)$.

b) *If n is even and if k is even, then* $[(M, \cdot)] \in \Omega_{2k-1}(B\mathbb{Z}_n, \xi_0)$.

c) *If n is even and if k is odd, then* $[(M, \cdot)] \in \Omega_{2k-1}(B\mathbb{Z}_n, \xi_1)$.

d) *If $2k - 1 > 1$, then M admits a metric of positive scalar curvature.*

e) *We have* $\eta(M)(\rho) = n^{-1}\Sigma_{\lambda^n=1,\lambda\neq 1} Tr(\rho)(\lambda)\psi(\vec{a})(\lambda)^{-1}$.

3.11. Lens space bundles. Let $n = 2^\ell > 2$. Let $H^{\otimes 2}$ be the tensor square of the Hopf line bundle over the sphere S^2 and let $\mathbb{1}$ be the trivial line bundle. Let $\mathcal{W}_k := H^{\otimes 2} \oplus (k-1)\,\mathbb{1}$ over S^2. We extend $\sigma(\vec{a})$ to act without fixed points on the fiber spheres of $\mathcal{S}(\mathcal{W}_k)$ and define $X^{2k+1}(n; \vec{a}) := \mathcal{S}(\mathcal{W}_k)/\sigma(\vec{a})(\mathbb{Z}_n)$. Lemma 3.10 generalizes to this setting; again, we refer to [8] for details.

3.12. Lemma. *Let* $M = X^{2k+1}(n; \vec{a})$.

a) *If k is even, then* $[(M, \cdot)] \in \Omega_{2k+1}(B\mathbb{Z}_n, \xi_0)$.

b) *If k is odd, then* $[(M, \cdot)] \in \Omega_{2k+1}(B\mathbb{Z}_n, \xi_1)$.

c) *If $2k + 1 \geq 3$, then M admits a metric of positive scalar curvature.*

d) *We have* $\eta(X^{2k+1})(\rho) = \Sigma_{\lambda^n=1,\lambda\neq 1} Tr(\rho)(\lambda)(1 + \lambda^{a_1})/(1 - \lambda^{a_1})\psi(\vec{a})(\lambda)^{-1}$.

3.13. Twisted products. To prove Theorem 1.13, we will need non-orientable examples. Let M be a closed orientable manifold of odd dimension m with cyclic fundamental group $\mathbb{Z}_n$ for $n = 2^\ell > 1$. We assume the universal cover $\tilde{M}$ of M is spin. Let the generator g_n of $\mathbb{Z}_n$ act on $\tilde{M}$ by the associated deck transformation. Let $\tilde{N} := \tilde{M} \times \tilde{M}$. We define a fixed point free action of $\mathbb{Z}_{2n}$ on $\tilde{N}$ by defining $g_{2n}(x, y) = (g_n y, x)$ and let $N := \tilde{N}/\mathbb{Z}_{2n}$. We refer to [13] for the proof of the following result.

3.14. Lemma. *Adopt the notation given above. There is a canonical pin$^-$ structure on N. If $m \equiv 3$ mod 4, let $\beta = 0$. If $m \equiv 1$ mod 4, let $\beta = n/2$. Let ρ_b be the determinant line bundle of the spin^c structure on M. Assume M has no harmonic spinors.*

a) *If $u = 2v - b + \beta$, then $\eta(N)(\rho_u) = \eta(M)(\rho_v)(\rho_0 - \rho_{n/2})$.*

b) *If $u = 2v - b + \beta + 1$, then $\eta(N)(\rho_u) = 0$.*

This theorem describes the eta invariant of the twisted product of two orientable odd dimensional manifolds; there is a corresponding formula for the twisted product of even dimensional manifolds; see Barrera–Yanez [4] for details.

3.15. Pontrjagin dual. If A is an Abelian group, the dual group A^* is the group of homomorphisms from A to $\mathbb{R}/\mathbb{Z}$. We define $\eta^*(M) \in R(G)^*$ or $\eta^*(M) \in R_0(G)^*$ by setting $\eta^*(M)(\rho) = \eta(M, \rho) \in \mathbb{R}/\mathbb{Z}$. We use Theorem 3.3 to see that:

(1) $\eta^* : \Omega_{4k+1}(B\mathbb{Z}_n) \to R(\mathbb{Z}_n)^*$.

(2) $\eta^* : \Omega_{4k-1}(B\mathbb{Z}_n) \to R_0(\mathbb{Z}_n)^*$.

Suppose n is even. If $m \equiv 3$ mod 8 and if $\xi = \xi_0$, let $\eta_0(M) = \eta(M)(\rho_0 - \rho_{n/2})$; if $m \equiv 2$ mod 8 and if $\xi = \xi_2$, let $\eta_0(M) = \eta(M)(\rho_0)$. We set $\eta^\# = \eta^* \oplus \eta_0$. We use Theorems 3.4 and 3.6 to see that:

(1) If $i = 0, 1$, then $\eta^* : ko_{4k+1}(B\mathbb{Z}_n, \xi_i) \to R(\mathbb{Z}_n)^*$.

(2) If $i = 0, 1$, then $\eta^* : ko_{4k-1}(B\mathbb{Z}_n, \xi_i) \to R_0(\mathbb{Z}_n)^*$.

(3) We have $\eta^\# : ko_{8k+3}(B\mathbb{Z}_n, \xi_0) \to R_0(\mathbb{Z}_n)^* \oplus \mathbb{R}/2\mathbb{Z}$.

(4) If $i = 2, 3$, then $\eta^* : ko_{2k}(B\mathbb{Z}_n, \xi_i) \to R(\mathbb{Z}_n)^*$.

(5) We have $\eta^\# : ko_{8k+2}(B\mathbb{Z}_n, \xi_2) \to R(\mathbb{Z}_n)^* \oplus \mathbb{R}/2\mathbb{Z}$.

3.16. Push forward. Let $\Xi : \pi \to \tilde{\pi}$ be a group homomorphism. Then Ξ induces a natural map $\Xi : B\pi \to B\tilde{\pi}$. If f gives M a π structure, then $\Xi \circ f$ gives M a $\tilde{\pi}$ structure. Let $\tilde{\xi}$ be a real vector bundle over $B\tilde{\pi}$ which admits a spin^c structure with a flat determinant line bundle. We have a natural map Ξ_* from $\Omega_m(B\pi; f^*\tilde{\xi})$ to $\Omega_m(B\tilde{\pi}; \tilde{\xi})$. Let Ξ^* be the pull-back from $R(\tilde{\pi})$ to $R(\pi)$. It is then immediate from the definition that $\eta(\Xi_* M)(\tilde{\rho}) = \eta(M)(\Xi^* \tilde{\rho})$. We remark that there is also a natural map $\Xi^* : \Omega_m(B\tilde{\pi}; \tilde{\xi}) \to \Omega_m(B\pi; \Xi^*\tilde{\xi})$ and that one can express $\eta(\Xi^* M)(\rho)$

in terms of the induced representation on $\tilde{\pi}$; we omit the details in the interests of brevity. Equivariant bordism and connective K-theory are Mackey functors and the eta invariant behaves nicely with respect to transfer and induction.

§4 Brief sketch of proofs

In this section, we will sketch the proofs of Theorems 1.5, 1.7, 1.9, 1.11, and 1.13. We shall omit all technical computations involving the eta invariant to give the broad outline of the proof and refer to [6, 8, 9, 10, 13] for details.

4.1. Proof of Theorem 1.5. Let n be odd and let $\pi = \mathbb{Z}_n$. By Corollary 2.3, we must show $\Omega_m^+(B\pi) = \ker(\hat{A}) \cap \Omega_m(B\pi)$. Since Theorem 1.3 establishes the Gromov–Lawson–Rosenberg conjecture in the simply connected case, we may work with the reduced theories instead. Note that the E^2 term in the Atiyah–Hirzebruch spectral sequence for $\tilde{\Omega}_m(B\pi)$ is given by $E^2_{p,q} := \tilde{H}_p(B\pi; \Omega_q)$; thus $\tilde{\Omega}_m(B\pi)$ is a finite group of odd order for any m and we may invert the prime 2. We can choose simply connected spin manifolds $X^{4j} \in \Omega_{4j} \otimes \mathbb{Z}[1/2]$ so that

$$\Omega_* \otimes \mathbb{Z}[1/2] = \mathbb{Z}[1/2, X^4, X^8, \dots].$$

We may let X^4 be the Kummer surface and choose the X^{4k} for $k > 1$ so that $\hat{A}(X^{4k}) = 0$. Thus all the X^{4k} admit metrics of positive scalar curvature except for X^4. Let $L^{2j-1} := S^{2j-1}/\pi$ where we use the diagonal embedding of π in $\mathrm{U}(j)$. The lenspaces admit metrics of positive scalar curvature for $2j - 1 > 1$. An elementary spectral sequence argument shows that the L^{2j-1} generate $\tilde{\Omega}_*(B\pi)$ as an Ω_* module. Consequently $\tilde{\Omega}_*(B\pi)$ is generated by products of the X^{4k} and the L^{2j-1}. If $M = M_1 \times M_2$ and if M_1 admits a metric of positive scalar curvature, then M admits a metric of positive scalar curvature. Thus the only manifolds which do not obviously admit metrics of positive scalar curvature in this generating set are the manifolds $S^1 \times (X^4)^\nu$. Theorem 1.5 will follow if we can show that $S^1 \times X^4$ is bordant to a manifold which admits a metric of positive scalar curvature. This reduces the proof of Theorem 1.5 to the special case $m = 5$. Let $\mathcal{P}_5 := \mathrm{span}_{\mathbb{Z}}\{L(n; 2, 1, 1), L(n; 2, 2, 2)\} \subset \tilde{\Omega}_5(B\pi)$. We define η^* as in §3.15 and use Lemma 3.10 to show that $|\eta^*(\mathcal{P}_5)| \geq n^2$ and thus $|\mathcal{P}| \geq n^2$. On the other hand, the Atiyah–Hirzebruch spectral sequence shows $|\tilde{\Omega}_5(B\pi)| \leq n^2$. This shows $\mathcal{P}_5 = \tilde{\Omega}_5(B\pi)$ so $\tilde{\Omega}_5^+(B\pi) = \tilde{\Omega}_5(B\pi)$. □

4.2. Proof of Theorems 1.7 and 1.11. Let $\pi = \mathbb{Z}_n$. We can work one prime at a time; if n is an odd prime power, the desired result follows from Theorem 1.5 so we suppose $n = 2^\ell$. We refer to Rosenberg and Stolz [27, 28, 29] for the case $n = 2$ and suppose therefore $n \geq 4$. Let $\xi = \xi_0$ or $\xi = \xi_1$. If m is even, we can use the

Adams spectral sequence to show that $\ker\alpha_\pi \cap ko_m(B\pi;\xi) = 0$ which completes the proof in this case. We therefore assume m odd. The Adams spectral sequence shows $|\ker\alpha_\pi \cap ko_m(B\pi,\xi)| \le \epsilon(m,n,\xi)$ where

(1) If $m = 8k+1$, $\epsilon(m,n,\xi_0) := (n/2)^{2k+1}$ and $\epsilon(m,n,\xi_1) := (2n)^{2k+1}$.

(2) If $m := 8k+3$, $\epsilon(m,n,\xi_0) := 2(2n)^{2k+1}$ and $\epsilon(m,n,\xi_1) := (n/2)^{2k+1}$.

(3) If $m := 8k+5$, $\epsilon(m,n,\xi_0) := (n/2)^{2k+2}$ and $\epsilon(m,n,\xi_1) := (2n)^{2k+2}$.

(4) If $m := 8k+7$, $\epsilon(m,n,\xi_0) := (2n)^{2k+2}$ and $\epsilon(m,n,\xi_1) := (n/2)^{2k+2}$.

We define

(1) $\mathcal{P}_{4k-1}(B\pi;\xi_0) := \text{span}_{\mathbb{Z}}\{L^{4k-1}(n;\vec{a})\} \subset ko^+_{4k-1}(B\pi;\xi_0)$

(2) $\mathcal{P}_{4k+1}(B\pi;\xi_0) := \text{span}_{\mathbb{Z}}\{X^{4k+1}(n;\vec{a})\} \subset ko^+_{4k+1}(B\pi;\xi_0)$

(3) $\mathcal{P}_{4k-1}(B\pi;\xi_0) := \text{span}_{\mathbb{Z}}\{X^{4k-1}(n;\vec{a})\} \subset ko^+_{4k-1}(B\pi;\xi_0)$

(4) $\mathcal{P}_{4k+1}(B\pi;\xi_0) := \text{span}_{\mathbb{Z}}\{L^{4k+1}(n;\vec{a})\} \subset ko^+_{4k+1}(B\pi;\xi_0)$

It is then a fairly straightforward if somewhat tedious computation using Lemmas 3.10 and 3.12 to see that $|\eta^*\mathcal{P}_m(B\pi;\xi)| \ge \epsilon(m,n,\xi)$ except if $m \equiv 3 \bmod 8$ and $\xi = \xi_0$ and that $|\eta^\#\mathcal{P}_m(B\pi;\xi)| \ge \epsilon(m,n,\xi)$ if $m \equiv 3 \bmod 8$ and $\xi = \xi_0$. This shows $\mathcal{P}_m(B\pi;\xi) = \ker\alpha_\pi \cap ko^+_m(B\pi;\xi)$. □

4.3. Proof of Theorem 1.13. The argument is similar to that given for the proof of Theorems 1.7 and 1.11. We use the twisted product construction discussed in §3.13. Let $\pi = \mathbb{Z}_{2n}$ where $n = 2^\ell > 1$. Let

$$\mathcal{P}_{4k-2} := \text{span}_{\mathbb{Z}}\{N(L^{2k-1}(n;\vec{a})), N(X^{2k-1}(n;\vec{a}))\} \subset ko^+_{4k-2}(B\pi;\xi_2)$$

be the span of all possible twisted products of $2k-1$ dimensional lens spaces and lens space bundles. We use Lemma 3.14 to express the eta invariant of a twisted product in terms of the eta invariant of the original manifold and then use Lemmas 3.10 and 3.12 to evaluate the resulting eta invariants η^* if $m \equiv 6 \bmod 8$ and the refined eta invariant $\eta^\#$ if $m \equiv 2 \bmod 8$ to establish a lower bound for the order of the groups $\mathcal{P}_m(B\pi;\xi_2)$. We use the Adams spectral sequence to establish an upper bound for the orders of the connective K-theory groups. The two bounds agree so $\mathcal{P}_m(B\pi;\xi_2) = \ker\alpha_\pi \cap ko_m(B\pi;\xi_2)$ if $m \equiv 2 \bmod 4$. This completes the proof. □

4.4. Remark. Let $n = 2^\ell$. We have shown that η^* or the refined eta invariant $\eta^\#$ completely detects $ko_{2\nu-1}(B\mathbb{Z}_n;\xi_0)$, $ko_{2\nu-1}(B\mathbb{Z}_n;\xi_1)$, and $ko_{4\nu-2}(B\mathbb{Z}_n;\xi_2)$; η^* and $\eta^\#$ provide the characteristic numbers of the theory. If m is odd, it is fairly easy to establish the Gromov–Lawson–Rosenberg conjecture for $\Omega_m(B\mathbb{Z}_n;\xi_2)$ and $\Omega_m(B\mathbb{Z}_n;\xi_3)$. The difficulty we have encountered in proving the remaining cases of the Gromov–Lawson–Rosenberg conjecture for m even and $\xi = \xi_2, \xi_3$ is to find suitable manifolds which (1) admit metrics of positive scalar curvature and which (2) permit the calculation of their eta invariants.

4.5. Proof of Theorem 1.9. Again, it suffices to work with the reduced groups. Let $\pi = \pi(n,k)$ where n is an odd prime power. We note that α_π vanishes on $\tilde{\Omega}_*(B\pi)$. Thus by Corollary 2.3, we must show show that if m is odd and if $k \leq 3$ or if m is even and $k \leq 2$, then the elements of $\tilde{\Omega}_m(B\pi)$ can all be represented by manifolds that admit metrics of positive scalar curvature. We proceed by induction. If $k = 1$, this follows from Theorem 1.5. Let $\pi = G_1 \oplus G_2$ be a non-trivial direct sum decomposition of π. Cartesian product defines a natural homomorphism $\tilde{\Omega}_*(BG_1) \otimes \tilde{\Omega}_*(BG_2) \to \tilde{\Omega}_*(B\pi)$. Let $\tilde{D}_m(BG)$ be the subgroup of $\tilde{\Omega}_*(B\pi)$ which is generated by the image of all such Cartesian products. Let $\tilde{T}_m(B\pi)$ be the subgroup of $\tilde{\Omega}_m(B\pi)$ generated by all geometrical $\mathbb{HP}^2$ fibrations as defined in §3.4. Let

$$\mathcal{A}_m(B\pi) := \tilde{\Omega}_m(B\pi)/[\tilde{T}_m(B\pi) + \tilde{D}_m(B\pi)].$$

The manifolds of $T_m(B\pi)$ all admit metrics of positive scalar curvature. Since G_1 and G_2 have lower rank than π, the manifolds of $\tilde{D}_m(B\pi)$ all admit metrics of positive scalar curvature by induction. Thus it suffices to show the elements of $\mathcal{A}_m(B\pi)$ can be represented by manifolds that admit metrics of positive scalar curvature. If m is even and if $k = 2$, then the Adams spectral sequence shows $\mathcal{A}_m(B\pi) = 0$. We therefore suppose m is odd. Let Ξ be an embedding of $\mathbb{Z}_n$ into π. We adopt the notation of §3.16. Let

$$\mathcal{P}_m := \text{span}_{\mathbb{Z}}\{\Xi_* L^m(n;\vec{a})\} \subset \mathcal{A}_m(B\pi)$$

where we take all possible embeddings and all possible lens spaces. We use the Adams spectral sequence and results of Johnson and Wilson [17] to obtain an upper estimate on $|\mathcal{A}_m(B\pi)|$. We use Lemma 3.10 and the observation made in §3.16 to obtain a lower estimate for $|\eta^*\mathcal{P}_m(B\pi)|$. This uses the action of the group of units in the field $\mathbb{Z}_p$ which is why we must impose the restriction $m \leq 2(p-1)$ if n is not prime. These two estimates coincide and the theorem is proved. We are not able to handle even dimensional manifolds in the rank 3 case; consequently, we are not able to handle the decomposable manifolds in odd dimensions in rank 4 and the proof stops at this point. □

References

[1] M. F. Atiyah, V. K. Patodi, and I. M. Singer, Spectral asymmetry and Riemannian geometry, I, Math. Proc. Cambridge Philos. Soc. 77 (1975), 43–69.

[2] ——, Spectral asymmetry and Riemannian geometry, II, Math. Proc. Cambridge Philos. Soc. 78 (1975), 405–432.

[3] —— Spectral asymetry and Riemannian geometry, III, Math. Proc. Cambridge. Philos. Soc. 79 (1976), 71–99.

[4] E. Barrera-Yanez, Ph. D. thesis, University of Oregon 1997.

[5] B. Botvinnik and P. Gilkey, The eta invariant and metrics of positive scalar curvature, Math. Ann. 302 (1995), 507–517.

[6] ——, The eta invariant, equivariant spin bordism, and metrics of positive scalar curvature, in: Partial Differential Operators and Mathematical Physics (M. Demuth and B. W. Schulze, eds.), Birkhäuser 1995, 141–152.

[7] ——, Metrics of positive scalar curvature on spherical space forms, Canadian Math J 48 (1996), 64–80.

[8] ——, The Gromov–Lawson–Rosenbeerg conjecture: the twisted case, Houston J. Math., to appear.

[9] ——, The eta invariant and the Gromov-Lawson conjecture for elementary Abelian groups of odd order, Topology Appl., to appear.

[10] B. Botvinnik, P. Gilkey, and S. Stolz, The Gromov–Lawson–Rosenberg conjecture for groups with periodic cohomology, J. Differential Geom., to appear.

[11] H. Donnelly, Eta invariant for G spaces, Indiana Univ. Math. J. 27 (1978), 889–918.

[12] P. Gilkey, The geometry of spherical space form groups, Ser. Pure Math. 7, World Scientific Press, Singapore 1989.

[13] —— The eta invariant of pin manifolds with cyclic fundamental groups, preprint.

[14] M. Gromov and H. B. Lawson, Jr., Spin and scalar curvature in the presence of a fundamental group I, Ann. of Math 111 (1980), 209–230.

[15] ——, The classification of simply connected manifolds of positive scalar curvature, Ann. of Math. 111 (1980), 423–434.

[16] N. Hitchin, Harmonic spinors, Adv. in Math. 14 (1974), 1–55.

[17] B. Johnson and S. Wilson, The BP homology of elementary p-groups, Amer. J. Math. 102 (1982), 427–454.

[18] M. Kreck and S. Stolz, HP^2-bundles and elliptic homology, Acta Math. 171 (1993), 231–261.

[19] S. Kwasik and R. Schultz, Positive scalar curvature and periodic fundamental groups, Comment. Math. Helv. 65 (1990), 271–286.

[20] ——, Fake spherical space forms of constant positive scalar curvature, Comment. Math. Helv. 71 (1996), 1–40.

[21] H. B. Lawson and M. L. Michelsohn, Spin geometry, Princeton Math. Ser. 38, Princeton Univ. Press, Princeton, NJ, 1989.

[22] A. Lichnerowicz, Spineurs harmoniques, C. R. Acad. Sci. Paris 257 (1963), 7–9.

[23] T. Miyazaki, On the existence of positive curvature metrics on non simply connected manifolds, J. Fac. Sci. Univ. Tokyo Sect IA 30 (1984), 549–561.

[24] J. Rosenberg, C*-algebras, positive scalar curvature, and the Novikov Conjecture I, Publ. Math. I. H. E. S. 58 (1983), 197–212.

[25] ——, C*-algebras, positive scalar curvature, and the Novikov Conjecture II, in: Geometric Methods in Operator Algebras, Pitman Res. Notes 123, 1986, 341–374.

[26,] ——, C*-algebras, positive scalar curvature, and the Novikov Conjecture III, Topology 25 (1986), 319–336.

[27] J. Rosenberg and S. Stolz, Manifolds of positive scalar curvature, Algebraic Topology and its applications (G. Carlsson, R. Cohen, W. C. Hsiang, and J. D. S. Jones, eds.), Math. Sci. Res. Inst. Publ. 27, Springer-Verlag, 1994, 241–267.

[28] ——, A stable version of the Gromov–Lawson conjecture, to appear in the Proceedings of the Centennial Czech conference.

[29] ——, A general approach to the Gromov–Lawson–Rosenberg conjecture, preprint.

[30] R. Schoen and S.-T. Yau, On the structure of manifolds with positive scalar curvature, Manuscripta Math. 28 (1969), 159–183.

[31] R. Schultz, Positive scalar curvature and odd order Abelian fundamental groups, preprint.

[32] S. Stolz, Simply connected manifolds of positive scalar curvature, Ann. of Math. 136 (1992), 511–540.

[33] ——, Splitting certain MSpin module spectra, Topology 33 (1994), 159–180.

Locally conformally Kählerian structures on complex manifolds and uniformization

Yoshinobu Kamishima

1991 Mathematics Subject Classification: 57S25, 51M10.

Introduction

A *locally conformally symplectic structure* on a $2n$-manifold W is a family of forms $\{U_\alpha, \Omega_\alpha\}_{\alpha\in\Lambda}$ consisting of an open cover $\{U_\alpha\}_{\alpha\in\Lambda}$ of W and a symplectic form Ω_α on each U_α which satisfies that if $U_\alpha \cap U_\beta \neq \emptyset$, then $\Omega_\beta = \lambda_{\beta\alpha}\Omega_\alpha$, where $\lambda_{\beta\alpha}$ is positive constant. In addition, if there exists a complex structure J on W such that each Ω_α is invariant under J, *i.e.*, Ω_α is a Kähler form on U_α, then the family $\{U_\alpha, \Omega_\alpha\}_{\alpha\in\Lambda}$ is called a *locally conformally Kählerian structure* on a complex manifold (W, J). A manifold equipped with this structure is said to be a *locally conformally symplectic* (*resp. Kählerian*) manifold. Examples of such (compact) manifolds are obviously Kähler manifolds. However, we are mainly interested in locally conformally Kählerian manifolds which do not admit any Kähler structure. For this purpose, we give a class of compact locally conformally Kählerian manifolds having the first Betti number b_1 odd. As a Kähler condition requires Betti numbers of odd dimensional cohomology groups to be even, such manifolds cannot admit the structure of Kähler manifolds. We prove the following.

Theorem A (Realization). *Given a standard pseudo-Hermitian manifold $\tilde{M}$ and a subgroup $\Gamma \subset \mathrm{Psh}(\tilde{M}) \times \mathbb{R}^+$ acting properly discontinuously and freely on $\tilde{W} = \tilde{M} \times \mathbb{R}^+$ with compact quotient, there exists a complex structure J on the orbit space $W = \tilde{W}/\Gamma$ for which the compact complex manifold (W, J) admits a locally conformally Kählerian structure.*

In particular, the product of a compact standard pseudo-Hermitian manifold M with S^1 admits a locally conformally Kählerian structure. The typical example of compact standard pseudo-Hermitian manifolds is the total space M of a circle bundle over a compact Hodge manifold B with nonzero euler class (cf. [1]). Using the Thom–Gysin exact sequence, $H^1(M;\mathbb{Q}) \approx H^1(B;\mathbb{Q})$, and the rank of $H^1(B;\mathbb{Q})$ is even by the Kähler condition. Thus $W = M \times S^1$ has the odd first Betti number so that W cannot be a Kähler manifold. Besides these, the Brieskorn homology sphere Σ^{2n-1} is known to be a compact simply connected standard pseudo-Hermitian manifold (cf. [9], [10]), so $\Sigma^{2n-1} \times S^1$ is an example of locally conformally Kählerian manifolds which is not a Kähler manifold. In particular, there exists a locally conformally Kählerian structure on a Hopf manifold $S^{2n-1} \times S^1$. The existence of locally conformally Kählerian structure on the Hopf manifold was shown by Vaisman ([21]) in the different method.

In the previous paper ([8]), we have classified compact Kähler manifolds with vanishing Bochner tensor. S. Bochner introduced the basic curvature tensor on Kähler manifolds, called *Bochner curvature tensor.* This tensor can be also derived from the Weyl's principle on unitary group representation theory (cf. [23], [2], [21]). When Hermitian manifolds are taken into account in the framework of conformal geometry, Tricerri and Vanhecke ([20]) have defined a *generalized Bochner curvature tensor* on Hermitian manifolds. This curvature tensor coincides with the original Bochner curvature tensor when a Hermitian manifold is Kähler. They observed that the generalized Bochner curvature tensor is a *conformal invariant* on Hermitian metrics on a Hermitian manifold. Here the complex structure is fixed. As a consequence, when a Hermitian manifold is a locally conformally Kählerian manifold, the (generalized) Bochner curvature tensor has the same formula as the (original) Bochner curvature tensor.

A *Bochner curvature flat Hermitian manifold* is defined to be a Hermitian manifold with vanishing (generalized) Bochner curvature tensor. It is unknown which compact Hermitian manifold supports a Bochner curvature flat metric. In addition, the classification of compact Bochner curvature flat Hermitian manifolds is far from valid as well as the classification of compact conformally flat manifolds.

From this observation, we restrict Hermitian manifolds to the class of locally conformally Kählerian manifolds. Then we prove the following uniformization theorem on compact locally conformally Kählerian manifolds: (Compare [8], [11].)

Theorem B (Uniformization). *Let (M, g, J) be a $2n$-dimensional compact locally conformally Kählerian manifold $(n > 1)$. If the Bochner curvature tensor vanishes, then M is conformally equivalent to one of the following locally conformally Kählerian manifolds*:

(i) *The complex projective space $\boldsymbol{CP}^n$.*

(ii) *A complex euclidean space form $T^n_{\mathbb{C}}/F$ $(F \subset U(n))$.*

(iii) *A complex hyperbolic space form $\mathbb{H}^n_{\mathbb{C}}/\Gamma$ $(\Gamma \subset \mathrm{PU}(n,1))$.*

(iv) *A fiber space $\mathbb{H}^m_{\mathbb{C}} \times_{\Gamma} CP^{n-m}$ $(\Gamma \subset \mathrm{PU}(m,1) \times \mathrm{PU}(n-m+1), m = 1, 2, \ldots, n-1)$.*

(v) *A Hopf manifold $S^{2n-1} \times_F S^1$ $(F \subset U(n) \times S^1)$.*

Here,

(1) *F is a finite group and* Γ *is a discrete cocompact subgroup which acts freely and properly discontinuously.*

(2) *The manifolds of the above* (i), (ii), (iii), (iv) *are Kähler manifolds. When* g *is a Kähler metric from the beginning,* (M, g, J) *is holomorphically isometric (up to a constant scalar multiple of the Kähler metric) to one of the above Kähler manifolds of* (i), (ii), (iii), (iv).

1. Symplectization of pseudo-Hermitian manifold

Given a locally conformally symplectic (Kählerian) structure $\{U_\alpha, \Omega_\alpha\}_{\alpha\in\Lambda}$, it is easy to check that

$$\lambda_{\beta\alpha}\lambda_{\alpha\gamma} = \lambda_{\beta\gamma} \text{ on } U_\alpha \cap U_\beta \cap U_\gamma.$$

Thus $\{\lambda_{\alpha\beta}\}_{\alpha,\beta\in\Lambda}$ is a 1-cocycle in $H^1(W; \mathbb{R}^+)$. As usual the cocycle $\{\lambda_{\alpha\beta}\}_{\alpha,\beta\in\Lambda}$ defines a line bundle E whose first Stiefel–Whitney class $w_1(E)$ coincides with the cocycle $\{\lambda_{\alpha\beta}\}_{\alpha,\beta\in\Lambda}$. Since $\lambda_{\alpha\beta}$ is positive, the bundle E is orientable, or $w_1(E) = 0$. So there is a family of functions $\{f_\alpha\}$ on the neighborhoods $\{U_\alpha\}$ such that $\delta^0 f = \lambda$ (*i.e.*, $f_\alpha \cdot (f_\beta)^{-1} = \lambda_{\beta\alpha}$). If $\Omega_\beta = \lambda_{\beta\alpha}\Omega_\alpha$, then $f_\beta \cdot \Omega_\beta = f_\alpha \cdot \Omega_\alpha$ on $U_\alpha \cap U_\beta$.

Proposition 1.1. *Given a locally conformally Kählerian structure* $\{U_\alpha, \Omega_\alpha\}_{\alpha\in\Lambda}$ *on a complex manifold* (W, J), *there exist a 2-form* Ω *on* W *such that* $\Omega|U_\alpha = f_\alpha \cdot \Omega_\alpha$ *and a Hermitian metric* $g(X, Y) = \Omega(X, JY)$.

The fundamental 2-form $\Omega(X, Y) = g(JX, Y)$ satisfies that n times $\Omega\wedge\cdots\wedge\Omega \neq 0$, but Ω is not necessarily closed from this propostion. Ω is not a Kähler form in general. Moreover $g_\alpha(X, Y) = \Omega_\alpha(X, JY)$ is a Kähler metric locally defined on each U_α such that $g|U_\alpha = f_\alpha \cdot g_\alpha$. At this stage, it is worthwhile to mention that Tricerri and Vanhecke defined a generalized Bochner curvature tensor $\mathcal{B}(R)$ on Hermitian manifolds which is a conformal invariant on Hermitian manifolds.

Theorem 1.2 (Theorem 11.1 of [20]). *Let* g, g' *be Hermitian metrics on a Hermitian manifold* M *and* R *(resp.* R'*) the corresponding curvatute tensor. If* $g' = \lambda \cdot g$ *for some positive function* λ, *then the Bochner curvature tensor* $\mathcal{B}(R')$ *of* g' *coincides with* $\mathcal{B}(R)$ *of* g.

By Proposition 1.1, we obtain that

Corollary 1.3. *Let* W *be a locally conformally Kählerian manifold. Then there is a Hermitian metric* g *such that* g *is locally conformal to a Kähler metric. In particular, the (generalized) Bochner curvature tensor of* (W, g) *coincides with that of* g_α, *restricted to each* U_α.

In order to prove Theorem A, we shall recall *pseudo-Hermitian structure* on odd dimensional manifolds. A pseudo-Hermitian structure on a $(2n+1)$-dimensional smooth manifold M consists of a pair (ω, J) for which ω is a *contact form* of M, and $\mathrm{Null}\,\omega = \{X \in TM \mid \omega(X) = 0\}$ is a codimension 1-subbundle of the tangent bundle TM, which is called a *contact subbundle.* J is a complex structure on $\mathrm{Null}\,\omega$. Namely, J is an almost complex structure on $\mathrm{Null}\,\omega$ which satisfies that if $\mathrm{Null}\,\omega \otimes \mathbb{C} = T^{1,0} \oplus T^{0,1}$ is the canonical splitting, then $[T^{1,0}, T^{1,0}] \subset T^{1,0}$. In addition a pseudo-Hermitian structure (or a *CR*-structure) is *strictly pseudoconvex* if the Hermitian pairing $Q : T^{1,0} \times T^{1,0} \to \mathbb{C}$ defined by $Q(X, Y) = d\omega(JX, \bar{Y})$ is positive definite (cf. [4], [22]). The pair $(\mathrm{Null}\,\omega, J)$ is called a *CR*-structure. (Compare §3.)

Proposition 1.4. *Let ω be a contact structure on a smooth manifold M. Then there exists a unique vector field ξ satisfying that;*

(i) $\omega(\xi) = 1$ *and* $d\omega(\xi, X) = 0$ *for all* $X \in TM$.

(ii) $L_\xi \omega = 0$ *and* $L_\xi d\omega = 0$.

In fact, if we note that $L_\xi = d \circ \iota + \iota \circ d$, then (ii) follows from (i). The vector field ξ is called a *characteristic vector* field for ω. Recall the following definition (cf. [12], [22], [18]).

Definition 1.5. Let (ω, J) be a pseudo-Hermitian structure on M for which $(\mathrm{Null}\,\omega, J)$ is a *CR*-structure on M.

(1) A diffeomorphism $f : M \to M$ is a pseudo-Hermitian transformation if $f_* \circ J = J \circ f_*$ and $f^*\omega = \omega$.

(2) A diffeomorphism $f : M \to M$ is a *CR* transformation if $f_* \circ J = J \circ f_*$ and the differential map f_* preserves $\mathrm{Null}\,\omega$ (*i.e.,* $f^*\omega = \lambda \cdot \omega$ for some positive function λ.)

(3) Let $\{\phi_t\}_{|t|<\epsilon}$ be a local 1-parameter group of transformations of M induced by the characteristic vector field ξ. If for each t, ϕ_t is a *CR* transformation of M, then ξ is called a characteristic *CR* vector field.

(4) A standard pseudo-Hermitian structure is a strictly pseudoconvex pseudo-Hermitian structure (ω, J) equipped with a characteristic *CR* vector field ξ.

A smooth manifold M is said to be a standard pseudo-Hermitian manifold if M admits a standard pseudo-Hermitian structure.

Let $\mathrm{Psh}(M)$ be the group of all pseudo-Hermitian transformations on a strictly pseudoconvex pseudo-Hermitian manifold M of dimension $2n-1$. We know from the theory of G-structure that $\mathrm{Psh}(M)$ is a Lie group with dimension less than or equal to n^2. $\mathrm{Psh}(M)$ is compact if M is compact. (Compare [22], [13].)

Now our method to produce a (compact) locally conformally Kählerian manifold is *symplectization* of a (compact) pseudo-Hermitian manifold. Let $(\tilde{\omega}, \tilde{J})$ be a strictly pseudoconvex pseudo-Hermitian structure on a smooth $(2n-1)$-manifold $\tilde{M}$. Consider the product $\tilde{W} = \tilde{M} \times \mathbb{R}^+$. Put $\tilde{\Omega} = d(t\tilde{\omega})$ where t is the coordinate of $\mathbb{R}^+$. Then $\tilde{\Omega}$ is a symplectic structure on $\tilde{W}$. Let ξ be a characteristic vector field on M for $\tilde{\omega}$. Recall that $T\tilde{W} = \mathrm{Null}\,\tilde{\omega} \oplus \xi \oplus \frac{d}{dt}$. Define $\tilde{J} : T(\tilde{W}) \to T(\tilde{W})$ by setting $\tilde{J}(\xi_{(x,t)}) = t\frac{d}{dt}$, $\tilde{J}(\frac{d}{dt}) = \frac{-1}{t}\xi_{(x,t)}$, and $\tilde{J}|\,\mathrm{Null}\,\tilde{\omega} = \tilde{I}$. Then $\tilde{J}$ is an almost complex structure on $\tilde{W}$.

Since $\tilde{\Omega}(\tilde{J}\xi, \tilde{J}\frac{d}{dt}) = \tilde{\Omega}(t\frac{d}{dt}, \frac{-1}{t}\xi) = \tilde{\Omega}(\xi, \frac{d}{dt})$ and by the properties (i), (ii) of (2.1), $\tilde{\Omega}$ is invariant under $\tilde{J}$.

Lemma 1.6. *If $\tilde{M}$ is a standard pseudo-Hermitian manifold, then $\tilde{J}$ is a complex structure on $\tilde{W}$. In particular, $\tilde{\Omega}$ is a Kähler form.*

Proof. If $\mathrm{Null}\,\tilde{\omega} \otimes \mathbb{C} = T^{1,0} \oplus T^{0,1}$ for $\tilde{I}$, then $T\tilde{W} \otimes \mathbb{C} = \{T^{1,0} + (\xi - i\tilde{J}\xi)\} \oplus \{T^{0,1} + (\xi + i\tilde{J}\xi)\}$. Put $S^{1,0} = T^{1,0} + (\xi - i\tilde{J}\xi)$. We prove that $[X, Y] \in S^{1,0}$ whenever $X, Y \in S^{1,0}$. Since $\tilde{I}$ is a complex structure on $\mathrm{Null}\,\tilde{\omega}$, this is true for $X, Y \in T^{1,0}$. For $X \in T^{1,0}$, we have $[X, \xi - i\tilde{J}\xi] = [X, \xi]$ because $\tilde{J}\xi = t\frac{d}{dt}$. Let $\{\phi_t\}_{|t|<\epsilon}$ be a local 1-parameter group induced by ξ. Since $[X, \xi] = \lim_{\theta\to 0} \frac{X - \phi_{\theta *}X}{\theta}$ and ξ is a characteristic *CR* vector field,

$$\tilde{J}[X, \xi] = \lim_{\theta\to 0} \frac{\tilde{I}X - \tilde{I} \circ \phi_{\theta *}X}{\theta} = \lim_{\theta\to 0} \frac{\tilde{I}X - \phi_{\theta *} \circ \tilde{I}X}{\theta} = [\tilde{I}X, \xi] = i[X, \xi].$$

Thus we obtain that $[X, \xi - i\tilde{J}\xi] \in S^{1,0}$. □

Let $\mathrm{Psh}(\tilde{M}) \times \mathbb{R}^+$ be the product group acting on $\tilde{M} \times \mathbb{R}^+$ by

$$(f, t_a)(x, t) = (f(x), at),$$

where $f \in \mathrm{Psh}(\tilde{M})$, $a \in \mathbb{R}^+$.

Lemma 1.7. *The group* $\mathrm{Psh}(\tilde{M}) \times \mathbb{R}^+$ *acts as homothetic transformations of $\tilde{W}$ with respect to $\tilde{\Omega}$. Moreover, each element of* $\mathrm{Psh}(\tilde{M}) \times \mathbb{R}^+$ *preserves the complex structure $\tilde{J}$ of $\tilde{W}$.*

Proof. By the definition each element of $\mathrm{Psh}(\tilde{M})$ preserves $\tilde{\Omega} = d(t\tilde{\omega})$ obviously. If $t_a \in \mathbb{R}^+$, then $t_a^*\tilde{\Omega} = d(at \cdot \omega) = a \cdot \tilde{\Omega}$. So, $\mathbb{R}^+$ acts as homothetic transformations. Recall that $\tilde{J}|\,\mathrm{Null}\,\tilde{\omega} = \tilde{I}$, $\tilde{J}(\xi, \frac{d}{dt}) = (t\frac{d}{dt}, \frac{-1}{t}\xi)$. If $f \in \mathrm{Psh}(\tilde{M})$, then by the definition $f_* \circ \tilde{J} = \tilde{J} \circ f_*$ on $\mathrm{Null}\,\tilde{\omega}$. Since $f^*\tilde{\omega} = \tilde{\omega}$ and (i) of (1.4), we obtain $f_*\xi = \xi$. As $f = id$ on $\mathbb{R}^+$-summand, it follows that $f_* \circ \tilde{J} = \tilde{J} \circ f_*$.

If $t_a \in \mathbb{R}^+$, then $t_{a*} \circ \tilde{J} = \tilde{J} \circ t_{a*}$ on Null $\tilde{\omega}$ because $t_{a*} = id$ on Null $\tilde{\omega}$. It is easy to see that $t_{a*}(\xi, \frac{d}{dt}) = (\xi, a\frac{d}{dt})$. So we have

$$\tilde{J}_{(x,at)} \circ t_{a*}(\xi, \frac{d}{dt}) = (at\frac{d}{dt}, \frac{-1}{t}\xi) = (a(t\frac{d}{dt}), \frac{-1}{t}\xi) = t_{a*} \circ \tilde{J}_{(x,t)}(\xi, \frac{d}{dt}).$$

□

Proof of Theorem A.

Let Γ be a subgroup of $\mathrm{Psh}(\tilde{M}) \times \mathbb{R}^+$ acting properly discontinuously and freely on $\tilde{W}$ with compact quotient. By Lemma 1.7, $\tilde{J}$ induces a complex structure J on the quotient space $W = \tilde{W}/\Gamma$. Moreover, W is homeomorphic to $M \underset{F}{\times} S^1$ where F is a finite subgroup of $\mathrm{Psh}(M) \times S^1$ acting freely on $M \times S^1$. Let $P : \tilde{W} \longrightarrow W$ be the covering map. First note that each element $h \in \Gamma$ satisfies that

$$h^*\tilde{\Omega} = a(h) \cdot \tilde{\Omega}$$

for some constant $a(h) \in \mathbb{R}^+$. Choose an evenly covered neighborhoods $\{U_\alpha\}_{\alpha\in\Lambda}$ on W. By the definition, there is a neighborhood $\tilde{U}_\alpha$ of $\tilde{W}$ such that $P : \tilde{U}_\alpha \rightarrow U_\alpha$ is homeomorphic, and $P^{-1}(U_\alpha) = \underset{\gamma\in\Gamma}{\cup}(\gamma \cdot \tilde{U}_\alpha)$. Furthermore $\gamma \cdot \tilde{U}_\alpha \cap \tilde{U}_\alpha \neq \emptyset$ if and only if $\gamma = 1$. Define a 2-form Ω_α on U_α by push-forward;

$$P^*\Omega_\alpha = \tilde{\Omega}|\tilde{U}_\alpha.$$

Obviously Ω_α is a Kähler form on (U_α, J). We obtain a family of local Kähler forms $\{U_\alpha, \Omega_\alpha\}_{\alpha\in\Lambda}$ on W. If $U_\alpha \cap U_\beta \neq \emptyset$, then $\tilde{U}_\beta \cap \gamma \cdot \tilde{U}_\alpha \neq \emptyset$ for some $\gamma \in \Gamma$. As above, $\gamma^*\tilde{\Omega} = a(\gamma) \cdot \tilde{\Omega}$ for some positive constant $a(\gamma)$. Put $a(\gamma) = a_{\beta\alpha}$. Then, on the intersection $\tilde{U}_\beta \cap \gamma \cdot \tilde{U}_\alpha$,

$$P^*\Omega_\beta = \tilde{\Omega}|\tilde{U}_\beta = \tilde{\Omega}|\tilde{U}_{\gamma\cdot\tilde{U}_\alpha} = \gamma^*\tilde{\Omega}|\tilde{U}_\alpha = a_{\beta\alpha} \cdot \tilde{\Omega}|\tilde{U}_\alpha = a_{\beta\alpha} \cdot P^*\Omega_\alpha.$$

Hence $\Omega_\beta = a_{\beta\alpha} \cdot \Omega_\alpha$. The family $\{U_\alpha, \Omega_\alpha\}_{\alpha\in\Lambda}$ is a locally conformally Kählerian structure on (W, J). □

Remark 1.8.

1. Recall the similarity geometry $(\mathbb{C}^n \rtimes (U(n) \times \mathbb{R}^+), \mathbb{C}^n)$ equipped with the canonical Kähler form Ω_0 and the complex structure J_0. The complex number space $\mathbb{C}^n$ with one point removed is homeomorphic to $S^{2n-1} \times \mathbb{R}^+$. We have the Hopf geometry $(U(n) \times \mathbb{R}^+, S^{2n-1} \times \mathbb{R}^+)$. As before choosing an infinite discrete subgroup Γ from $U(n) \times \mathbb{R}^+$ which acts freely, we have a Hopf manifold $W = S^{2n-1} \underset{\Gamma}{\times} \mathbb{R}^+$. Since $U(n) \times \mathbb{R}^+$ is a subgroup of holomorphic transformations of $\mathbb{C}^n$, the Hopf manifold W is a complex manifold. Note that W is an S^1-bundle $S^{2n-1} \underset{F}{\times} S^1$ over a spherical orbifold S^{2n-1}/F.

On the other hand, the sphere S^{2n-1} equipped with the spherical CR-structure is the standard pseudo-Hermitian manifold with automorphism group $U(n)$. By Theorem A, the Hopf manifold (W, J) is a locally conformally Kählerian manifold. In this case, our Kähler structure $(\tilde{\Omega}, \tilde{J})$ obtained above is exactly the Kähler structure (Ω_0, J_0) restricted to $\mathbb{C}^n - \{0\} = S^{2n-1} \times \mathbb{R}^+$.

2. Similar to Hopf's construction, Calabi–Eckmann, Morimoto ([16], [13]) have constructed a complex structure $\bar{J}$ on the product $M_1 \times M_2$, where M_i $(i = 1, 2)$ is a standard pseudo-Hermitian manifold. Moreover, there is a Hermitian metric on the complex manifold $M_1 \times M_2$. In particular, Calabi–Eckmann manifolds $S^{2k+1} \times S^{2\ell+1}$ are obtained as above when we take an odd dimensional sphere as a standard pseudo-Hermitian manifold (spherical CR-manifold) M_i. It is unknown which $(M_1 \times M_2, g, \bar{J})$ admits a locally conformally symplectic structure.

2. Uniformization of L. C. Kählerian manifolds

We collect some facts on CR-geometry. Let (ω, J) be a pseudo-Hermitian structure on a smooth manifold N^{2n+1} for which $\mathrm{Null}\,\omega = \{X \in TN | \omega(X) = 0\}$ is the contact subbundle. Denote by $[\omega]$ the conformal class of ω, *that is*, $\omega' \in [\omega]$ if and only if $\omega' = u \cdot \omega$ for a positive function u. Note that ω' is also a contact form of N and $\mathrm{Null}\,\omega' = \mathrm{Null}\,\omega$.

A *CR-structure* on N^{2n+1} is a pair $([\omega], J)$. A smooth manifold N^{2n+1} with a CR-structure is said to be a *CR-manifold.* The CR-geometry has been extensively developed by S. S. Chern, J. Moser, S. Webster and others. By making use of the structure equations of a real hypersurface in $\mathbb{C}^{n+1}$, Chern and Moser ([4]) found a fourth-order curvature tensor $S = (S_{\alpha\beta\bar{\rho}\bar{\sigma}})$ on N^{2n+1} which is a CR invariant.

In view of Weyl's principle, it is conceivable that $S = (S_{\alpha\beta\bar{\rho}\bar{\sigma}})$ plays the same role as the Weyl curvature tensor in CR geometry. For this, among the numerical invariants of pseudo-Hermitian manifolds introduced by Webster ([22]), there is a fourth-order curvature tensor $R = (R_{\alpha\bar{\rho}\beta\bar{\sigma}})$ which behaves like the Riemannian curvature tensor. The Webster curvature tensor R is in general not a CR invariant on N^{2n+1}. However, just like the formula of Weyl curvature tensor, the Chern–Moser curvature tensor $S = (S_{\alpha\beta\bar{\rho}\bar{\sigma}})$ may be related with the Webster curvature tensor $R = (R_{\alpha\bar{\rho}\beta\bar{\sigma}})$ on N^{2n+1}. In fact he gave the formula of S which is the same as the Bochner curvature tensor.

Theorem 2.1 (Webster).

$$\begin{aligned} S_{\beta\rho}{}^{\alpha}{}_{\bar{\sigma}} = {} & R_{\beta}{}^{\alpha}{}_{\rho\bar{\sigma}} - \frac{1}{n+2}(R_{\beta}{}^{\alpha} g_{\rho\bar{\sigma}} + R_{\rho}{}^{\alpha} g_{\beta\bar{\sigma}} + \delta_{\beta}{}^{\alpha} R_{\rho\bar{\sigma}} + \delta_{\rho}{}^{\alpha} R_{\beta\bar{\sigma}}) \\ & + \frac{R}{2(n+1)(n+2)}(\delta_{\beta}{}^{\alpha} g_{\rho\bar{\sigma}} + \delta_{\rho}{}^{\alpha} g_{\beta\bar{\sigma}}) \end{aligned}$$

where $R_{\rho\bar{\sigma}} = R_{\alpha}{}^{\alpha}{}_{\rho\bar{\sigma}}$ *and* $\frac{1}{2}R = g^{\rho\bar{\sigma}} R_{\rho\bar{\sigma}}$ *is the Ricci tensor and the scalar curvature respectively.*

When the sphere S^{2n+1} is viewed as a hyperquadric in $\mathbb{C}^{n+1}$, S^{2n+1} is a *CR*-manifold whose curvature tensor S vanishes identically.

The complex analogue of conformal geometry is that if the Chern–Moser curvature tensor S of a *CR*-manifold N^{2n+1} vanishes, then N^{2n+1} is locally *CR* equivalent to the sphere S^{2n+1} $(n > 1)$. (Compare [4], [22], [7].) In this case, N^{2n+1} is said to be a *spherical CR-manifold.* Let $\mathrm{Aut}_{CR}(S^{2n+1})$ be the group of all *CR* automorphisms of the sphere S^{2n+1}. Note that $\mathrm{Aut}_{CR}(S^{2n+1})$ is isomorphic to the unitary Lorentz group $\mathrm{PU}(n+1, 1)$.

Then the pair $(\mathrm{Aut}_{CR}(S^{2n+1}), S^{2n+1})$ is called *spherical CR-geometry.* By the theorem of Frobenius, the *uniformization* with respect to the spherical *CR*-structure states that N is locally modelled on S^{2n+1} with coordinate changes lying in $\mathrm{Aut}_{CR}(S^{2n+1})$. By the usual monodromy argument (cf. [14], [7], [6]), there is a *developing pair* :

$$(\rho, \mathrm{dev}) : (\mathrm{Aut}_{CR}(\tilde{N}), \tilde{N}) \to (\mathrm{PU}(n+1, 1), S^{2n+1}).$$

Here $\mathrm{Aut}_{CR}(\tilde{N})$ is the group of *CR* automorphisms of the universal covering space $\tilde{N}$ of N. The developing map dev is a *CR* immersion which is *equivariant* with respect to a *holonomy homomorphism* ρ. In general, the developing map is very complicated (cf. [17]). However, we know that if there exists an invariant domain of S^{2n+1}, the *symmetry* imposes a strong restriction on the *CR*-structure on its domain.

Lemma 2.2 (Geometric invariant domain). *Let H be a connected abelian subgroup of* $\mathrm{PU}(n+1, 1)$ *acting invariantly on a domain X of S^{2n+1}. If H has no fixed point on X, then X is contained in one of the following mutually distinct spherical CR-manifolds: $X \subset S^{2n+1}$ for which the closure of H is a proper subgroup of $U(n+1)$, or $X \subset S^{2n+1} - \{\infty\}$, $X \subset S^{2n+1} - S^{2m-1}$ $(m = 1, 2, \dots, n)$.*

See [8], [11] for the proof. Moreover, X is a spherical *CR*-manifold which is a fibration over a Kähler manifold. In fact, let S^{2n+1} be the standard sphere in $\mathbb{C}^{n+1}$. Put $S^1 = \mathcal{Z}(n+1)$, the center of $U(n+1)$. We have an equivariant fibration:

$$\text{(I)} \qquad (S^1, S^1) \to (U(n+1), S^{2n+1}) \xrightarrow{p} (\mathrm{PU}(n+1), CP^n).$$

Here CP^n carries the Fubini–Study metric with the group of isometries $\mathrm{PU}(n+1)$.

Recall that the sphere S^{2n+1} with one point $\{\infty\}$ removed is identified with the Heisenberg nilpotent Lie group $\mathcal{N}$ as a *CR*-structure (cf. [7], [6]). It has the central extension $1 \to \mathcal{R} \to \mathcal{N} \longrightarrow \mathbb{C}^n \to 1$ for which $\mathrm{Aut}_{CR}(\mathcal{N}) = \mathcal{N} \rtimes (U(n) \times \mathbb{R}^+)$. The subgroup $\mathbb{E}_{CR}(\mathcal{N}) = \mathcal{N} \rtimes U(n)$ is the group of isometries of the left invariant metric on $\mathcal{N}$. Note that $\mathcal{R}$ is the center of $\mathbb{E}_{CR}(\mathcal{N})$. Since $\mathcal{R}$ is a normal subgroup of $\mathrm{Aut}_{CR}(\mathcal{N})$, we have an equivariant principal bundle:

$$\text{(II)} \qquad (\mathcal{R}, \mathcal{R}) \to (\mathcal{N} \rtimes (U(n) \times \mathbb{R}^+), \mathcal{N}) \xrightarrow{p} (\mathbb{C}^n \rtimes (U(n) \times \mathbb{R}^+), \mathbb{C}^n).$$

Here $\mathbb{C}^n$ is the complex n-number space with euclidean metric.

Suppose that G has the fixed point set S^{2m-1} for each $m = 1, 2, \ldots, n$. The complement $S^{2n+1} - S^{2m-1}$ is CR equivalent to the quotient of the product of the Lorentz hyperbolic space form with the sphere, $P(V_{-1}^{2m+1} \times S^{2(n-m)+1})$, and $\mathrm{Aut}_{CR}(S^{2n+1} - S^{2m-1}) = P(U(m,1) \times U(n-m+1))$ which has the center $S^1 = P(\mathcal{Z}(m,1) \times \mathcal{Z}(n-m+1))$. We have an equivariant principal bundle:

(III)
$$(S^1, S^1) \to (\mathrm{Aut}_{CR}(S^{2n+1} - S^{2m-1}), S^{2n+1} - S^{2m-1})$$
$$\xrightarrow{p} (\mathrm{PU}(m,1) \times \mathrm{PU}(n-m+1), \mathbb{H}_{\mathbb{C}}^m \times CP^{n-m}).$$

In particular when $m = n$, $S^{2n+1} - S^{2n-1} = P(V_{-1}^{2n+1} \times S^1) = V_{-1}^{2n+1}$ and

$$(S^1, S^1) \to (U(n,1), V_{-1}^{2n+1}) \xrightarrow{p} (\mathrm{PU}(n,1), \mathbb{H}_{\mathbb{C}}^n)$$

is a principal bundle over the complex hyperbolic space. Note that the complex hyperbolic metric on $\mathbb{H}_{\mathbb{C}}^n$ is the Bergman metric on the unit ball of $\mathbb{C}^n$. And the Kähler metric on $\mathbb{H}_{\mathbb{C}}^m \times CP^{n-m}$ is the product of Bergman metric and Fubini–Study metric.

The canonical Kähler metric defines a Kähler form Ω_Y on $Y_{\mathbb{C}}^n = CP^n$, $\mathbb{C}^n$ or $\mathbb{H}_{\mathbb{C}}^m \times CP^{n-m}$ equipped with the canonical complex structure J_Y respectively. There exists a pseudo-Hermitian structure (ω_X, J_X) which represents the spherical CR-structure on $X = S^{2n+1}$, $\mathcal{N}$ or $S^{2n+1} - S^{2m-1}$. Then the principal fibration of (I), (II), (III) satisfies that

(*)
$$d\omega_X = p^*\Omega_Y, \quad p_* \circ J_X = J_Y \circ p_*.$$

In this case note that $\mathrm{Psh}(X, \omega_X) = U(n+1)$, $\mathcal{N} \rtimes U(n)$ or $\mathrm{PU}(m,1) \times \mathrm{PU}(n-m+1)$ according as (I), (II) or (III). (Compare [9], [1] for this construction.)

A geometry $(\mathcal{G}, X)$ consists of a finite dimensional Lie group $\mathcal{G}$ with finitely many components and an n-dimensional homogeneous space X from $\mathcal{G}$. We shall recall *the uniformization* on smooth n-manifolds with respect to $(\mathcal{G}, X)$. (Compare [14],[19].)

Suppose that there exists a maximal collection of charts $\{U_\alpha, \phi_\alpha\}_{\alpha \in \Lambda}$ of M;

$$M = \bigcup_{\alpha \in \Lambda} U_\alpha,$$
$$\phi_\alpha : U_\alpha \to \phi_\alpha(U_\alpha) \ (\subset X) \text{ is a homeomorphism,}$$

satisfying that if $U_\alpha \cap U_\beta \neq \emptyset$, then the coordinate change $g_{\alpha\beta} = \phi_\beta \circ \phi_\alpha^{-1}$ extends to an element of $\mathcal{G}$.

Then the collection of charts is said to give a *uniformization* on M. An equivalence class of uniformizations (by refinement) is called a $(\mathcal{G}, X)$-*structure* on M (or simply, *geometric structure*).

Thus, a $(\mathcal{G}, X)$-structure on a smooth n-manifold M is a geometric structure locally modelled on X with coordinate changes lying in $\mathcal{G}$. A manifold M equipped with a $(\mathcal{G}, X)$-structure is said to be a $(\mathcal{G}, X)$-manifold.

Let $\tilde{M}$ be the universal covering space of M and $\pi_1(M)$ the fundamental group of M. $\mathrm{Aut}(M)$ denotes the group of all $(\mathcal{G}, X)$-automorphisms of M.

A $(\mathcal{G}, X)$-structure on M induces a *developing pair:* (cf. [14])

$$(\rho, \mathrm{dev}) : (\mathrm{Aut}(\tilde{M}), \tilde{M}) \rightarrow (\mathcal{G}, X),$$

where $\pi_1(M) \subset \mathrm{Aut}(\tilde{M})$ and the developing map dev is a $(\mathcal{G}, X)$-structure preserving immersion, equivariant with respect to a holonomy homomorphism ρ. Moreover, the pair (ρ, dev) is unique up to an element of $\mathcal{G}$, *i.e.*, the developing pairs (ρ', dev'), (ρ, dev) define the same structure on M if and only if one is a multiple of the other by an element of $\mathcal{G}$, $\mathrm{dev}' = g \circ \mathrm{dev}$ for some $g \in \mathcal{G}$. It turns out that the developing pair is a geometric invariant on $(\mathcal{G}, X)$-structures on M. We prove the following uniformization of locally conformally Kählerian manifolds of dimension $2n$ $(n > 1)$. (Compare [8], [11].)

Theorem 2.3. *Let (W, g, J) be a locally conformally Kählerian manifold of dimension $2n$ $(n > 1)$. Suppose that the Bochner curvature tensor B vanishes with respect to the Hermitian metric g. Then W is uniformized over $Y^n_{\mathbb{C}}$ with respect to $\mathcal{G}$. Here $Y^n_{\mathbb{C}}$ is a connected simply connected Kähler manifold with a canonical Kähler metric g_Y, equipped with a transitive group of biholomorphic transformations $\mathcal{G}$. More precisely, $(\mathcal{G}, Y^n_{\mathbb{C}}, g_Y)$ is one of the following geometries:*

(1) *A projective geometry* $(\mathrm{PU}(n+1),\ \mathbb{C}P^n;$ *Fubini–Study metric*$)$.

(2) *A similarity geometry* $(\mathbb{C}^n \rtimes (U(n) \times \mathbb{R}^+), \mathbb{C}^n;$ *Euclidean metric*$)$.

(3) *A hyperbolic geometry* $(\mathrm{PU}(n, 1), \mathbb{H}^n_{\mathbb{C}};$ *Hyperbolic metric*$)$.

(4) *A projective-hyperbolic geometry* $(\mathrm{PU}(m, 1) \times \mathrm{PU}(n-m+1), \mathbb{H}^m_{\mathbb{C}} \times \mathbb{C}P^{n-m};$ *Product metric*$)$. $(m = 1, 2, \dots, n-1)$.

Moreover, the uniformization $\{U_\alpha, \phi_\alpha\}_{\alpha\in\Lambda}$ of (W, g, J) has the following property that:

$$c_\alpha \cdot g = \phi^*_\alpha g_Y \quad \text{for some constant } c_\alpha \text{ on } U_\alpha,$$
$$\phi_{\alpha *} \circ J = J_Y \circ \phi_{\alpha *},$$

where g_Y is the Kähler metric of $Y^n_{\mathbb{C}}$ as above and J_Y is the canonical complex structure on $Y^n_{\mathbb{C}}$.

Remark 2.4. The above geometries (1), (2) and (3) are subgeometries of the projective geometry $(\mathrm{PGL}(n+1, \mathbb{C}), \mathbb{C}P^n)$. Except for (2), the group $\mathcal{G}$ is a transitive group of $Y^n_{\mathbb{C}}$ preserving the canonical Kähler structure. The euclidean group $\mathbb{E}_{\mathbb{C}}(n) = \mathbb{C}^n \rtimes U(n)$ preserves the Kähler structure of $\mathbb{C}^n$ as well. We recall the following fact (cf. [24]) that *a conformal transformation preserving complex structures between Kähler manifolds (M^{2n}, g, Ω, J), $(N^{2n}, g', \Omega', J')$ is necessarily homothetic.* Namely, if $h^*g' = \lambda \cdot g$ and $h_* \circ J = J' \circ h_*$ (equivalently, $h^*\Omega' = \lambda \cdot \Omega$, $h_* \circ J = J' \circ h_*$), then λ is a constant, provided that $\dim M = 2n \geq 4$.

The basic tool to proving Theorem 2.3 is *contactization* of Kähler manifolds with exact forms. The contactization produces a pseudo-Hermitian manifold which is the trivial $\mathbb{R}$-bundle over the Kähler manifold.

Proof of Theorem 2.3.

Suppose that a $2n$-dimensional complex manifold (W, g, J) admits a locally conformally Kählerian structure $\{U_\alpha, \Omega_\alpha\}_{\alpha\in\Lambda}$, consisting of an open cover $\{U_\alpha\}_{\alpha\in\Lambda}$ of W and a Kähler form Ω_α on U_α. By Proposition 1.1, $g(X, Y) = \Omega(X, JY)$ is a Hermitian metric and $\Omega|U_\alpha = f_\alpha \cdot \Omega_\alpha$. We may assume that each U_α is homeomorphic to a ball. Then Ω_α is exact, so there is a 1-form θ_α such that $d\theta_\alpha = \Omega_\alpha$. Let $M_\alpha = \mathbb{R}\times U_\alpha$ be the product. We apply the contactization to define a pseudo-Hermitian structure $(\omega_\alpha, J_\alpha)$ on M_α. Let $p : M_\alpha \longrightarrow U_\alpha$ be the projection.

Put

$$\omega_\alpha = dt + p^*\theta_\alpha,$$

where t is the coordinate of $\mathbb{R}$. Then we have

(i) ω_α is $\mathbb{R}$-invariant.

(ii) $\omega_\alpha(\xi) = 1$ for the vector field ξ $(= \dfrac{d}{dt})$ induced by the $\mathbb{R}$-action.

(iii) ω_α is a contact form and $d\omega_\alpha = p^*\Omega_\alpha$.

Let $\mathrm{Null}\,\omega_\alpha = \{X \in TM_\alpha \mid \omega_\alpha(X) = 0\}$ be as before. Since $p_* : \mathrm{Null}\,\omega_\alpha \to TU_\alpha$ is an isomorphism at each point, we define an almost complex structure J_α on $\mathrm{Null}\,\omega_\alpha$ to be the pullback of J by p_*,

$$J_\alpha(X) = p_*^{-1} \circ J \circ p_*(X) \quad \text{for } X \in \mathrm{Null}_{(t,x)}\,\omega_\alpha.$$

As in the argument of §1, J_α is a complex structure on $\mathrm{Null}\,\omega_\alpha$. In this case, ξ is a characteristic *CR* vector field and $(\omega_\alpha, J_\alpha, \xi)$ is a standard pseudo-Hermitian structure by Definition 1.5. Now $(\mathrm{Null}\,\omega_\alpha, J_\alpha)$ is a *CR*-structure on M_α.

Let $g_\alpha(X, Y) = \Omega_\alpha(X, JY)$ be the Kähler metric on U_α. Note that the Webster curvature tensor R on M_α coincides with the Riemannian curvature tensor $\hat{R}$ of U_α with respect to the Kähler metric g_α (cf. [11]). In view of the formula of the Bochner curvature tensor B of U_α, the Webster's theorem 2.1 implies that the Bochner curvature tensor of $(U_\alpha, \Omega_\alpha, g_\alpha)$ coincides with the fourth-order Chern–Moser curvature tensor of $(M_\alpha, (\mathrm{Null}\,\omega_\alpha, J_\alpha))$. By Proposition 1.1, the Hermitian metric g on W satisfies that $g|U_\alpha = f_\alpha \cdot g_\alpha$, and the Bochner curvature tensor of (W, g, J) restricted to U_α is equal to that of (U_α, g_α, J) by Corollary 1.3.

Suppose that the Bochner curvature tensor of W vanishes with respect to g. Then the Chern–Moser curvature tensor of $(M_\alpha, (\mathrm{Null}\,\omega_\alpha, J_\alpha))$ vanishes, which turns out that M_α is a spherical *CR*-manifold. There exists a developing pair:

$$(\rho_\alpha, \mathrm{dev}_\alpha) : (\mathbb{R}, M_\alpha) \longrightarrow (\mathrm{PU}(n+1, 1), S^{2n+1})$$

where $\mathbb{R} \subset \mathrm{Aut}_{CR}(M_\alpha)$. Let G_α be the closure of the holonomy $\rho_\alpha(\mathbb{R})$ in $\mathrm{PU}(n+1, 1)$. Since $\mathbb{R}$ acts freely on M_α and dev_α is an immersion, $\rho_\alpha(\mathbb{R})$ has no fixed point in $\mathrm{dev}_\alpha(M_\alpha)$. Thus Lemma 2.2 implies that

$$\mathrm{dev}_\alpha(M_\alpha) \subset X,$$

where $X = S^{2n+1}$ in which G_α is proper in $U(n+1)$, or $X = \mathcal{N}$, $S^{2n+1} - S^{2m-1}$ $(m = 1, 2, \ldots, n)$. Moreover there is the principal bundle of pseudo-Hermitian manifold (X, ω_X, J_X) of (I),(II), (III):

$$(**) \qquad (T, T) \to (\mathcal{H}, X) \xrightarrow{p} (\mathcal{G}, Y^n_{\mathbb{C}}),$$

where $T = \mathcal{R}$ or S^1. Since $\mathcal{H}$ is $U(n+1)$, $\mathrm{Aut}_{CR}(\mathcal{N})$ or $\mathrm{Aut}_{CR}(S^{2n+1} - S^{2m-1})$, T is normal in $\mathcal{H}$.

$\mathcal{G}$ is a transitive group of isometries of $Y^n_{\mathbb{C}}$ with respect to g_Y, except in (II). For this case, $\mathbb{E}_{\mathbb{C}}(n)$ is a group of isometries of $\mathbb{C}^n$, and the subgroup $\mathbb{R}^+$ acts as homotheties.

Since $\mathcal{H}$ $(\subset \mathrm{Aut}_{CR}(X))$ acts transitively on X, we can assume that $\mathrm{dev}_\alpha(0 \times U_\alpha)$ is transverse to the fiber T. Then we can define a map $\varphi_\alpha : U_\alpha \to Y^n_{\mathbb{C}}$ by setting

$$\varphi_\alpha(u) = p \circ \mathrm{dev}_\alpha(0, u) \qquad \text{for} \quad u \in U_\alpha.$$

Obviously φ_α is a homeomorphism onto its image. We have the following commutative diagram:

$$\begin{array}{ccc} M_\alpha & \xrightarrow{\mathrm{dev}_\alpha} & X \\ {\scriptstyle p}\downarrow & & \downarrow{\scriptstyle p} \\ U_\alpha & \xrightarrow{\varphi_\alpha} & Y^n_{\mathbb{C}}. \end{array}$$

Since dev_α is a CR immersion,

$$\begin{aligned} (\mathrm{dev}_\alpha)^*\omega_X &= \lambda_\alpha \cdot \omega_\alpha \text{ for some positive function } \lambda_\alpha, \\ (\mathrm{dev}_\alpha)_* \circ J_\alpha &= J_X \circ (\mathrm{dev}_\alpha)_*. \end{aligned}$$

Lemma 2.5.

(1) ${\varphi_\alpha}^*\Omega_Y = c_\alpha \cdot \Omega_\alpha$ *for some constant* c_α.

(2) $(\varphi_\alpha)_* \circ J = J_Y \circ (\varphi_\alpha)_*$, *i.e.,* φ_α *is a holomorphic immersion into* $Y^n_{\mathbb{C}}$.

Proof. We first prove that λ_α is constant on M_α. Differentiate $(\mathrm{dev}_\alpha)^*\omega_X = \lambda_\alpha \cdot \omega_\alpha$;

$$d(\mathrm{dev}_\alpha)^*\omega_X = (\mathrm{dev}_\alpha)^* d\omega_X = (\mathrm{dev}_\alpha)^* \circ p^*\Omega_Y = p^* \circ {\varphi_\alpha}^*\Omega_Y.$$

As $d\omega_\alpha = p^*\Omega_\alpha$,

$$(***) \qquad p^* \circ {\varphi_\alpha}^*\Omega_Y = d\lambda_\alpha \wedge \omega_\alpha + \lambda_\alpha \cdot p^*\Omega_\alpha.$$

Since $p_* : \mathrm{Null}_{(t,x)}\, \omega_\alpha \to T_x U_\alpha$ is an isomorphism for arbitrary $t \in \mathbb{R}$, the equation $(***)$ implies that

$$({\varphi_\alpha}^*\Omega_Y)_x(X, Y) = \lambda_\alpha(t, x) \cdot (\Omega_\alpha)_x(X, Y)$$

for $X, Y \in T_x U_\alpha$ such that $p_* \tilde{X} = X$, $p_* \tilde{Y} = Y$ with $\tilde{X}, \tilde{Y} \in \mathrm{Null}_{(t,x)}\ \omega_\alpha$. This holds for arbitrary t, so λ_α is $\mathbb{R}$-invariant, $\lambda_\alpha(x) = \lambda_\alpha(t, x)$.

Substitute $\{\frac{d}{dt}, \tilde{X}\}$ into $(***)$ for $\tilde{X} \in \mathrm{Null}_{(t,x)}\ \omega_\alpha$. Using $\omega_\alpha(\frac{d}{dt}) = 1$, we have

$$0 = d\lambda_\alpha \wedge \omega_\alpha(\frac{d}{dt}, \tilde{X}) = -d\lambda_\alpha(\tilde{X}) = -\tilde{X}\lambda_\alpha.$$

Since $TM_\alpha = \mathrm{Null}_{(t,x)}\ \omega_\alpha \oplus \frac{d}{dt}$, $V\lambda_\alpha = 0$ for all $V \in TM_\alpha$, λ_α is a constant c_α.

On the other hand,

$$\begin{aligned}(\varphi_\alpha)_* \circ J(p_*(\tilde{X})) &= (\varphi_\alpha)_* \circ p_* \circ J_\alpha(\tilde{X}) = p_* \circ (\mathrm{dev}_\alpha)_*(J_\alpha(\tilde{X})) \\ &= p_* \circ J_X \circ (\mathrm{dev}_\alpha)_*(\tilde{X}) = J_Y \circ p_* \circ (\mathrm{dev}_\alpha)_*(\tilde{X}) \\ &= J_Y \circ (\varphi_\alpha)_*(p_*(\tilde{X})) \text{ by the commutativity.}\end{aligned}$$

□

By using a chart $(U_\alpha, \varphi_\alpha)$, we construct a uniformization with respect to one of the geometries $(\mathcal{G}, Y^n_{\mathbb{C}})$. Let (U_β, φ_β) be a chart. (Note that the corresponding space $Y^n_{\mathbb{C}}$ for (U_β, φ_β) is not necessarily the same one as that of $(U_\alpha, \varphi_\alpha)$.) For this, we need the following. (Compare [8].)

Lemma 2.6. *Given two developing pairs $(\rho_\alpha, \mathrm{dev}_\alpha)$, $(\rho_\beta, \mathrm{dev}_\beta)$, if $U_\alpha \cap U_\beta \neq \emptyset$, then they define the same CR-structure, i.e., there is an element $h \in \mathrm{PU}(n+1, 1)$ such that $\mathrm{dev}_\beta = h \circ \mathrm{dev}_\alpha$ on $M_\alpha \cap M_\beta$.*

Proof. As $\Omega_\beta = \lambda_{\beta\alpha} \cdot \Omega_\alpha$ on $U_\alpha \cap U_\beta$ where $\lambda_{\beta\alpha}$ is constant, we have

$$d(\lambda_{\beta\alpha} \cdot \theta_\alpha - \theta_\beta) = \lambda_{\beta\alpha} \cdot \Omega_\alpha - \Omega_\beta = 0.$$

We may assume that $U_\alpha \cap U_\beta$ is contractible so that there is a smooth map $\chi : U_\alpha \cap U_\beta \to \mathbb{R}$ with $\lambda_{\beta\alpha} \cdot \theta_\alpha - \theta_\beta = d\chi$. Noting that $M_\alpha \cap M_\beta = \mathbb{R} \times (U_\alpha \cap U_\beta)$, define a diffeomorphism $f : M_\alpha \cap M_\beta \to M_\alpha \cap M_\beta$ to be

$$f(t, x) = (\lambda_{\beta\alpha} \cdot t + \chi(x), x).$$

Then $f_*(\frac{d}{dt}) = \lambda_{\beta\alpha} \frac{d}{dt}$. Since $\chi \circ p(t, x) = \chi(x)$ and $p \circ f = p$, it follows that $f^*\omega_\beta = \lambda_{\beta\alpha} \cdot \omega_\alpha$. Moreover, as $p_* \circ J_\alpha = J \circ p_*$ (the same is true for J_β), we have $p_*(f_* \circ J_\alpha) = J \circ p_* = p_*(J_\beta \circ f_*)$. Thus, $f_* \circ J_\alpha = J_\beta \circ f_*$. Therefore $f_* : (\mathrm{Null}\, \omega_\alpha, J_\alpha) \to (\mathrm{Null}\, \omega_\beta, J_\beta)$ is a *CR* diffeomorphism.

As a consequence, they define the same *CR*-structure on $M_\alpha \cap M_\beta$.

Since $(\rho_\alpha, \mathrm{dev}_\alpha)$, $(\rho_\beta, \mathrm{dev}_\beta)$ are developing pairs, by the developing theorem (cf. [14]) there exists an element $h \in \mathrm{PU}(n+1, 1)$ such that:

$$\begin{aligned}\mathrm{dev}_\beta &= h \circ \mathrm{dev}_\alpha \quad \text{on } M_\alpha \cap M_\beta, \\ \rho_\beta(t) &= h \cdot \rho_\alpha(t) \cdot h^{-1} \quad \text{for } t \in \mathbb{R}.\end{aligned}$$

□

From this lemma, the closure G_α of $\rho_\alpha(\mathbb{R})$ (resp. G_β) satisfies that $G_\beta = h \cdot G_\alpha \cdot h^{-1}$.

If $U_\alpha \cap U_\beta \neq \emptyset$, then both dev_α and dev_β map into the same space X (and so both φ_α and φ_β map into the same $Y^n_{\mathbb{C}}$). In particular, $h \in \mathrm{Aut}_{CR}(X)$.

On the other hand, recall that $\mathrm{Aut}_{CR}(X) = \mathcal{H}$ for $X = \mathcal{N}$, $S^{2n+1} - S^{2m-1}$ of (II), (III). If $X = S^{2n+1}$ of (I), then both G_α and G_β are proper subgroups of $U(n+1)$, which means that G_α (resp. G_β) fixes the unique point at the origin of the complex hyperbolic space $\mathbb{H}^{n+1}_{\mathbb{C}}$. The above relation $G_\beta = h \cdot G_\alpha \cdot h^{-1}$ implies that h fixes the origin, *i.e.*, $h \in U(n+1) = \mathcal{H}$. So in each case, h lies in $\mathcal{H}$. As is noted in (**) that T is a normal subgroup of $\mathcal{H}$, h induces an element $\hat{h} \in \mathcal{G}$. Calculate

$$\varphi_\beta \circ p = p \circ \mathrm{dev}_\beta = p \circ h \circ \mathrm{dev}_\alpha = \hat{h} \circ p \circ \mathrm{dev}_\alpha = \hat{h} \circ \varphi_\alpha \circ p.$$

So $\varphi_\beta = \hat{h} \circ \varphi_\alpha$ on $U_\alpha \cap U_\beta$, hence $\varphi_\beta \circ \varphi_\alpha^{-1} = \hat{h} | \varphi_\alpha(U_\alpha \cap U_\beta)$ for $\hat{h} \in \mathcal{G}$. Therefore the collection of charts $\{U_\alpha, \varphi_\alpha\}_{\alpha \in \Lambda}$ gives a uniformization of (W, Ω, J) with respect to $(\mathcal{G}, Y^n_{\mathbb{C}})$ of (I), (II), (III). This proves Theorem 2.3. □

3. Conformal equivalence of locally Kählerian structure

Now we prove Theorem B by using Theorem 2.3. Suppose (W, g, J) is compact. Given a uniformization for (W, g, J) (cf. [14]), there exists a developing pair:

$$(\rho, \mathrm{dev}) : (\pi_1(W), \tilde{W}) \longrightarrow (\mathcal{G}, Y^n_{\mathbb{C}}).$$

Put $\rho(\pi_1(W)) = \Gamma$. For brevity, we put $Y^n_{\mathbb{C}} = Y$.

Recall that $\mathcal{G}$ acts as isometries of (Y, g_Y) except in case (II). Then it is easy to see that dev is a covering map onto Y because W is compact. As Y is simply connected, dev is a homeomorphism. The map dev induces a homeomorphism $\hat{\mathrm{dev}} : W \to Y/\Gamma$ where $Y/\Gamma = \boldsymbol{CP}^n$, $\mathbb{H}^n_{\mathbb{C}}/\Gamma$ or $\mathbb{H}^m_{\mathbb{C}} \times_\Gamma \boldsymbol{CP}^{n-m}$ $(m = 1, 2, \ldots, n-1)$. There is the commutative diagram:

$$\begin{array}{ccc} \tilde{W} & \xrightarrow{\ \mathrm{dev}\ } & Y \\ {\scriptstyle p}\downarrow & & \downarrow{\scriptstyle p} \\ W & \xrightarrow{\ \hat{\mathrm{dev}}\ } & Y/\Gamma. \end{array}$$

By the construction of developing, if $p : \tilde{U}_\alpha \to U_\alpha$ is homeomorphic, then

$$\varphi_\alpha \circ p = \mathrm{dev}\,|\tilde{U}_\alpha.$$

The commutative diagram implies that

$$(*) \qquad p \circ \varphi_\alpha = \hat{\mathrm{dev}}|U_\alpha.$$

Consider Case (II). If we note that $\mathcal{G} = \mathbb{C}^n \rtimes (U(n) \times \mathbb{R}^+) \subset \mathrm{Sim}(\mathbb{R}^{2n})$, which is the group of similarity transformations, then M is a similarity manifold. It follows from the result of Fried ([5]) that

$$(\rho, \mathrm{dev}) : (\pi_1(W), \tilde{W}) \longrightarrow (\mathcal{G}', Y') \text{ is homeomorphic}$$

where

$$(\mathcal{G}', Y') = \begin{cases} (\mathbb{E}_{\mathbb{C}}(n), \mathbb{C}^n), \\ (U(n) \times \mathbb{R}^+, S^{2n-1} \times \mathbb{R}^+). \end{cases}$$

Here $\mathbb{E}_{\mathbb{C}}(n) = \mathbb{C}^n \rtimes U(n)$ is the euclidean group endowed with the canonical Kähler form as in (II), and $\mathbb{C}^n - \{0\} = S^{2n-1} \times \mathbb{R}^+$ with the group of similarities $U(n) \times \mathbb{R}^+$.

As a consequence, the induced developing map $\hat{\mathrm{dev}}$ maps W homeomorphically onto a complex euclidean space form $\mathbb{C}^n / \Gamma$ or a Hopf manifold $S^{2n-1} \underset{F}{\times} S^1$.

Note 3.1. When (W, g, J) admits a Kähler structure (for example, Ω itself Kähler), $S^{2n-1} \times S^1$ is ruled out.

For the cases (I), (III) or $(\mathrm{II})_a$: $(\mathbb{E}_{\mathbb{C}}(n), \mathbb{C}^n)$, $\mathcal{G}$ is the group of holomorphic isometries of (Y, g_Y), *i.e.*, it preserves the canonical Kähler form Ω_Y. If $p : Y \rightarrow Y/\Gamma$ is a covering map, then Ω_Y (resp. g_Y) induces a Kähler form $\Omega_{Y/\Gamma}$ (resp. $g_{Y/\Gamma}$) on Y/Γ such that $p^*\Omega_{Y/\Gamma} = \Omega_Y$. We have the canonical Kähler manifold $(Y/\Gamma, \Omega_{Y/\Gamma}, J_{Y/\Gamma}, g_{Y/\Gamma})$.

Let (W, Ω, J, g) be a compact locally conformally Kählerian manifold with vanishing Bochner curvature tensor. For the cases (I), (III) or $(\mathrm{II})_a$, we must show that

(i) (W, Ω, J, g) is conformally equivalent to $(Y/\Gamma, \Omega_{Y/\Gamma}, J_{Y/\Gamma}, g_{Y/\Gamma})$.

(ii) If (W, Ω, J, g) is a Kähler manifold (Ω is a Kähler form), then (W, Ω, J, g) is holomorphically isometric to $(Y/\Gamma, \Omega_{Y/\Gamma}, J_{Y/\Gamma}, g_{Y/\Gamma})$ up to rescaling the metric by a constant.

First recall from Lemma 2.5 that ${\varphi_\alpha}^*\Omega_Y = c_\alpha \cdot \Omega_\alpha$ for some constant c_α ($\alpha \in \Lambda$). If $U_\alpha \cap U_\beta \neq \emptyset$, then the argument of proof of Lemma 2.6 shows that $\varphi_\beta = \hat{h} \circ \varphi_\alpha$ for $\hat{h} \in \mathcal{G}$. Since $\hat{h}$ preserves Ω_Y for the above cases, ${\varphi_\beta}^*\Omega_Y = {\varphi_\alpha}^*\hat{h}^*\Omega_Y = {\varphi_\alpha}^*\Omega_Y = c_\alpha \cdot \Omega_\alpha$. As ${\varphi_\beta}^*\Omega_Y = c_\beta \cdot \Omega_\beta$ and $\Omega_\beta = \lambda_{\beta\alpha} \cdot \Omega_\alpha$ on $U_\alpha \cap U_\beta$ (cf. §1), it follows that $c_\beta \lambda_{\beta\alpha} = c_\alpha$. Since $\lambda_{\beta\alpha} = f_\alpha \cdot f_\beta^{-1}$, we obtain that $c_\alpha f_\alpha^{-1} = c_\beta f_\beta^{-1}$ on $U_\alpha \cap U_\beta$.

So there is a global function $\tau : W \longrightarrow \mathbb{R}^+$ such that $\tau | U_\alpha = c_\alpha f_\alpha^{-1}$. If we note that $\Omega | U_\alpha = f_\alpha \cdot \Omega_\alpha$, then

$$\begin{aligned} \hat{\mathrm{dev}}^* \Omega_{Y/\Gamma} | \, U_\alpha &= {\varphi_\alpha}^* p^* \Omega_{Y/\Gamma} | \, U_\alpha = {\varphi_\alpha}^* \Omega_Y | \, U_\alpha \text{ by } (*) \\ &= c_\alpha \cdot \Omega_\alpha = c_\alpha f_\alpha^{-1} \Omega | \, U_\alpha = \tau \cdot \Omega | \, U_\alpha. \end{aligned}$$

This holds for every U_α, so $\hat{\mathrm{dev}}^* \Omega_{Y/\Gamma} = \tau \cdot \Omega$.

Moreover, on TU_α,

$$\hat{\mathrm{dev}}_* \circ J = p_* \circ (\varphi_\alpha)_* \circ J = p_* \circ J_Y \circ (\varphi_\alpha)_* = J_{Y/\Gamma} \circ p_* \circ (\varphi_\alpha)_* = J_{Y/\Gamma} \circ \hat{\mathrm{dev}}_*.$$

If Ω is Kähler (*i.e.*, $f_\alpha \equiv$ constant), then $\tau = c$, a constant.

In summary:

(1) $\widehat{\mathrm{dev}}^{*}\Omega_{Y/\Gamma} = \tau \cdot \Omega$ for some positive function τ.

(2) If Ω is Kähler, $\widehat{\mathrm{dev}}^{*}\Omega_{Y/\Gamma} = c \cdot \Omega$ for some constant c.

(3) $\widehat{\mathrm{dev}}_{*} \circ J = J_{Y/\Gamma} \circ \widehat{\mathrm{dev}}_{*}$.

We now consider Case (II)$_b$: Hopf geometry $(\mathcal{G}', Y') = (U(n) \times \mathbb{R}^+, S^{2n-1} \times \mathbb{R}^+)$ endowed with the Kähler form $\Omega_{Y'} = \Omega_Y | Y'$.

Given a locally conformally Kählerian structure $\{U_\alpha, \Omega_\alpha\}_{\alpha\in\Lambda}$ on W, put $V_\alpha = \widehat{\mathrm{dev}}(U_\alpha)$. Then $\{V_\alpha\}_{\alpha\in\Lambda}$ is an open covering of $Y'/\Gamma = S^{2n-1} \underset{F}{\times} S^1$.

We have already constructed a locally conformally Kählerian structure on the Hopf manifold Y'/Γ. (Compare Remark 1.8.) Passing to a refinement if necessary, we may assume that our local Kähler form Θ_α is defined on each V_α. So we have a family $\{V_\alpha, \Theta_\alpha\}_{\alpha\in\Lambda}$ and a 2-form Θ on $(Y'/\Gamma, J_{Y'/\Gamma})$ satisfying that;

$$\begin{aligned} \Theta | V_\alpha &= h_\alpha \cdot \Theta_\alpha \quad \text{for a positive function } h_\alpha, \\ \Theta_\beta &= a_{\beta\alpha}\Theta_\alpha \quad \text{for a positive constant}, \\ a_{\beta\alpha} &= h_\alpha \cdot h_\beta^{-1}. \end{aligned}$$

As before, we have the commutative diagram:

$$\begin{array}{ccc} \tilde{W} & \xrightarrow{\ \mathrm{dev}\ } & Y' \\ {\scriptstyle p}\downarrow & & \downarrow{\scriptstyle p} \\ W & \xrightarrow{\ \widehat{\mathrm{dev}}\ } & Y'/\Gamma, \end{array}$$

Put $\tilde{V}_\alpha = \mathrm{dev}(\tilde{U}_\alpha)$. By the commutativity, $p : \tilde{V}_\alpha \longrightarrow V_\alpha$ is a homeomorphism. As in the construction of Theorem A, recall that $\Omega_{Y'}|\tilde{V}_\alpha = p^*\Theta_\alpha$. Moreover, if $\varphi_\alpha : U_\alpha \to Y'$ is a chart, then

$$\begin{aligned} p \circ \varphi_\alpha &= \widehat{\mathrm{dev}}|U_\alpha, \\ \varphi_\beta &= \hat{h} \circ \varphi_\alpha \text{ on } U_\alpha \cap U_\beta \text{ for some } \hat{h} \in \mathcal{G}', \end{aligned}$$

where each element of $\mathcal{G}'$ acts as homothetic transformations with respect to $\Omega_{Y'}$.

Then, for each $\alpha \in \Lambda$,

$$\widehat{\mathrm{dev}}^{*}(\Theta_\alpha | V_\alpha) = (p \circ \varphi_\alpha)^*(\Theta_\alpha | V_\alpha) = \varphi_\alpha{}^*(\Omega_{Y'}|\tilde{V}_\alpha) = \varphi_\alpha{}^*\Omega_Y = c_\alpha \cdot \Omega_\alpha$$

by Lemma 2.5. So if $U_\alpha \cap U_\beta \neq \emptyset$, then

$$\begin{aligned} \widehat{\mathrm{dev}}^{*}\Theta_\beta &= c_\beta \cdot \Omega_\beta = c_\beta \cdot \lambda_{\beta\alpha}\Omega_\alpha \\ &= \widehat{\mathrm{dev}}^{*}(a_{\beta\alpha}\Theta_\alpha) = a_{\beta\alpha}\widehat{\mathrm{dev}}^{*}\Theta_\alpha = a_{\beta\alpha} \cdot c_\alpha\Omega_\alpha. \end{aligned}$$

We obtain that $c_\beta \cdot \lambda_{\beta\alpha} = a_{\beta\alpha} \cdot c_\alpha$ on $U_\alpha \cap U_\beta$.

Recall that h_α is a positive function on $V_\alpha = \hat{\mathrm{dev}}(U_\alpha)$. Put

$$\hat{h}_\alpha = h_\alpha \circ \hat{\mathrm{dev}} = (\hat{\mathrm{dev}})^* h_\alpha : U_\alpha \to \mathbb{R}^+.$$

Note that $a_{\beta\alpha} = h_\alpha \cdot h_\beta^{-1} = \hat{h}_\alpha \cdot \hat{h}_\beta^{-1}$ on $U_\alpha \cap U_\beta$.

As $\lambda_{\beta\alpha} = f_\alpha \cdot f_\beta^{-1}$, we have $c_\beta \cdot \hat{h}_\beta \cdot f_\beta^{-1} = c_\alpha \cdot \hat{h}_\alpha \cdot f_\alpha^{-1}$ on $U_\alpha \cap U_\beta$. Thus there is a global function $\mu : W \to \mathbb{R}^+$ such that $\mu | U_\alpha = c_\alpha \cdot \hat{h}_\alpha \cdot f_\alpha^{-1}$.

Now

$$\begin{aligned}(\hat{\mathrm{dev}}^* \Theta)|\, U_\alpha &= ({\varphi_\alpha}^* p^*(h_\alpha \cdot \Theta_\alpha))|U_\alpha = ({\varphi_\alpha}^*(p^* h_\alpha \cdot \Omega_{Y'}|\, \tilde{V}_\alpha))|U_\alpha \\ &= ({\varphi_\alpha}^* p^* h_\alpha \cdot {\varphi_\alpha}^* \Omega_{Y'}|\, \tilde{V}_\alpha)|U_\alpha = \hat{\mathrm{dev}}^* h_\alpha \cdot c_\alpha \Omega_\alpha | U_\alpha \\ &= \hat{\mathrm{dev}}^* h_\alpha \cdot c_\alpha \cdot f_\alpha^{-1} \Omega|\, U_\alpha = \hat{h}_\alpha \cdot c_\alpha \cdot f_\alpha^{-1} \Omega|\, U_\alpha \\ &= \mu \cdot \Omega|\, U_\alpha.\end{aligned}$$

Since this is true for all U_α, we have $\hat{\mathrm{dev}}^* \Theta = \mu \cdot \Omega$. Similarly as above, $\hat{\mathrm{dev}}_* \circ J = J_{Y'/\Gamma} \circ \hat{\mathrm{dev}}_*$. Hence (W, Ω, J, g) is conformally equivalent to the Hopf manifold $(Y'/\Gamma, \Theta, J_{Y'/\Gamma}, g_{Y'/\Gamma})$, where $g_{Y'/\Gamma}(X, Y) = \Theta(X, J_{Y'/\Gamma} Y)$. This completes the proof of Theorem B.

References

[1] D. E. Blair, Contact manifolds in Riemannian geometry, Lecture Notes in Math. 509, Springer-Verlag, 1976.

[2] S. Bochner, Curvature and Betti numbers. II, Ann. of Math. 50 (1949), 77–93.

[3] S. Bochner and K. Yano, Curvature and Betti numbers, Ann. of Math. Studies 32, Princeton Univ. Press, 1953.

[4] S. S. Chern and J. Moser, Real hypersurfaces in complex manifolds, Acta Math. 133 (1974), 48–69.

[5] D. Fried, Closed similarity manifolds, Comment. Math. Helv. 55 (1980), 576–582.

[6] W. Goldman, Complex hyperbolic geometry, Oxford Mathematical Monographs, Oxford Univ. Press, to appear.

[7] Y. Kamishima and T. Tsuboi, *CR*-structures on Seifert manifolds, Invent. Math. 104 (1991), 149–163.

[8] Y. Kamishima, Uniformization of Kähler manifolds with vanishing Bochner tensor, Acta. Math. 172 (1994), 299–308.

[9] Y. Kamishima, Standard pseudo-Hermitian structure and Seifert fibration on *CR* manifold, Ann. Global Anal. Geom. 12 (1994), 261–289.

[10] Y. Kamishima, Geometric flows on compact manifolds and Global rigidity, Topology 35 (1996), 439–450.

[11] Y. Kamishima, A geometry of vanishing Bochner tensor on complex manifolds with a geometric structure, preprint.

[12] S. Kobayashi, Transformation Groups in Differential Geometry, Ergeb. Math. Grenzgeb. 70, Springer-Verlag, 1970.

[13] S. Kobayashi and K. Nomizu, Foundations of differential geometry I, II, Interscience John Wiley & Sons, New York, 1969.

[14] R. Kulkarni, On the principle of uniformization, J. Differential Geom. 13 (1978), 109–138.

[15] R. Kulkarni, Proper actions and pseudo-Riemannian space forms, Adv. Math. 40 (1981), 10–51.

[16] A. Morimoto, On normal contact structures, J. Math. Soc. Japan 15 (4) (1963), 420–436.

[17] D. Sullivan and W. Thurston, Manifolds with canonical coordinates: Some examples, Enseign. Math. 29 (1983), 15–25.

[18] S. Tanno, Sasakian manifolds with constant ϕ-holomorphic sectional curvature, Tôhoku Math. J. 21 (1969), 501–507.

[19] W. Thurston, The geometry and topology of 3-manifolds, Lecture Notes, The Geometry Center, Minnesota, 1992.

[20] F. Tricerri and L. Vanhecke, Curvature tensors on almost Hermitian manifolds, Trans. Amer. Math. Soc. 267 (1981), 365–398.

[21] I. Vaisman, On locally conformal almost Kähler manifolds, Israel J. Math. 24 (1976), 338–351.

[22] S. Webster, Pseudo-Hermitian structures on a real hypersurface, J. Differential Geom. 13 (1978), 25–41.

[23] H. Weyl, Classical groups, their invariants and representations, Princeton Univ. Press, 1946.

[24] K. Yano, Differential geometry on complex and almost complex spaces, Pergamon Press, New York 1965.

A topological method for finding non-absolute minima for the Yang–Mills functional with Dirichlet data

Antonella Marini

Abstract. This lecture presents a theory by T. Isobe and myself, which is the generalization to Yang–Mills equations in four dimensions of the existence theory of large harmonic maps from the two dimensional disk to $\mathbb{R}^3$.

In the most general situation, there is a countable family of solutions for the Dirichlet problem for Yang–Mills on the four-dimensional cell B^4. The solutions are labelled by an integral index which is determined by the value of $\int_{B^4} F \wedge F$ (where F is the curvature of a connection A) and has topological meaning.

Here we highlight the analogy between our results and the results due to Brezis, Coron, and Jost (independently) on harmonic maps and prove that, under slightly more restrictive hypotheses, there exists a solution in every topological class. This comes essentially from Morse theory.

1991 Mathematics Subject Classification: 35Q99

1. Introduction

We generalize to the Yang–Mills equations a well known result by Brezis and Coron [3], and Jost (independently) [4, 5] on harmonic maps. They proved existence of "large" harmonic maps in dimension two, from the disk to the sphere.

The main theorems presented in this lecture are Theorem 2.2, and the more specific versions 3.1, 3.2. They are due to T. Isobe and myself. For the complete proofs, as well as all details involved in the theory, the reader is referred to [2].

In this paper we highlight the similarities between Yang–Mills equations in dimension four and harmonic maps in dimension two, and give a more qualitative discussion of the result in [2]. We also prove that a minimizer can be found in every topological class, under slightly more restrictive hypothesis. Our results can be understood from the perspective of Morse theory.

In Section 2, we describe the result on harmonic maps and define our problem for

Yang–Mills carrying out the analogy between the two problems.

In Section 3, we define the meaning of "topologically distinct" for solutions of a boundary value problem, for Yang–Mills and for harmonic maps. We also describe the dichotomy of bevaviour related to different assumptions on the boundary data, which pertains to Yang–Mills fields (cf. Theorems 3.1, 3.2).

In Section 4, we describe the key ingredients of our theory and analyse more closely the situations that may occur in the minimizing procedure. We analyse the phenomenon of "bubbling" and also prove some related results. We explore the relations between the absolute minimum and the existence of solutions. In particular, we show that when the absolute minimum carries low energy, then minima are reached in every topological class.

2. Generalizing a result on harmonic maps

The result we generalize is known as the existence of "large" harmonic maps from the two dimensional disk D^2 to the sphere S^2.

This is summarized by the following theorem (Theorem 1 in [3]).

Theorem 2.1 (Brezis and Coron, 1983; Jost, 1984, independently). *Let $\gamma : \partial D^2 \sim S^1 \to S^2$ be a map which extends to a map $\gamma \in H^1(D^2;\ S^2)$. Let $E(u) \equiv \int_{D^2} |\nabla u|^2$ with $u \in H^1(D^2;\ R^3)$ be the classical Dirichlet functional. Let us consider the space $\mathcal{S} \equiv \{u \in H^1(D^2;\ S^2)\ :\ u = \gamma \ \text{ on } \partial D^2\}$.*

If γ is non-constant, then there exists a critical point $u \neq \bar{u}$ of the Dirichlet functional over $\mathcal{S}$, where $\bar{u}$ is the absolute minimum.

We consider principal bundles $P \to B^4$, the four-dimensional disk, with group structure a compact Lie group G. Connections over these bundles are Lie algebra valued one-forms $A \in T^*P \otimes su(2)$.

The Dirichlet boundary value problem is obtained minimizing the pure Yang–Mills functional

$$\mathcal{Y}.\mathcal{M}.(A) \equiv \int_{B^4} |F_A|^2 * 1,$$

and is given by the following system (cf. [6])

$$\mathcal{D}: \quad \begin{cases} D^* F_A = 0\,, \\ i^* A \sim A_0 \ \text{ on } \partial B^4 = S^3. \end{cases}$$

Here $\sim$ means gauge equivalent and i^* is the pull-back of A via the inclusion $i : S^3 \hookrightarrow B^4$.

Our hypotheses on the regularity of A are $A \in H^1$ and also $i^*A \in H^1$ at the boundary. We take A_0 to be a smooth admissible boundary data.

In [2] we examine the Dirichlet boundary value problem for Yang–Mills connections on the four-dimensional disk and prove the following main theorem.

Theorem 2.2. *Let A_0 be non-flat. Then there exists a solution $A \neq A_{\min}$, where $A_{\min}$ is an absolute $\mathcal{Y}.\mathcal{M}.$ minimizer.*

(For the existence of an absolute minimum cf. [6].)

Notice the similarity between Theorem 2.1 on harmonic maps in dimension two and the above theorem on Yang–Mills equations in dimension four. The methods used in the two cases are also similar. Both are of topological nature as seen in Section 3. The general idea is to associate topological invariants to families of functions (or connections in the case of Yang–Mills), which have in common the prescribed boundary data, then trying to carry out a direct minimizing procedure within any fixed topological class. This does not work in a straight forward manner in either case, with more complexity in the case of Yang–Mills.

In the next section, we take a closer look at the topological methods used in both cases, and give a more detailed description of what happens for Yang–Mills in dimension four.

3. The topological method

In the case of Yang–Mills there is dichotomy of behaviour depending on whether the boundary value A_0 satisfies or not the following condition $(\mathcal{SD})$.

$(\mathcal{SD})$: *There exists a self-dual connection A_s on the four dimensional sphere S^4 such that $j^*A_s = A_0$, where j^* is the pull back via the inclusion $j : S^3 \equiv \partial B^4 \hookrightarrow S^4$.*

(Here we identified B^4 with the southern hemisphere S^- of S^4.)

The following theorems hold:

Theorem 3.1. *Let A_0 be the non-flat boundary data assigned for the Dirichlet problem satisfying $(\mathcal{SD})$.*

Then there exist (at least) as many topologically distinct solutions of the Dirichlet problem for Yang–Mills as $c_2(A_s)$ (here $c_2(A_s)$ is the second Chern number of the self-dual connection A_s described above).

Theorem 3.2. *Let A_0, which does not satisfy $(\mathcal{SD})$, be given as boundary data for the Dirichlet problem for Yang–Mills.*

Then there exist countably many topologically distinct solutions of the Dirichlet problem for Yang–Mills connections.

Both theorems are proved in [2].

In both cases, it is natural to ask whether there is more than one solution (not necessarily minimizing) in each topological class.

For a flat boundary data A_0 there is proof of uniqueness ([1]), thus only the flat connection solves the Dirichlet problem and there are no solutions in the non-trivial classes. It is important to notice that these theorems regard the full Yang–Mills equations and not their self dual or anti-self dual versions.

In the following we examine the meaning of "topologically distinct" in the context of harmonic maps and Yang–Mills equations.

In the case of harmonic maps, Brezis and Coron introduce the functional

$$\mathcal{Q}(u) \equiv \frac{1}{4\pi} \int_D u \cdot u_x \wedge u_y ,$$

where $u \in L^\infty$ and $u_x, u_y \in L^2(D, \mathbb{R}^3)$.

If u_1, u_2 agree at the boundary of the disk, then $\mathcal{Q}(u_1) - \mathcal{Q}(u_2)$ is an integer. In fact, in this case u_1 and u_2 glue well along $\partial D \equiv S^1 \equiv$ equator of S^2 via stereographic projection, making a harmonic map on the sphere S^2, and the number above is a second cohomology class of S^2. Once a solution is fixed (for example the absolute minimizer $\bar{u}$), one can associate the integer $\mathcal{Q}(\bar{u}) - \mathcal{Q}(u)$ to each function u in the space $\mathcal{S}$ of Theorem 2.1. This way $\mathcal{S}$ is subdivided in classes which are topologically distinct, and one can try to minimize the Dirichlet functional, restricted to a given topological class.

For the harmonic map problem with non-zero boundary data, Brezis and Coron prove that a minimizer (different from the absolute minimum) exists either in the class corresponding to the integer 1, or in the class corresponding to the integer -1.

For proof of existence of "large" Yang–Mills connections which solve $\mathcal{D}$, we introduce the functional

$$\mathcal{C}(A) \equiv \int_{B^4} F \wedge F$$

on the space

$$\mathcal{A}_0 \equiv \{A \in T^*P \otimes su(2) : A \in H^1,\ i^*A \in H^1(S^3),\ i^*A \sim A_0\}.$$

Again, connections in the space $\mathcal{A}$, glue well along the equator, via stereographic projection, with the absolute minimizer A_{min} found in [6]. This can be done with enough regularity; i.e. yielding connections which are of class H^1 on a L^2_2 bundle over S^4.

Once A_{min} has been fixed (note that there is no uniqueness result in [6]), one associates to a connection $A \in \mathcal{A}_0$ a connection $\hat{A}$ on S^4, thus a number $c_2(\hat{A}) = \mathcal{C}(A_{\text{min}}) - \mathcal{C}(A)$, which is the second Chern class of $\hat{A}$. The space $\mathcal{A}_0$ is subdivided this way into classes of topologically distinct connections. This subdivision carries down to gauge-equivalent classes. Thus, one can define

$$\alpha_k \equiv \{A \in \mathcal{A}_0 \ s.t.\ \mathcal{C}(A) - \mathcal{C}(A_{\text{min}}) = k\}/\mathcal{G},$$

where $\mathcal{G}$ is the gauge group. The space α_0 is not empty ($A_{\text{min}} \in \alpha_0$), thus α_k is not empty $\forall k$, by Taubes gluing construction [8].

As for the harmonic map problem, these classes are not closed under weak H^1-

convergence, thus the minimizing procedure carried out in a given topological class may hand a limiting connection in a different class, due to bubbling.

4. Key ingredients. Minimization and topological results

Here we briefly outline the minimizing procedure carried out in [2] and highlight the main ideas. We prove some more results which also come as corollaries of the theory developed in [2].

In order to find solutions different from the absolute minimum, one carries out minimization as follows. Let

$$\epsilon_k \equiv \inf_{A \in \alpha_k} \mathcal{Y}.\mathcal{M}.(\mathcal{A})$$

and $\{A_i\} \subset \alpha_k$ be a minimizing sequence (that is, satisfying $\mathcal{Y}.\mathcal{M}.(A_i) \to \epsilon_k$). By weak-compactness, one can extract a minimizing subsequence $A_i \rightharpoonup A_\infty$, weakly converging in H^1. Since α_k is not closed a priori under H^1-convergence, we cannot say that A_∞ belongs to the class α_k for the same k. Nonetheless, we can say that A_∞ satisfies the Dirichlet problem $\mathcal{D}$. This can be seen from general arguments as in [2]. Here (cf. Corollary 4.1) we prove that if $A_\infty \in \alpha_l$, then $\mathcal{Y}.\mathcal{M}.(A_\infty) = \epsilon_l$, a simple consequence of the three key inequalities below.

A preliminary estimate known for the energy of A_∞ comes from lower semicontinuity, this is $\mathcal{Y}.\mathcal{M}.(A_\infty) \leq \lim_{i\to\infty} \mathcal{Y}.\mathcal{M}.(A_i)$. In order to prove anything about the multiplicity of solutions, one needs to estimate to which class A_∞ belongs.

For this purpose, one uses the following three key inequalities.

Inequality #1. This is consequence of a similar inequality which holds for closed manifolds and is due to Sedlacek [7]. It shows that, for the Chern class to change in the minimizing process, there needs to be a jump in energy of at least $8\pi^2$:

$$8\pi^2 |C(A_i) - C(A_\infty)| \leq \mathcal{Y}.\mathcal{M}.(A_i) - \mathcal{Y}.\mathcal{M}.(A_\infty) + o(1) \tag{1}$$

Inequality #2. This is due to Taubes (from the gluing construction in [8]):

$$\epsilon_l - \epsilon_k \leq 8\pi^2 |l - k| \text{ for all } l, k \in \mathbb{Z}. \tag{2}$$

Inequality #3. This is also due to Taubes and holds under the hypotheses that $l > k$, and there is a connection $A \in \alpha_k$ such that $\mathcal{Y}.\mathcal{M}.(A) = \epsilon_k$ (i.e. the minimum is attained in α_k), with A not anti-self dual:

$$\epsilon_l - \epsilon_k < 8\pi^2 (l - k), \quad l > k. \tag{3}$$

Likewise, if the minimum attained in α_k is not self dual, the following holds:

$$\epsilon_l - \epsilon_k < 8\pi^2 (l - k), \quad l < k. \tag{3'}$$

These last estimates come from the construction in Proposition 6-2 in [8], which shows that if the self-dual projection of a connection $A_k \in \alpha_k$ is non-zero at a point $\bar{z} \in B^4$, then one can modify the connection in a neighborhood of $\bar{z}$ in such a way to change the Chern number, while the energy changes by less than $8\pi^2$; i.e.

$$\mathcal{Y}.\mathcal{M}.(A_{k+1}) < \mathcal{Y}.\mathcal{M}.(A_k) + 8\pi^2,$$

where $A_{k+1} \in \alpha_{k+1}$ and $A_k \in \alpha_k$ coincide in a neighborhood of $\bar{z}$.

An application of the inequalities above is

Corollary 4.1. *The connection $A_\infty \in \alpha_l$ obtained via minimization in α_k is an absolute minimum in α_l.*

Proof. By 1 and 2,

$$8\pi^2|k-l| \leq \epsilon_k - \mathcal{Y}.\mathcal{M}.(A_\infty) \leq \epsilon_k - \epsilon_l \leq 8\pi^2|k-l|,$$

thus $\mathcal{Y}.\mathcal{M}.(A_\infty) = \epsilon_l$. □

The following lemma is of fundamental importance.

Lemma 4.1. *Let $\{A_i\} \subset \alpha_k$, be a minimizing sequence (i.e. $\mathcal{Y}.\mathcal{M}.(A_i) \to \epsilon_k$), and $A_i \rightharpoonup A_\infty$ (in H^1) with $A_\infty \in \alpha_l$. Then the following holds:*

If $l < k$, the limiting connection A_∞ is anti-self dual;

If $l > k$, the limiting connection A_∞ is self dual.

Proof. Let $A_\infty \in \alpha_l$ for $l < k$ and let us assume that A_∞ be not anti-self dual. By 1 and 3,

$$8\pi^2|k-l| \leq \epsilon_k - \epsilon_l < 8\pi^2|k-l|,$$

thus a contradiction. Similarly, using 1 and 3′, one shows that A_∞ must be self dual if $l > k$. □

This lemma, together with the unique continuation principle for self dual (anti-self dual) connections, accounts for the dichotomy of behaviour announced in Section 3.

In fact, if the boundary data prescribed satify ($\mathcal{SD}$), then there exist both a self dual and an anti-self dual solution of $\mathcal{D}$ on $B^4 \sim S^-$ (the southern hemisphere). These must be unique. Otherwise, by gluing, one could construct distinct self dual (or anti-self dual) connections on S^4 which coincide on one hemisphere. This is not possible by the unique continuation principle which holds for the self dual (anti-self dual) equations.

The number of topologically distinct solutions, which one has when $\mathcal{SD}$ is satisfied, comes from the following construction. One identifies the southern hemisphere of S^4 with B^4 ($S^- \sim B^4$) via stereographic projection, then glues together the (unique) self dual and anti-self dual connections, A_{sd} and A_{asd}, by reflecting A_{sd} along the equator.

This way, one obtains an anti-self dual connection

$$\hat{A} \equiv \begin{cases} r^* A_{\text{sd}} & \text{on } S^+ \\ A_{\text{asd}} & \text{on } S^- \sim B^4 , \end{cases} \tag{4}$$

with Chern number $c_2(\hat{A}) \equiv -h$. (Here r^* is reflection across the equator.)

It is not difficult to see that if one starts with a minimizing sequence in α_k, bubbling off is not allowed within a closed range of values of k (i.e. A_∞ belongs to α_k for the same k). Outside that range, an instanton (or anti-instanton) bubbles off and the resulting limit A_∞ coincides either with A_{sd} or with A_{asd}. In [2] we find at least h distinct solutions, with $h \equiv -c_2(\hat{A})$, the second Chern number of $\hat{A}$. (We warn the reader that we used slightly different constructions in [2], depending on whether $\mathcal{SD}$ were satisfied or not.)

Let us now prove some results not contained in [2], and also analyse more closely what may happen in the minimizing procedure. Once $A_{\min}$ is fixed, $c_0 \equiv \mathcal{C}(A_{\min})$ and $\epsilon_0 \equiv \mathcal{Y}.\mathcal{M}.(A_{\min})$ are also fixed. For simplicity of notation, let $8\pi^2 c_k \equiv \mathcal{C}(A)$ for $A \in \alpha_k$. (This is topological, thus it does not depend on the choice of $A \in \alpha_k$.) With this definition, $c_k \equiv c_0 - k$. Let also $\epsilon_0 \equiv 8\pi^2(c_0 + a)$ with $a \geq 0$. Let us further assume that the absolute minimum $A_{\min}$ be not anti-self dual.

The fact that $\mathcal{C}(A)$ is a topological lower bound for $\mathcal{Y}.\mathcal{M}.(A)$, together with 3 yields the following estimates for all values of $k > 0$,

$$8\pi^2 |c_0 - k| \leq \epsilon_k < 8\pi^2(c_0 + k + a) . \tag{5}$$

Likewise, if $A_{\min}$ is not self dual,

$$8\pi^2 |c_0 + k| \leq \epsilon_{-k} < 8\pi^2(c_0 + k + a) \qquad (k > 0). \tag{6}$$

Once we hit an anti-self dual solution for a strictly positive value of k, say $A_{\bar{k}}$, with $\mathcal{Y}.\mathcal{M}.(A_{\bar{k}}) = \epsilon_{\bar{k}}$, then for all $l > \bar{k}$ the eventual minima must also be anti-self dual. In fact, $\epsilon_l + 8\pi^2 c_{\bar{k}} = \epsilon_l - \epsilon_{\bar{k}} \leq 8\pi^2(l - \bar{k})$, thus $\epsilon_l \leq 8\pi^2(l - \bar{k} - c_{\bar{k}}) = -8\pi^2 c_l$. Likewise, once we hit a self dual solution for a negative k, then all the ones below must also be self dual.

Thus we have proved that

Proposition 4.1. *Let $A_{\min}$ be not anti-self dual. If an anti-self dual minimum is reached in $\alpha_{\bar{k}}$, then eventual minima in α_k, for $k > \bar{k}$ must also be anti-self dual.*

Let us now give sufficient conditions under which the minimum is attained in every class. Clearly, if there are neither self dual nor anti-self dual solutions for the boundary data A_0, Lemma 4.1 guarantees that the minimum is attained for all values of k.

Starting from a minimizing sequence $\{A_i\} \subset \alpha_k$, if $A_\infty \in \alpha_l$ is not anti-self dual, then $l \geq k$ and inequality 1 implies that $\epsilon_l \leq \epsilon_k$ $(l \geq k)$. To investigate further, we prove the following proposition.

Proposition 4.2. *Let us assume there are no anti-self dual solutions and an absolute minimum is reached in α_{k-1} for k strictly positive. Let $\epsilon_{k-1} \leq \epsilon_l$, for $l > k$. Then the minimum is attained in α_k.*

Proof. Let us take a minimizing sequence $\{A_i\} \in \alpha_k$ and let the limiting connection $A_\infty \in \alpha_l$. From 1, 3, and by our hypothesis, if $l > k$,

$$8\pi^2|k-l| \leq \epsilon_k - \epsilon_l < \epsilon_{k-1} + 8\pi^2 - \epsilon_l \leq 8\pi^2.$$

Thus a contradiction if $k \neq l$. □

If the energy of the minimum ϵ_0 is small, ϵ_{k+1} and ϵ_{k-1} range in disjoint sets (by 5, 6) and $\epsilon_{k-1} \leq \epsilon_{k+1}$. In this case, the minima are attained. If instead there is bubbling for the sequence in α_k (still in the hypothesis that there are no anti-self dual solutions), $\epsilon_{k-1} > \epsilon_l$ and A_∞ must be self dual. This implies $c_l \equiv c_0 - l \geq 0$.

The following propositions yield conditions for the Chern integral $c_0 \equiv \mathcal{C}(A_{\min})$ under which the minimum is attained in every class.

Proposition 4.3. *Let $\{A_i\} \subset \alpha_k$ satisfy $\mathcal{Y}.\mathcal{M}.(A_i) \to \epsilon_k$ and $A_\infty \in \alpha_l$ be a limiting connection (as in the minimizing procedure). The following holds:*

If $c_0 > k$, then $l \geq k$ (in this case, A_∞ is self dual and $l \in [k+1, c_0]$, or $l = k$);

if $c_0 < k$, then $l \leq k$ (in this case, A_∞ is anti-self dual and $l \in [c_0, k-1]$, or $l = k$);

if $c_0 = k$, then $l = k$.

Proof. From Lemma 4.1, if $l < k$, then A_∞ is anti-self dual and $c_l \equiv c_0 - l \leq 0$. Thus $c_0 \leq l < k$. Similarly for the other cases. □

Thus, the value of c_0 determines whether or not a given class α_k is closed under the minimizing procedure. As a consequence, we have

Proposition 4.4. *If there are no anti-self dual solutions and $c_0 < 1$, the minimum is attained in every class α_k with k positive. (If there are no self dual solutions and $c_0 > -1$ the minimum is attained in every class α_{-k} with $-k$ negative.)*

Concluding remarks. From this discussion we see how existence results are related to properties of $A_{\min}$. If $A_{\min}$ has a small topological lower bound for energy, the classes α_k end up being closed under the minimizing procedure, while whenever there is enough energy to allow bubbling, the limiting connection is self dual or anti-self dual.

Also, the Yang–Mills equations in dimension four can be regarded as analogous to harmonic maps in dimension two, whereas the self-dual (anti-self dual) equations can be regarded as analogous to holomorphic (anti-holomorphic) maps. This plays a role in the dichotomy of behaviour described in Section 3, because of the unique continuation principle which holds for self dual (anti-self dual) connections.

References

[1] T. Isobe, Non-Existence and Uniqueness Results for Boundary Value Problems for Yang–Mills Connections, Proc. Amer. Math. Soc., to appear.

[2] T. Isobe, A. Marini, On Topologically Distinct Solutions of the Dirichlet Problem for Yang–Mills Connections, Calculus of Variations, to appear.

[3] H. Brezis, J. M.Coron, Large solutions for harmonic maps in two dimensions, Comm. Math. Phys. 92 (1983), 203–215.

[4] J. Jost, The Dirichlet problem on harmonic maps from a surface with boundary onto a 2-sphere with non constant boundary values, J. Differential Geom. 19 (1984), 393–401.

[5] J. Jost, Harmonic maps between surfaces, Lecture Notes in Math. 1062, Springer-Verlag, Berlin 1984.

[6] A. Marini, Dirichlet and Neumann boundary problems for Yang–Mills connections, Comm. Pure. Appl. Math. XLV (1992), 1015–1050.

[7] S. Sedlacek, A direct method for minimizing the Yang–Mills functional over four dimensional manifold, Comm Math. Phys. 86 (1982), 515–528.

[8] C. H. Taubes, Path-connected Yang–Mills moduli spaces, J. Differential Geom. 19 (1984), 337–392.

Some remarks on equivariant harmonic and non-holomorphic maps in Riemannian manifolds

Caio J. C. Negreiros

Abstract. In this paper we show that the method described in [16] for obtaining new examples of tori in Riemannian manifolds cannot be adapted to produce harmonic and non-holomorphic 2-spheres in flag manifolds. The method in question uses separation of variables and is based on Uhlenbeck's method as described in [18].

1. Introduction

The modern study of harmonic surfaces in Riemannian symmetric spaces started with Calabi [6] and Chern [7] and, after Uhlenbeck [19], is now very well understood. An important step was the complexification of the problem, as done by Eells and Wood [11], Din and Zakarewski [8] and Glaser and Stora [12]. Much less attention has been given to the case where the target is an homogeneous non-symmetric space like G/T (i.e. the quotient of a semi-simple, compact Lie group by a maximal torus). Some of the first attempts in this direction were described in [13] and [15]. See also Black's book [3] for an account of this case, as well as for a generalization to the closely related case of twistor spaces (see also [14]). The interest in such cases relies heavily on their similarities with the variational approach to problems in low dimensional topology. For related material see, for example, [2] and [9].

In this paper we show that the method used in [16] to obtain new tori in Riemannian manifolds (based on Uhlenbeck's method as described [18]) cannot be simply adapted so as to produce harmonic and non-holomorphic 2-spheres into flag manifolds.

The main results are the following ones:

5.4. Theorem. *Let* $\tilde{\varphi} = (\tilde{\pi}_i)_{i=1}^n : \mathbb{R} \times \mathbb{R} \to (F(n), g_{a=(a^{ij})})$ *be an equivariant map defined in the following way:* $\tilde{\pi}_i = \exp(Ax + Bf(y)) \cdot E_i \cdot \exp(-Ax - Bf(y))$, *where* A *and* B *are in* $u(n)$, $[A, B] = 0$, *and* $f : \mathbb{R} \to \mathbb{R}$ *is an arbitrary non-constant function at least* C^2. *Furthermore we assume that* $\operatorname{diag}(A) = \operatorname{diag}(B) = 0$. *Then* $\tilde{\varphi}$ *is harmonic if and only if* $f''(y) = 0 \ \forall y \in \mathbb{R}$.

5.5. Corollary. *Let $\tilde{\varphi} = (\tilde{\pi}_i)_{i=1}^n : \mathbb{R}\times\mathbb{R} \to (F(n), g_{a=(a^{ij})})$ be defined so as to satisfy all the hypothesis of Theorem 5.4. Then A and B can be chosen so that $\tilde{\varphi}$ induces $\varphi : T^2 = S^1 \times S^1 \to (F(n), g_{a=(a^{ij})})$ producing the harmonic nonholomorphic T^2 in $(F(n), g_{a=(a^{ij})})$ described in §4.*

Acknowledgements. I am grateful to the referee for all the suggestions. I want sincerely to thank Professor Karen Uhlenbeck for all her enormous support throughout these years.

2. General facts concerning the geometry of G/T

In this paper we will focus our attention on the homogeneous space G/T where G is an arbitrary compact semi-simple Lie group and T is a maximal torus of G. We will consider a fixed representation of G into $U(n)$ and the derived homogeneous space $F(n) = U(n)/T$ where $T = \underbrace{U(1) \times \ldots \times U(1)}_{n\text{-times}}$.

Note that it is well known that $U(n)/T$ is closely related with the twistor space of $\mathbb{C}P^{n-1}$. See, for example, [14] for more information.

As is well known, the trace form of $U(n)$ gives rise to an inner product $\langle\, ,\, \rangle$ on the Lie algebra $u(n) = p \oplus \underbrace{\mathbb{R} \oplus \ldots \oplus \mathbb{R}}_{n\text{-times}}$ where p is identified with $T(F(n))_{(T)}$. Since we can easily find A, B, and C in p such that $[A, [B, C]] \notin p$ we see for a result of E. Cartan that $F(n)$ is not a symmetric space. In general $p = \sum_{\alpha\in\Lambda} E_\alpha$ where $\Lambda \subseteq \mathcal{N}^*$ is the set of roots, E_α is the root space corresponding to $\alpha \in \Lambda$ and $\mathcal{N}$ is the Lie algebra of T.

A T-invariant almost complex structure on $U(n)/T$ corresponds to a T-invariant endomorphism J of p with $J^2 = -I$. Such endomorphisms correspond to a decomposition $\Lambda = \Lambda^+ \oplus \Lambda^-$ with the property $\Lambda^- = \{-\alpha, \alpha \in \Lambda^+\}$, via the decomposition $p_{\mathbb{R}} \otimes \mathbb{C} = p_{(1,0)} \oplus p_{(0,1)}$ into (1,0) and (0,1) parts given by: $p_{(1,0)} = \sum_{\alpha\in\Lambda^+} E_\alpha$ and $p_{(0,1)} = \sum_{\alpha\in\Lambda^-} E_\alpha$.

In [4], Borel studied the left invariant metrics on $F(n)$: let A and B in p and consider the following inner products $\langle A, B\rangle_{a=(a^{ij})} = \sum_{i,j} \operatorname{tr}(a^{ij} E_i \cdot A \cdot E_j \cdot B^*)$, where E_i is the matrix which has 1 in the (i, i)-position and zero elsewhere and $a^{ij} = a^{ji} > 0$ for any $1 \le i \ne j \le n$.

3. Harmonic and holomorphic map equations in $U(n)/T$

As is well known, we can view $F(n)$ geometrically as the set of n-tuples $(L_1, \dots, L_n)$, where L_i is a 1-dimensional complex subspace of $\mathbb{C}^n$, L_i is perpendicular to L_j if $i \neq j$ and $L_1 \oplus \dots \oplus L_n = \mathbb{C}^n$. There are then tautologously defined vector bundles on $F(n)$ in which the fibers over $(L_1, \dots, L_n)$ are the vector spaces $L_1, \dots, L_n$ respectively.

We identify a smooth map $\varphi : M^2 \to \mathbb{C}P^{n-1}$ with a subbundle $\tilde{\varphi}$ of $\tilde{\mathbb{C}}^n = M^2 \times \mathbb{C}^n$ with rank one which has fiber at $x \in M$ given by: $\tilde{\varphi}_x = T_{\varphi(x)}$ where T is the tautological line bundle over $\mathbb{C}P^{n-1}$.

Any subbundle $\tilde{\varphi}$ of $\tilde{\mathbb{C}}^n$ inherits a metric denoted by $\langle\, , \,\rangle_{\tilde{\varphi}}$, and connection denoted by D_φ, induced from the canonical flat metric and connection ∂ on $\mathbb{C}^n$. More explicitly $\langle V, W\rangle_{\tilde{\varphi}} = \langle V, W\rangle$ for any $V, W \in \tilde{\varphi}_x$, $x \in M^2$ and $(D_{\tilde{\varphi}})_Z W = \pi_{\tilde{\varphi}}(\partial_Z W)$, $W \in \Gamma(\tilde{\varphi})$, $Z \in T(M)^{(1,0)}$ where $\pi_{\tilde{\varphi}} : \tilde{\mathbb{C}}^n \to \tilde{\varphi}$ denotes the Hermitian projection on the subbundle $\tilde{\varphi}$.

$F(n)$ embeds naturally in $\underbrace{\mathbb{C}P^{n-1} \times \dots \times \mathbb{C}P^{n-1}}_{n\text{-times}}$ so that any $\varphi : M^2 \to F(n)$ can be described as $\varphi = (\pi_1, \dots, \pi_n)$, where $\pi_i : M^2 \to \mathbb{C}P^{n-1}$ and $\pi_i \cdot \pi_j = \delta_{ij}\pi_i$. Thus if $\varphi = (\pi_1, \dots, \pi_n) : M^2 \to F(n)$ is a smooth map, then φ determines the tautologously defined vector bundles $\pi_1, \dots, \pi_n$ over M^2. Let $\partial_{\partial/\partial x}\pi_i = \dfrac{\partial \pi_i}{\partial x}$ be the covariant derivative of π_i with respect to x and A_x^i the projection of $\dfrac{\partial \pi_i}{\partial x}$ onto $\pi_i^\perp$. See [15] for more details.

We define the partial second fundamental forms of $\varphi = (\pi_1, \dots, \pi_n) : M^2 \to F(n)$ to be the maps $A_x^{ij} = \pi_i(A_x^j) = \pi_i \dfrac{\partial \pi_j}{\partial x}$ for $i \neq j$. Of course $A_x^{ii} = 0$ for $1 \le i \le n$. Note that $A_x^{ij} \in \operatorname{Hom}(\pi_j, \pi_i)$ and $\sum_{j=1}^{n} A_x^{ij}$ is the second fundamental form of the span of π_i in $\mathbb{C}^n$. We can clearly define A_y^{ij} in the same way.

As usual, we define:

$$\frac{\partial \pi_i}{\partial z} = \frac{1}{2}\left(\frac{\partial \pi_i}{\partial x} - \sqrt{-1}\,\frac{\partial \pi_i}{\partial y}\right) \quad \text{and} \quad \frac{\partial \pi_i}{\partial \bar{z}} = \frac{1}{2}\left(\frac{\partial \pi_i}{\partial x} + \sqrt{-1}\frac{\partial \pi_i}{\partial y}\right).$$

For simplicity we denote $\pi_i \dfrac{\partial \pi_j}{\partial \mu}$ by A_μ^{ij} when $\mu = z$ or $\bar{z}$. We also define the matrix A_μ where the (i, j)-entry is A_μ^{ij} for $1 \le i, j \le n$. If $F(n)$ is equipped with the left invariant metric $g_{a=(a^{ij})}$ we denote $a^{ij}A_\mu^{ij}$ by $(A_\mu^a)^{(ij)}$.

3.1. Proposition. *We have* $\dfrac{\partial}{\partial z}(A_{\bar{z}}^a) - \dfrac{\partial}{\partial \bar{z}}(A_z^a) = [A_z^a, A_{\bar{z}}^a] + [A_z^a, A_{\bar{z}}^a]_h$ *where* $[A_z^a, A_{\bar{z}}^a]_h$ *denotes the diagonal part of the matrix* $[A_z^a, A_{\bar{z}}^a]$.

Proof. See [15].

We define the energy of a map $\varphi = (\pi_i)_{i=1}^n : M^2 \to (F(n), g_a)$ by:

$$E(\varphi) = \frac{1}{2}\sum_{i=1}^{n}\int_{M^2}\left(\langle\frac{\partial\pi_i}{\partial z}, \frac{\partial\pi_i}{\partial z}\rangle_{g_a} + \langle\frac{\partial\pi_i}{\partial\bar{z}}, \frac{\partial\pi_i}{\partial\bar{z}}\rangle_{g_a}\right)v_g.$$

We have the following result.

3.2. Proposition. *Let $\varphi : M^2 \to (F(n), g_a)$ be a smooth map. Then φ is harmonic if and only if* $\frac{\partial}{\partial z}(A^a_{\bar{z}}) + \frac{\partial}{\partial\bar{z}}(A^a_z) = 0 = \frac{\partial}{\partial x}(A^a_x) + \frac{\partial}{\partial y}(A^a_y)$.

Proof. We first notice that

$$(\delta E)\ (\delta\varphi(q)) = -2\,\mathrm{Re}\left\{\sum_{i,j=1}^{n}\int_{M^2}\langle(A^a_\mu)^{ij}, \frac{\partial q}{\partial\mu}\rangle v_g\right\} = -2\,\mathrm{Re}\left\{\left(\langle\langle A^a_\mu, \frac{\partial q}{\partial\mu}\rangle\rangle\right)\right\}$$

where $\mu = z$ or $\bar{z}$. In fact:

$$\begin{aligned}\delta E &= -2\,\mathrm{Re}\left\{\sum_{i,j=1}^{n}\int_{M^2}\langle a^{ij}A^{ij}_\mu, \delta(A^{ij}_\mu)\rangle v_g\right\}\\ &= 2\,\mathrm{Re}\left\{\sum_{i,j=1}^{n}a^{ij}\int_{M^2}\langle A^{ij}_\mu, [A^{ij}_{\mu,q}] - \pi_i\frac{\partial q}{\partial\mu}\rangle v_g\right\}\\ &= -2\,\mathrm{Re}\left\{\sum_{i,j=1}^{n}\int_{M^2}\langle(A^a_\mu)^{ij}, \frac{\partial q}{\partial\bar{\mu}}\rangle v_g\right\}\\ &= -2\,\mathrm{Re}\left\{\langle\langle A^a_\mu, \frac{\partial q}{\partial\mu}\rangle\rangle\right\}\end{aligned}$$

since $\mathrm{Re}\{a^{ij}\langle\langle[A^{ij}_\mu, A^{ji}_{\bar{\mu}}], q\rangle\rangle\} = 0$.

But according to Nöether's theorem (see [1]), we conclude that φ is harmonic if and only if $\mathrm{Re}\left\{\frac{\partial}{\partial\bar{z}}(A^a_z)\right\} = \frac{\partial}{\partial x}(A^a_x) + \frac{\partial}{\partial y}(A^a_y) = 0$. Equivalently, we have $\frac{\partial}{\partial\bar{z}}(A^a_z) + \frac{\partial}{\partial z}(A^a_{\bar{z}}) = 2\left\{\frac{\partial}{\partial x}(A^a_x) + \frac{\partial}{\partial y}(A^a_y)\right\} = 0.$

Now we will take $[1, n] = \{x \in I\!N; 1 \le x \le n\}$ consider $D = \{(i.i); 1 \le i \le n\}$, and S^+ to be a $\frac{(n^2 - n)}{2}$ elements such that if $(i, j) \in S^+$ then $(j, i) \notin S^+$ in $([1, n] \times [1, n] - D)$. We call S^+ a positive system in $[1, n]$.

Let E^o and $\overline{E}$ denote the ∂ and $\overline{\partial}$ energy respectively, defined by

$$E^o_{S^+}(\varphi) = \sum_{(i,j)\in S^+} \int_{M^2} |A^{ij}_z|^2_{v_g} \quad \text{and} \quad \overline{E}_{S^+}(\varphi) = \sum_{(i,j)\in S^+} \int_{M^2} |A^{ij}_{\bar z}|^2_{v_g}.$$

Then a map $\varphi = (\pi_1, \ldots, \pi_n) : M^2 \to F(n)$ is holomorphic with respect to the almost complex structure determined by S^+ if and only if

$$\overline{E}_{S^+}(\varphi) = \sum_{(i,j)\in S^+} \int_{M^2} |A^{ij}_{\bar z}|^2_{v_g} = 0$$

i.e. $A^{ij}_z \cdot A^{ji}_{\bar z} = 0 \ \forall (i,j) \in S^+$.

4. Harmonic and non-holomorphic tori in $U(n)/T$

In this paragraph we describe a large class of the new harmonic and non-holomorphic tori in $(F(n), g_{a=(a^{ij})})$, as obtained in [16]. Our method there was based on Uhlenbeck's method as described in [18]. The tori thus obtained are harmonic with respect to any metric $g_{a=(a^{ij})}$ on $F(n)$ and so are equiharmonic according to the definition in [3].

We define $\varphi = (\pi_1, \ldots, \pi_n) : \mathbb{R}^2 \to (F(n), g_{a=(a^{ij})})$, $n \geq 4$ by $\pi_i(x,y) = \exp(Ax + By) \cdot E_i \cdot \exp(-Ax - By)$, where A and B are in $u(n)$ and $[A, B] = 0$.

After some computation we obtain that φ is harmonic if and only if

$$\sum_{i,j=1}^{n} a^{ij} E_i \cdot ([A, \operatorname{diag} A] + [B, \operatorname{diag} B]) \cdot E_j = 0.$$

By using the holomorphic and harmonic map equation obtained we can prove the following result (see [16] for more details).

4.1. Theorem. *Let $\tilde{\varphi} = (\tilde{\pi}_i)_{i=1}^n \frac{\mathbb{R}^2}{\alpha\mathbb{Z}\oplus\beta\mathbb{Z}} \to (F(n), g_{a=(a^{ij})})$ be a map defined as above (α, β are fixed real numbers) and assume that $E_i \cdot A \cdot E_j \neq 0$ or $E_i \cdot B \cdot E_j \neq 0$ for some $1 \leq i \neq j \leq n$ and that $\sum_{i,j=1}^{n} a^{ij} E_i \cdot ([A, \operatorname{diag} A] + [B, \operatorname{diag} B]) \cdot E_j = 0$. Then $\tilde{\varphi}$ is harmonic with respect to the metric $g_{a=(a^{ij})}$ but it is not holomorphic with respect to any almost complex structure on $F(n)$ defined in* §3.

This result allows us to construct several examples of harmonic and non-holomorphic tori in $(U(n)/T, g_a)$. For example, $A, B \in u(n)$ given by

$$A = \begin{pmatrix} 0 & \alpha\sqrt{-1} & 0 \dots 0 \\ \alpha\sqrt{-1} & 0 & 0 \dots 0 \\ 0 & 0 & 0 \dots 0 \\ \vdots & \vdots & \vdots \quad \vdots \\ 0 & 0 & 0 \quad 0 \end{pmatrix} \quad \text{and} \quad B = \begin{pmatrix} 0 & 0 & 0 & 0 & \dots & 0 \\ 0 & 0 & 0 & 0 & \dots & 0 \\ 0 & 0 & 0 & \beta\sqrt{-1} & \dots & 0 \\ 0 & 0 & \beta\sqrt{-1} & 0 & \dots & 0 \\ \vdots & \vdots & \vdots & \vdots & & \vdots \\ 0 & 0 & 0 & 0 & \dots & 0 \end{pmatrix}$$

where α and β are fixed nonzero real numbers.

Then, for example,

$$\exp(At) = \begin{pmatrix} \cos(\alpha t) & 0 & 0 \dots 0 \\ 0 & \cos(\alpha t) & 0 \dots 0 \\ 0 & 0 & 0 \dots 0 \\ \vdots & \vdots & \vdots \quad \vdots \\ 0 & 0 & 0 \quad 0 \end{pmatrix} + \sqrt{-1} \begin{pmatrix} 0 & \sin(\alpha t) & 0 \dots 0 \\ \sin(\alpha t) & 0 & 0 \dots 0 \\ 0 & 0 & 0 \dots 0 \\ \vdots & \vdots & \vdots \quad \vdots \\ 0 & 0 & 0 \quad 0 \end{pmatrix}$$

and similarly for $\exp(Bt)$.

So we now consider $\tilde{\varphi} : \mathbb{R}^2 \to U(n)$ given by $\tilde{\varphi}(x, y) = \exp(Ax + By)$. Then $\tilde{\varphi}$ induces a map: $\varphi : \dfrac{\mathbb{R}^2}{(\frac{2\pi}{\alpha})\mathbb{Z} \oplus (\frac{2\pi}{\beta})\mathbb{Z}} \to F(n)$ given by:

$$\varphi(x, y) = \tilde{\varphi}(x, y)(E_1, \dots, E_n) \cdot \tilde{\varphi}^* = \exp(Ax + By) \cdot (E_1, \dots, E_n) \cdot \exp(-Ax - By).$$

Since $[A, B] = 0$ and $\operatorname{diag}(A) = \operatorname{diag}(B) = 0$, according to Theorem 4.1 φ is harmonic with respect to any left-invariant metric on $F(n)$ but not holomorphic with respect to any almost complex structure defined in §3, since $E_1 \cdot A \cdot E_2 = E_2 \cdot A \cdot E_1 = \alpha\sqrt{-1} \neq 0$.

Another family of equiharmonic but not holomorphic tori is obtained considering

$$A = \begin{pmatrix} 0 & \alpha\sqrt{-1} & 0 & 0 & \dots & 0 \\ \alpha\sqrt{-1} & 0 & 0 & 0 & \dots & 0 \\ 0 & 0 & 0 & \beta\sqrt{-1} & \dots & 0 \\ 0 & 0 & \beta\sqrt{-1} & 0 & \dots & 0 \\ \vdots & \vdots & \vdots & \vdots & & \vdots \\ 0 & 0 & 0 & 0 & \dots & 0 \end{pmatrix}$$

and

$$B = \begin{pmatrix} 0 & \beta\sqrt{-1} & 0 & 0 & \dots 0 \\ \beta\sqrt{-1} & 0 & 0 & 0 & \dots 0 \\ 0 & 0 & 0 & \alpha\sqrt{-1} & \dots 0 \\ 0 & 0 & \alpha\sqrt{-1} & 0 & \dots 0 \\ \vdots & \vdots & \vdots & \vdots & \vdots \\ 0 & 0 & 0 & 0 & \dots 0 \end{pmatrix},$$

where α and β are fixed nonzero real numbers such that $\alpha/\beta \in Q$. Then A and B are in $u(n)$, $[A, B] = 0$, and furthermore there exists $\gamma \in \mathbb{R}$ such that $\alpha \cdot \gamma$ and $\beta \cdot \gamma$ are integers. Then we consider $\varphi : \dfrac{\mathbb{R}^2}{2\pi\gamma(\mathbb{Z} \oplus \mathbb{Z})} = T^2 \to F(n)$ given by:

$$\varphi(x, y) = \exp(Ax + By) \cdot (E_1, \dots, E_n) \cdot \exp(-Ax - By).$$

See [16] for many more examples. See also [5] for some important related ideas.

5. On harmonic and non-holomorphic maps into $F(n)$

We will now generalize the construction of §4 in an attempt to obtain new solutions of our harmonic map equations. More explicitly: we will consider

$$\tilde{\varphi} = (\tilde{\pi}_1, \dots, \tilde{\pi}_n) : \mathbb{R}^2 \to F(n), \quad n \geq 4,$$

with $\pi_i(x, y) = \exp(Ax + Bf(y)) \cdot E_i \cdot \exp(-Ax - Bf(y))$, where $A, B \in u(n)$ satisfy $[A, B] = 0$ and $f : \mathbb{R} \to \mathbb{R}$ is an arbitrary non-constant function at least C^2.

One reason for such interest is the following: by considering the above maps we will naturally generate maps $\varphi : S^1 \times \mathbb{R} \to F(n)$ (for a lot of functions f). But it is well known that $S^1 \times \mathbb{R}$ is conformally equivalent to $S^2 - \{p, q\}$ and, since the energy functional is conformally invariant, by using the Theorem of Sacks and Uhlenbeck [17], we could remove these isolated singularities (under additional hypotheses on f to ensure that $E(\varphi) < \infty$). So in this way we would produce harmonic and non-holomorphic maps of S^2 into $U(n)/T$. Another reason for such study relies on its similarities with variational aspects of the Yang–Mills functional. See for example [2] or [9] for related material and explanations on these similarities

We shall show in this section that we obtain no new solutions beyond those obtained in §4 and so, in particular, no solutions of finite energy. I am very thankful to Francis Burstall for pointing out to me a basic mistake in an earlier version of this paper.

5.1. Lemma. *Let* $\tilde{\varphi} = (\tilde{\pi}_i)_{i=1}^n : \mathbb{R}^2 \to (F(n), g_{a=(a^{ij})})$ *be defined as above. Then:*

$$A_x^a = \sum_{i,j=1}^n a^{ij} \exp(Ax + Bf(y)) \cdot E_i \cdot A \cdot E_j \cdot \exp(-Ax - Bf(y))$$

$$A_y^a = \sum_{i,j=1}^n a^{ij} f'(y) \exp(Ax + Bf(y)) \cdot E_i \cdot B \cdot E_j \cdot \exp(-Ax - Bf(y)).$$

Proof. We have the following:

$$\begin{aligned} a^{ij} A_x^{ij} &= a^{ij} \tilde{\pi}_i \frac{\partial \tilde{\pi}_j}{\partial x} = a^{ij} \tilde{\pi}_i \{A \cdot \exp(Ax + Bf(y)) \cdot E_j \cdot \exp(-Ax - Bf(y)) \\ &\qquad + \exp(Ax + Bf(y)) \cdot E_j \cdot (-A) \cdot \exp(-Ax - Bf(y))\} \\ &= a^{ij} \exp(Ax + Bf(y)) \cdot E_i \cdot \exp(-Ax - Bf(y)) \cdot \{A \exp(Ax + Bf(y)) \cdot E_j \\ &\qquad \times \exp(-Ax - Bf(y)) - \exp(Ax + Bf(y)) \cdot E_j \cdot A \cdot \exp(-Ax - Bf(y))\} \\ &= a^{ij} \exp(Ax + Bf(y)) \cdot E_i \cdot A \cdot E_j \cdot \exp(-Ax - Bf(y)) \end{aligned}$$

since $A \cdot B = B \cdot A$ and $E_i \cdot E_j = 0$ if $i \neq j$.

On the other hand:

$$\begin{aligned} a^{ij} A_y^{ij} &= a^{ij} \tilde{\pi}_i \frac{\partial \tilde{\pi}_j}{\partial y} = a^{ij} \tilde{\pi}_i \{B \cdot \exp(Ax + Bf(y)) \cdot E_j \cdot \exp(-Ax - Bf(y)) \\ &\qquad + \exp(Ax + Bf(y)) \cdot E_j \cdot (-Bf'(y)) \cdot \exp(-Ax - Bf(y))\} \\ &= a^{ij} \exp(Ax + Bf(y)) \cdot E_i \cdot \exp(-Ax - Bf(y)) \cdot \{Bf'(y) \cdot \exp(Ax + Bf(y)) \cdot E_j \\ &\qquad \times \exp(-Ax - Bf(y)) - \exp(Ax + Bf(y)) \cdot E_j \cdot B \cdot f'(y) \cdot \exp(-Ax - Bf(y))\} \\ &\underset{[A,B]=0}{\overset{\text{since}}{=}} a^{ij} \exp(Ax + Bf(y)) \cdot E_i \cdot E_j \cdot B \cdot f'(y) \cdot \exp(-Ax - Bf(y)) \\ &= a^{ij} \exp(Ax + Bf(y)) \cdot E_i \cdot E_j \cdot B \cdot f'(y) \cdot \exp(-Ax - Bf(y)) \\ &= a^{ij} \exp(Ax + Bf(y)) \cdot E_i \cdot B \cdot E_j \cdot f'(y) \cdot \exp(-Ax - Bf(y)). \end{aligned}$$

5.2. Proposition. *We consider* $\tilde{\varphi} = (\pi_i)_{i=1}^n : \mathbb{R}^2 \to (U(n)/T, g_{a=(a^{ij})})$ *a smooth map defined as above i.e.* $\tilde{\pi}_i = \exp(Ax + Bf(y)) \cdot E_i \cdot \exp(-Ax - Bf(y))$, *where* A *and* B *are in* $u(n)$ *and satisfy* $[A, B] = 0$. *Then* $\tilde{\varphi}$ *is harmonic with respect to the metric* $g_{a=(a^{ij})}$ *if and only if*

$$\exp(Ax + Bf(y)) \cdot \left\{ \left(\sum_{i,j} a^{ij} E_i \cdot [A, \operatorname{diag} A] \cdot E_j + f''(y) \left(\sum_{i,j} a^{ij} E_i \cdot B \cdot E_j \right) \right. \right.$$
$$\left. \left. + (f'(y))^2 \left(\sum_{i,j} a^{ij} E_i \cdot [B, \operatorname{diag} B] \cdot E_j \right) \right) \right\} \exp(-Ax - Bf(y)) = 0,$$

where $\operatorname{diag}(A)$ *and* $\operatorname{diag}(B)$ *denote the diagonal part of* A *and* B*, respectively.*

Proof. We know by using Proposition 3.2 that $\varphi : M^2 \to (F(n), g_{a=(a^{ij})})$ is harmonic if and only if $\frac{\partial}{\partial x}(A^a_x) + \frac{\partial}{\partial y}(A^a_y) = 0$. We will now compute explicitly the Euler–Lagrange equations and so be able to find the formulas claimed:

$$\begin{aligned}
\frac{\partial}{\partial x}(A^a_x) &\overset{\text{Lemma 5.1}}{=\!=\!=} \frac{\partial}{\partial x}\Big(\sum_{i,j} a^{ij}\exp(Ax + Bf(y))\Big)\cdot E_i \cdot A \cdot E_j \cdot \exp(-Ax - Bf(y))\\
&= \sum_{i,j} a^{ij} A \cdot \exp(Ax + Bf(y)) \cdot E_i \cdot A \cdot E_j \cdot \exp(-Ax - Bf(y))\\
&\quad + \sum_{i,j} a^{ij}\exp(Ax + Bf(y)) \cdot E_i \cdot A \cdot E_j \cdot A \cdot \exp(-Ax - Bf(y))\\
&\overset{\text{since}}{\underset{[A,B]=0}{=\!=\!=}} \sum_{i,j} a^{ij}\exp(Ax + Bf(y)) \cdot A \cdot E_i \cdot A \cdot E_j \cdot \exp(-Ax - Bf(y))\\
&\quad + \sum_{i,j} a^{ij}\exp(Ax + Bf(y)) \cdot (A \cdot E_i \cdot A \cdot E_j - E_i \cdot A \cdot E_j \cdot A)\exp(-Ax - Bf(y))\\
&= \exp(Ax + Bf(y)) \cdot ([A, \sum_{i,j} a^{ij} E_i \cdot A \cdot E_j]) \cdot \exp(-Ax - Bf(y))\\
&= \sum_{i,j=1}^{n} a^{ij}\exp(Ax + Bf(y)) \cdot E_i \cdot [A, \operatorname{diag}(A)] \cdot E_j \cdot \exp(Ax + Bf(y)). \qquad (*)
\end{aligned}$$

On the other hand we have:

$$\begin{aligned}
\frac{\partial}{\partial y}(A^a_y) &= \frac{\partial}{\partial y}\sum_{i,j} a^{ij} f'(y)\exp(Ax + Bf(y)) \cdot E_i \cdot B \cdot E_j \cdot \exp(-Ax - Bf(y)))\\
&= \sum_{i,j} a^{ij} f''(y)\exp(Ax + Bf(y)) \cdot E_i \cdot B \cdot E_j \cdot \exp(-Ax - Bf(y)) \qquad (**)\\
&\quad + \sum_{i,j} a^{ij} f'(y)Bf'(y)) \cdot \exp(Ax + Bf(y)) \cdot E_i \cdot B \cdot E_j \cdot \exp(-Ax - Bf(y))\\
&\quad + (f'(y))^2\exp(Ax + Bf(y))) \cdot \Big(\sum_{i,j} a^{ij} E_i \cdot [B, \operatorname{diag} B] \cdot E_j \cdot \exp(-Ax - Bf(y)\Big).
\end{aligned}$$

Adding up $(*)$ and $(**)$ we conclude that $\varphi = (\pi_i)_{i=1}^n : \mathbb{R}^2 \to (F(n), g_{a=(a^{ij})})$ is harmonic if and only if

$$\begin{aligned}
&\exp(Ax + Bf(y)) \cdot \Big\{\Big(\sum_{i,j} a^{ij} E_i \cdot [A, \operatorname{diag} A] \cdot E_j + f''(y)\Big(\sum_{i,j} a^{ij} E_i \cdot B \cdot E_j\Big)\\
&\quad + (f'(y))^2\Big(\sum_{i,j} a^{ij} E_i \cdot [B, \operatorname{diag} B] \cdot E_j\Big)\Big\}\exp(-Ax - Bf(y)) = 0.
\end{aligned}$$

We will now make a very natural hypothesis namely that $\mathrm{diag}(A) = \mathrm{diag}(B) = 0$ i.e. we are considering matrices in $u(n)$ far from a maximal tori (basically we are considering variations in $T(F(n))$).

5.3. Corollary. *Let $\tilde{\varphi} = (\tilde{\pi}_1, \ldots, \tilde{\pi}_n) : \mathbb{R}^2 \to (F(n), g_{a=(a^{ij})})$ be a map which admits an $\mathbb{R}$-action defined as above i.e. $\tilde{\pi}_i = \exp(Ax + B(f(y)) \cdot E_i \cdot \exp(-Ax - Bf(y))$ with A, B in $u(n)$ satisfying $[A, B] = 0$. We assume that* $\mathrm{diag}(A) = \mathrm{diag}(B) = 0$. *Then $\tilde{\varphi}$ is harmonic if and only if $f''(y)(\sum_{i,j=1}^{n} a^{ij} E_i \cdot B \cdot E_j) = 0$.*

Proof. Just apply the equations in Proposition 4.2.

5.4. Theorem. *Let $\tilde{\varphi} = (\tilde{\pi}_1, \ldots, \tilde{\pi}_n) : \mathbb{R}^2 \to (F(n), g_{a=(a^{ij})})$ be an equivariant map defined in the following way: $\tilde{\pi}_i = \exp(Ax + Bf(y)) \cdot E_i \cdot \exp(-Ax - Bf(y))$ where A and B are in $u(n)$, $[A, B] = 0$, and $f : \mathbb{R} \to \mathbb{R}$ is an arbitrary non-constant function at least C^2. Furthermore we assume that* $\mathrm{diag}(A) = \mathrm{diag}(B) = 0$ *and $n \geq 4$. Then $\tilde{\varphi}$ is harmonic if and only if $f''(y) = 0$, $\forall y \in \mathbb{R}$.*

Proof. If $f''(y) = 0$, $\forall y \in \mathbb{R}$ then by using Corollary 5.3 we conclude that $\tilde{\varphi}$ is harmonic with respect to the metric $g_{a=(a^{ij})}$.

Reciprocally, if $\tilde{\varphi}$ is harmonic again according to Corollary 5.3 we have

$$f''(y)\left(\sum_{i,j=1}^{n} a^{ij} E_i \cdot B \cdot E_j\right) = 0.$$

But $B \neq 0$ so there is i_0 and j_0 such that $E_{i_0} B E_{j_0} \neq 0$ $(1 \leq i_0, j_0 \leq n)$. But $a^{i_0 j_0} > 0$ so we must have $f''(y) = 0$, $\forall y \in \mathbb{R}$.

5.5. Corollary. *Let $\tilde{\varphi} = (\tilde{\pi}_i)_{i=1}^{n} : \mathbb{R} \times \mathbb{R} \to (F(n), g_{a=(a^{ij})})$ be defined so as to satisfy all the hypothesis of Theorem 5.4. Then A and B can be chosen so that $\tilde{\varphi}$ induces $\varphi : T^2 = S^1 \times S^1 \to (F(n), g_{a=(a^{ij})})$ producing the harmonic non-holomorphic T^2 in $(F(n), g_{a=(a^{ij})})$ described in §4.*

Proof. According Theorem 5.4 we must have $f''(y) = 0$, $\forall y \in \mathbb{R}$ i.e. $f(y) = ay + b$, $a, b \in \mathbb{R}$, $a \neq 0$ since f is non-constant.

5.6. Remark. It remains a nice open problem to determine if we can apply some variant of Uhlenbeck's method as described in [18] to obtain new examples of 2-spheres in $F(n)$ which are harmonic and non-holomorphic (with respect to any almost complex structure defined in §3). The function f may satisfy some additional relation.

References

[1] M. Abraham and J. Marsden, Foundations of Mechanics, Benjamin Cummings, 1987.

[2] M. Atiyah, Instantons in two and four manifolds, Commun. Math. Phys. 93 (1984), 437–452.

[3] M. Black, Harmonic maps into homogeneous spaces, Pitman Res. Notes Math. Ser. 255, 1991.

[4] A. Borel, Kählerian coset spaces of semi-simple Lie groups, Proc. Nat. Acad. Sci. USA 40 (1954), 1147–1151.

[5] F. Burstall, D. Ferus, F. Pedit & U. Pinkall, Harmonic tori in symmetric spaces and commutative Hamiltonian systems on loop algebras, Ann. of Math. 138 (1993), 173–212.

[6] E. Calabi, Minimal immersions of surfaces in Euclidean spheres, J. Differential Geom. 1 (1967), 111–125.

[7] S. S. Chern, Minimal surfaces in a Euclidean space of N dimensions, Diff. Comb. Topology, Princeton, (1965), 187–198.

[8] A. Din and W. Zakarewski, General classical solution in the $\mathbb{C}P^{n-1}$ model, Nuclear Phys. B 174 (1980), 397–401.

[9] S. Donaldson, Instantons and geometric invariant theory, Commun. Math. Phys. 93 (1984), 453–460.

[10] J. Eells and L. Lemaire, Another report on harmonic maps, Bull. London Math. Soc. 20 (1988), 385–524.

[11] J. Eells and J. C. Wood, Harmonic maps from surfaces to complex projective spaces, Adv. in Math. 49 (1983), 217–263.

[12] V. Glaser and R. Stora, Regular solutions of the $\mathbb{C}P^n$ model and further generalizations, preprint, Cern 1980.

[13] M. Guest, The geometry of maps between generalized flag manifolds, J. Diff. Geom. 25 (1987), 223–250.

[14] H. Lawson, Surfaces minimales et la construction de Calabi–Penrose, Sém. Bourbaki 624 (1984).

[15] C. Negreiros, Some remarks about harmonic maps into flag manifolds, Indiana Univ. Math. J. 37 (1988), 617–636.

[16] C. Negreiros, Harmonic maps into periodic flag manifolds and into loop groups, J. Geom. Phys. 7 (1990), 339–361.

[17] J. Sacks and K. Uhlenbeck, The existence of minimal immersion of 2-spheres, Ann. of Math. 113 (1981), 1–24.

[18] K. Uhlenbeck, Equivariant harmonic maps into spheres, Lecture Notes in Math. 949, Springer-Verlag, 1981.

[19] K. Uhlenbeck, Harmonic maps into Lie groups (Classical solutions of the Chiral model), J. Differential Geom. 30, 1–50 (1989).

Focal decomposition in geometry, arithmetic and physics

Mauricio M. Peixoto

Abstract. It is remarked that the focal decomposition associated to the 2-point boundary value problem for $\ddot{x} = f(t, x, \dot{x})$ is a prerequisite for the semiclassical quantization of this equation via the Feynman path integral method. Besides the concept of Brillouin zones is extended to arbitrary Riemannian manifolds. Also a convolution formula is given involving the number of solutions of the Diophantine equation $Ax^2 + Bxy + Cy^2 = N$, an extension of a previous work on $Ax^2 + Cy^2 = N$.

1991 Mathematics Subject Classification: 34B10, 11B09, 81Q30.

§1. Introduction

Roughly speaking we deal here with the way the trajectories of a second order differential equation starting at a given point will meet again, i.e. focus afterwards. The concept of focal decomposition formalizes this intuitive idea.

This concept was introduced in [1] in the context of the 2-point boundary value problem for a second order ordinary differential equation (see §2 below). It was further considered by Peixoto and Thom in [2] where the possibility of a general, analytic theory became clear. Also in that paper the fundamental example of the pendulum equation $\ddot{x} + \sin x = 0$ was worked out in detail. See also [3].

Afterwards the concept of focal decomposition was put by Kupka and Peixoto [4] in the context of geodesics of a Riemannian manifold M. When M is a flat torus the corresponding focal decomposition, surprizingly, bumps into the Brillouin zones of a crystal, to which it bears a close formal relationship. It also leads naturally to relationships with the arithmetic of positive definite quadratic forms [5]. In [2], [4], [5] the expression "sigma decomposition" was used for what we call here "focal decomposition".

In this paper we make a survey of above developments, adding some new insights.

One of them is the remark that the knowledge of the focal decomposition relative to a second order ordinary differential equation may be considered as a prerequisite for its semiclassical quantization via the Feynman path integral method.

Another insight concerns the possibility of extending the concept of Brillouin zones to general Riemannian manifolds.

We end the present paper with the theorem of §6, an application of the concept of focal decomposition. There we extend to the Diophantine equation $Ax^2 + Bxy + Cy^2 = N$ the convolution formula derived in [5] for the equation $Ax^2 + Cy^2 = N$.

For the convenience of the reader the present paper can be read independently of [5]. In [5] we proved the following proposition. Let A, C, p be distinct odd primes, $A \equiv C \pmod 4$, and consider the four Diophantine equations

$$Ax^2 + Cy^2 = p \text{ or } 2p, \quad ACx^2 + y^2 = p \text{ or } 2p.$$

Then either none of these equations is solvable or else exactly one is solvable and the other three are not. Although I could not find a reference for this in the literature, J.W.S. Cassels pointed out to the author in a private communication that the above proposition can be proved via the theory of composition of ambiguous binary quadratic forms. Now all forms $Ax^2 + Cy^2$ are ambiguous but among the forms considered here infinitely many may well be not ambiguous. This makes natural the study of the forms we consider. See a remark at the end of §6.

The author thanks C. A. A. de Carvalho for discussions.

§2. Focal decomposition in the 2-point boundary value problem

We consider the 2-point boundary value problem

$$\ddot{x} = f(t, x, \dot{x}), \quad x(t_1) = x_1, \quad x(t_2) = x_2, \quad t, x, \dot{x} \in \mathbb{R}, \tag{1}$$

the simplest and oldest of all boundary value problems. Accordingly there is a vast literature, mostly in the context of Applied Mathematics, where one uses frequently the methods of Functional Analysis.

Here we adopt a different point of view: we look for the number of solutions of problem (1), if any, and how this number varies with the endpoints.

The concept of focal decomposition expresses precisely this point of view and can be formalized as follows.

Let $\mathbb{R}^4(t_1, x_1; t_2, x_2)$ be the set of all pairs of points of the (t, x)-plane and to each pair associate the integer $i(t_1, x_1; t_2, x_2) = i$ defined as the number of solutions of problem (1). The possible values of i are $0, 1, \ldots, \infty$. When $t_1 = t_2$ the index i is defined as being 0 if $x_1 \neq x_2$ and ∞ if $x_1 = x_2$.

Call $\Sigma_i \subset \mathbb{R}^4$ the set of points of $\mathbb{R}^4$ to which the index i has been assigned. Clearly the Σ_i define a partition of $\mathbb{R}^4$ i.e. $\mathbb{R}^4$ is the disjoint union of the Σ_i,

$$\mathbb{R}^4 = \Sigma_0 \cup \Sigma_1 \cup \cdots \cup \Sigma_\infty . \tag{2}$$

This is called the *focal decomposition* associated to problem (1).

The fundamental problem is: to describe the sets Σ_i and how they decompose $\mathbb{R}^4$.

To get a feeling about what can be done about this general, seemingly hopeless problem, consider the special case where one of the end points in (1) is fixed, say (t_1, x_1). Then the sets Σ_i induce a decomposition of $\mathbb{R}^2(t_2, x_2)$ by the sets $\sigma_i = \Sigma_i \cap (t_1, x_1) \times \mathbb{R}^2(t_2, x_2))$. The *restricted problem* with base point (t_1, x_1) consists in finding the corresponding focal decomposition of

$$\mathbb{R}^2 = \sigma_0 \cup \sigma_1 \cup \cdots \cup \sigma_\infty . \tag{3}$$

Now to deal with the restricted problem we associate to the equation (1) the first order equation

$$\dot{t} = 1, \quad \dot{x} = u, \quad \dot{u} = f(t, x, u), \qquad t, x, u \in \mathbb{R} \tag{1'}$$

Call $\mathcal{F}$ the foliation of $\mathbb{R}^3(t, x, u)$ defined by the trajectories of (1′) and $\mathcal{F}(t, x, u)$ the leaf through (t, x, u).

Define the *star* $\Omega(t_1, x_1)$ of a given point $(t_1, x_1) \in \mathbb{R}^2$ as the union of all leaves $\mathcal{F}(t_1, x_1, u)$, $-\infty < u < \infty$.

Clearly $\Omega(t_1, x_1)$ is a two dimensional manifold with the same class of differentiability as f. It is also clear that denoting by $\pi : \mathbb{R}^3 \to \mathbb{R}^2$ the projection parallel to the u-axis, $\pi(t, x, u) = (t, x)$ then with respect to base point (t_1, x_1)

$$(t_2, x_2) \in \sigma_i \quad \Leftrightarrow \quad \operatorname{card}\{(\pi \mid \Omega(t_1, x_1))^{-1} (t_2, x_2)\} = i .$$

So the focal decomposition with base point (t_1, x_1) appears naturally in the description of the projection of the star $\Omega(t_1, x_1)$ on $\mathbb{R}^2(t, x)$.

Now if the equation (1) is analytic so is the star $\Omega(t_1, x_1)$. From results of Hironaka and Thom, see [6, pp. 40–43] and [4, p. 243], one has the following

Existence Theorem. *If f is analytic and $\pi \mid \Omega(t_1, x_1)$ is proper on $\mathbb{R}^2 \backslash \delta$ where $\delta = \{(t, x) \mid t = t_1\}$, then there is a Whitney stratification of $\mathbb{R}^2 \backslash \delta$ such that each $\sigma_i \backslash \delta$ is the union of strata of this stratification.*

The result of all this is that the analyticity of f and the properness of $\pi \mid \Omega(t_1, x_2)$ guarantee that each σ_i is the locally finite union of nice, well behaved analytic manifolds. For instance, no σ_i can be a Cantor set. Of course the properness condition is in general a touchy one to verify. This because it involves the whole star $\Omega(t_1, x_1)$, which stands for the totality of all trajectories of (1), passing through (t_1, x_1).

The above theorem extends to the general case of the focal decomposition, the star being substituted by the graph of the foliation $\mathcal{F}$. See [4, p. 243].

§3. The Feynman path integral quantization and focal decomposition

Given the second order equation (1), which we assume is the Euler equation of some action integral S, there is a well defined mathematical operation, involving eigenfunctions and spectra, called its quantization. And one of the methods of quantization is the Feynman path integral method. In this method a crucial step is the calculation of a certain amplitude called the propagator $K(P_1, P_2)$ relative to a pair of points $P_1(t_1, x_1)$ and $P_2(t_2, x_2)$. This $K(P_1, P_2)$ is obtained by making a sum over all continuous curves connecting P_1 and P_2.

Consider now the case where the data of the problem, such as lenghts, masses, times, etc are so large that S representing the classical action then $S/\hbar$ is very big. This is called the semiclassical quantization.

In this case, to calculate the propagator $K(P_1, P_2)$ we have only to consider the curves joining P_1 and P_2 which are actual solutions of the equation (1). See [7, p. 29].

Therefore the knowledge of the focal decomposition associated to the problem (1) appears as a preliminary problem that one encounters before doing the semiclassical quantization of the equation (1).

Now a good candidate for the semiclassical quantization via path integral is the pendulum, whose motion is described by the equation

$$\ddot{x} + \sin x = 0. \tag{4}$$

Its focal decomposition was determined in [2].

Consider the case of base point $(0, 0)$ which is typical of what happens at other base points $(0, x_0)$, $|x_0| < \pi$.

The focal decomposition at base point $(0, 0)$ is described by Fig. 1 below.

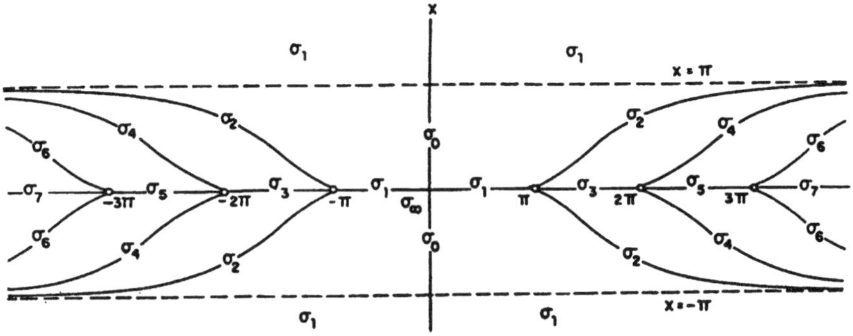

Figure 1

It should be remarked that in a related context, the one of tunneling in Quantum Statistical Mechanics we found recently the appearance of the above Fig. 1, in the work of C. A. A. Carvalho and R. M. Cavalcanti [8]. See Fig. 5 of [8]. But in Fig. 5 the axis where the cusps are situated stands for absolute temperature so that Fig. 5 is only half of the above Fig. 1.

§4. Focal decomposition on Riemannian manifolds

In [4], joint work with Kupka, the concept of focal decomposition is extended to Riemannian manifolds. This is quite natural since to every Riemannian manifold one associates in a canonical way a second order differential equation, namely the differential equation of the geodesics. Observe that in this equation the independent "time" variable is substituted by the arc lenght variable. But instead of working directly with the equations of the geodesics it is better to work with the exponential map.

Let M be a complete Riemannian manifold and TM be its tangent bundle. Let $v \in TM$, $v \neq 0$, $p = \pi(v)$ where $\pi\colon TM \to M$ is the canonical projection. Call $I(v)$ the number of geodesic arcs of lenght $||v||$ connecting p to $\exp_p(v)$ and define

$$\Sigma_i = \{v \in TM \mid I(v) = i\}, \qquad i = 1, 2, \dots, \infty.$$

The zero section, $M \subset TM$ by definition belongs to Σ_1. Since M is complete, $\Sigma_0 = \emptyset$.

The *focal decomposition of* TM is the partition of TM into the sets Σ_i, $i = 1, 2, \dots, \infty$. Clearly the focal decomposition always exists and depends only on the Riemannian metric of M. If $p \in M$ and $\sigma_i = \Sigma_i \cap T_pM$ the study of the partition of T_pM by the sets σ_i constitutes the *restricted problem with base point p*. In [4, p.243] we proved the following

Existence Theorem. *If M is analytic and complete then there is an analytic Whitney stratification of* TM *such that each* Σ_i *is the union of strata of this stratification. Similarly for the restricted problem with base point* $p \in M$: *there is an analytic Whitney stratification of* T_pM *such that every set* σ_i *is the union of strata of this stratification.*

So this existence theorem assumes only that the Riemannian metric is analytic and complete. In the differential equation case considered in §2, to get the corresponding existence theorem it is necessary to assume that a certain map is proper. So analyticity alone in the differential equation case may eventually lead to a very complicated focal decomposition. Because of this, we may say that the natural place to study focal decomposition is a complete analytic Riemannian manifold.

The determination of the focal decomposition relative to a complete analytical Riemannian manifold is, in general, a difficult problem because it requires a very precise knowledge of all geodesics passing through the base point.

The flat torus. Consider now what seems to be the simplest of all non trivial examples of focal decomposition: the two dimensional flat torus $T^2 = \mathbb{R}^2/\mathbb{Z}^2$, with the usual metric, induced by the canonical metric of $\mathbb{R}^2$. This metric is canonically associated to the quadratic form $x^2 + y^2$.

Observe that the exponential map is identified with the covering map $\pi : \mathbb{R}^2 \to T^2$. We take the base point as the origin in the tangent space.

We want to find out the index $I(x, y)$ of an arbitrary point (x, y). It clearly is the cardinality of the pair of integers $(m, n) \in \mathbb{Z}^2$ such that

$$(x+m)^2 + (y+n)^2 = x^2 + y^2. \tag{5}$$

Clearly if $(x, y) \in \mathbb{Z}^2$, $I(x, y)$ coincides with the number of integral solutions (X, Y) of the Diophantine equation

$$X^2 + Y^2 = x^2 + y^2 = N \in \mathbb{Z}.$$

This number is usually designated by $R(N)$ and is expressed by a classical formula of Gauss in terms of the prime factorization of N.

In general if x, y are not both integers, $I(x, y)$ is the number of solutions of (5), considered as a Diophantine equation in (m, n), so that $I(x, y)$ is the number of lattice points $(m, n) \in \mathbb{Z}^2$ on the circle centered at $(-x, -y)$ and passing through the origin. Now the determination of $I(x, y)$ by some explicit formula analogue to the above mentioned formula of Gauss seems to be a hopeless problem.

To get the desired $I(x, y)$ we approach this question geometrically as follows. In a way we "invert" the problem, considering in (5), $(m, n) \in \mathbb{Z}^2 \backslash (0, 0)$ as given and look for the corresponding x, y. Now (5) is equivalent to

$$L(m, n) \colon 2mx + 2ny + m^2 + n^2 = 0 \tag{5'}$$

so that this equation represents the straight line $L(m, n)$, the perpendicular bisector to the lattice vector connecting the origin to the lattice point $(-m, -n)$. Assume we have drawn all lines $L(m, n)$. Clearly, from (5), the index $I(x, y)$ equals 1 plus the number of lines $L(m, n)$ passing through (x, y). So the rule set for determining the index $I(x, y)$ automatically determines the whole focal decomposition on the tangent space $\mathbb{R}^2$. See Fig. 1 of [4] for a drawing of the lines $L(m, n)$.

The above can be extended to a flat torus in N dimensions with a metric defined by any positive definite quadratic form in N variables. The resulting focal decomposition of the tangent space $\mathbb{R}^N$ being defined by hyperplanes, the analogue of the lines $L(m, n)$,i.e. the analogues of the (5) with m, n held fixed.

Actually we can forget about the geodesic interpretation and say that any positive definite quadratic form in N variables determines a focal decomposition in $\mathbb{R}^N$. See [5, p. 458] for a generalization of the concept of focal decomposition along these lines.

§5. Brillouin zones

We go back to the above family of lines $L(m, n)$, corresponding to the quadratic form $x^2 + y^2$. Call

$$\mathcal{L} = \{L(m, n)\}, (m, n) \in \mathbb{Z}^2 \backslash (0, 0).$$

The Brillouin zones of a cubic crystal correspond to a certain way of assembling the connected components of the complement of $\mathcal{L}$. See [9, p. 126]. For latter use we give a formal presentation of this concept as follows.

Definition 5.1. The Brillouin index $B(x, y)$ of a point $(x, y) \in \mathbb{R}^2$ is 1 plus the number of lines $L(m, n)$ meeting the closed segment connecting $(0, 0)$ to (x, y).

If B is one connected component of the complement of $\mathcal{L}$ then all points of B have the same Brillouin index, called the Brillouin index of B. This is clear because no line of $\mathcal{L}$ passes through the origin, and B is convex.

Definition 5.2. The Brillouin zone of order k, $k \geq 1$, is the interior, $\overset{\circ}{B}_k$, of the set B_k of all points with Brillouin index k.

Clearly $\overset{\circ}{B}_1$ is the interior of the square bounded by the lines $x = \pm 1/2$, $y = \pm 1/2$, $\overset{\circ}{B}_2$ is formed by the interior of the four equilateral triangles determined by the four sides of $\overset{\circ}{B}_1$. In general $\overset{\circ}{B}_k$ is formed by the interiors of a certain number of convex polygons joined by the vertices and forming a kind of collar arround the origin.

Brillouin zones are a classical concept of Solid State Physics and were introduced some sixty years ago in a context quite different from the purely geometrical one given above. Yet these two concepts, focal decomposition and Brillouin zones are formally almost the same thing in the sense that both are completely and immediately determined by the lines $L(m, n)$.

A remarkable fact discovered by Brillouin on physical grounds (conservation of energy) and latter proved by Bieberbach [10] is that all Brillouin zones have the same area. This was done by showing that A_{k_j}, $j = 1, 2, \ldots.$ representing the connected components of the zone $\overset{\circ}{B}_k$, then for each j there is a translation of A_{k_j} by a vector of integral components such that these translated of these connected components are disjoint an fill completely the first zone $\overset{\circ}{B}_1$, with the exception of the boundary lines.

The above fact was proved by G. A. Jones in [11] this author being apparently unaware of the earlier work of Bieberbach in [10]. But Jone's proof is independent of Bieberbach's and besides he proves a number of interesting new facts about Brillouin zones.

Although we are not going to discuss here Brillouin zones from the physical point of view it is appropriate to point out that the lines $L(m, n)$ do have a very down-to-earth physical interpretation, having to do with waves.

In fact, $L(m, n)$ is the locus of all points (x, y) satisfying (5) with m, n held fixed. But then $L(m, n)$ appears as the locus of all real points common to the two circles

$$x^2 + y^2 = \lambda, \quad (x + m)^2 + (y + n)^2 = \lambda, \quad \lambda \geq 0, \tag{5''}$$

of equal radii $\lambda^{1/2}$, centered at $(0, 0)$ and $(-m, -n)$ respectively.

So if two stones strike simultaneously the surface of a pool at (0,0) and $(-m, -n)$, the locus of the interferences of the wave fronts thus formed is a good image of the line $L(m, n)$. Physicists call them Bragg lines.

It is worthwile to point out here that although equations (5), (5') and (5") are obviously equivalent to each other they "point out" in different directions.

So far we considered Brillouin zones associated to the quadratic form $x^2 + y^2$.

But one can extend the concept to the case of any positive definite quadratic form in any number of variables in essentially the same way. Below we propose a definition of the Brillouin zones in T_pM which seems to be natural enough and raises interesting questions.

Brillouin zones in Riemannian manifolds. Let M be an analytic, complete Riemannian manifolds and $O \in M$. The interest of defining Brillouin zones in T_OM lies in the fact that this would give some information of a global nature about the exponential map $\exp_0 \colon T_OM \to M$. Assuming of course that we manage to get some understanding of what these zones are.

The existence theorem of §4 suggests that in the analytical Whitney stratification of T_pM associated to the sets σ_i, one might use the codimension 1 strata of this stratification to build the "walls" of the zones, in analogy with the lines $L(m, n)$ above. But the information given by the existence theorem is too scanty to allow one to go any further.

So Definition 5.1 of the Brillouin index, as it is, does not generalizes to T_0M. But an equivalent form of this definition does. In fact to say that $L(m, n)$ meets the segment connecting $(0, 0)$ and (x, y) is the same thing as saying that the point $(-m, -n)$ lies on the disc determined by the circle centered at (x, y) and passing through $(0, 0)$. So m, n must satisfy

$$(x + m)^2 + (y + n)^2 \leq x^2 + y^2 \tag{6}$$

and the Brillouin index $B(x, y)$ is 1 plus the cardinality of the pair of integers $(m, n) \in \mathbb{Z}^2 \backslash (0, 0)$ satisfying (6). Observe that the combination $m = n = 0$ does satisfy (6). So the Brillouin index $B(x, y)$ is the cardinality of all points (X, Y) on the disc

$$X^2 + Y^2 \leq x^2 + y^2$$

equivalent mod 1 to (x, y); whereas to get the index $I(x, y)$ one has only to change the inequality sign into an equality one.

It is now clear how to define the Brillouin index and zones in T_0M, extending Definition 5.1 and Definition 5.2, corresponding to the flat torus case.

Definition 5.3. (i) Let $p \in T_0M$, then the Brillouin index $B(p)$ is

$$B(p) = \#\{q \in T_0M \mid \|q\| \leq \|p\|,\ \exp_0 q = \exp_0 p\}.$$

(ii) For every integer $k \geq 1$ the k-th Brillouin zone is the interior, $\overset{\circ}{B}_k$, of the set $B_k = \{q \in T_0M \mid B(q) = k\}$ of all points with Brillouin index k.

From this definition a natural question arises: how does the stratification that appears in the existence theorem of §4 fit into the picture of the zones?

We now mention two simple examples of the application of Definition 5.3 in which we have a situation quite similar to the one of the flat torus, the codimension 1 strata — the walls — bounding the zones.

(i) $M = S^n$, $n \geq 1$, with the usual metric. The first zone is an open disc, all the other are bounded by concentric $(n-1)$-spheres, the walls. A ray joining the origin to a point on the k-th zone, crosses $(k-1)$ walls.

(ii) $M = S^1 \times \mathbb{R}$, the cylinder, with the usual metric. If the x-axis corresponds to S^1, the y-axis corresponds to $\mathbb{R}$ and S^1 has length 2, then the lines $x = \pm 1, x = \pm 2, \dots, x = \pm k, \dots$ all belong to σ_2 and are the walls defining the zones. Then B_1 is the strip $-1 < x < 1$ and B_k, $k > 1$, is the disconnected strip

$$k - 1 < x < k, \quad -k < x < -k + 1.$$

Again a line joining the origin to any point in B_k crosses $(k-1)$ walls.

§6. Arithmetic

We consider the Diophantine equations

$$Ax^2 + Bxy + Cy^2 = N, \quad \Delta = 4AC - B^2 > 0 \tag{7}$$

$$\Delta x^2 + y^2 = N \tag{8}$$

and assume that

$$A, C \text{ odd}, \tag{9}$$

$$A \equiv C \pmod 4 \tag{10}$$

$$B \equiv 0 \pmod 4 \tag{11}$$

$$\overline{\Delta} = \Delta/4 \text{ is square free and odd.} \tag{12}$$

Call $R(N)$ the number of solutions of (7), $s(N)$ the number of primitive solutions of (8) and $s^+(N)$ the number of those primitive solutions for which $x \geq 0$, $y \geq 1$.

Theorem. *Under the above assumptions we have*

$$\text{(I)}\quad \tfrac{1}{2}R(N)^2 = \textstyle\sum_{d|4\Delta N} s^+(d)\, R\left(\frac{4\Delta N}{d}\right), \tag{13}$$

or equivalently

$$\text{(II)}\quad R(N)^2 = R(\Delta N) - R(N) + \tfrac{1}{2}\textstyle\sum_{d|4\Delta N} s(d)\, R\left(\frac{4\Delta N}{d}\right). \tag{14}$$

The proof of this theorem will be preceeded by an explanation of the very simple geometric considerations that underlies it, by some definitions and by several lemmas.

We consider the focal decomposition associated to the quadratic form (7).

Repeating the argument of §4 the index of a point (x, y), $I(x, y)$, is the number of pairs $(m, n) \in \mathbb{Z}^2$ such that

$$A(x+m)^2 + B(x+m)(y+n) + C(y+n)^2 = Ax^2 + Bxy + Cy^2 \tag{15}$$

a difficult question to settle. If we fix $(m, n) \in \mathbb{Z}^2\backslash(0,0)$, (15) becomes the equation of a straight line

$$(2Am + Bn)x + (Bm + 2Cn)y + Am^2 + Bmn + Cn^2 = 0, \tag{16}$$

passing through the base point $(-m/2, -n/2)$. Call this line $L(m, n; A, B, C) \doteq L(m, n)$ and $\mathcal{L} = \{L(m, n)\}$ their totality. Clearly the index $I(x, y)$ equals 1 plus the number of lines of $\mathcal{L}$ passing through (x, y).

A trivial computation shows that a line $L(m, n)$, at its base point $(-m/2, -n/2)$, is tangent to the ellipse (7) passing through it ,i.e. for which $N = (Am^2 + Bmn + Cn^2)/4$.

Now, concerning the family $\mathcal{L}$ of lines one of the simplest questions one can ask about $\mathcal{L}$ is: how many lines $\mathcal{L}$ passes through a lattice point $(x_0, y_0) \neq (0, 0)$? The answer, we just saw above is $R(Ax_0^2 + Bx_0y_0 + Cy_0^2) - 1$.

Another question: how many lines of $\mathcal{L}$ meet the ellipse (7) at some lattice point?

Definition 6.1. A line $L(m, n)$ of the family $\mathcal{L}$ is said to be allowable if and only if it meets the ellipse (7) at some lattice point.

The proof of the theorem consists in calculating the number of allowable lines in two different ways and equating them. Actually the statement of the theorem was obtained in answering the above question.

In Lemma 1 below the number of allowable lines is calculated in an obvious way.

No restriction is necessary on A, B, C, N except that they are positive integers and satisfy the inequality in (7).

Lemma 1. *The number of allowable lines is* $\frac{1}{2}R(N)^2$.

Proof. Let $(x_1, y_1), \ldots, (x_k, y_k)$, $R = R(N)$, be the lattice points on the ellipse (7). Now from the very definition of the lines $L(m, n)$ it is clear that $L(-x_1 - x_i, -y_1 - y_i)$, for $(x_i, y_i) \neq (x_1, y_1), (-x_1, -y_1)$, passes through both (x_1, y_1) and (x_i, y_i). So, passing through (x_1, y_1) there are $R(N) - 2$ such lines. Also as it was seen above, the tangent to the ellipse at (x_1, y_1), which is $L(-2x_1, -2y_1)$, should also be counted. So, altogether there are $R(N)[R(N) - 2]/2$ lines of $\mathcal{L}$ that correspond to the lines joining the lattice points, to which one should add the $R(N)$ tangents. Adding up one gets $R(N)^2/2$ allowable lines, as stated. □

Now consider another way to compute the number of allowable lines. Of course one needs to find a necessary and sufficient condition for $L(m, n)$ to be allowable. This will be done piecemeal, in several lemmas, so that one keeps track of the assumptions on A, B, C, m, n that will be introduced on the way.

First of all remark that $L(m, n) \in \mathcal{L}$ meets the ellipse (7) somewhere — not necessarily a lattice point — if and only if

$$Am^2 + Bmn + Cn^2 = E \leq 4N. \tag{17}$$

All lines $L(m, n)$ corresponding to the same E are tangent at their base point to the same ellipse in the x, y plane which is contained in the original ellipse (7); this smaller ellipse is the homothetic replica of ratio 1/2 of the ellipse (17) in the m, n plane.

Now one has to distinguish among the lines $L(m, n)$ that meet the ellipse (7), the ones that meet it at some lattice point ,i.e. the allowable ones. Of course these are to be found among the lines $L(m, n)$ that contain some lattice point.

Lemma 2. *Assuming* (10), (11), *then* $L(m, n)$ *contains some lattice point if and only if* m, n *have the same parity,*

$$m \equiv n \pmod 2. \tag{18}$$

Proof. We first show that $m \not\equiv n \pmod 2$ implies that $L(m, n)$ contains no lattice point. In fact, from the equation (16) of $L(m, n)$ we see that the coefficients of x, y are both even whereas the term independent of x, y is odd, contradiction.

It remains to show that (18) implies that $L(m, n)$ contains some lattice point. If m, n are both even, the base point $(-m/2, -n/2)$ is a lattice point in $L(m, n)$. If m, n are both odd, from (16) it follows that a parametric representation of $L(m, n)$ is

$$x = -\frac{m}{2} - (Bm + 2Cn)t, \quad y = -\frac{n}{2} + (2Am + Bn)t$$

so that for $t = 1/4$ we get a lattice point in $L(m, n)$. Lemma 2 is proved. □

Another lemma is necessary.

Lemma 3. *Assuming conditions* (9), (10), (11), *let* x, y *be a solution of* (7). *Then*

(i) x, y *are both even if and only if* $N \equiv 0 \pmod 4$,

(ii) x, y *are both odd if and only if* $N \equiv 0 \pmod 2$.

Proof. (i) We need to prove the equivalence of (i1) $x, y \equiv 0 \pmod 2$ and (i2) $N \equiv 0 \pmod 4$.

Clearly (i1) $\Rightarrow$ (i2). To show that (i2) $\Rightarrow$ (i1) suppose that it does not, and get a contradiction. We then have to show that (i2) is in contradiction with each of (i3) $x \equiv 0 \pmod 2$, $y \equiv 1 \pmod 2$, (i4) $x \equiv 1 \pmod 2$, $y \equiv 0 \pmod 2$ or (i5) $x \equiv 1 \pmod 2$, $y \equiv 1 \pmod 2$. Since A, C are odd, (i3) and (i4) contradict (i2). It remains to show that (i5) contradicts (i2). This is so because we are also assuming (10).

In fact (i5) implies that $x^2, y^2 \equiv 1 \pmod 4$ and condition (10) implies that either $A, C \equiv 1 \pmod 4$ or else $A, C \equiv 3 \pmod 4$, and in either case $N \equiv 2 \pmod 4$ contradicting (i2).

This proves part (i) of Lemma 3.

(ii) We have to prove that (ii1) $x, y \equiv 1 \pmod 2$ is equivalent to (ii2) $N \equiv 2 \pmod 4$.

That (ii1) $\Rightarrow$ (ii2) is clear by the argument just used at the end of the proof of part (i).

Now show that (ii2) $\Rightarrow$ (ii1). If not we would have $N \equiv 2 \pmod 4$ consistent with one of the possibilities (ii3) $x, y \equiv 0 \pmod 2$, (ii4) $x \equiv 1 \pmod 2$, $y \equiv 0 \pmod 2$ or (ii5) $x \equiv 0 \pmod 2$, $y \equiv 1 \pmod 2$. This is certainly not the case because (ii3) implies $N \equiv 0 \pmod 4$ and both (ii4) and (ii5) imply $N \equiv 1 \pmod 2$. So (ii2) $\Rightarrow$ (ii1) proving part (ii) of Lemma 3. Lemma 3 is proved. □

We now have the following characterization of the allowable lines

Lemma 4. *Assume conditions* (9), (10), (11) *are satisfied. A necessary and sufficient condition for $L(m, n)$ to be allowable is that m, n satisfy conditions* (17), (18) *and that there exist integers $p \geq 0$, $q \geq 1$ with $(p, q) = 1$ such that*

$$\rho = \left(\frac{4N - E}{\Delta E}\right)^{\frac{1}{2}} = \frac{p}{q}, \tag{19}$$

and

$$q \mid Bm + 2Cn, \quad q \mid 2Am + Bn. \tag{20}$$

Proof. (i) Necessity. If $L(m, n)$ is allowable then (17) is verified because it means that $L(m, n)$ meets the ellipse (7); (18) is verified because of Lemma 2. Consider now the intersection of the ellipse (7) with $L(m, n)$ defined by (16). Solving for x, y we get the intersection points

$$x_1 = \frac{1}{2}[-m + (Bm + 2Cn)\rho], \quad y_1 = [-n - (2Am + Bn)\rho]$$

and (21)

$$x_2 = \frac{1}{2}[-m - (Bm + 2Cn)\rho], \quad y_2 = [-n + (2Am + Bn)\rho]$$

where

$$\rho = \left(\frac{4N - E}{\Delta E}\right)^{\frac{1}{2}}, \quad \Delta = 4AC - B^2, \quad E = Am^2 + Bmn + Cn^2.$$

Since we are assuming that say x_1, y_1 are both integers, ρ must be some rational $\rho = \frac{p}{q}$ and this means that condition (19) must be satisfied. Besides, x_1, y_1 integers require that

$$P = (Bm + 2Cn)\frac{p}{q}, \quad Q = -(2Am + Bn)\frac{p}{q} \tag{22}$$

be both integers so that condition (20) must be satisfied. Necessity is proved.

(ii) Sufficiency. Suppose now that m, n satisfy (17), (18), (19), (20) and we have to show that $L(m, n)$ is allowable. In fact, because of (17), $L(m, n)$ meets the ellipse (7), because of (18) and Lemma 2, $L(m, n)$ contains lattice points. Because of (19) and (20), P, Q given by (22) are integers. This is not enough to show that the intersection points (21) are lattice points ,i.e. that $L(m, n)$ is allowable. To get that we have to show that P has the parity of m and Q has the parity of n. From (22) one gets

$$AP^2 + BPQ + CQ^2 = 4N - E. \tag{23}$$

Because of (18) we have to distinguish two cases: (ii1) m, n even; (ii2) m, n odd.

In case (ii1), $E \equiv 0 \pmod 4$ and so $4N - E \equiv 0 \pmod 4$ so that from Lemma 3 applied to (23) one gets that P, Q are even. In case (ii2) one get that $E \equiv 2 \pmod 4$ and thus $4N - E \equiv 2 \pmod 4$ and Lemma 3 tell us that P, Q are both odd. In all cases P has the parity of m and Q the parity of n.

So $L(m, n)$ is allowable, proving sufficiency and Lemma 4. □

We now give another form of Lemma 4,

Lemma 5. *Assume conditions* (9), (10), (11) *are satisfied. A necessary and sufficient condition for* $L(m, n)$ *to be allowable is that* m, n *satisfy* (17), (18) *and that there are integers* $p \geq 0$, $q \geq 1$, $(p, q) = 1$ *and integers* μ, ν *such that*

$$E = Am^2 + Bmn + Cn^2 = \frac{4Nq^2}{\Delta p^2 + q^2}, \tag{24}$$

$$2Am + Bn = \mu q, \qquad Bm + 2Cn = \nu q. \tag{25}$$

Proof. In fact (24) is equivalent to (19) and (25) is an equivalent way of expressing (20), taking into account the definition of E in (17). □

Now we get the final form of the characterization of the allowable lines, which will allow us to calculate their number.

Lemma 6. *A necessary and sufficient condition for $L(m, n)$ to be allowable is that m, n satisfy* (17), (18) *and that besides there exist integers p, q, μ, ν such that*

$$(\Delta p^2 + q^2)(C\mu^2 - B\mu\nu + A\nu^2) = 4\Delta N, \quad p \geq 0,\ q \geq 1,\ (p, q) = 1 \tag{26}$$

$$m = \frac{q}{\Delta}(-2C\mu - B\nu), \quad n = \frac{q}{\Delta}(2A\nu - B\mu). \tag{27}$$

Proof. In fact (27) is (25) solved for m, n, and (26) is obtained by substituting these values of m, n into (24). □

Lemma 6 is a characterization of the allowable lines in a form that is amenable to the computation of their number.

In fact from (26) we see that by just considering divisors of $4\Delta N$ one can find p, q, μ, ν verifying (26) from which (27) gives the desired m, n. But from Lemma 2 the only quadruples p, q, μ, ν that are valid are those for which the m, n obtained in (27) are integers and of the same parity. Also we need the different quadruples p, q, μ, ν go into different pairs m, n and consequently into different lines $L(m, n)$. This injectivity property is expressed by

Lemma 7. *Let d_i, δ_i, $i = 1, 2$ be divisors of $4\Delta N$ such that*

$$d_i\, \delta_i = 4\Delta N. \tag{28}$$

Let p_i, q_i, $p_i \geq 0$, $q_i \geq 1$ be such that

$$\Delta\, p_i^2 + q_i^2 = d_i \tag{29}$$

and μ_i, ν_i be integers satisfying

$$C\mu_i^2 - B\mu_i\nu_i + A\nu_i^2 = \delta_i\,. \tag{30}$$

If we assume that

$$\begin{aligned} m_1 &= \frac{q_1}{\Delta}(2C\mu_1 - B\nu_1) = \frac{q_2}{\Delta}(2C\mu_2 - B\nu_2) = m_2, \\ n_1 &= \frac{q_1}{\Delta}(2A\nu_1 - B\mu_1) = \frac{q_2}{\Delta}(2A\nu_2 - B\mu_2) = n_2 \end{aligned} \tag{31}$$

then

$$p_1 = p_2\,, \quad q_1 = q_2\,, \quad \mu_1 = \mu_2\,, \quad \nu_1 = \nu_2\,. \tag{32}$$

Proof. Since the q's are positive, from (31) we get that

$$\begin{vmatrix} 2C\mu_1 - B\nu_1 & 2C\mu_2 - B\nu_2 \\ 2A\nu_1 - B\mu_1 & 2A\nu_2 - B\mu_2 \end{vmatrix} = 0$$

Cancelling the factor Δ we get

$$\mu_1\nu_2 - \mu_2\nu_1 = 0.$$

If one of these numbers, say $\mu_1 = 0$ then necessarily $\nu_1 \neq 0$, because they satisfy (30), and then from the above relation $\mu_2 = 0$, which requires $\nu_2 \neq 0$. In all cases we see that there is a constant $k \neq 0$ such that

$$\mu_2 = k\mu_1, \quad \nu_2 = k\nu_1. \tag{33}$$

From (28) we get

$$d_1\delta_1 = d_2\delta_2 \tag{28'}$$

which, because of (29), (30) becomes

$$(\Delta p_1^2 + q_1^2)(C\mu_1^2 - B\mu_1\nu_1 + A\nu_1^2) = (\Delta p_2^2 + q_2)(C\mu_2^2 - B\mu_2\nu_2 + A\nu_2^2). \tag{34}$$

From (33) this gives after cancelling a common factor,

$$\Delta p_1^2 + q_1^2 = k^2(\Delta p_2^2 + q_2^2). \tag{35}$$

On the other hand substituting (33) into (31) we get

$$q_1(2C\mu_1 - B\nu_1) = k\,q_2(2C\mu_1 - B\nu_1)$$
$$q_1(2A\nu_1 - B\mu_1) = k\,q_2(2A\nu_1 - B\mu_1).$$

Since it is impossible to have simultaneously $2C\mu_1 - B\nu_1 = 0$, $-B\mu_1 + 2A\nu_1 = 0$, we get

$$q_1 = k\,q_2. \tag{36}$$

Therefore from (35) we get

$$p_1^2 = k^2 p_2^2$$

from which we conclude, since $p_1, p_2 \geq 0$ that

$$p_1 = k\,p_2 \tag{37}$$

so that also $k > 0$. Then from (36), (37) we get

$$1 = (p_1, q_1) = (kp_2, kq_2) = k(p_1, p_2) = k.$$

Then (33), (36), (37) give (32), proving Lemma 7. □

Lemma 8. *Assume that* (9)–(12) *are satisfied. Let* $d \mid 4\Delta N$ *and* p, q *with* $p \geq 0$, $q \geq 1$, $(p, q) = 1$ *be such that*

$$\Delta p^2 + q^2 = d. \tag{38}$$

Call μ, ν *any integers such that*

$$C\mu^2 - B\mu\nu + A\nu^2 = \frac{4\Delta N}{d} = \frac{16\overline{\Delta}N}{d}. \tag{39}$$

Define m, n *by*

$$m = \frac{q}{\Delta}(2C\mu - B\nu), \quad n = \frac{q}{\Delta}(2A\nu - B\mu) \tag{27}$$

Then m, n *are integers with the same parity.*

Proof. It consists of two parts:

(i) $m, n \in \mathbb{Z}$,

(ii) $m \equiv n \pmod 2$.

Consider (i). From (12) $\Delta = 4\overline{\Delta}$ with $\overline{\Delta}$ odd and square free. Putting

$$U = (2C\mu - B\nu), \qquad V = (2A\nu - B\mu) \tag{40}$$

(27) becomes

$$m = \frac{qU}{4\overline{\Delta}}, \qquad n = \frac{qV}{4\overline{\Delta}}. \tag{27'}$$

We now prove that $m \in \mathbb{Z}$. Since $\overline{\Delta}$ is odd it is enough to prove (i1) $\overline{\Delta} \mid qU$ and (i2) $4 \mid qU$.

To prove (i1) observe that from (38) and (39) we get

$$(\Delta p^2 + q^2)(C\mu^2 - B\mu\nu + A\nu^2) = 4\Delta N \tag{41}$$

so that

$$q^2(C\mu^2 - B\mu\nu + A\nu^2) = 4\Delta N - \Delta p^2\delta = 4\overline{\Delta}(4N - p^2\delta). \tag{42}$$

We now make use of the following obvious identity: if

$$f(x, y) = Ax^2 + Bxy + Cy^2$$

then

$$4Af(x, y) = \left(\frac{\partial f}{\partial x}\right)^2 + \Delta y^2 \qquad \Delta = 4AC - B^2. \tag{43}$$

Then multiplying both members of (42) by $4C$ we get

$$q^2 4C(C\mu^2 - B\mu\nu + A\nu^2) = q^2[(2C\mu - B\nu)^2 + \Delta\nu^2] = 16C\overline{\Delta}(4N - p^2\delta) \tag{44}$$

From (40) we get

$$q^2(U^2 + \Delta\nu^2) = 16C\overline{\Delta}(4N - p^2\delta). \tag{45}$$

Let $\overline{\Delta} = s_1 s_2 \ldots s_j \ldots$ be the prime factorization of $\overline{\Delta}$ so that for every j

$$s_j \mid q^2(U^2 + \Delta\nu^2).$$

But then either $s_j \mid q^2$, i.e. $s_j \mid q$ or else $s_j \mid U^2 + \Delta\nu^2$ and in this case $s_j \mid U$ because $s_j \mid \Delta$. So $s_j \mid qU$ which implies $\overline{\Delta} \mid qU$, proving (i1).

We now prove (i2). To prove that $4 \mid qU$, i.e. that $4 \mid q(2C\mu - B\nu)$ we have to look at equations (38), (39) where the variables q, μ, ν, come from. Any number d we choose in (38) has to be of one of the three mutually exclusive types:

(i2.1) d odd,

(i2.2) $d \equiv 2 \pmod 4$,

(i2.3) $d \equiv 0 \pmod 4$.

In case (i2.1), q is necessarily odd, because of (38). On the other hand, because of Lemma 3, if μ, ν satisfy (39), μ, ν must be even. Applying Lemma 3 again we get $\mu, \nu \equiv 0 \pmod 4$. Clearly then $q \mid qU$, proving that m is an integer. Since $qU \equiv 0 \pmod 8$ we proved that m is even.

Consider now case (i2.2). The corresponding Diophantine equation (38) has no solution.

Consider case (i2.3). Then q is even and repeating the previous argument we get that m is even. So in all cases m is even. Now the same argument shows that n is also even. Lemma 8 is proved. □

Proof of the theorem. As indicated before we have to equate the number of allowable lines $L(m, n)$ given by Lemma 1 with that same number obtained in a more elaborate way by intercepting the ellipse (7) with $L(m, n)$. The crucial point is Lemma 6 which allows us to find the m, n which will generate the allowable lines. Lemma 7 implies that there is a bijection between the allowable lines and the quadruplies p, q, μ, ν given by Lemma 6. As for condition (17), that means that $L(m, n)$ intercepts the ellipse (7), observe that substituting m, n given by (27) into $E = Am^2 + Bmn + Cn^2$ we get (24) which implies $E \leq 4N$, i.e. (17).

Definition 6.2. If $d \mid 4\Delta N$, call $s^+(d)$ the number of solutions p, q of

$$\Delta p^2 + q^2 = d \tag{38}$$

which are primitive, i.e. $(p, q) = 1$, and besides $p \geq 0$, $q \geq 1$.

To any such solution p, q associate any solution μ, ν of

$$C\mu^2 - B\mu\nu + A\nu^2 = \frac{4\Delta N}{d}\cdot \tag{39}$$

All allowable lines are then obtained by associating to a quadruple p, q, μ, ν the line $L(m, n)$ where m, n are defined by (27).

Now the number of solutions μ, ν of (39) is easily seen to be $R(4\Delta N \mid d)$, R being defined with respect to (7). So corresponding to $d \mid 4\Delta N$ we have

$$s^+(d) R\left(\frac{4\Delta N}{d}\right)$$

allowable lines and then the grand total of all allowable lines is

$$\sum_{d \mid 4\Delta N} s^+(d) R\left(\frac{4\Delta N}{d}\right).$$

From Lemma 1 we get

$$\frac{1}{2} R(N)^2 = \sum_{d \mid 4\Delta N} s^+(d) R\left(\frac{4\Delta N}{d}\right). \tag{13}$$

which proves part (I) of the theorem.

To get a formula involving $s(d)$ instead of $s^+(d)$ we proceed as follows.

Observe first that the only primitive solutions of (38) for which one of the variables p, q is zero are $p = 0$, $q = \pm 1$, corresponding to $d = 1$ and $p = \pm 1$, $q = 0$, corresponding to $d = \Delta$. Then we have

$$s(1) = 2, \quad s^+(1) = 1, \quad s(\Delta) = 2, \quad s^+(\Delta) = 0. \tag{46}$$

We also have that whenever $d \neq 1, \Delta$,

$$s^+(d) = s(d)/4, \tag{47}$$

because every point on the ellipse (38) outside the coordinate axis has quadruple symmetry. Because of (46) and (47), (13) can be written

$$\frac{1}{2}R(N)^2 = R(4\Delta N) + \frac{1}{4}\sum_{d\neq 1,\Delta} s(d)R\left(\frac{4\Delta N}{d}\right) \tag{48}$$

where the sum is extended over all divisors of $4\Delta N$, except 1 and Δ. Because of Lemma 3, $R(4\Delta N) = R(\Delta N)$ so that (48) becomes

$$R(N)^2 = 2R(\Delta N) + \frac{1}{2}\sum_{d\neq 1,\Delta} s(d)R\left(\frac{4\Delta N}{d}\right). \tag{49}$$

Now adding and subtracting to the right hand side of this equation the expression

$$\frac{1}{2}[S(1)R(4\Delta N) + S(\Delta)R(4N)] = R(\Delta N) + R(N)$$

we get rid of the exceptional divisors in (49) and obtain

$$R(N)^2 = R(\Delta N) - R(N) + \frac{1}{2}\sum_{d|4\Delta N} s(d)R\left(\frac{4\Delta N}{d}\right) \tag{14}$$

which is (14) and proves part (II) of our theorem.

Remark 1. If $B = 0$ it is fairly simple to see that $R(\Delta N) = R(N)$ so that $R(N)^2$ is expressed as a "pure" convolution, as in [5].

Remark 2. If $k > 1$, the form

$$5x^2 + 4xy + (4k+1)y^2$$

satisfies all conditions (9)–(12), $\overline{\Delta} = 20k + 1$. From the Dirichlet theorem on primes on an arithmetical progression, there are infinitely many values of k making $\overline{\Delta}$ prime.

Besides the above form is not ambiguous [12, p. 68]. An ambiguous form is one which is invariant through a unimodular transformation of determinant -1.

References

[1] M. M. Peixoto, On end point boundary value problems, J. Differential Equations 44 (1982), 273–280.

[2] M. M. Peixoto et R. Thom, Le point de vue énumératif dans les problèmes aux limites pour les équations différentielles ordinaires, C. R. Acad. Sc. Paris I 303 (1986), 629–632; 307 (1988), 197–198 (Erratum); II, C. R. Acad. Sc. Paris I 303 (1986), 693–698.

[3] M. M. Peixoto and A. R. da Silva, Focal decomposition and some results of Bernstein on the 2-point boundary value problem, to appear.

[4] M. M. Peixoto and I. Kupka, On the enumerative geometry of geodesics, in: From Topology to Computation (M. W. Hirsch, J. E. Marsden & M. Shub, eds.), Proceedings of the Smalefest, Springer-Verlag, Heidelberg–Berlin–New York 1993, 243–253.

[5] M. M. Peixoto, Sigma décomposition et arithmétique de quelques formes quadratiques définies positives, in: R. Thom Festschrift volume: Passion des Formes (M. Porte, ed.), ENS Editions, Fonteny-St. Cloud 1994, 455–479.

[6] M. Goresky and R. MacPherson, Stratified Morse Theory, Ergeb. Math. Grenzgeb. (3) 14, Springer-Verlag, Heidelberg–Berlin–New York 1988.

[7] R. P. Feynman and A. R. Hibbs, Quantum Mechanics and Path Integrals, McGraw Hill, New York 1965.

[8] C. A. de Carvalho and R. M. Cavalcanti, Tunnelings as Catastrophes, preprint, hep–th 9511005.

[9] L. Brillouin, Wave propagation in periodic structures, Electric filters and crystal lattices, 2nd ed., Dover, New York 1953.

[10] L. Bieberbach, Über die Inhaltgleichheit der Brillouinischen Zonen, Monatsh. Math. und Phys. 48 (1939), 508–515.

[11] G. A. Jones (1984), Geometric and asymptotic properties of Brillouin zones, Bull. London Math. Soc. 16 (1984), 241–263.

[12] D. E. Flath, Introduction to Number Theory, John Willey, New York 1989.

Simple type is not a boundary phenomenon

Lorenzo Sadun†

Abstract. This is an expository article, explaining recent work by D. Groisser and myself [GS] on the extent to which the boundary region of moduli space contributes to the "simple type" condition of Donaldson theory. The presentation is intended to complement [GS], presenting the essential ideas rather than the analytical details. It is shown that the boundary region of moduli space contributes $6/64$ of the homology required for simple type, regardless of the topology or geometry of the underlying 4-manifold. The simple type condition thus reduces to a statement about the interior of moduli space, namely that the interior of the $k+1$st ASD moduli space, intersected with two representatives of (4 times) the point class, be homologous to 58 copies of the k-th moduli space. This is peculiar, since the only known embeddings of the k-th moduli space into the $k+1$st involve Taubes patching, and the image of such an embedding lies entirely in the boundary region.

In this paper I discuss some recent work of David Groisser and myself [GS] on how the "simple type" condition of Donaldson theory is related to the geometry of the moduli spaces of anti-self-dual connections over a given 4-manifold. But before I begin, I must answer the obvious nagging question: *Haven't the Seiberg–Witten equations made all of Donaldson theory obsolete?* I obviously think not. While SW theory has eclipsed much of Donaldson theory, it has actually made other uses of Donaldson invariants *more* practical. This paper, I hope, will serve as an example of how to extract insight from Donaldson theory in the post-Seiberg–Witten era.

A historical digression

Once upon a time, the Yang–Mills (YM) equations were proposed and were studied without regard to topological consequences. The motivation was from physics: the equations accurately describe physics at the subnuclear scale, and the "standard model" is an $SU(3) \times SU(2) \times U(1)$ gauge theory. In the 1970s and 1980s, people began to

† Research supported in part by an NSF Mathematical Sciences Postdoctoral Fellowship and Texas ARP Grant 003658-037

write down solutions to the YM equations. Of particular interest were anti-self-dual (ASD) connections, which automatically satisfy the YM equations. People began to look at "moduli spaces" of ASD connections, both over manifolds of physical interest ($\mathbb{R}^4$, or equivalently S^4) and over more general 4-manifolds. In addition to aiding our understanding of the connections themselves, these moduli spaces were of interest in their own right, especially over complex manifolds, where ASD connections correspond to holomorphic vector bundles.

Donaldson's brilliant insight was that the topology of moduli spaces (and in particular how they sit in the larger space of all connections modulo gauge transformations) could tell a lot about the differential topology of the underlying 4-manifolds. Donaldson invariants capture the essential topological information about the moduli spaces, and proved very useful for classifying smooth 4-manifolds. In fact, they were *so* successful that the previous quest, to understand the YM equations and the ASD moduli spaces for their own sake, was largely neglected.

By 1990, then, mathematicians primarily studied the YM equations in order to understand moduli spaces, studied moduli spaces in order to understand Donaldson invariants, and studied Donaldson invariants in order to aid in the classification of smooth 4-manifolds. This is schematically shown in Figure 1. This was a grand and difficult project, with a huge number of practitioners making gradual progress.

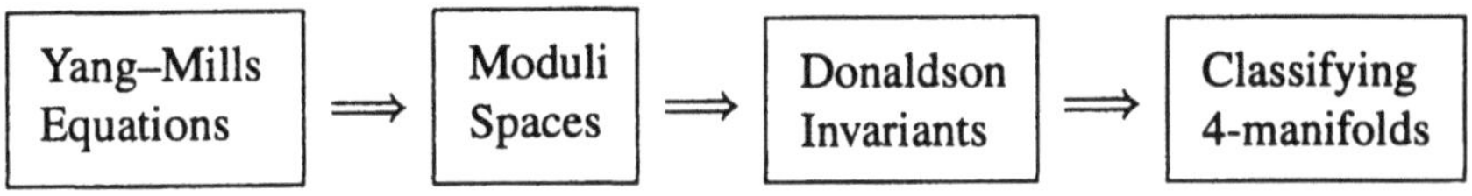

Figure 1. The traditional flow of ideas

The Seiberg–Witten revolution

In late 1994, this project was largely made irrelevant by the advent of the Seiberg–Witten equations [W]. The SW invariants are far easier to compute than Donaldson invariants, and are generally believed to carry exactly the same information. Witten's formula [W] for the Donaldson invariants in terms of the SW invariants is almost universally believed, although as of this writing a mathematical proof is still lacking. As far as classifying smooth 4-manifolds goes, just about anything that can be done with Donaldson theory can be done, far more easily, with SW invariants.

The last arrow in Figure 1 must therefore be abandoned. Indeed, the middle arrow is also largely superceded, as (modulo a proof of the Witten conjecture) the Donaldson invariants are most easily computed by first computing the SW invariants and then applying Witten's formula. The flow of ideas in Figure 1 is simply obsolete.

$SU(2)$ gauge theory, however, is *not* obsolete, if we remember the original interest in the YM equations and the moduli spaces. We just have to reverse the arrows in Figure 1!

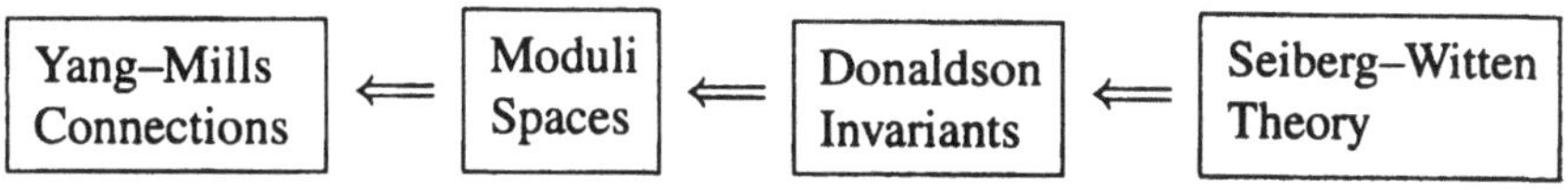

Figure 2. The new paradigm

We can use SW theory to gain insight into the structure of Donaldson invariants. We can use Donaldson invariants to tell us about moduli spaces. Finally, we can use moduli spaces to tell us about solutions to the YM equations. Perhaps this is not a "politically correct" program; nothing points to a classification of 4-manifolds! But the structure of moduli spaces can be extremely interesting, so let's get to work using our new-found tools.

Our results

This paper is an exercise along the lines of Figure 2. I will take as given that a manifold has simple type (as all smooth orientable 4-manifolds with $b_+ > 1$ are believed to have), and see what that says about the structure of its moduli spaces. The results are quite surprising!

Simple type says that the $k + 1$st moduli space $\mathcal{M}_{k+1}$, intersected with certain varieties, has the homology of 64 copies of the k-th moduli space $\mathcal{M}_k$. I will show that the portion of (a small perturbation of) $\mathcal{M}_{k+1}$ near the boundary, cut down, looks like exactly 6 copies of $\mathcal{M}_k$, regardless of the topology or geometry of the underlying 4-manifold. Simple type thus implies that the *interior* of $\mathcal{M}_{k+1}$, intersected with certain varieties, has the homology of 58 copies of $\mathcal{M}_k$. This is quite surprising, since the only known embeddings of $\mathcal{M}_k$ into $\mathcal{M}_{k+1}$ involve Taubes patching, and have images near the boundary of $\mathcal{M}_{k+1}$. Nobody has the slightest idea of what $\mathcal{M}_k$ has to do with the interior of $\mathcal{M}_{k+1}$, yet for all known 4-manifolds with $b_+ > 1$, they appear to be closely related. This is a mystery that warrants further investigation.

Here is an outline for the rest of the paper. First I review the definitions of the Donaldson invariants, and of simple type, and state the result precisely. Then I sketch the proof, which has three essential ingredients. First there is a choice of the geometric representative of the point class. Then there is an approximate formula for the curvature of a connection in the boundary region of $\mathcal{M}_{k+1}$. Finally there is a very naive calculation that illustrates clearly why the boundary region of $\mathcal{M}_{k+1}$ contributes exactly 6 copies of $\mathcal{M}_k$, rather than any other number. It takes quite a bit of analysis to thoroughly justify the approximate formula and the naive calculation; all that can be found in [GS]. Here I will concentrate on presenting the essential ideas as clearly as possible, making simplifying assumptions, as needed, along the way.

What is simple type, anyway?

Let X be an oriented 4-manifold, let $G = SU(2)$ or $SO(3)$ and let $\mathcal{B}_k^*$ be the space of irreducible connections (up to gauge equivalence) on P_k, the principal G bundle of instanton number k over X. Let $\mathcal{M}_k \subset \mathcal{B}_k$ be the space of irreducible connections on P_k with anti-self-dual curvature, modulo gauge transformations.

Donaldson [D1, D2] defined a map $\mu : H_i(X, Q) \to H^{4-i}(\mathcal{B}_k^*, Q)$, i =0, 1, 2, 3, whose image freely generates the rational cohomology of $\mathcal{B}_k^*$. Donaldson invariants are defined by pairing the fundamental class of $\mathcal{M}_k$ with products of μ of the homology classes of X, where k is chosen so that the dimensions match. Formally, for elements $[\Sigma_1], \ldots, [\Sigma_n] \in H_*(X)$, we write

$$D([\Sigma_1]\ldots[\Sigma_n]) = \mu([\Sigma_1]) \smile \cdots \smile \mu([\Sigma_n])[\mathcal{M}_k]. \tag{1}$$

Now let x be the point class in $H_0(X)$, and let ω be any formal product of classes in $H_*(X)$. The simple type condition is that, for all ω,

$$D(x^2\omega) = 4D(\omega). \tag{2}$$

Of course, the "fundamental class of $\mathcal{M}_k$" is usually not well defined, as $\mathcal{M}_k$ is typically not compact. The usual way to make sense of (1) and (2) is with geometric representatives. One finds finite-codimension varieties V_Σ in $\mathcal{B}_k^*$ that are Poincaré dual to $\mu([\Sigma])$. One then counts points, with sign, in $V_{\Sigma_1} \cap \cdots \cap V_{\Sigma_n} \cap \mathcal{M}_k$. To make a topological invariant one must show that the number of intersection points is independent of auxiliary data, such as the metric and the choice of representatives. This requires careful analysis of the bubbling-off phenomena that make $\mathcal{M}_k$ noncompact.

Unfortunately, $\mu(x)$ is not an integral class in $H^4(\mathcal{B}^*)$. However, $-4\mu(x)$ *is* an integral class. Let ν_1 and ν_2 be two (generic) geometric representatives of $-4\mu(x)$. The simple type condition can be rewritten as

$$\#(\mathcal{M}_{k+1} \cap \nu_1 \cap \nu_2 \cap V_\omega) = 64\#(\mathcal{M}_k \cap V_\omega), \tag{3}$$

where ω is an arbitrary formal product of homology cycles of X, and V_ω is a geometric representative of $\mu(\omega)$. Still more formally, one can write

$$[\mathcal{M}_{k+1} \cap \nu_1 \cap \nu_2] = 64[\mathcal{M}_k]. \tag{4}$$

I will show you that, with the right choice of ν_i, the boundary region of (a perturbation of) $\mathcal{M}_{k+1}$, intersected with ν_1 and ν_2, looks like 6, not 64, copies of $\mathcal{M}_k$.

The geometric representative ν_p

Our first step is to find appropriate geometric representatives of $\mu(x)$. The geometric representatives we will use are closely tied to the notion of reducibility. An $SU(2)$ connection is said to be reducible if the gauge group reduces to a proper subgroup of $SU(2)$. The only proper, nontrivial connected Lie subgroup of $SU(2)$ is $U(1)$, so the curvature of a reducible connection lives in the Lie Algebra of $U(1)$, which is 1-dimensional. Thus all the components of the curvature must be colinear. In fact, one can show that, on a contractible set, an $SU(2)$ connection is reducible if and only if, at each point, the components of the curvature are colinear. We therefore define "reducible at a point" to mean that the components of the curvature are colinear at that point.

Let p be any point in X, and let

$$\nu_p = \{[A] \in \mathcal{B}_k^* | F_A^- \text{ is reducible at } p\}. \tag{5}$$

Here $F_A^- = (F_A - *F_A)/2$ is the anti-self-dual part of the curvature F_A. In [S] I proved the following theorem, which applies equally well to $SU(2)$ and $SO(3)$ bundles.

Theorem 1. *ν_p is a geometric representative of $-4\mu(x)$.*

Sketch of Proof. We first need a geometric representative for the universal $SO(3)$ bundle, and then pull it back to a gauge theory setting. Let V be any real vector space, and let S_V be the Stiefel manifold of linearly independent triples of vectors in V. $SO(3)$ acts freely on S_V, with a quotient we denote G_V. Topologically, G_V is $\mathbb{R}^6\times$ the Grassmannian of oriented 3-planes in V. If V is infinite-dimensional, then S_V is contractible, and $S_V \to G_V$ is the universal $SO(3)$ bundle. Let $\pi : V \to \mathbb{R}^3$ be any linear surjection. If $\dim(V) > 7$, then the first Pontryagin class of the bundle $S_V \to G_V$ is represented by

$$\nu_\pi = \{[v_1, v_2, v_3] \in G_V | \pi(v_1), \pi(v_2), \text{ and } \pi(v_3) \text{ are colinear.}\} \tag{6}$$

A proof, using Schubert cycles, may be found in [S], but this result was almost certainly known to Pontryagin.

We are now able to construct μ of the point class. Let p be a point on the manifold X, let D be a geodesic ball around p, let $\mathcal{A}_D$ be the $SU(2)$ (or $SO(3)$) connections on D within the Sobolev space L_k^q (the choice of q and k is not important), let $\mathcal{G}^0$ be the gauge transformations in L_{k+1}^q that leave the fiber at p fixed, and let $\mathcal{G}$ be all gauge transformations in L_{k+1}^q. Define $\mu_D(p)$ to be $-\frac{1}{4}p_1$ of the $SO(3)$ bundle $\mathcal{A}_D/\mathcal{G}^0 \to \mathcal{A}_D/\mathcal{G}$. $\mu_D(p)$ is a cohomology class in $H^*(\mathcal{B}(D))$. Let $R : \mathcal{B}(X) \to \mathcal{B}(D)$ be the map obtained by restricting connections on a bundle over X to a bundle over D, and define $\mu(x) = R^*\mu_D(p)$. $\mu(x)$ is then a class in $H^*(\mathcal{B}(X))$, which turns out not to depend on the choice of point p or neighborhood D. For more about the topology of the μ map, see Chapter 5 of [DK].

Note that, when the gauge group is $SU(2)$, the bundle $\mathcal{A}_D/\mathcal{G}^0 \to \mathcal{A}_D/\mathcal{G}$ is a principal $SO(3)$ bundle, not a principal $SU(2)$ bundle. The reason is that $SU(2)$ does not act freely on $\mathcal{A}_D/\mathcal{G}^0$. A gauge transformation by ± 1 leaves a connection fixed, so the typical fiber of our bundle is $SU(2)/Z_2 \sim SO(3)$.

Although the restriction of the original $SU(2)$ (or $SO(3)$) bundle to D is trivial, the bundle $\mathcal{A}_D/\mathcal{G}^0 \to \mathcal{A}_D/\mathcal{G}$ is highly nontrivial. Indeed, it is essentially the universal $SO(3)$ bundle. The space $\mathcal{A}_D/\mathcal{G}^0$ is isomorphic to the set of connections in radial gauge with respect to the point p. In such a gauge the connection form A vanishes in the radial direction but is otherwise unconstrained. In particular, $A(p) = 0$, and the curvature at p, $F_A(p) = dA(p) + A(p) \wedge A(p) = dA(p)$, is a linear function of A.

Let V be the space of (scalar valued) 1-forms with no radial component. A connection in radial gauge is defined by a triple of elements of V, one for each direction in the Lie Algebra. Deleting the infinite-codimension set for which these elements are linearly dependent we get S_V. Thus $\mu_D(p)$ is $-1/4p_1$ of $S_V \to G_V$, which we have already computed. Now let $\pi(\alpha)$ be $d^-(\alpha)$ evaluated at p. π is a linear map of V onto the 3-dimensional space of ASD 2-forms at p. $-4\mu_D(p)$ is then represented by ν_π, which is the set of connections on D for which $F^-(p)$ is reducible. Pulling $-4\mu_D(p)$ back by the restriction map we get the connections on X for which $F_A^-(p)$ is reducible, i.e. ν_p. □

The main result

Pick geodesic normal coordinates about some point in X, and let p and q be the points $(\pm L, 0, 0, 0)$, with L small. Our problem is to count the points on the left hand side of (3) that lie near the boundary of $\mathcal{M}_{k+1}$. This is tantamount to answering the basic question: *Given a connection $A_0 \in \mathcal{M}_k$, how many ways are there to glue a small charge-1 bubble onto A_0 so as to make the resulting curvatures reducible at both p and q?* This is a local calculation, and yields an extremely simple answer:

Theorem 2. *For a generic background connection A_0, and for any $\alpha \in (0, 2)$, there are exactly six ways to glue in a bubble of size $O(L^\alpha)$ so as to make $F^-(p)$ and $F^-(q)$ both reducible. All six solutions have bubbles of size $O(L^2)$, and all six have positive orientation. This answer is independent of the global topology and geometry of X, and in particular is independent of whether X has simple type.*

The approximate curvature formula

Suppose we have a background connection A_0, in a radial gauge with respect to the origin, and glue in a bubble with center at the origin, size λ, and gluing angle m to get a new connection A. What is the curvature F_A of A? Remarkably, there is an extremely simple approximate formula:

$$F_A(x) \approx F_{A_0}(x) + F_{\text{std}}(x), \tag{7}$$

as long as $\lambda \ll |x| \ll 1$. Here F_{std} is the curvature of a standard $k = 1$ instanton, centered at the origin with size λ, in a gauge that is radial, singular at the origin, and regular at ∞. This gauge is not unique; the choice of this gauge is essentially our gluing angle m. If a bubble is to be glued in at a point $y \neq 0$, then formula (7) still applies, except that the relevant gauges are radial with respect to y, not to the origin.

The reason for the formula is this. Let A_{std} be the connection form for the standard instanton. In the appropriate gauge, $|A_{\text{std}}(x)|$ is of order $\lambda^2/|x|^3$, while $|A_0(x)| = O(|x|)$. In the relevant region, the connection form for A is essentially $A_{\text{std}} + A_0$, and so the curvature is $F_0 + F_{\text{std}} + A_{\text{std}} \wedge A_0 + A_0 \wedge A_{\text{std}}$. The last two terms have norms of order $\lambda^2/|x|^2$, and so may be ignored for $|x| \gg \lambda$.

For the remainder of this paper, we will pretend that (7) is an equality, rather than an approximation. The error terms really do *not* matter, although it takes a fair bit of work [GS] to prove it.

Thanks to formula (7), our problem reduces to finding gluing data such that $F_0 + F_{\text{std}}$ is reducible at p and q. This involves two steps: finding what values of $F_{\text{std}}(p)$ and $F_{\text{std}}(q)$ are required, and counting the sets of gluing data that yield those values. Both steps require the following notational tool.

Expressing curvatures as 3 x 3 real matrices

Relative to the standard oriented basis of $\Lambda^2_- T^*\mathbb{R}^4$ (namely $\omega_1 = dx^0dx^1 - dx^2dx^3$, $\omega_2 = dx^0dx^2 - dx^3dx^1$, $\omega_3 = dx^0dx^3 - dx^1dx^2$), an anti-self-dual curvature form F has, at each point, 3 Lie-algebra-valued components. F can thus be viewed as a triple of 3-vectors, or equivalently a 3×3 real matrix that we (momentarily) denote $\text{Mat}(F)$. Reducibility at a point means that this matrix has a rank of 1 (or 0) there. More precisely, $\text{Mat}(F)$ is constructed as follows. The first, second and third columns of $\text{Mat}(F)$ are half the ω_1, ω_2 and ω_3 components of F. The first, second and third entries of each column refer to the i, j and k directions in $su(2)$, the Lie Algebra of $SU(2)$. Of course, this construction is dependent on gauge and a choice of basis for TX. A gauge transformation is a change of basis in $su(2)$, and thus changes $\text{Mat}(F)$ by left-multiplication by an element of $SO(3)$. A change of basis in TX changes $\text{Mat}(F)$ by right-multiplication by an element of $SO(3)$. Thus the singular values of $\text{Mat}(F)$, and in particular the rank of $\text{Mat}(F)$, are gauge- and basis-independent.

Now let's compute the matrix of a standard $k = 1$ instanton. Think of $SU(2)$ as the unit quaternions, with $su(2)$ as the imaginary quaternions. The connection form of a standard instanton of scale size 1, centered at the origin, is $A_{\text{std}_0} = Im(\bar{x}dx/(1+|x|^2))$.

The curvature of this connection is

$$F_{\text{std}_0} = \frac{d\bar{x}dx}{(1+|x|^2)^2} = \frac{2i\omega_1 + 2j\omega_2 + 2k\omega_3}{(1+|x|^2)^2} \tag{8}$$

Note that the matrix $\text{Mat}(F_{\text{std}_0})$ is $1/(1+|x|^2)^2$ times the identity matrix.

Unfortunately, that is in the wrong gauge, in which $A \sim \phi^{-1}d\phi$ as $|x| \to \infty$, where $\phi(x) = x/|x|$. We do a gauge transformation by ϕ^{-1}, to get a radial gauge in which $A = O(|x|^{-3})$ as $|x| \to \infty$ (and A is singular at the origin). We then do a further gauge transformation by an arbitrary constant g_0 to get the most general radial gauge with this property. F_{std} is the curvature form in this gauge. Since $F_{\text{std}} = g_0^{-1}\phi F_{\text{std}_0}\phi^{-1}g_0$, $\text{Mat}(F_{\text{std}}) = \rho(g_0^{-1})\rho(\phi)\,\text{Mat}(F_{\text{std}_0})$, where ρ is the standard double covering map from $SU(2)$ to $SO(3)$; the three columns of $\rho(\phi)$ are $\phi i\phi^{-1}$, $\phi j\phi^{-1}$, and $\phi k\phi^{-1}$. The matrix $\rho(g_0)$ is our gluing angle m.

Now suppose that we have a $k = 1$ instanton, centered at a point y, with scale size λ. The curvature matrix, expressed in the exterior radial gauge of gluing angle m, is then

$$\text{Mat}(F_{\text{std}}) = \frac{\lambda^2}{(\lambda^2 + |x-y|^2)^2}\, m^{-1}\rho\left(\frac{x-y}{|x-y|}\right) \tag{9}$$

Note that $\text{Mat}(F_{\text{std}})$ is a positive multiple of an $SO(3)$ matrix. The multiple is determined by λ and $|x-y|$, while the $SO(3)$ matrix is determined by m and $(x-y)/|x-y|$.

From now on, we will identify curvatures with their matrices, and will omit the explicit function "Mat".

A linear algebra lemma

Recall that the singular values $\sigma_1 \geq \sigma_2 \geq \sigma_3 \geq 0$ of a 3×3 real matrix M are the square roots of the eigenvalues of $M^T M$. For M generic, these are distinct and positive. The non-generic cases are as follows: Matrices in a codimension-1 set have $\sigma_3 = 0$. Matrices in a codimension-2 set either have $\sigma_1 = \sigma_2$ or $\sigma_2 = \sigma_3$. Matrices in a codimension-4 set have $\sigma_2 = \sigma_3 = 0$; these matrices have rank 1 or 0. Matrices in a codimension-5 set have $\sigma_1 = \sigma_2 = \sigma_3$; these are all scalar multiples of $SO(3)$ matrices. Only the zero matrix (codimension-9) has $\sigma_1 = \sigma_2 = \sigma_3 = 0$.

Lemma. *Let P be a 3 by 3 real matrix with singular values $\sigma_1 \geq \sigma_2 \geq \sigma_3 \geq 0$. If these singular values are all distinct, then there are exactly two pairs $(s, M) \in (0, \infty) \times SO(3)$ for which $P + sM$ has rank 1 (and no pairs (s, M) for which $P + sM = 0$). In both cases $s = \sigma_2(P)$. If exactly two of the singular values of P are the same and nonzero, then the two solutions (s, M) coalesce to a double root.*

The proof, although straightforward, is not especially enlightening, so we'll skip it here. You may want to attempt it as an exercise, or even assign it to an advanced linear algebra class. One of the many possible proofs can be found in [GS].

This means that, for any generic background connection A_0 and a generic point z, there is one size s_z and two $SO(3)$ matrices $M_1(z)$ and $M_2(z)$ such that $F_0(z)+s_zM_i(z)$ is reducible. Note that these are completely determined by $F_0(z)$, which is a smooth function of z. Since p and q are separated by a distance $O(L)$, s_p is within $O(L)$ of s_q, and $M_i(p)$ is within $O(L)$ of $M_i(q)$.

Solving for the magnitude of F_{std}

According to formula (7), in order to have $F(p)$ and $F(q)$ reducible, we must solve $F_{\text{std}}(p) = s_pM_i(p)$ and $F_{\text{std}}(q) = s_qM_j(q)$. We do this in two steps, first solving for the magnitudes and then for the $SO(3)$ matrices. We will pretend that $s_p = s_q$. The actual $O(L)$ difference between them complicates the algebra, but does not make any qualitative difference.

The condition for the standard curvature F_{std} to have magnitude s_p at p is

$$\frac{\lambda^2}{(|y-p|^2+\lambda^2)^2} = s_p, \tag{10}$$

or equivalently

$$\lambda^2 + |y-p|^2 = \lambda/\sqrt{s_p}. \tag{11}$$

Similarly, we need

$$\lambda^2 + |y-q|^2 = \lambda/\sqrt{s_q}. \tag{12}$$

Since $s_p = s_q$, this implies that $|y-p| = |y-q|$, so $y = (y_0, y_I)$ must lie in the plane half-way between p and q. That is, $y_0 = 0$. As long as this is the case, any solution to (11) is also a solution to (12).

As long as $|y-p| < 1/2\sqrt{s_p}$ there are two solutions to (11), while for $|y-p| > 1/2\sqrt{s_p}$ there are none. When $|y-p| < 1/2\sqrt{s_p}$, one solution has $\lambda > 1/2\sqrt{s_p}$, which is not $O(L^\alpha)$ for L small. The other solution qualifies as small if $|y-p|$ is small enough, and, for $|y-p| \ll 1/\sqrt{s_p}$, is approximately $\lambda = |y-p|^2\sqrt{s_p} = (L^2+|y_I|^2)\sqrt{s_p}$.

Section Solving for the gluing angle

We still have to get the $SO(3)$ matrices right. By eq. (9), this means simultaneously solving the equations

$$m^{-1}\rho((y-p)/|y-p|) = M_i(p) \tag{13}$$

and

$$m^{-1}\rho((y-q)/|y-q|) = M_j(q) \tag{14}$$

for m. If a solution exists it is obviously unique. A solution exists if and only if $\rho((y-p)/|y-p|)^{-1}\rho((y-q)/|y-q|) = M_i(p)^{-1}M_j(q)$. Let

$$g(y) = \frac{(\bar{y}-\bar{p})(y-q)}{|(y-p)(y-q)|}. \tag{15}$$

We must count the points on our 3-disk (of small solutions to (11) and (12)) for which the $SO(3)$-valued function $\rho(g(y))$ equals $M_i(p)^{-1}M_j(q)$. Note that

$$g(y) = -I + 2y_I/L + O((|y_I|/L)^2) \quad \text{for } |y_I| \ll L, \tag{16}$$

while

$$g(y) = I - 2Ly_I/|y_I|^2 + O((L/|y_I|)^2) \quad \text{for } |y_I| \gg L. \tag{17}$$

Pick a constant $K > 0$ and let $R_{K,\alpha}$ be such that $|y_I| < R_{K,\alpha}$ implies $\lambda \leq KL^\alpha$. For L small we have $R^2_{K,\alpha} \sim KL^\alpha/\sqrt{s_m} - L^2 \sim KL^\alpha/\sqrt{s_m}$, since $\alpha < 2$. $L/R_{K,\alpha}$ is $O(L^{1-\alpha/2})$ and hence goes to zero as $L \to 0$. On the 3-disk of admissible y_I, the map g covers all of $SU(2)$ except for a ball of radius $\sim L^{1-\alpha/2}$ around the origin. Since ρ is a 2-1 map, $\rho(g(y))$ hits all of $SO(3)$ twice, except for a ball of radius $\sim L^{1-\alpha/2}$ around the origin, which is only hit once. The number of solutions to our problem depends on whether, for small L, $M_i^{-1}(p)M_j(q)$ is in this ball or not.

As $L \to 0$, $M_1(p)^{-1}M_2(q)$ and $M_2(p)^{-1}M_1(q)$ are bounded away from the identity, but $M_1(p)^{-1}M_1(q)$ and $M_2(p)^{-1}M_2(q)$ are within $O(L)$ (and hence within $o(L^{1-\alpha/2})$) of the identity. Thus we have two sets of parameters (y, λ, m) that give $F_{\text{std}}(p) = s_pM_1(p)$ and $F_{\text{std}}(q) = s_qM_2(q)$, two that give $F_{\text{std}}(p) = s_pM_2(p)$ and $F_{\text{std}}(q) = s_qM_1(q)$, one that gives $F_{\text{std}}(p) = s_pM_1(p)$ and $F_{\text{std}}(q) = s_qM_1(q)$ and one that gives $F_{\text{std}}(p) = s_pM_2(p)$ and $F_{\text{std}}(q) = s_qM_2(q)$. A total of six solutions in all.

Peeking under the carpet

If you accept all the simplifying assumptions I have made, then the proof of Theorem 2 is finished. I reduced the central question to counting the solutions to some explicit (and simple!) algebraic equations, and I not only counted the solutions, but actually showed you how to find them.

The skeptics among you, however, may be worried that my simplifying assumptions are unrealistic, or hide some deep problems. To ease your fears, here is a list of the shortcuts I have taken, and how these issues are dealt with in [GS].

I assumed that $s_p = s_q$ when, in fact, s_p and s_q differ by $O(L)$. The set of y for which both (11) and (12) can be solved for λ is typically not a disk in the plane $y_0 = 0$; rather, it is an ellipsoid that passes between p and q. The portion of the ellipsoid that results in λ being small is a topological 3-disk located between p and q. As $L \to 0$, this topological 3-disk approaches a geometric 3-disk in the plane $y_0 = 0$ sufficiently rapidly that the previous discussion goes through essentially unchanged.

I showed that there are 6 solutions, but did not show that they all give intersection number +1. This involves two steps. First we show that the intersection numbers are, by continuity, independent of $M_i(p)$ and $M_j(q)$. A similar argument shows that the intersection numbers for the two solutions for a given $M_i(p)$ and $M_j(q)$ are equal. We then compute the intersection number at $(y = 0, \lambda, m = I)$ for $M_i(p) = M_j(q) = I$, which is indeed +1.

In order to apply formula (7), A_0 must be in radial gauge with respect to y. However, we computed $M_i(p)$ and $M_j(q)$ from A_0 in radial gauge with respect to the origin, not with respect to y. In reality, $M_i(p)$ and $M_j(q)$ should really be viewed as functions of y. The extent of this y-dependence can be estimated, and we show that the derivative of $M_i(p)^{-1}M_j(q)$ with respect to y is too small to change the count.

Formula (7) is itself an approximation, not an equality. The error terms may be treated as a perturbation to F_0. By estimating the dependence of (y, λ, m) on F_0, and the dependence of the error terms of (y, λ, m), we show that any solution to $F_0(p) + F_{\mathrm{std}}(p) =$ reducible and $F_0(q) + F_{\mathrm{std}}(q) =$ reducible can be perturbed to a solution to $F(p) =$ reducible and $F(q) =$ reducible, and vice versa. These estimates are, technically, the most difficult part of the whole problem.

Finally, formula (7) applies not to a true ASD connection, but to a connection obtained by an explicit grafting formula. The set of such connections is an L^2-small perturbation $\tilde{\mathcal{M}}_{k+1}$ of the true ASD moduli space $\mathcal{M}_{k+1}$. Theorem 2 does not directly relate $\mathcal{M}_k$ to $\mathcal{M}_{k+1} \cap \nu_p \cap \nu_q$. Rather, it relates $\mathcal{M}_k$ to $\tilde{\mathcal{M}}_{k+1} \cap \nu_p \cap \nu_q$.

Ideally, one would like to interpolate from $\tilde{\mathcal{M}}_{k+1} \cap \nu_p \cap \nu_q$ to $\mathcal{M}_{k+1} \cap \nu_p \cap \nu_q$. This is quite difficult, as ν_p and ν_q are defined by pointwise conditions. I know of no pointwise estimates relating the curvature of an almost-ASD connection to that of a nearby ASD connection. In order to make use of the integral estimates available in the literature one would have to replace ν_p and ν_q by geometric representatives defined by integral conditions. While this is possible (Cliff Taubes once showed me such an extended representative), it is well beyond the scope of this work.

A differential forms approach

There is a quite different approach to measuring the importance of the boundary region of $\mathcal{M}_{k+1}$ to simple type. One can use differential form representatives of $\mu(\cdot)$, and integrate these forms over $\mathcal{M}$ to obtain Donaldson invariants. In that setting, our problem is to integrate $\mu_{dR}(p) \wedge \mu_{dR}(q)$ over the 8-dimensional space of gluing parameters. Here $\mu_{dR}(p)$ is a de Rham representative of $\mu(x)$ based on connections near p, much the way that ν_p is a geometric representative. Let B_{λ_0} be the set of gluing data (y, λ, m) for which $\lambda < \lambda_0$. In [GS] we prove that

$$\lim_{\lambda_0 \to 0} \lim_{L \to 0} \int_{B_{\lambda_0}} \mu_{dR}(p) \wedge \mu_{dR}(q) = 1/2. \tag{18}$$

Surprisingly, this is a different answer than obtained from the geometric representative calculation (1/8 of what is required for simple type, as opposed to 6/64). Moreover, the bulk of the integral (18) is from λ being of order L, while the geometric representative calculation had all the intersection points having λ of order L^2. This is not a contradiction. Although the Donaldson invariants are topological, hence independent of a choice of representatives, the contribution of the boundary region is geometric, and can definitely depend on a choice of representatives.

Conclusions

While the two approaches disagree on the exact contribution of the boundary region, and on just how close to the boundary we should consider, they agree on the central theme of this paper. Simple type is *not* a boundary phenomenon. Simple type implies that the features of each moduli space $\mathcal{M}_k$ are duplicated in the structure of the interior of $\mathcal{M}_{k+1}$. This duplication is not at all explained by our present understanding of moduli spaces; perhaps the explanation lies in quantum duality.

Acknowledgements. I gratefully acknowledge the collaboration of David Groisser, and the many helpful discussion I have had with Dan Freed, Rob Kusner, Tom Parker, Cliff Taubes and Karen Uhlenbeck. This work was partially supported by an NSF Mathematical Sciences Postdoctoral Fellowship and Texas ARP Grant 003658-037.

References

[D1] S. K. Donaldson, Connections, cohomology and the intersection forms of four manifolds, J. Differential Geom. 24 (1986), 275–341.

[D2] S. K. Donaldson, Polynomial invariants for smooth 4-manifolds, Topology 29 (1990), 257–315.

[DK] S. K. Donaldson and P. B. Kronheimer, The Geometry of Four-Manifolds, Oxford University Press, Oxford 1990.

[GS] D. Groisser and L. Sadun, Simple type and the boundary of moduli space, preprint, 1997.

[S] L. Sadun, A simple geometric representative of μ of a point, Commun. Math. Phys. 178 (1996), 107–113.

[W] E. Witten, Monopoles and four-manifolds, Math. Res. Lett. 1 (1994), 769–796.

On the homology cobordism group of homology 3-spheres

Nikolai Saveliev *

Abstract. In this paper we present our results on the homology cobordism group $\Theta^3_{\mathbb{Z}}$ of oriented integral homology 3-spheres. We specially emphasize the role played in the subject by the gauge theory including Floer homology and invariants by Donaldson and Seiberg–Witten.

1991 Mathematics Subject Classification: 57R80, 57M25, 57R55, 57R57.

A closed oriented 3-manifold Σ is said to be an integral homology sphere if it has the same integral homology as S^3. Two homology spheres Σ_0 and Σ_1 are homology cobordant, if there is a smooth compact oriented 4-manifold W with $\partial W = \Sigma_0 \cup -\Sigma_1$ such that $H_*(W, \Sigma_0; \mathbb{Z}) = H_*(W, \Sigma_1; \mathbb{Z}) = 0$. The set of all homology cobordism classes forms an abelian group $\Theta^3_{\mathbb{Z}}$ with the group operation defined by connected sum. Here, the zero element is homology cobordism class of S^3, and the additive inverse is obtained by a reverse of orientation.

The Rohlin invariant μ is an epimorphism $\mu : \Theta^3_{\mathbb{Z}} \to \mathbb{Z}_2$ defined by the formula

$$\mu(\Sigma) = \frac{1}{8} \text{ sign}(W) \mod 2,$$

where W is a smooth simply connected parallelizable compact manifold with $\partial W = \Sigma$. This invariant is well-defined due to the Rohlin theorem [R] which states that the signature $\text{sign}(V)$ of any smooth simply connected *closed* parallelizable manifold V is divisible by 16.

We will focus our attention on the following problem concerning the group $\Theta^3_{\mathbb{Z}}$ and the homomorphism μ : does there exist an element of order two in $\Theta^3_{\mathbb{Z}}$ with non-trivial Rohlin invariant ? This is one of the R. Kirby's problems [K], Problem 4.4. According to [GS] and [Ma], a positive solution to this problem would imply, in particular, that all closed topological n-manifolds are simplicially triangulable if $n \geq 5$ (this is not true in dimension 4 where a counterexample is due to A. Casson and M. Freedman, see e.g. [AM]).

*Supported in part by the NSF. Research at MSRI was supported in part by NSF grant DMS-9022140.

We study this problem by trying to lift the Rohlin homomorphism from $\mathbb{Z}_2$ to the integers. If we succeeded in doing this, then all elements of finite order would have to lie in the kernel of μ, and the Kirby problem would have a negative solution. In fact, our goal is less ambitious—we define a lift of μ on a certain subgroup of $\Theta^3_{\mathbb{Z}}$. This implies that there are no elements in this subgroup solving the problem. This is done in Section 1 with the help of the $\bar{\mu}$-invariant introduced by W. Neumann and L. Siebenmann in 1978, see [N] and [Sb], for the so called plumbed homology spheres.

Section 2 is devoted to an interesting observation about Floer homology of homology 3-spheres homology cobordant to zero. We manage to prove a two-periodicity in the Floer homology groups of such homology spheres in a series of examples. This leads us to defining a new non-trivial integral invariant ν, see (4), which vanishes for all these examples.

In Section 3 we prove that the invariants $\bar{\mu}$ and ν coincide for Seifert fibered homology spheres. Thus, the invariant ν provides an extension of the $\bar{\mu}$-invariant to *all* integral homology spheres. Another result in Section 3 relates Floer homology to the Jones polynomial.

1. Homology cobordisms of plumbed homology 3-spheres

We first recall the definition of the invariant $\bar{\mu}$ by W. Neumann [N]. Note that our definition differs from the original one by a factor of $1/8$.

Let Γ be a plumbing graph, that is a (not necessarily connected) graph with no cycles, each of whose vertices carries an integer weight e_i , $i = 1, \dots, s$. The matrix $A(\Gamma) = (a_{ij})_{i,j=1,\dots,s}$ with the entries

$$a_{ij} = \begin{cases} e_i, & \text{if } i = j \\ 1, & \text{if } i \text{ is connected to } j \text{ by an edge} \\ 0, & \text{otherwise} \end{cases}$$

is the intersection matrix of the 4-dimensional manifold $P(\Gamma)$, obtained by plumbing D^2-bundles over 2-spheres according to Γ. This manifold is simply connected.

If Γ is a plumbing graph as above, then $M(\Gamma) = \partial P(\Gamma)$ is an integral homology sphere if and only if $\det A(\Gamma) = \pm 1$. For example, all Seifert fibered homology spheres $\Sigma(a_1, \dots, a_n)$ are of the form $\partial P(\Gamma)$ where Γ is a star-shaped graph, see [NR].

If $M(\Gamma)$ is a homology sphere, there exists a unique homology class $w \in H_2(P(\Gamma); \mathbb{Z})$ satisfying the following two conditions. First, w is *characteristic*, that is (dot represents intersection number)

$$w.x \equiv x.x \mod 2 \quad \text{for all } x \in H_2(P(\Gamma); \mathbb{Z}), \tag{1}$$

and second, all coordinates of w are either 0 or 1 in the natural basis of $H_2(P(\Gamma); \mathbb{Z})$. We call w *the integral Wu class* for $P(\Gamma)$. It follows from [N] that the integer $\operatorname{sign} P(\Gamma) - w.w$ only depends on $M(\Gamma)$ and not on Γ. This integer is divisible by 8, see [Sr], so one can define the Neumann-Siebenmann invariant by the formula

$$\bar{\mu}(M(\Gamma)) = \frac{1}{8}(\operatorname{sign} P(\Gamma) - w.w). \qquad (2)$$

The class w is spherical, so its modulo 2 reduction is the usual Rohlin invariant μ, see [NR].

W. Neumann conjectured in [N] that the number $\bar{\mu}$ is in fact a homology cobordism invariant. This conjecture has been checked for all known examples of plumbed homology spheres homology cobordant to zero, including ones from the lists of A. Casson and J. Harer [CH] and R. Stern [St]. The list in [CH] includes Seifert fibered homology spheres $\Sigma(p, ps \pm 1, ps \pm 2)$, $p \geq 3$ odd, $s \geq 1$, and $\Sigma(p, ps-1, ps+1)$, $p \geq 2$ even, $s \geq 1$ odd. All the Stern's examples are of the form $\Sigma(a_1, a_2, 2a_1a_2 \pm a_3)$ for certain $\Sigma(a_1, a_2, a_3)$ in the Casson–Harer list. Now, the gauge theory enables us to prove some results of general nature which weigh for a positive answer to the conjecture.

Algebraic links. Our first result is concerned with the links of algebraic singularities. It was shown in [EN] that a plumbed homology sphere Σ is the link of an algebraic singularity if and only if there exists a plumbing graph Γ such that the intersection form of the manifold $P(\Gamma)$ with $\Sigma = \partial P(\Gamma)$ is negative definite. The simplest case is the Seifert fibered case: any Seifert fibered homology sphere $\Sigma(a_1, \dots, a_n)$ is the link of the singularity of $f^{-1}(0)$ where $f : \mathbb{C}^n \to \mathbb{C}^{n-2}$ is a map of the form

$$f(z_1, \dots, z_n) = \left(\sum_{k=1}^{n} b_{1,k} z_k^{a_k}, \dots, \sum_{k=1}^{n} b_{n-2,k} z_k^{a_k}\right)$$

with sufficiently general coefficient matrix $(b_{i,j})$, see [NR]. For instance,

$$\Sigma(p, q, r) = \{z \in \mathbb{C}^3 \mid \|z\| = \varepsilon \text{ and } z_1^p + z_2^q + z_3^r = 0\}$$

for $\varepsilon > 0$ small enough.

The next simplest case is the following: if p, q, r are pairwise relatively prime integers, as are p', q', r', then the homology sphere Σ obtained by splicing $\Sigma(p, q, r)$ and $\Sigma(p', q', r')$ along the singular fibers of degrees r and r', is the link of singularity if and only if $rr' > pp'qq'$, see [NW], § 4.

Theorem 1. *Let a homology sphere Σ be the link of an algebraic singularity. If Σ is homology cobordant to zero then $\bar{\mu}(\Sigma) \geq 0$.*

Proof. Since Σ is an algebraic link, one may assume that Σ is the boundary of a plumbed 4-manifold $P(\Gamma)$ with the negative definite intersection form. Suppose that Σ bounds a smooth homology ball M. Let us consider the manifold $W = P(\Gamma) \cup_{\Sigma} (-M)$. This is a smooth closed oriented manifold whose intersection form is naturally isomorphic to the intersection form of $P(\Gamma)$, in particular, is negative definite. By S. Donaldson's Theorem 1 from [D], this form is diagonalizable over the integers.

We use this fact to evaluate $\bar{\mu}(\Sigma)$. Since sign $P(\Gamma) = -s$ where s is the number of vertices in the graph Γ, we only need to find the Wu class w. In the standard basis associated with the plumbing, the matrix A of the intersection form of $P(\Gamma)$ is of the form $A = U^t(-E)U$ where $U \in \mathrm{SL}_s(\mathbb{Z})$ and E is the identity matrix. The defining relation (1) translates into

$$w^t U^t(-E)Ux \equiv x^t U^t(-E)Ux \mod 2 \quad \text{for all } x \in H_2(P(\Gamma); \mathbb{Z}),$$

or, equivalently,

$$(Uw)^t(-E)y \equiv y^t(-E)y \mod 2 \quad \text{for all } y \in H_2(P(\Gamma); \mathbb{Z}).$$

Therefore, Uw is characteristic for $-E$, in particular, all the coordinates of Uw are odd. Now, we have

$$w.w = w^t U^t(-E)Uw = -(Uw)^t(Uw)$$

which is equal to minus the square of the Euclidean length of the vector Uw. Since all coordinates of Uw are odd, $w.w \leq -s$, and therefore $\bar{\mu}(\Sigma) \geq 0$. □

Corollary. *If a plumbed homology sphere Σ is an algebraic link and $\bar{\mu}(\Sigma) < 0$, then Σ has infinite order in the group $\Theta^3_{\mathbb{Z}}$.*

Example. For any relatively prime integers $p, q > 0$, one can see that $\bar{\mu}(\Sigma(p, q, pq-1)) < 0$. Therefore, all these homology spheres are of infinite order in $\Theta^3_{\mathbb{Z}}$. In particular, Poincaré homology sphere $\Sigma(2, 3, 5)$ is of infinite order.

Example. Let Σ be a splice [EN] of $\Sigma(4, 7, 9)$ and $\Sigma(2, 3, 25)$ along singular fibers of degrees 9 and 25. This manifold is an algebraic link though is not Seifert fibered. By using the additivity of $\bar{\mu}$ proven in [Du], see also [S3], we find that

$$\bar{\mu}(\Sigma) = \bar{\mu}(\Sigma(4, 7, 9)) + \bar{\mu}(\Sigma(2, 3, 25)) = -2 + 0 < 0.$$

Therefore, Σ has infinite order in $\Theta^3_{\mathbb{Z}}$.

Many Seifert homology spheres $\Sigma(a_1, \dots, a_n)$ having infinite order in the group $\Theta^3_{\mathbb{Z}}$ by the corollary above, can also be detected by the Fintushel-Stern invariant

$$R(a_1, \dots, a_n) = \frac{2}{a} - 3 + n + \sum_{i=1}^{n} \frac{2}{a_i} \sum_{k=1}^{a_i - 1} \cot\left(\frac{\pi a k}{a_i^2}\right) \cot\left(\frac{\pi k}{a_i}\right) \sin^2\left(\frac{\pi k}{a_i}\right)$$

where $a = a_1 \cdot \ldots \cdot a_n$. A theorem in [FS2] says that if $\Sigma(a_1, \dots, a_n)$ bounds a homology ball then $R(a_1, \dots, a_n) < 0$. There are however Seifert spheres which are not homology cobordant to zero and which can be detected by $\bar{\mu}$ and not by R, and vice versa.

Example. Both Seifert spheres $\Sigma(2, 11, 19)$ and $\Sigma(3, 5, 7)$ are not homology cobordant to zero. As to the former one, this follows from the fact that $\bar{\mu}(\Sigma(2, 11, 19)) = -1$ is negative (though $R(2, 11, 19) = -1$); the latter one has $\bar{\mu}(\Sigma(3, 5, 7)) = 0$ and the result follows from $R(3, 5, 7) = 1$.

Let $P(\Gamma)$ be a resolution of a singularity whose link Σ is a homology 3-sphere. Let K be the Poincaré dual of the canonical class of a complex structure on $P(\Gamma)$, so K is characteristic. If Σ is homology cobordant to zero then the same argument as in Theorem 1 shows that the number sign $P - K.K$ should be non-negative (in fact, this is true for *any* characteristic class K, not necessarily canonical).

On the other hand, one can compute sign $P - K.K$ easily in terms of the Milnor fiber M of the singularity. According to [Mo],

$$(3\,\text{sign} + 2e)(M) - K'.K' = (3\,\text{sign} + 2e)(P) - K.K$$

where e stands for the Euler characteristic, and K' is the canonical class of M. Since the intersection form of the manifold P is negative definite, we get

$$\text{sign}\, P - K.K = (3\,\text{sign} + 2e)(M) - K'.K' - 2.$$

One can get more precise result for Seifert manifolds.

Theorem 2. *If a Seifert fibered homology sphere $\Sigma(a_1, \dots, a_n)$ is homology cobordant to zero then*

$$\lambda(\Sigma(a_1, \dots, a_n)) \le \frac{1}{12} \prod_{i=1}^{n} (a_i - 1)$$

where λ is the Casson invariant.

Proof. In the special case of $\Sigma(a_1, \dots, a_n)$, the Milnor fiber M is simply-connected and its canonical class K' vanishes, see [Dv]. In addition,

$$\text{rk}\, H_2(M) = \prod_{i=1}^{n} (a_i - 1), \text{ and } \text{sign}\, M = -8\lambda(\Sigma(a_1, \dots, a_n)).$$

The first formula is due to Milnor [M], and the second one to Fintushel–Stern [FS1] and Neumann–Wahl [NW]. □

Example. For any positive integer k, $\lambda(\Sigma(2, 3, 6k-1)) = k$. Therefore, the homology sphere $\Sigma(2, 3, 6k - 1)$ is not homology cobordant to zero due to Theorem 2 though its $\bar{\mu}$-invariant vanishes if k is even. In fact, this can be also proven with the help of the Fintushel–Stern R-invariant.

More constraints from even plumbing. Now, we have to deal with those plumbed homology spheres whose $\bar{\mu}$-invariant is positive. Theorem 1 does not prohibit such homology spheres from bounding a homology ball. A good example of this kind is $\Sigma(2,3,7)$ with $\bar{\mu} = 1$. This homology sphere bounds a four-manifold with the following plumbing description:

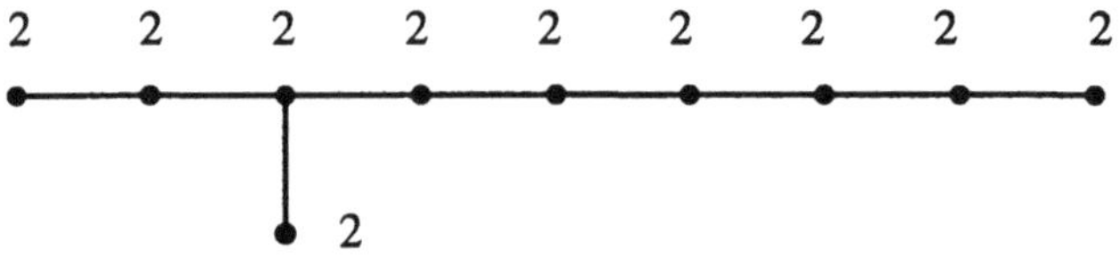

Its intersection form is even and indefinite, hence is isomorphic over the integers to the form $E_8 \oplus H$ where E_8 is the positive definite E_8-form, and H is the hyperbolic 2×2-form, see [Sr]. Suppose that a double of $\Sigma(2,3,7)$ is homology cobordant to zero. Then, by gluing the homology ball and a double of the plumbing along $\Sigma(2,3,7)\#\Sigma(2,3,7)$, we get a closed smooth manifold with the intersection form $2E_8 \oplus 2H$ which is prohibited by S. Donaldson's Theorem 2, see [D]. Therefore, $\Sigma(2,3,7)$ is not of the second order in $\Theta^3_{\mathbb{Z}}$.

The Seiberg–Witten gauge theory has led M. Furuta [Fu], see also [A], to a proof of the 10/8-conjecture saying that if the intersection form Q of a simply-connected closed smooth 4-manifold is even then

$$\frac{\operatorname{rank} Q}{|\operatorname{sign} Q|} > \frac{10}{8}.$$

One can use this result to show that in fact none of $\Sigma(2,3,7)$ multiples can bound a contractible manifold. Because if any of them did, there would exist a closed smooth simply-connected manifold with rank/sign = 10/8.

One can use a similar argument with even plumbings for many other Seifert fibered homology spheres with $\bar{\mu}$ positive. Unfortunately, the construction does not work for *all* homology spheres because the number of hyperbolics in the even plumbing may be too large, and in order to get an even form with the desired quotient rank/sign one may need to surger some of them out (at the expense of losing the plumbing structure). In our paper [S2] we do this for Seifert fibered homology spheres $\Sigma(p,q,pqk \pm 1)$, $k > 0$ odd, which can be thought of as the manifolds obtained by the $(-1/k)$-surgery along a (p,q)-torus knot.

Theorem 3. *Let $p, q > 1$ be relatively prime integers, and $k > 0$ an odd integer. Then neither $\Sigma(p,q,pqk+1)$ nor any of its multiples can bound a smooth contractible 4-manifold.*

Note that $\bar{\mu}(\Sigma(p,q,pqk+1)) > 0$ for all p, q, k as in the theorem, and that a similar result for $\Sigma(p,q,pqk-1)$ has followed from Theorem 1. Theorem 3 can not be generalized for $\Sigma(p,q,pqk+1)$ with k even; for instance, $\Sigma(2,3,13)$ is known to bound a contractible manifold [AK].

We refer for a complete proof of Theorem 3 to [S2], and only draw the final surgery diagrams in the simplest case of $p = 2$. If $q \equiv 3 \bmod 4$, then $\Sigma(2, q, 2qk + 1)$ is the boundary of the 4-manifold obtained by plumbing according to the following diagram:

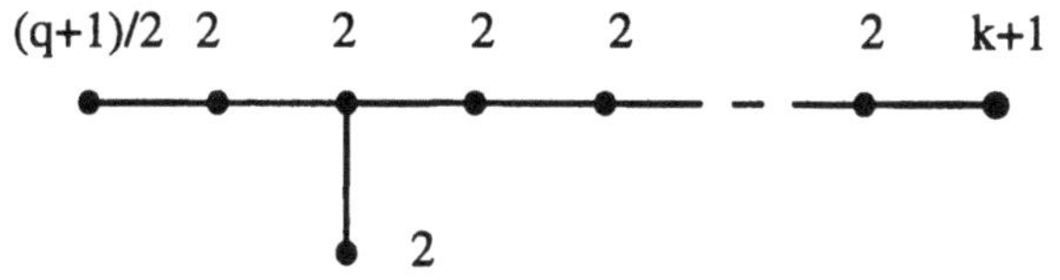

The intersection form of this manifold is isomorphic to $\frac{q+1}{4} \cdot E_8 \oplus H$. If $q \equiv 1$ modulo 4, some hyperbolics have to be surgered out. The homology sphere $\Sigma(2, q, 2qk + 1)$ is then a surgery on the link shown in Figure 1.

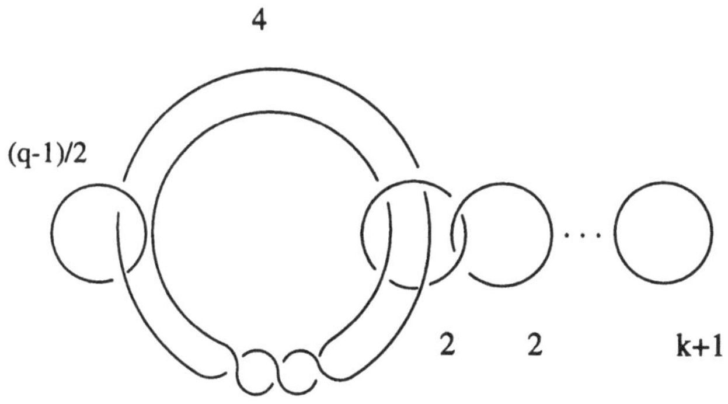

Figure 1

The intersection form of the corresponding 4-manifold is isomorphic over the integers to $\frac{q-1}{4} \cdot E_8 \oplus H$.

2. Floer homology and invariants of homology cobordism

The Floer homology groups $I_n(\Sigma)$, $n = 0, \ldots, 7$, are abelian groups, intrinsically associated to an oriented $\mathbb{Z}$-homology 3-sphere Σ. The definition of Floer homology in [F1] makes essential use of gauge theory on 3- and 4-dimensional manifolds, so the groups $I_n(\Sigma)$ are generally difficult to compute.

We use the Floer exact triangle [F2] to compute $I_*(\Sigma)$ for several classes of integral homology spheres Σ unified by one property—they all are homology cobordant to zero. Among these homology spheres are: Mazur homology spheres [AK]; Seifert fibered homology spheres which bound smooth contractible manifolds [CH] and [St]; homology spheres obtained by (± 1)-Dehn surgery on certain slice knots. In short, the main result of our computations can be stated as follows.

Theorem 4. *For all integral homology spheres Σ from the list above, the Floer homology groups are 2-periodic, that is, $I_j(\Sigma) = I_{j+2}(\Sigma)$ for any j.*

Below we will give a more specific description of the homology spheres in Theorem 4, but first we would like to use this theorem to justify introduction of a new invariant. For every oriented integral homology sphere Σ, we will define a number $\nu(\Sigma)$ keeping in mind the following conditions we would like ν ideally to satisfy:

(1) ν is a $\mathbb{Z}$-homology cobordism invariant,

(2) ν is additive with respect to connected sums of $\mathbb{Z}$-homology spheres,

(3) ν is a lifting of the Rohlin homomorphism μ , i.e. $\nu(\Sigma) \equiv \mu(\Sigma) \bmod 2$.

Recall that according to C. Taubes [T] the Casson's λ-invariant equals one-half of the Euler characteristic of Floer homology,

$$\lambda(\Sigma) = \frac{1}{2}\sum_{n=0}^{7}(-1)^n \operatorname{rk}_{\mathbb{Q}} I_n(\Sigma). \tag{3}$$

We define our invariant ν in a manner similar to that of (3), i.e. as an algebraic sum of the ranks of the Floer homology groups. In doing so, we observe that there is essentially only one choice of the signs, besides that in (3), that makes the invariant sensitive to the orientation (as it is required by the conditions (1) and (2) above). We use this choice to define our invariant ν as follows:

$$\nu(\Sigma) = \frac{1}{2}\sum_{n=0}^{7}(-1)^{\frac{(n+1)(n+2)}{2}} \operatorname{rk}_{\mathbb{Q}} I_n(\Sigma). \tag{4}$$

This is an invariant of orientation preserving homeomorphism just because so is Floer homology. The ν-invariant is not identically zero, for instance, $\nu(\Sigma(2,3,5)) = -1$. The most important problem, therefore, is to check its homology cobordism invariance. Theorem 4 gives some evidence towards that.

Corollary. *The ν-invariant vanishes on all the integral homology spheres homology cobordant to zero listed in Theorem* 4.

Concerning the condition (3), we mention that $\nu(\Sigma)$ obviously equals $\mu(\Sigma)$ modulo 2 for those homology spheres Σ whose Floer homology is 4-periodic. Since the 4-periodicity does not hold in general, it is not known if $\nu(\Sigma) = \mu(\Sigma) \bmod 2$ for all homology spheres Σ. In [S3] we address the condition (2). We use Fukaya's spectral sequence in Floer homology [Fk] to show that $\nu(\Sigma)$ is additive with respect to connected sums under some restrictions on Floer homology of the factors.

Mazur homology spheres. By a Mazur homology sphere we mean an integral homology 3-sphere Σ obtained by the following construction. Let W be a compact contractible 4-manifold obtained by attaching a two-handle to $S^1 \times B^3$ along its boundary as in Figure 2.

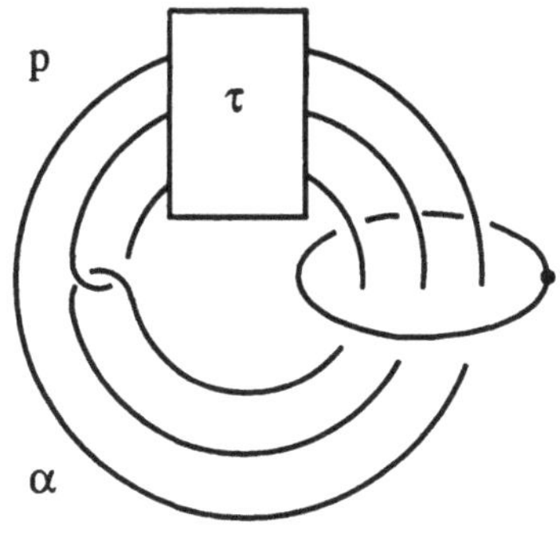

Figure 2

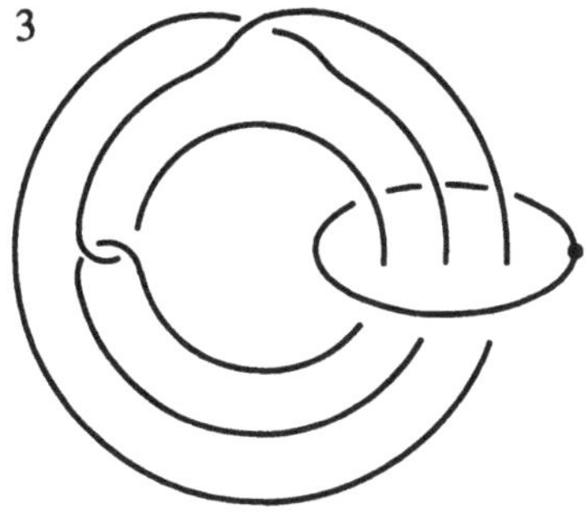

Figure 3

Here, the framing p is an integer number, and τ is an arbitrary tangle such that the component α of the link is connected. The manifold W is a straight generalization of the well known Mazur manifold shown in Figure 3. The homology spheres Σ in question are the boundaries of such 4-manifolds. For example, the boundary of the manifold in Figure 3 is homeomorphic to Seifert fibered homology sphere $\Sigma(2, 5, 7)$, see [AK]. The following result is proven in [S4].

Theorem 5. *Let $\Sigma = \partial W$ be the boundary of a manifold W in Figure 2. Then, for any tangle τ and for any p, the Floer homology groups $I_*(\Sigma)$ are two-periodic, that is $I_i(\Sigma) = I_{i+2}(\Sigma)$ for any i. Moreover, the groups $I_*(\Sigma)$ are independent of the framing p.*

This theorem implies in particular that the ν-invariant vanishes for all the Mazur homology spheres. One can compute the groups $I_*(\Sigma)$ completely in the case when the tangle τ in Figure 2 is a braid, see [S3].

Theorem 6. *If the tangle τ in Figure 2 is in fact a braid then for every p, the group $I_i(\Sigma)$ is trivial if i is odd, and $I_i(\Sigma) = \mathbb{Z}$ if i is even, or vice versa, depending on the orientation.*

Casson–Harer and Stern homology spheres. Mazur manifolds are examples of contractible manifolds having a handlebody decomposition with one 0-handle and one pair of 1- and 2-handles. The attaching circle for the 2-handle passes geometrically three times through the 1-handle. In general, this number may be greater than three (of course, the algebraic intersection still needs to be ± 1). This possibility is realized by the Casson–Harer homology spheres [CH]. The Stern homology spheres [St] bound contractible manifolds with more than one pair of 1- and 2-handles.

In [S3], we prove that the Floer homology groups of all the Seifert fibered homology spheres from [CH] and [St] known to bound contractible manifolds are 2-periodic. In our proof, we apply the Floer exact triangle to some special surgery on plumbed homology spheres which we introduce in [S3]. We use the results of R. Fintushel and R. Stern [FS1] on Floer homology of Seifert manifolds as well. Here are some explicit formulae following from our computations.

Theorem 7. *Let p be an odd integer greater than 1, and $s \geq 1$ an arbitrary integer. Then the Floer homology group $I_j(\Sigma(p, ps \pm 1, ps \pm 2))$ is trivial for j odd, and isomorphic to a free abelian group of rank $s(p^2-1)(ps \pm 3)/48$ for any even j. In particular, $\nu(\Sigma(p, ps \pm 1, ps \pm 2)) = 0$.*

Slice knots. In [S3] we also compute Floer homology of the homology spheres obtained by Dehn surgery on the knots $T(p,q)\#T(p,q)^*$ in S^3 where $T(p,q)$ is the left-handed (p,q)-torus knot, and $T(p,q)^*$ its mirror image. In the particular case of $(p,q) = (2, 2n+1)$, the Floer homology groups have been computed by A. Stipsicz and Z. Szabó in [SS] by different methods using the results of [KKR] on splitting spectral flow.

It is worth mentioning that for any relatively prime integers (p,q), the knot $T(p,q)\#T(p,q)^*$ is slice, and that both $(+1)$- and (-1)-surgeries on it bound contractible manifolds. As it follows from Theorem 8 below, the ν-invariant vanishes in both cases.

Theorem 8. *Let $k = T(p,q)\#T(p,q)^*$ be a connected sum of the (p,q)-torus knot $T(p,q)$ and its mirror image $T(p,q)^*$. Let $M_n(p,q)$ be the homology sphere obtained by $(-1/n)$-surgery of S^3 along k, $n = 1, 2, \ldots$. Then the group $I_j(M_n(p,q))$ is trivial if j is odd, and isomorphic to a free abelian group of rank $n(p^2-1)(q^2-1)/24$ if j is even.*

In our proof, we first use the Floer exact triangle to discover some symmetries in Floer homology of amphicheiral knots, and then apply a description of the SU(2)-representation space of the fundamental group of a splice of Seifert fibered homology spheres along singular fibers, see [S3].

3. Comparing the invariants

Our computations of both the ν- and $\bar{\mu}$-invariants suggest that these invariants coincide for plumbed homology spheres. In comparing the two invariants, we use the well-known fact, see [EN] and [Sb], that any plumbed homology sphere can be obtained by pasting together graph links, which are Seifert fibered homology spheres with a collection of Seifert fibers. This operation is called splicing [EN]. As a special case, for any $2 \leq j \leq n-2$, a Seifert fibered homology sphere $\Sigma(a_1, \dots, a_n)$ is a splice of Seifert fibered homology spheres $\Sigma(a_1, \dots, a_j, p)$ and $\Sigma(q, a_{j+1}, \dots, a_n)$ along the singular fibers of degrees p and q, where the integers $q = a_1 \cdots a_j$ and $p = a_{j+1} \cdots a_n$ are the products of the first j and the last $(n - j)$ Seifert invariants, respectively. In this case,

$$\bar{\mu}(\Sigma(a_1, \dots, a_n)) = \bar{\mu}(\Sigma(a_1, \dots, a_j, p)) + \bar{\mu}(\Sigma(q, a_{j+1}, \dots, a_n)),$$

$$\nu(\Sigma(a_1, \dots, a_n)) = \nu(\Sigma(a_1, \dots, a_j, p)) + \nu(\Sigma(q, a_{j+1}, \dots, a_n)).$$

The first formula follows from [Du], see also [S3], the second from [S1]. Therefore, the invariants must coincide for Seifert fibered homology spheres with an arbitrary number of singular fibers if they do for Seifert fibered homology spheres with just three singular fibers. With the latter proven in [S5], we get the following result.

Theorem 9. *For any Seifert fibered homology sphere $\Sigma(a_1, \dots, a_n)$, the ν-invariant and the $\bar{\mu}$-invariant coincide,*

$$\nu(\Sigma(a_1, \dots, a_n)) = \bar{\mu}(\Sigma(a_1, \dots, a_n)).$$

Another formulation of this result establishes links between Floer homology and the Jones polynomial. It is well-known that any Seifert fibered homology sphere $\Sigma(a_1, \dots, a_n)$ is a double cover over S^3 branched over a Montesinos knot $k(a_1, \dots, a_n)$, see [Mt]. Note that the $\bar{\mu}$-invariant can be defined in terms of the signature of this knot, as it was originally done by L. Siebenmann [Sb]. The result below follows from Theorem 9 and D. Mullins [Mu] formula for the Casson invariant of two-fold branched covers.

Theorem 10. *Let $\Sigma(a_1, \dots, a_n)$ be a Seifert fibered homology sphere then*

$$\operatorname{rank} I_0(\Sigma(a_1, \dots, a_n)) + \operatorname{rank} I_4(\Sigma(a_1, \dots, a_n)) = -\frac{1}{6} \cdot \frac{d}{dt}\bigg|_{t=-1} \ln V_k(t)$$

where V_k is the Jones polynomial of the Montesinos knot $k = k(a_1, \dots, a_n)$.

In conclusion, it is worth mentioning that the $\bar{\mu}$-invariant is additive with respect to splicing of *arbitrary* graph links, see [Du] and [S3]. It is not known whether the same is true in general for the ν-invariant.

References

[A] S. Akbulut, Lectures on Seiberg–Witten Invariants, Turkish J. Math. 20 (1996), 95–118.

[AK] S. Akbulut, R. Kirby, Mazur manifolds, Michigan Math. J. 26 (1979), 259–284.

[AM] S. Akbulut, J. McCarthy, Casson's Invariant for Oriented Homology 3-Spheres. An exposition, Math. Notes 36, Princeton Univ. Press, Princeton, NJ, 1990.

[CH] A. Casson, J. Harer, Some homology lens spaces which bound rational homology balls, Pacific J. Math. 96 (1981), 23–36.

[Dv] I. Dolgachev, Weighted projective varieties, Lecture Notes in Math. 956, Springer-Verlag, Berlin–New-York 1982, 34–71.

[D] S. Donaldson, The orientation of Yang-Mills moduli spaces and 4-manifold topology, J. Differential Geom. 26 (1987), 397–428.

[Du] N. Duchon, Involutions of plumbed manifolds, Ph.D. Thesis, University of Maryland, College Park 1982.

[EN] D. Eisenbud, W. Neumann, Three-dimensional Link Theory and Invariants of Plane Curve Singularities, Ann. of Math. Stud. 110, Princeton Univ. Press, Princeton, NJ, 1985.

[FS1] R. Fintushel, R. Stern, Instanton homology groups of Seifert fibered homology three spheres, Proc. London Math. Soc. 61 (1990), 109–137.

[FS2] R. Fintushel, R. Stern, Pseudofree orbifolds, Ann. Math. 122 (1985), 335–364.

[F1] A. Floer, An instanton-invariant for 3-manifolds, Comm. Math. Phys. 118 (1988), 215–240.

[F2] A. Floer, Instanton homology and Dehn surgery, The Floer memorial volume, Birkhäuser, Basel 1995, 77–97.

[Fk] K. Fukaya, Floer homology of connected sum of homology 3-spheres, preprint, University of Tokyo.

[Fu] M. Furuta, Monopole equation and the 11/8 conjecture, preprint.

[GS] D. Galewski, R. Stern, Classification of simplicial triangulations of topological manifolds, Ann. Math. 111 (1980), 1–34.

[K] R. Kirby, Problems in low dimensional topology, Proc. Sympos. Pure Math. 32 (1978), 273–312.

[KKR] P. Kirk, E. Klassen, D. Ruberman, Splitting the spectral flow and the Alexander matrix, Comm. Math. Helv. 69 (1994), 375–416.

[Ma] T. Matumoto, Triangulation of manifolds, Proc. Sympos. Pure Math. 32 (1978), 3–6.

[M] J. Milnor, Singular Points of Complex Hypersurfaces, Ann. of Math. Stud. 91, Princeton Univ. Press, Princeton 1968.

[Mt] J. Montesinos, Variedades de Seifert que son recubridores ciclicos ramificados de dos hojas, Bol. Soc. Mat. Mexicana 18 (1973), 1–32.

[Mo] S. Morita, Almost complex manifolds and Hirzebruch invariants for isolated singularities in complex space, Math. Ann. 211 (1974), 245–260.

[Mu] D. Mullins, The generalized Casson invariant for 2-fold branched covers of S^3 and the Jones polynomial, Topology 32 (1993), 419–438.

[N] W. Neumann, An invariant of plumbed homology spheres, Lecture Notes in Math. 788, Springer-Verlag, Berlin–New York 1980, 125–144.

[NR] W. Neumann, F. Raymond, Seifert manifolds, plumbing, μ-invariant and orientation reversing maps, Lecture Notes in Math. 664, Springer-Verlag, Berlin–New York 1978, 163–196.

[NW] W. Neumann, J. Wahl, Casson invariant of links of singularities, Comm. Math. Helv. 65 (1990), 58–78.

[R] V. Rohlin, New results in the theory of 4-dimensional manifolds, Dokl. Akad. Nauk SSSR 84 (1952), 221–224.

[S1] N. Saveliev, Adding relations to instanton homology groups of Seifert fibered homology spheres, Mat. Sbornik 183 (1992), 125–140; English translation in: Russian Acad. Sci. Sb. Math. 77 (1994), 497–510.

[S2] N. Saveliev, Dehn surgery along torus knots, Topology Appl., to appear 1997.

[S3] N. Saveliev, Floer homology and 3-manifold invariants, Ph.D. Thesis, University of Oklahoma, 1995.

[S4] N. Saveliev, Floer homology and invariants of homology cobordism, submitted.

[S5] N. Saveliev, Floer homology of Brieskorn homology spheres: solution to Atiyah's problem, MSRI Preprint No. 1997-006.

[S6] N. Saveliev, Notes on homology cobordisms of plumbed homology 3-spheres, Proc. Amer. Math. Soc., to appear.

[Sr] J.-P. Serre, A Course in Arithmetic, Springer-Verlag, New-York–Heidelberg–Berlin 1973.

[Sb] L. Siebenmann, On vanishing of the Rohlin invariant and nonfinitely amphicheiral homology 3-spheres, Lecture Notes in Math. 788, Springer-Verlag, Berlin–New York 1980, 172–222.

[St] R. Stern, Some more Brieskorn spheres which bound contractible manifolds, Notices Amer. Math. Soc. 25 (1978), A448.

[SS] A. Stipsicz, Z. Szabó, Floer homology groups of certain algebraic links, Conf. Proc. Lecture Notes Geometry and Topology, International Press, Cambridge, MA, 1994, 173–185.

[T] C. Taubes, Casson's invariant and gauge theory, J. Differential Geom. 31 (1990), 547–599.

θ-vacua in the light front quantized Schwinger model*

Prem P. Srivastava

Abstract. The light-front (LF) quantization of the bosonized Schwinger model is discussed in the *continuum formulation*. The proposal, successfully used earlier for describing the spontaneous symmetry breaking (SSB) on the LF, of separating first the scalar field into the dynamical condensate and the fluctuation fields before employing the standard Dirac method works here as well. The condensate variable, however, is now shown to be a q-number operator in contrast to the case of SSB where it was shown to be a c-number or a background field. The condensate or θ-vacua emerge straightforwardly together with their continuum normalization which avoids the violation of the cluster decomposition property in the theory. Some topics on the *front form* theory are summarized in the appendices and attention is drawn to the fact that the theory quantized, say, at equal x^+ seems already to carry information on equal x^- commutators as well.

1. Introduction

Dirac [4] in 1949 pointed out the advantages of studying the relativistic quantum dynamics of physical system on the hyperplanes of the light-front: $x^0 + x^3 = \text{const}.$, the *front form*. The LF or light-cone coordinates, in place of (x^0, x^1, x^2, x^3), are then convenient to use; they are defined by $(x^+, x^-, x^\perp)$ where $x^\pm = (x^0 \pm x^3)/\sqrt{2}$ $= x_\mp$ and $x^\perp = (x^1, x^2)$. The metric tensor for the indices $\mu = +, -, 1, 2$ may be read from the Lorentz invariant expression $A^\mu B_\mu = g_{\mu\nu} A^\mu B^\nu = g^{\mu\nu} A_\mu B_\nu =$ $A^+ B^- + A^- B^+ - A^\perp . B^\perp$. We make the *convention* to regard‡ $x^+ \equiv \tau$ as the LF *time* coordinate while $x^- \equiv x$ is the longitudinal spatial coordinate and we study the evolution in τ of the dynamical system. The LF components of any four-vector or tensor are similarly defined. We note that the separation of two points x and y on

* Invited talk given at the *First Brazil/USA Workshop on Geometry, Topology and Physics*, UNICAMP, Campinas, SP, Brasil, July '96 and at the *Intl. Workshop on Light-Front Quantization and Non-Perturbative QCD*, Iowa State University, Ames, Iowa, USA, March '96.

‡ We can of course make the convention with the role of x^+ and x^- interchanged. In Appendix D we illustrate by an example how the equal-x^+ quantized theory does seem to contain information on the equal-x^- commutators as well.

equal-τ plane is also spacelike. It becomes lightlike when $x^\perp = y^\perp$ but, unlike in the equal-time case, the points need not be coincident since $(x^- - y^-)$ may take any value. The *microcausality principle* leads to locality requirement only in $x^\perp$ and the appearence of any nonlocality in the longitudinal coordinate in the theory would not be unexpected [15].

The transformation from the conventional to LF components is, however, *not a Lorentz transformation* and the structure of the *LF phase space* is notably different when compared with the conventional phase space. For example, the momentum four-vector is $(k^-, k^+, k^\perp)$ where k^- is the LF energy while $k^\perp$ and k^+ indicates the transverse and the longitudinal components of the momentum. A massive particle on the mass shell, $k^- = (m^2 + {k^\perp}^2)/(2k^+)$, has positive definite values for $k^\pm$ in contrast to $-\infty \le k^{1,2,3} \le \infty$ for the usual components. An immediate consequence is that the *vacuum* in the LF *quantized theory* may become simpler than the one in the conventional (equal-time) theory and in many cases the interacting theory vacuum on the LF may be the same as the perturbation theory vacuum. For example, the conservation of the total longitudinal momentum would not permit the excitations of particle-antiparticle pairs by the LF vacuum (having $\sum k^+ = 0$). The SSB on the LF, for example, is described [15] in a way different from the conventional one even though the physical outcome, as expected, is the same (Apppendix C).

An important advantage pointed out by Dirac is that *seven* out of the ten Poincaré generators are *kinematical* on the LF while in the conventional theory constructed on the hyperplanes $x^0 = const.$, the *instant form*, only *six* have this property. Also the notions of spin on the LF for massive and massless particles seem to get unified (Appendices A and B).

Another notable feature of a relativistic theory in the *front form* is that it gives rise to a singular Lagrangian, e.g., a constrained dynamical system [5]. It leads in general to a reduction in the number of independent field operators on the corresponding phase space. The vacuum structure may then become more tractable and the computation of the physical observables may become simpler. It is illustrated below by the detailed study of the Schwinger model [14]. In the conventional framework, for example, the QCD vacuum is quite complex due to the *infrared slavery* and it contains also the gluonic and fermionic condensates. There seems also to exist a contradiction between the Standard Quark Model and the QCD containing a sea of partons (quarks, anti-quarks and gluons); the *front form* theory may throw light on such problems.

The LF field theory was rediscovered in 1966 by Weinberg [20] in his Feynman rules adapted for infinite momentum frame. Later it was demonstrated [7] that these rules correspond to the quantization of the theory on the LF. The recent revival [3, 21, 13] of the interest in LF quantization owes, say, to the difficulties encountered in the computation of nonperturbative effects in the *instant form* QCD or in the study of the relativistic bound states of light fermions [21, 3] which can not be handled, say, by the Lattice gauge theory. The LF coordinates have proved also very useful in the study of the theories of (super-) strings and membranes as well.

The purpose of the present work is to show how the quantization of the massless Schwinger model on the LF leads in a straightforward way to the θ-vacua known to emerge [9, 1] in the *instant form* theory. The important feature of the continuum normalization of these states arises naturally on the LF. This is in contrast to the discussions in the conventional Lorentz or the Coulomb gauge framework where we have to invoke arguments to impose it, so as to avoid [9, 1, 8] the violation of the the cluster decomposition property. The discussion on the LF is more transparent due to a reduced number of independent fields. It is obtained here by making use of the method proposed earlier in connection with the *front form* description of the SSB (and the tree level Higgs mechanism) [16, 15]. The scalar field (of the equivalent bosonized Schwinger model) is separated, based on physical considerations, into the dynamical bosonic condensate and the quantum fluctuation fields. The Dirac procedure [5] is then followed in order to construct the Hamiltonian formulation and the quantized theory.

After a brief discussion in Section 2 of the Schwinger model its bosonized version is quantized on the LF. The *condensate* or θ-vacua are described in Section 3 while the conclusions are summarized in Section 4. Appendices contain some material related to the *front form* field theory.

2. LF quantization of the Schwinger model

The field theory model is simply the two dimensional massless quantum electrodynamics with the Lagrangian

$$\mathcal{L} = \bar{\psi}\, i\gamma^{\mu}\partial_{\mu}\psi - \frac{1}{4}F^{\mu\nu}F_{\mu\nu} - e\bar{\psi}\,\gamma^{\mu}\psi A_{\mu}, \tag{1}$$

where ψ is a two-component spinor field† and A_{μ} is the $U(1)$ gauge field. The Lagrangian is invariant under the global $U(1)_5$ chiral transformations $\psi \to e^{(i\gamma_5\alpha)}\psi$ apart from under the usual $U(1)$ gauge transformations. The model is exactly solvable and its physical spectrum consists solely of massive vector field [14]. This was made explicit by Lowenstein and Swieca [9, 1] through their operator solution of the model in the *instant form* framework in the Lorentz gauge and where the complex structure of the ground state was also studied. It has also been studied in the Coulomb gauge [8] and on the light-cone in the discretized formulation [11] where there are many subtleties in the fermionic version (1). We will work here directly in *the continuum formulation* (Appendix C) on the LF so that the spurious finite volume effects get automatically suppressed. For the purposes of studying the vacuum structure it is also convenient to study the *equivalent* bosonized version.

† $\gamma^0 = \sigma_1$, $\gamma^1 = i\sigma_2$, $\gamma_5 = -\sigma_3$, $\eta^{00} = -\eta^{11} = 1$, $\epsilon^{-+} = \epsilon^{01} = +1$, and $\gamma_5\gamma^{\mu} = \epsilon^{\mu\nu}\gamma_{\nu}$ etc.

The LF coordinates are now (x^+, x^-) where $x^{\pm} = (x^0 \pm x^1)/\sqrt{2} = x_{\mp}$. The conserved Noether currents defined by $j_\mu = \bar\psi\gamma_\mu\psi$ and $j_{5\mu} = \bar\psi\gamma_5\gamma_\mu\psi$ satisfy the relation $j_5^\mu = \epsilon^{\mu\nu} j_\nu$ which is particular to the two dimensional theory. Written explicitly, $j_5^+ = -j^+ = -\sqrt{2}\psi_2{}^\star\psi_2$, $j_5^- = +j^- = \sqrt{2}\psi_1{}^\star\psi_1$. The classical conservation of both the current leads to $j^+ = j^+(x^-)$ and $j^- = j^-(x^+)$ like in the case of the free Dirac field. In the quantized theory, however, the currents must be defined, say, by the introduction of the of the point-splitting of operators. When the gauge coupling is also present we are required to construct gauge invariant currents together with the point-splitting. The divergence of the axial current is then found anomalous: $\partial_\mu j_5^\mu = (e/(2\pi))\epsilon_{\mu\nu}F^{\mu\nu}$ and the eqs. of motion lead to $(\partial_\mu\partial^\mu + e^2/\pi)\epsilon^{\mu\nu}F_{\mu\nu} = 0$ implying dynamical mass generation [14] for the gauge field in the quantized theory. The same results may also be obtained by employing functional integal methods [1].

The abelian gauge theory under consideration has been extensively studied and its exact solvability [14, 9] derives from the remarkable property of one-dimensional fermion systems, viz, that they can equivalently be described in terms of canonical one-dimensional boson fields [10]. The operator solution [9] mentioned above in a sense amounts to bosonization. Some of the relevant correspondences in abelian bosonization are $\bar\psi\psi = K : \cos 2\sqrt{\pi}\,\phi :$, $\bar\psi\gamma_5\psi = K : \sin 2\sqrt{\pi}\,\phi :$, $\bar\psi\gamma_5\gamma_\mu\psi = \partial_\mu\phi/\sqrt{\pi}$, $\bar\psi\gamma_\mu\psi = \epsilon_{\mu\nu}\partial^\nu\phi/\sqrt{\pi}$, $\bar\psi\, i\gamma.\partial\psi = \frac{1}{2}\partial_\mu\phi\partial^\mu\phi$ where ϕ is a bosonic scalar field and K is a constant. The fermionic condensate $\langle\bar\psi\psi\rangle_0$, for example, may then be expressed in terms of the value of the bosonic condensate. The bosonized theory can also be constructed with the use of the functional integral method [1]. The original fermionic and the bosonized theories are *equivalent* in the sense that they have the same current commutation relations and the energy-momentum tensor is the same when expressed in terms of the currents.

The Lagrangian density of the *bosonized* massless Schwinger Model takes the following form

$$\mathcal{L} = \frac{1}{2}\partial_\mu\phi\partial^\mu\phi - gA_\mu\epsilon^{\mu\nu}\partial_\nu\phi - \frac{1}{4}F^{\mu\nu}F_{\mu\nu}, \tag{2}$$

where $g = e/\sqrt{\pi}$. It carries in it all the symmetries of the original fermionic model including the information on the mass generation. Under the $U(1)$ gauge field transformation the scalar field is left invariant while under the $U_5(1)$ chiral transformations, in view of the correspondences above, it suffers a translation by a constant. This is crucial in obtaining the so called θ- or *condensate* vacua. The quantization of the bosonized theory would allow us to describe the vacuum structure of the original theory and to compute, say, the fermionic condensate.

We first make the *separation*, proposed in Ref. [16], in the scalar field (a generalized function): $\phi(\tau, x^-) = \omega(\tau) + \varphi(\tau, x^-)$, where $\omega(\tau)$ is the (dynamical) bosonic *condensate* variable and the field φ represents the quantum fluctuations. This enabled us to give [16] a description on the LF of the SSB and (tree level) Higgs mechanism and where the variable ω was shown to be a c-number, e.g., a background field. In the Schwinger model, on the contrary, it will be shown below to turn out as a q-number operator and its eigenvalues would label the vacuum states. We will set $\int dx^-\varphi(x^+,x^-)=0$ so that the entire zero-momentum mode of ϕ is represented by the condensate variable. The *chiral transformations* would be defined [8] as: $\omega\to\omega+\text{const.}$, $\varphi\to\varphi$, and $A_\mu\to A_\mu$ so that the boundary conditions at infinity on φ are left unaltered (see Section 3). The bosonized Lagrangian then becomes

$$L=\int_{-R/2}^{R/2}dx^-\Big[\dot\varphi\varphi'+g(A_+\varphi'-A_-\dot\varphi)+\frac{1}{2}(\dot A_- - A_+{}')^2\Big]-g\dot\omega\int_{-R/2}^{R/2}dx^-A_-(\tau,x^-). \tag{3}$$

Here $R\to\infty$, an overdot (a prime) indicates the partial derivative with respect to $x^+\equiv\tau$ $(x^-\equiv x)$, and φ is assumed to satisfy the conditions required for the existence of the Fourier transform in the spatial variable x^-.

The last term of (3) shows that the light-cone gauge $A_-=0$ is not convenient to adopt in the present case. It is suggested and in fact will be shown below that the zero-momentum mode of A_-, viz, $h(\tau)\equiv\int dx^-A_-$ and ω form a canonically conjugate pair. We would instead adopt the gauge [6] $\partial_-A_-=0$ which is shown to be accessible on the phase space. The canonical momenta defined from (3) are

$$\begin{aligned}\pi&=\frac{\delta\mathcal{L}}{\delta\dot\varphi}=\varphi'-gA_-\\ E^+&=\frac{\delta\mathcal{L}}{\delta\dot A_+}=0\\ E^-&=\frac{\delta\mathcal{L}}{\delta\dot A_-}=(\dot A_- - A_+{}')\\ p&=\frac{\partial L}{\partial\dot\omega}=-gh(\tau)\equiv-\pi_\omega.\end{aligned} \tag{4}$$

The primary constraints [5] are thus $\chi\equiv\pi-\varphi'+gA_-\approx0$, $E^+\approx0$, and $T(\tau)\equiv(-\pi_\omega+gh)\approx0$, where $\approx$ indicates the weak equality [5]. The canonical Hamiltonian is found to be

$$H_c=\int dx^-\Big[\frac{1}{2}E^{-2}+E^-A_+{}'-gA_+\varphi'\Big], \tag{5}$$

and we take the preliminary [5] Hamiltonian as

$$H'=H_c+\int dx^-\Big[u_+E^++u\chi\Big]+\lambda(\tau)T(\tau), \tag{6}$$

where u_+, u, and λ are the Lagrange multiplier fields. We assume initially the standard equal-τ Poisson brackets: $\{\pi_\omega(\tau),\omega(\tau)\}=+1$, $\{E^\mu(\tau,x^-),A_\nu(\tau,y^-)\}=-\delta^\mu_\nu\delta(x^--y^-)$, $\{\pi(\tau,x^-),\varphi(\tau,y^-)\}=-\delta(x^--y^-)$ etc. from which it follows, for example, that $\{E^\mu(\tau,x^-),h(\tau)\}=-\delta^\mu_-$. On requiring the persistency of the constraints in τ employing $df/d\tau=\{f,H'\}+\partial f/\partial\tau$ we derive the secondary constraints $K(\tau)\equiv\int dx^-[E^-+A'_+]\approx0$ and $\Omega\equiv\partial_-(E^-+g\varphi)\approx0$. We go over now to an extended Hamiltonian including these constraints as well and repeat the procedure.

No new constraints are shown to be generated since only the equations which would determine the Lagrange multiplier fields are left. The constraints Ω and E^+ are easily shown to be first class [5] while T, K, and χ are second class [3]. From the eqs. of motion we show that we may determine the multiplier fields such that $\partial_- A_- \approx 0$ and $(E^- + A'_+) \approx 0$ along with their persistency conditions are satisfied. We may thus add to the theory these ones as the (external) gauge-fixing constraints, corresponding to the two first class constraints found above. The whole set of constraints then becomes second class and we may proceed to construct Dirac brackets [5], which would replace the Poisson ones, such that the weak equalities may be replaced by the strong equalities (even) inside them.

It is straightforward to construct the Dirac bracket iteratively. We first handle the pair $T \approx 0$, $K \approx 0$ with $\{T, K\} \approx gR$, $\{T, T\} \approx 0$, and $\{K, K\} \approx 0$. The corresponding modified equal-τ bracket is easily constructed:‡

$$\{f, k\}_1 = \{f, k\} + \frac{1}{gR}\Big[\{f, T\}\{K, k\} - \{f, K\}\{T, k\}\Big]. \tag{7}$$

We verify, for example, $\{f, T\}_1 \equiv 0$, $\{K, f\}_1 \equiv 0$ etc.. The modified bracket differs from the Poisson one only when the variables ω, E^-, E^+, or A_- are to be found in the functionals f and k. For example, $\{\omega, h\}_1 = -(1/g)$, $\{\omega, E^+\}_1 = 0$, $\{E^-, A_-\}_1 = \{E^-, A_-\} + (1/R)$, $\{f, \chi\}_1 = \{f, \chi\} - (1/R)\{f, T\}$, $\{\chi, \chi\}_1 = \{\chi, \chi\} = -2\partial_-^x \delta(x^- - y^-)$.

A further modification

$$\{f, k\}_2 = \{f, k\}_1 + \frac{1}{4}\iint dudv\ \{f, \chi\}_1 \epsilon(u - v)\{\chi, k\}_1 \tag{8}$$

allows us to set $\chi = 0$ also as strong equality. We find, for example, $\{\omega, \omega\}_2 = 0$, $\{E^-, E^-\}_2 = (-g^2/4)[\epsilon(x - y) + (1/R)\int du\ \{\epsilon(u - x) - \epsilon(u - y)\}]$, $\{\omega, E^-\}_2 = -(g/(4R))\int du\, \epsilon(u - y)$, $\{\varphi, \pi\}_2 = (1/2)\delta(x - y)$, $\{\varphi, \varphi\}_2 = (-1/4)\epsilon(x - y)$, $\{\varphi, E^-\}_2 = (g/4)\epsilon(x - y)$, $\{E^+, \Omega\}_2 = 0$, $\{\Omega, \Omega\}_2 = 0$ etc., and that E^+ and Ω continue to remain first class even with respect to $\{,\}_2$.

Next the *gauge-fixing* constraint $\chi_3 \equiv \partial_- A_- \approx 0$ along with $\Omega \approx 0$ are implemented. The constraint matrix has elements $\{\chi_3, \chi_3\}_2 = 0$, $\{\chi_3, \Omega\}_2 = -\partial_x^2 \delta(x - y)$, and $\{\Omega, \Omega\}_2 = 0$ and its inverse is easily found to be $i\sigma_2\, \partial_x^{-2} \delta(x - y)$. Hence we define

$$\{f, k\}_3 = \{f, k\}_2 + \int du\ \Big[\{f, A_-\}_2\{E^- + g\varphi, k\}_2 - \{f, E^- + g\varphi\}_2\{A_-, k\}_2\Big]. \tag{9}$$

We find $\{A_-, \omega\}_3 = 1/(gR)$ and $\{A_-, f\}_3 \to 0$ when $R \to \infty$ for $f \neq \omega$. Some of the others are $\{\omega, E^-\}_3 = (-g/(4R))\int du\, \epsilon(u - y)$, $\{h, E^-\}_3 = 0$, $\{\omega, h\}_3 = -1/g$, $\{E^-, E^-\}_3 = (-g^2/4)\epsilon(x - y)$.

---

‡ We make the convention that the first variable in an equal-τ bracket $\{f, k\}$ refers to longitudinal coordinate x^- while the second one to y^-. We remind that we are working *in the continuum formulation* and that $R \equiv \int_{-R/2}^{R/2} \to \infty$ at the end of the computation.

The remaining first class constraint $E^+ \approx 0$ may be taken care of by adding still another *gauge-fixing* constraint $\Phi \equiv E^- + A'_+ \approx 0$. The two dimensional constraint matrix $C(x, y)$ here has the elements: $C_{11} = \{E^+, E^+\}_3 = 0$, $C_{12} = C_{21} = \{E^+, \Phi\} = \{E^+, A'_+\} = \partial_x \delta(x-y)$, $C_{22} = \{\Phi, \Phi\}_3 = -g^2\epsilon(x-y)/4$. An inverse matrix C^{-1} is shown to have the elements $C^{-1}_{12} = C^{-1}_{21} = (1/2)\epsilon(x-y)$ and $C^{-1}_{22} = 0$ while C^{-1}_{11} satisfies $\partial_x^2 C^{-1}_{11}(x, y) = (g^2/4)\epsilon(x-y)$. The final Dirac bracket which implements all the constraints is thus constructed to be

$$\begin{aligned}\{f, k\}_D &= \{f, k\}_3 \\ &- \frac{1}{2}\iint du dv \Big[\{f, \Phi\}_3 \epsilon(u-v)\{E^+, k\}_3 - \{f, E^+\}_3 \epsilon(u-v)\{\Phi, k\}_3 \\ &+ 2\{f, E^+\}_3\, C^{-1}_{11}(u, v)\, \{E^+, k\}_3\Big].\end{aligned} \tag{10}$$

We find $\{E^-, E^-\}_D = -g^2\epsilon(x-y)/4$, $\{\varphi, \varphi\}_D = -\epsilon(x-y)/4$, $\{A_-, k\}_D = 0$ for $k \neq \omega$, $\{A_-, \omega\}_D = 1/(gR)$, $\{\pi_\omega, \omega\}_D = 1$, $\{A_+, A_+\}_D = C^{-1}_{11}(x, y)$, and $\{\omega, E^-\}_D = -g \int du\, \epsilon(u-y)/(4R)$ among the others. All the constraints can now be written as strong equalities and we are left behind with the independant variables $\varphi, \omega, \pi_\omega$ and $E^- = -\partial_- A_+ = -g\varphi$. The Hamiltonian density is effectively given by $\mathcal{H}_D = E^{-2}/2 + \partial_-(E^- A_+)$ and $H_D = \int dx\, E^{-2}/2 = g^2 \int dx\, \varphi^2/2$ corresponding to a free scalar field φ of mass e^2/π. It may be checked that for the self-consistency with the Lagrangian eqs. we require that A_+ should satisfy the periodic boundary conditions at infinity in x^-.

3. θ-vacua

We discuss now the vacuum state in the LF quantized theory. The *ultimate physical conclusions* would of course coincide with those following from the *instant form* theory. The *quantized theory* is obtained through the correspondence [5] $i\{f, k\}_D \to [f, k]$ with the commutators of the field operators. We find $[\omega, \omega] = 0$, $[\pi_\omega, \omega] = i$, $[\omega, E^-] = [\omega, \varphi] = [\pi_\omega, \varphi] = [\omega, H_D] = [\pi_\omega, H_D] = 0$, and the well known LF commutator $[\varphi, \varphi] = -i\epsilon(x-y)/4$ which leads to $2\pi[j^+(\tau, x), j^+(\tau, y)] = i\partial_-\delta(x-y)$, where the right hand side is the Schwinger term. The condensate field ω turns out to be a q-number (operator) in the present model in contrast to the case of the LF quantized scalar theory where it is shown (Appendix C) to be a c-number. This is expected since the chiral symmetry in the model is realized as translation of the value of the condensate. The latter must then be represnted by an operator ω in order that its eigenvalues may be shifted by another operator. The $\omega(\tau)$ in our discussion has been treated as a dynamical variable and we let the Dirac procedure to determine if it is a c- or q-number in the quantized theory.

The original fermionic Schwinger model is invariant under *global* $U(1)_5$ chiral transformations $\psi \to \exp(i\alpha\gamma_5)\,\psi$, $\bar\psi\psi \to \bar\psi\,\exp(i2\alpha\gamma_5)\,\psi$. From $\exp(i\alpha\gamma_5) = (\cos\alpha + i\gamma_5\sin\alpha)$ it is clear that the $U(1)_5$ like $U(1)$ is a compact group with $\alpha = \alpha_0 + 2\pi n$, $n = 0, \pm 1, \pm 2, \ldots$, and, for example, $n = 0$ with $0 \le \alpha_0 \le 2\pi$ would enumerate all the distinct elements of the group. For $\alpha = n\pi$, $n = 0, \pm 1, \pm 2, \ldots$, $\bar\psi\psi$ and $\bar\psi\gamma_5\psi$ are clearly left invariant under chiral transformations. From the correspondence $\bar\psi\,\psi \leftrightarrow K : \cos(2\sqrt{\pi}\phi) :$ implied in the construction of the bosonized theory it follows that the chiral transformation on the bosonic field ϕ is realized by $\omega \to \omega + \beta/\sqrt{\pi}$, $\varphi \to \varphi$ with $\beta = \beta_0 + n\pi$, $n = 0, \pm 1, \pm 2, \ldots$, $0 \le \beta_0 \le \pi$. The ω is therefore an angular (coordinate or) variable with the period $\sqrt{\pi}$. In the quantized field theory the chiral transformations are generated by the unitary operator $Q_5(\beta)$ such that‡ $Q_5^\dagger(\beta)\,\omega\,Q_5(\beta) = \omega + \beta/\sqrt{\pi}$, $Q_5^\dagger(\beta)\,\varphi\,Q_5(\beta) = \varphi$ and $Q_5(\pi) = Q_5(0)$. The commutation relations given above result in $\exp(-ia\pi_\omega)\,\varphi\,\exp(ia\pi_\omega) = \varphi$, $\exp(-ia\pi_\omega)\,\omega\,\exp(ia\pi_\omega) = \omega + a$. The unitary field theory operator which generates the chirality transformation may then be defined as $Q_5(\beta) \equiv \exp(i\pi_\omega\,\beta/\sqrt{\pi})$.

In view of the commutation relations above the space of states may be built as a tensor product, written as $|\varphi) \otimes |n\}$, of a conventional Fock space for the massive field φ and a space spanned by the eigenvalues of π_ω. From $Q_5(\pi) = \exp(i\pi_\omega\sqrt{\pi}) = Q_5(0) = I$, where I is identity operator, it follows that π_ω has discrete eigenvalues: $\pi_\omega|n\} = 2n\sqrt{\pi}|n\}$ where $n = 0, \pm 1, \pm 2, \ldots$ and $\{m|n\} = \delta_{mn}$. With the help of $\exp(ia\omega)\,\pi_\omega\,\exp(-ia\omega) = \pi_\omega + a$ we may construct the ladder operators $d^\pm = \exp(\mp 2i\sqrt{\pi}\omega)$ such that $d^\pm|n\} = |(n\pm 1)\}$. The Hamiltonian contains only the field φ and it commutes with the chirality operator $Q_5(\beta)$ and $d^\pm$. We have infinite degeneracy corresponding to the *chiral* vacuum states $|0) \otimes |n\}$ with $n = 0, \pm 1, \pm 2, \ldots$ since ω and π_ω are absent from the LF Hamiltonian. There are also no transitions between these vacua characterized by the differnt values of n. In the original Schwinger solution the vacuum state was chosen to be the chirally symmetric state $|0) \otimes |n = 0\}$ which leads to the violation of the cluster decomposition property. This may be avoided by choosing instead the (alternative) vacuum state for the model to be an eigenstate of the condensate operator. The state vectors which are degenerate with respect to the eigenvalue ω' of the condensate operator are easily constructed:§

$$|\omega'\} \otimes |\varphi) = \frac{1}{\pi^{1/4}} \sum_{n=-\infty}^{\infty} e^{2i\sqrt{\pi}\,n\,\omega'}\,|n\} \otimes |\varphi) = e^{i\pi_\omega\omega'}|\omega' = 0\} \otimes |\varphi), \tag{11}$$

where $0 \le \omega' \le \sqrt{\pi}$ or $0 \le \theta' \equiv 2\sqrt{\pi}\omega' \le 2\pi$ specify the physical values of the condensate. These states have the continuum normalization $\{\omega''|\omega'\} = \delta(\omega'' - \omega')$ and it comes out naturally in our discussion of the LF quantized theory. This is in contrast

‡ Here ω, φ, and π_ω are field operators while β, a, and n are c-no. constants.

§ In the coordinate representation $\omega|\omega'\} = \omega'|\omega'\}$ and π_ω is represented by $i\partial/\partial\omega'$. We recall also the *Poisson summation formula* of the distribution theory: $\delta(x) = \sum_{-\infty}^{\infty} e^{i2\pi nx}$ for $-1 \le x \le 1$ and note that $\{\omega'|\pi_\omega|\omega''\} = i\partial\delta(\omega' - \omega'')/\partial\omega'$ etc.

to the arguments required in the equal-time formulation to impose it so as to avoid the violation of the cluster decomposition property and other inconsistencies in the theory [9, 1, 8]. The *condensate-* or *θ-vacuum* in the theory under consideration is the state $|\omega'\} \otimes |0)$ with a fixed given value for θ' and we note that $Q_5(\beta)|\omega'\} = |\omega' + \beta/\sqrt{\pi}\}$. The vacuum state is infinitely degenerate and there are no transitions among the states with different values of θ'. The chiral symmetry is thus spontaneously broken. The corresponding generator π_ω also does not annihilate the vacuum state. This feature here is different from the one found in connection with the description on the LF of the SSB (and Higgs mechanism). There the LF vacuum is annihilated by all of the symmetry generators (in contrast to the case of the equal-time formulation) and the broken symmetry is manifested in the quantized theory Hamiltonian (Appendix C). The fermionic condensate in the Schwinger model may be computed as

$$\begin{aligned}(0| \otimes \{\omega''| : \bar{\psi}\psi : |\omega'\} \otimes |0) &= K(0| \otimes \{\omega''| : \cos 2\sqrt{\pi}(\omega + \varphi) : |\omega'\} \otimes |0) \\ &= K(0| \otimes \{\omega''| : \cos(2\sqrt{\pi}\varphi) : |\omega'\} \otimes |0) \cos(2\sqrt{\pi}\omega') \\ &= K \cos\theta' \, \delta(\omega'' - \omega').\end{aligned} \tag{12}$$

5. Conclusions

On the LF the θ-vacua in the massless Schwinger model are obtained in straightforward fashion on quantizing the equivalent bosonized theory. Self-consistent Hamiltonian formulation on the LF may be built by first seperating the scalar field into the dynamical condensate and quantum fluctuation fields and then following the Dirac procedure. The same procedure allowed earlier to describe also the SSB and (tree level) Higgs mechanism. The integrability of QCD_2 has been conjectured [2] recently from the studies in the conventional framework employing the equivalent bosonic description. It would be interesting to study the vacua here in the *front form* theory and also to find an alternative (simpler) proof [19] of the conjecture. The present work adds to the previous experience in that the physical results following from the discussion in the conventional *instant form* and the *front form quantum dynamics*, in spite of the notable differences in the structure of the corresponding phase spaces, come out to be the same though the mechanism to arrive at them may be different.

Acknowledgements. Acknowledgements with thanks are due to James Vary, Stan Brodsky, Heinrich Leutwyler, Maria Cristina Abdalla, Elcio Abdalla, and Luiz Alvarez-Gaume for constructive suggestions. Thanks are due also to Waldyr Rodrigues Jr. and the Organizers of the Workshop at the UNICAMP for the invitation at the very stimulating interdisciplinary meeting. The hospitality offered at the Theory Division of CERN, at the Workshop at Iowa State University by Prof. James Vary, and at CBPF/LAFEX by Profs. A. Santoro and F. Caruso is gratefully acknowledged.

Appendix A. Poincaré generators on the LF

The Poincaré generators in coordinate system (x^0, x^1, x^2, x^3), satisfy $[M_{\mu\nu}, P_\sigma] = -i(P_\mu g_{\nu\sigma} - P_\nu g_{\mu\sigma})$ and $[M_{\mu\nu}, M_{\rho\sigma}] = i(M_{\mu\rho}g_{\nu\sigma} + M_{\nu\sigma}g_{\mu\rho} - M_{\nu\rho}g_{\mu\sigma} - M_{\mu\sigma}g_{\nu\rho})$ where the metric is $g_{\mu\nu} = \mathrm{diag}(1, -1, -1, -1)$, $\mu = (0, 1, 2, 3)$ and we take $\epsilon_{0123} = \epsilon_{-+12} = 1$. If we define $J_i = -(1/2)\epsilon_{ikl}M^{kl}$ and $K_i = M_{0i}$, where $i, j, k, l = 1, 2, 3$, we find $[J_i, F_j] = i\epsilon_{ijk}F_k$ for $F_l = J_l, P_l$ or K_l while $[K_i, K_j] = -i\epsilon_{ijk}J_k$, $[K_i, P_l] = -iP_0 g_{il}$, $[K_i, P_0] = iP_i$, and $[J_i, P_0] = 0$.

The LF generators are P_+, P_-, P_1, P_2, $M_{12} = -J_3$, $M_{+-} = -K_3$, $M_{1-} = -(K_1 + J_2)/\sqrt{2} \equiv -B_1$, $M_{2-} = -(K_2 - J_1)/\sqrt{2} \equiv -B_2$, $M_{1+} = -(K_1 - J_2)/\sqrt{2} \equiv -S_1$ and $M_{2+} = -(K_2 + J_1)/\sqrt{2} \equiv -S_2$. We find $[B_1, B_2] = 0$, $[B_a, J_3] = -i\epsilon_{ab}B_b$, $[B_a, K_3] = iB_a$, $[J_3, K_3] = 0$, $[S_1, S_2] = 0$, $[S_a, J_3] = -i\epsilon_{ab}S_b$, $[S_a, K_3] = -iS_a$ where $a, b = 1, 2$ and $\epsilon_{12} = -\epsilon_{21} = 1$. Also $[B_1, P_1] = [B_2, P_2] = iP^+$, $[B_1, P_2] = [B_2, P_1] = 0$, $[B_a, P^-] = iP_a$, $[B_a, P^+] = 0$, $[S_1, P_1] = [S_2, P_2] = iP^-$, $[S_1, P_2] = [S_2, P_1] = 0$, $[S_a, P^+] = iP_a$, $[S_a, P^-] = 0$, $[B_1, S_2] = -[B_2, S_2] = -iJ_3$, $[B_1, S_1] = [B_2, S_2] = -iK_3$. For $P_\mu = i\partial_\mu$, and $M_{\mu\nu} \to L_{\mu\nu} = i(x_\mu\partial_\nu - x_\nu\partial_\mu)$ we find $B_a = (x^+P^a - x^aP^+)$, $S_a = (x^-P^a - x^aP^-)$, $K_3 = (x^-P^+ - x^+P^-)$ and $J_3 = (x^1P^2 - x^2P^1)$. Under the conventional *parity* operation $\mathcal{P}$: $(x^\pm \leftrightarrow x^\mp, x^{1,2} \to -x^{1,2})$ and $(p^\pm \leftrightarrow p^\mp, p^{1,2} \to -p^{1,2})$, we find $\vec{J} \to \vec{J}$, $\vec{K} \to -\vec{K}$, $B_a \to -S_a$ etc.. The six generators P_l, M_{kl} leave $x^0 = 0$ hyperplane invariant and are called [4] *kinematical* while the remaining P_0, M_{0k} the *dynamical* ones. On the LF there are *seven* kinematical generators: P^+, P^1, P^2, B_1, B_2, J_3 and K_3 which leave the LF hyperplane, $x^0 + x^3 = 0$, invariant and the three *dynamical* ones S_1, S_2 and P^- form a mutually commuting set. We note that each of the set $\{B_1, B_2, J_3\}$ and $\{S_1, S_2, J_3\}$ generates an $E_2 \simeq SO(2) \otimes T_2$ algebra; this will be shown below to be relevant for defining the *spin* for massless particle. Including K_3 in each set we find two subalgebras each with four elements. Some useful identities are $e^{i\omega K_3}P^\pm e^{-i\omega K_3} = e^{\pm\omega}P^\pm$, $e^{i\omega K_3}P^\perp e^{-i\omega K_3} = P^\perp$, $e^{i\bar{v}.\bar{B}}P^- e^{-i\bar{v}.\bar{B}} = P^- + \bar{v}.\bar{P} + \frac{1}{2}\bar{v}^2P^+$, $e^{i\bar{v}.\bar{B}}P^+ e^{-i\bar{v}.\bar{B}} = P^+$, $e^{i\bar{v}.\bar{B}}P^\perp e^{-i\bar{v}.\bar{B}} = P^\perp + v^\perp P^+$, $e^{i\bar{u}.\bar{S}}P^+ e^{-i\bar{u}.\bar{S}} = P^+ + \bar{u}.\bar{P} + \frac{1}{2}\bar{u}^2P^-$, $e^{i\bar{u}.\bar{S}}P^- e^{-i\bar{u}.\bar{S}} = P^-$, $e^{i\bar{u}.\bar{S}}P^\perp e^{-i\bar{u}.\bar{S}} = P^\perp + u^\perp P^-$ where $P^\perp \equiv \bar{P} = (P^1, P^2)$, $v^\perp \equiv \bar{v} = (v_1, v_2)$ and $(v^\perp.P^\perp) \equiv (\bar{v}.\bar{P}) = v_1P^1 + v_2P^2$ etc. Analogous expressions with P^μ replaced by X^μ can be obtained if we use $[P^\mu, X_\nu] \equiv [i\partial^\mu, x_\nu] = i\delta^\mu_\nu$.

Appendix B. LF spin operator. Hadrons in LF Fock basis

The Casimir generators of the Poincaré group are: $P^2 \equiv P^\mu P_\mu$ and W^2, where $W_\mu = (-1/2)\epsilon_{\lambda\rho\nu\mu}M^{\lambda\rho}P^\nu$ defines the Pauli–Lubanski pseudovector. It follows from $[W_\mu, W_\nu] = i\epsilon_{\mu\nu\lambda\rho}W^\lambda P^\rho$, $[W_\mu, P_\rho] = 0$ and $W.P = 0$ that in a representation charactarized by particular eigenvalues of the two Casimir operators we may simultaneously diagonalize P^μ along with just one component of W^μ. We have $W^+ = -[J_3P^+ + B_1P^2 - B_2P^1]$, $W^- = J_3P^- + S_1P^2 - S_2P^1$, $W^1 = K_3P^2 + B_2P^- - S_2P^+$, and $W^2 = -[K_3P^1 + B_1P^- - S_1P^+]$ and it shows that W^+ *has a special place* since it contains only the kinematical generators [17]. On the LF we define $\mathcal{J}_3 = -W^+/P^+$ as the *spin operator* [17]. It may be shown to commute with P_μ, B_1, B_2, J_3, and K_3. For $m \neq 0$ we may use the parametrizations $p^\mu : (p^- = (m^2 + p^{\perp 2})/(2p^+), p^+ = (m/\sqrt{2})e^\omega, p^1 = -v_1p^+, p^2 = -v_2p^+)$ and $\tilde{p}^\mu : (1,1,0,0)(m/\sqrt{2})$ in the rest frame. We have $P^2(p) = m^2I$ and $W(p)^2 = W(\tilde{p})^2 = -m^2[J_1^2 + J_2^2 + J_3^2] = -m^2s(s+1)I$ where s assumes half-integer values. Starting from the rest state $|\tilde{p}; m, s, \lambda, \ldots\rangle$ with $J_3\,|\tilde{p}; m, s, \lambda, \ldots\rangle = \lambda\,|\tilde{p}; m, s, \lambda, \ldots\rangle$ we may build an arbitrary eigenstate of P^+, $P^\perp$, $\mathcal{J}_3$ (and P^-) on the LF by

$$|p^+, p^\perp; m, s, \lambda, \ldots\rangle = e^{i(\bar{v}.\bar{B})}e^{-i\omega K_3}|\tilde{p}; m, s, \lambda, \ldots\rangle.$$

If we make use of the following *identity* [17]

$$\mathcal{J}_3(p) = J_3 + v_1B_2 - v_2B_1 = e^{i(\bar{v}.\bar{B})}\; J_3\; e^{-i(\bar{v}.\bar{B})}$$

we find $\mathcal{J}_3\,|p^+, p^\perp; m, s, \lambda, \ldots\rangle = \lambda\,|p^+, p^\perp; m, s, \lambda, \ldots\rangle$. Introducing also the operators $\mathcal{J}_a = -(\mathcal{J}_3P^a + W^a)/\sqrt{P^\mu P_\mu}$, $a = 1,2$, which do, however, contain dynamical generators, we verify that $[\mathcal{J}_i, \mathcal{J}_j] = i\epsilon_{ijk}\mathcal{J}_k$.

For $m = 0$ case when $p^+ \neq 0$ a convenient parametrization is $p^\mu : (p^- = p^+v^{\perp 2}/2,\; p^+, p^1 = -v_1p^+,\; p^2 = -v_2p^+)$ and $\tilde{p} : (0, p^+, 0^\perp)$. We have $W^2(\tilde{p}) = -(S_1^2 + S_2^2)p^{+2}$ and $[W_1, W_2](\tilde{p}) = 0$, $[W^+, W_1](\tilde{p}) = -ip^+W_2(\tilde{p})$, $[W^+, W_2](\tilde{p}) = ip^+W_1(\tilde{p})$ showing that W_1, W_2 and W^+ generate the algebra SO(2) $\otimes$ T_2. The eigenvalues of W^2 are hence not quantized and they vary continuously. This is contrary to the experience so we impose that the physical states satisfy in addition $W_{1,2}|\,\tilde{p};\, m = 0, \ldots\rangle = 0$. Hence $W_\mu = -\lambda P_\mu$ and the invariant parameter λ is taken to define as the *spin* of the massless particle. From $-W^+(\tilde{p})/\tilde{p}^+ = J_3$ we conclude that λ assumes half-integer values as well. We note that $W^\mu W_\mu = \lambda^2P^\mu P_\mu = 0$ and that on the LF the definition of the spin operator appears unified for massless and massive particles. A parallel discussion based on $p^- \neq 0$ may also be given.

As an illustration consider the three particle state on the LF with the total eigenvalues p^+, λ and $p^\perp$. In the *standard frame* with $p^\perp = 0$ it may be written as $(|x_1p^+, k_1^\perp; \lambda_1\rangle|x_2p^+, k_2^\perp; \lambda_2\rangle|x_3p^+, k_3^\perp; \lambda_3\rangle\,)$ with $\sum_{i=1}^3 x_i = 1$, $\sum_{i=1}^3 k_i^\perp = 0$, and $\lambda = \sum_{i=1}^3 \lambda_i$. Aplying $e^{-i(\tilde{p}.\bar{B})/p^+}$ on it we obtain $(|x_1p^+, k_1^\perp + x_1p^\perp; \lambda_1\rangle|x_2p^+, k_2^\perp + x_2p^\perp; \lambda_2\rangle|x_3p^+, k_3^\perp + x_3p^\perp; \lambda_3\rangle\,)$ now with $p^\perp \neq 0$. The x_i and $k_i^\perp$ indicate relative (invariant) parameters and do not depend upon the reference frame. The x_i is the fraction of the total longitudinal momentum carried by the ith particle while $k_i^\perp$ its transverse momentum. The state of a pion with momentum $(p^+, p^\perp)$, for example, may be expressed as an expansion over the LF Fock states constituted by the different number of partons [3]

$$|\pi : p^+, p^\perp\rangle =$$

$$\sum_{n,\lambda} \int \bar{\Pi}_i \frac{dx_i d^2k^\perp{}_i}{\sqrt{x_i}\, 16\pi^3} |n : x_i p^+, x_i p^\perp + k^\perp{}_i, \lambda_i\rangle\, \psi_{n/\pi}(x_1, k^\perp{}_1, \lambda_1; x_2, \dots,),$$

where the summation is over all the Fock states n and spin projections λ_i, with $\bar{\Pi}_i dx_i = \Pi_i dx_i\ \delta(\sum x_i - 1)$, and $\bar{\Pi}_i d^2k_i^\perp = \Pi_i dk_i^\perp\ \delta^2(\sum k_i^\perp)$. The wave function of the parton $\psi_{n/\pi}(x, k^\perp)$ indicates the probability amplitude for finding inside the pion the partons in the Fock state n carrying the 3-momenta $(x_i p^+, x_i p^\perp + k_i^\perp)$. The Fock state of the pion is also off the energy shell: $\sum k_i^- > p^-$.

The *discrete symmetry* transformations may also be defined on the LF Fock states [17]. For example, under the conventional parity $\mathcal{P}$ the spin operator $\mathcal{J}_3$ is not left invariant. We may rectify this by defining *LF Parity operation* by $\mathcal{P}^{lf} = e^{-i\pi J_1}\mathcal{P}$. We find then $B_1 \to -B_1, B_2 \to B_2, P^\pm \to P^\pm, P^1 \to -P^1, P^2 \to P^2$ etc. such that $\mathcal{P}^{lf}|p^+, p^\perp; m, s, \lambda, \dots\rangle \simeq |p^+, -p^1, p^2; m, s, -\lambda, \dots\rangle$. Similar considerations apply for charge conjugation and time inversion. For example, it is straightforward to construct the free *LF Dirac spinor* $\chi(p) = [\sqrt{2}p^+\Lambda^+ + (m - \gamma^a p^a)\,\Lambda^-]\tilde{\chi}/\sqrt{\sqrt{2}p^+ m}$ which is also an eigenstate of $\mathcal{J}_3$ with eigenvalues $\pm 1/2$. Here $\Lambda^\pm = \gamma^0\gamma^\pm/\sqrt{2} = \gamma^\mp\gamma^\pm/2 = (\Lambda^\pm)^\dagger$, $(\Lambda^\pm)^2 = \Lambda^\pm$, and $\chi(\tilde{p}) \equiv \tilde{\chi}$ with $\gamma^0\tilde{\chi} = \tilde{\chi}$. The conventional (equal-time) spinor can also be constructed by the procedure analogous to that followed for the LF spinor and it has the well known form $\chi_{con}(p) = (m + \gamma.p)\tilde{\chi}/\sqrt{2m(p^0 + m)}$. Under the conventional parity operation $\mathcal{P}$: $\chi'(p') = c\gamma^0\chi(p)$ (since we must require $\gamma^\mu = L^\mu{}_\nu\, S(L)\gamma^\nu S^{-1}(L)$, etc.). We find $\chi'(p) = c[\sqrt{2}p^-\Lambda^- + (m - \gamma^a p^a)\,\Lambda^+]\,\tilde{\chi}/\sqrt{\sqrt{2}p^- m}$. For $p \neq \tilde{p}$ it is not proportional to $\chi(p)$ in contrast to the result in the case of the usual spinor where $\gamma^0\chi_{con}(p^0, -\vec{p}) = \chi_{con}(p)$ for $E > 0$ (and $\gamma^0\eta_{con}(p^0, -\vec{p}) = -\eta_{con}(p)$ for $E < 0$). However, applying parity operator twice we do show $\chi''(p) = c^2\chi(p)$ hence leading to the usual result $c^2 = \pm 1$. The LF parity operator over spin 1/2 Dirac spinor is $\mathcal{P}^{lf} = c\,(2J_1)\,\gamma^0$ and the corresponding transform of χ is shown to be an eigenstate of $\mathcal{J}_3$.

Appendix C. SSB Mechanism. Continuum limit of discretized LF quantized theory. Nonlocality of LF Hamiltonian

The existence of the continuum limit [18, 15] of the Discretized Light Cone Quantized (DLCQ) [12] theory, the nonlocal nature of the LF Hamiltonian, and the description of the SSB on the LF were clarified [16] only recently.

Consider first the two dimensional case with $\mathcal{L} = [\dot{\phi}\phi' - V(\phi)]$. The eq. of motion, $\dot{\phi}' = (-1/2)\delta V(\phi)/\delta\phi$, shows that ϕ = const. is a possible solution. We write $\phi(x,\tau) = \omega(\tau) + \varphi(x,\tau)$. The general case of $\omega = \omega(\tau)$ can be made [15] but to make the discussion here short we assume that ω is a constant so that $\mathcal{L} = \dot{\varphi}\varphi' - V(\phi)$. Dirac procedure is applied now to construct Hamiltonian field theory which may be quantized. We may avoid using distribuitions if we restrict x to a finite interval from $-R/2$ to $R/2$. The *physical limit to the continuum* ($R \to \infty$), however, must be taken later to remove the spurious finite volume effects. Expanding φ by Fourier series we obtain $\phi(\tau,x) \equiv \omega + \varphi(\tau,x) = \omega + \frac{1}{\sqrt{R}}q_0(\tau) + \frac{1}{\sqrt{R}}\sum'_{n\neq 0} q_n(\tau)\, e^{-ik_n x}$ where $k_n = n(2\pi/R)$, $n = 0, \pm 1, \pm 2, \ldots$ and the *discretized theory* Lagrangian becomes $i\sum_n k_n\, q_{-n}\, \dot{q}_n - \int dx\; V(\phi)$. The momenta conjugate to q_n are $p_n = ik_n q_{-n}$ and the canonical LF Hamiltonian is found to be $\int dx\, V(\omega + \varphi(\tau,x))$. The primary constraints are thus $p_0 \approx 0$ and $\Phi_n \equiv p_n - ik_n q_{-n} \approx 0$ for $n \neq 0$. We follow the *standard* Dirac procedure [5] and find three weak constraints $p_0 \approx 0$, $\beta \equiv \int dx\, V'(\phi) \approx 0$, and $\Phi_n \approx 0$ for $n \neq 0$ on the phase space and they are shown to be second class. We find for $n \neq 0$ and $m \neq 0$: $\{\Phi_n, p_0\} = 0$, $\{\Phi_n, \Phi_m\} = -2ik_n\delta_{m+n,0}$, $\{\Phi_n, \beta\} = \{p_n, \beta\} = -(1/\sqrt{R})\int dx\;[V''(\phi) - V''([\omega + q_0]/\sqrt{R})]\, e^{-ik_n x} \equiv -\alpha_n/\sqrt{R}$, $\{p_0, p_0\} = \{\beta, \beta\} = 0$, $\{p_0, \beta\} = -(1/\sqrt{R})\int dx\;\, V''(\phi) \equiv -\alpha/\sqrt{R}$. Implement first the pair of constraints $p_0 \approx 0$, $\beta \approx 0$ by modifying the Poisson brackets to the star bracket $\{\}^*$ defined by $\{f,g\}^* = \{f,g\} - [\{f,p_0\}\,\{\beta,g\} - (p_0 \leftrightarrow \beta)](\alpha/\sqrt{R})^{-1}$. We may then set $p_0 = 0$ and $\beta = 0$ as strong equalities. We find by inspection that the brackets $\{\}^*$ of the remaining variables coincide with the standard Poisson brackets except for the ones involving q_0 and p_n $(n \neq 0)$: $\{q_0, p_n\}^* = \{q_0, \Phi_n\}^* = -(\alpha^{-1}\alpha_n)$. For example, if $V(\phi) = (\lambda/4)\,(\phi^2 - m^2/\lambda)^2$, $\lambda \geq 0, m \neq 0$ we find $\{q_0, p_n\}^*\, [\{3\lambda\,(\omega + q_0/\sqrt{R})^2 - m^2\}R + 6\lambda(\omega + q_0/\sqrt{R})\int dx\varphi + 3\lambda\int dx\,\varphi^2] = -3\lambda\,[2(\omega + q_0/\sqrt{R})\sqrt{R}q_{-n} + \int dx\,\varphi^2\, e^{-ik_n x}]$.

Implement next the constraints $\Phi_n \approx 0$ with $n \neq 0$. We have $C_{nm} = \{\Phi_n, \Phi_m\}^* = -2ik_n\delta_{n+m,0}$ and its inverse is given by ${C^{-1}}_{nm} = (1/2ik_n)\delta_{n+m,0}$. The Dirac bracket which takes care of all the constraints is then given by

$$\{f,g\}_D = \{f,g\}^* - \sum\nolimits'_n \frac{1}{2ik_n}\{f, \Phi_n\}^*\,\{\Phi_{-n}, g\}^*,$$

where we may now in addition write $p_n = ik_n q_{-n}$. It is easily shown that $\{q_0, q_0\}_D = 0$, $\{q_0, p_n\}_D = \{q_0, ik_n q_{-n}\}_D = \frac{1}{2}\{q_0, p_n\}^*$, $\{q_n, p_m\}_D = \frac{1}{2}\delta_{nm}$.

The limit to the continuum, $R \to \infty$ is taken as usual: $\Delta = 2(\pi/R) \to dk$, $k_n = n\Delta \to k$, $\sqrt{R}q_{-n} \to \lim_{R\to\infty}\int_{-R/2}^{R/2} dx\,\varphi(x)\,e^{ik_n x} \equiv \int_{-\infty}^{\infty} dx\,\varphi(x)\,e^{ikx} = \sqrt{2\pi}\tilde{\varphi}(k)$ for all n, $\sqrt{2\pi}\varphi(x) = \int_{-\infty}^{\infty} dk\,\tilde{\varphi}(k)e^{-ikx}$, and $(q_0/\sqrt{R}) \to 0$. From $\{\sqrt{R}\,q_m, \sqrt{R}\,q_{-n}\}_D = R\,\delta_{nm}/(2ik_n)$ following from $\{q_n, p_m\}_D$ for $n, m \neq 0$ we derive, on using $R\delta_{nm} \to \int_{-\infty}^{\infty} dx e^{i(k-k')x} = 2\pi\delta(k-k')$, that $\{\tilde{\varphi}(k), \tilde{\varphi}(-k')\}_D = \delta(k-k')/(2ik)$ where $k, k' \neq 0$. If we use the integral representation of the sgn function the well known LF Dirac bracket $\{\varphi(x,\tau), \varphi(y,\tau)\}_D = -\frac{1}{4}\epsilon(x-y)$ is obtained. The expressions of $\{q_0, p_n\}_D$ (or $\{q_0, \varphi'\}_D$) show that the DLCQ is harder to work with here. The continuum limit of the *constraint eq.* $\beta = 0$ is

$$\lim_{R\to\infty} \frac{1}{R}\int_{-R/2}^{R/2} dx\, V'(\phi) \equiv$$

$$\omega(\lambda\omega^2 - m^2) + \lim_{R\to\infty} \frac{1}{R}\int_{-R/2}^{R/2} dx\Big[\,(3\lambda\omega^2 - m^2)\varphi + \lambda(3\omega\varphi^2 + \varphi^3)\,\Big] = 0$$

while that for the LF Hamiltonian is ($P^- \equiv H^{l.f.}$)

$$P^- = \int dx\left[\omega(\lambda\omega^2 - m^2)\varphi + \frac{1}{2}(3\lambda\omega^2 - m^2)\varphi^2 + \lambda\omega\varphi^3 + \frac{\lambda}{4}\varphi^4\right].$$

These results follow immediately if we worked directly in the continuum formulation; we do have to handle generalized functions now. In the LF Hamiltonian theory we have an additional new ingredient in the form of the *constraint equation*. Elimination of ω using it would lead to a *nonlocal LF Hamiltonian* corresponding to the local one in the equal-time formulation. At the tree or classical level the integrals appearing in in the constraint eq. are convergent and when $R \to \infty$ it leads to $V'(\omega) = 0$. In equal-time theory this is essentially *added* to it as an external constraint based on physical considerations. In the renormalized theory [16] the constraint equation describes the high order quantum corrections to the tree level value of the condensate.

The quantization is performed via the correspondence $i\{f, g\}_D \to [f, g]$. Hence $\varphi(x,\tau) = (1/\sqrt{2\pi})\int dk\, \theta(k)\, [a(k,\tau)\, e^{-ikx} + a^\dagger(k,\tau)e^{ikx}]/(\sqrt{2k})$, were $a(k,\tau)$ and $a^\dagger(k,\tau)$ satisfy the canonical equal-τ commutation relations, $[a(k,\tau), a(k',\tau)^\dagger] = \delta(k-k')$ etc.. The vacuum state is defined by $a(k,\tau)|vac\rangle = 0$, $k > 0$ and the tree level description of the *SSB* is given as follows. The values of $\omega = \langle|\phi|\rangle_{vac}$ obtained from $V'(\omega) = 0$ the different vacua in the theory. Distinct Fock spaces corresponding to different values of ω are built as usual by applying the creation operators on the corresponding vacuum state. The $\omega = 0$ corresponds to a *symmetric phase* since the Hamiltonian is then symmetric under $\varphi \to -\varphi$. For $\omega \neq 0$ this symmetry is violated and the system is in a *broken or asymmetric phase.*

The extension to 3+1 dimensions and to global continuous symmetry is straightforward [16, 15]. Consider real scalar fields $\phi_a (a = 1, 2, \ldots, N)$ which form an isovector of global internal symmetry group $O(N)$. We now write $\phi_a(x, x^\perp, \tau) = \omega_a + \varphi_a(x, x^\perp, \tau)$ and the Lagrangian density is $\mathcal{L} = [\dot\varphi_a\varphi_a' - (1/2)(\partial_i\varphi_a)(\partial_i\varphi_a) - V(\phi)]$, where $i = 1, 2$ indicate the transverse space directions. The Taylor series expansion of the constraint equations $\beta_a = 0$ gives a set of coupled eqs. $R\, V_a'(\omega) + V_{ab}''(\omega)\int dx\varphi_b + V_{abc}'''(\omega)\int dx\varphi_b\varphi_c/2 + \cdots = 0$. Its discussion at the tree level leads to the conventional theory results. The LF symmetry generators are found to be $G_\alpha(\tau) = -i\int d^2x^\perp dx\varphi_c'(t_\alpha)_{cd}\varphi_d = \int d^2k^\perp\, dk\, \theta(k)\, a_c(k, k^\perp)^\dagger (t_\alpha)_{cd} a_d(k, k^\perp)$ where $\alpha, \beta = 1, 2, \ldots, N(N-1)/2$, are the group indices, t_α are hermitian and antisymmetric generators of $O(N)$, and $a_c(k, k^\perp)^\dagger$ ($a_c(k, k^\perp)$) is creation (destruction) operator contained in the momentum space expansion of φ_c. These are to be contrasted with the generators in the equal-time theory, $Q_\alpha(x^0) = \int d^3x\, J^0 = -i\int d^3x(\partial_0\varphi_a)(t_\alpha)_{ab}\varphi_b - i(t_\alpha\omega)_a\int d^3x(d\varphi_a/dx_0)$. All the symmetry generators thus annihilate the LF vacuum and the SSB is now seen in the broken symmetry of the quantized theory Hamiltonian. The criteria for the counting of the number of Goldstone bosons on the LF is found to be the same as in the conventional theory. In contrast, the first term on the right hand side of $Q_\alpha(x^0)$ does annihilate the conventional theory vacuum but the second term gives now non-vanishing contributions for some of the (broken) generators. The symmetry of the conventional theory vacuum is thereby broken while the quantum Hamiltonian remains invariant. The physical content of SSB in the *instant form* and the *front form*, however, is the same though achieved by different descriptions. Alternative proofs on the LF, in two dimensions, can be given of the Coleman's theorem related to the absence of Goldstone bosons and of the pathological nature of massless scalar theory; we are unable to implement the second class constraints over the phase space.

We remark that the simplicity of the LF vacuum is in a sense compensated by the involved nonlocal Hamiltonian. The latter, however, may be treatable using advance computational techniques. In a recent work [16] it was also shown that renormalized theory may be constructed without the need of first solving the constraint eq. for ω. Instead we may perform renormalization and obtain a renormalized constraint equation.

Appendix D. Commutators for equal-x^-

The LF formulation is symmetrical with respect to x^+ and x^- and it is a matter of convention that we take the plus component as the LF *time* while the other as a spatial coordinate. The theory quantized at $x^+ =$ const . hyperplanes, however, does seem to already incorporate in it the information from the equal-x^- quantized theory.

For illustration we consider the two dimensional massive free scalar theory. The LF quantization, assuming x^+ as the LF *time* coordinates, leads to $\omega = 0$ and the equal-x^+ commutator $[\varphi(x^+, x^-), \varphi(x^+, y^-] = -i\epsilon(x^- - y^-)/4$. The commutator can be realized in the momentum space through the expansion (Appendix C)

$$\varphi(x^+, x^-) = \frac{1}{\sqrt{2\pi}}\int_{k^+>0}^{\infty}\frac{dk^+}{\sqrt{2k^+}}\Big[a(k^+)e^{-i(k^+x^-+k^-x^+)} + a^\dagger(k^+)e^{i(k^+x^-+k^-x^+)}\Big],$$

where $[a(k^+), a(l^+)^\dagger] = \delta(k^+ - l^+)$ etc. and $2k^+k^- = m^2$. It is then easy to show

$$[\varphi(x^+, x^-), \varphi(y^+, x^-)] = \frac{1}{2\pi}\int_{k^+>0}^{\infty}\frac{dk^+}{2k^+}\Big[e^{ik^-(y^+-x^+)} - e^{-ik^-(y^+-x^+)}\Big].$$

We may change the integration variable to k^- by making use of $k^-dk^+ + k^+dk^- = 0$. Hence on employing the integral representation $\epsilon(x) = (i/\pi)\mathcal{P}\int_{-\infty}^{\infty}(d\lambda/\lambda)\,e^{(-i\lambda x)}$

we arrive at the equal-x^- commutator

$$[\varphi(x^+, x^-), \varphi(y^+, x^-)] = -\frac{i}{4}\epsilon(x^+ - y^+).$$

The above field expansion on the LF, in contrast to the equal-time case, does not involve the mass parameter m and the same result follows in the massless case also if we assume that $k^+ = l^+$ implies $k^- = l^-$. Defining the right and the left movers by $\varphi(0, x^-) \equiv \varphi^R(x^-)$, and $\varphi(x^+, 0) \equiv \varphi^L(x^+)$ we obtain $[\varphi^R(x^-), \varphi^R(y^-)] = (-i/4)\epsilon(x^- - y^-)$ while $[\varphi^L(x^+), \varphi^L(y^+)] = (-i/4)\epsilon(x^+ - y^+)$.

References

[1] See for example, Abdalla, E., Abdalla, M. C., Rothe, K., Non-Perturbative Methods in Two Dimensional Quantum Field Theory, World Scientific, Singapore 1991, and references therein.

[2] Abdalla, E., Abdalla, M. C.. Internat. J. Moderen Phys. A10 (1995), 1611.

[3] Brodsky, S. J., Pauli, H. C., Light-cone Quantization and QCD, Lecture Notes in Phys. 396 (H. Mitter et. al., eds.), Springer-Verlag, Berlin 1991; Brodsky, S. J., Lepage, G. P., in: Perturbative Quantum Chromodynamics (A. H. Mueller, ed.), World Scientific, Singapore 1989.

[4] Dirac, P. A. M., Rev. Modern Phys. 21 (1949), 392.

[5] Dirac, P. A. M., Lectures in Quantum Mechanics, Benjamin, New York 1964; Sudarshan, E. C. G., Mukunda, N., Classical Dynamics: a modern perspective, Wiley, New York 1974; Hanson, A., Regge T., Teitelboim, C., Constrained Hamiltonian Systems, Atti Accad. Naz. Lincei Mem. Cl. Sci. Fis. Mat. Natur. Sez., Roma 1976.

[6] See for example, 't Hooft, G., in: New Phenomena in Subnuclear Physics 13A (A. Zichichi, ed.), Plenum, New York 1977.

[7] Kogut, J. B., Soper, D. E., Phys. Rev. D1 (1970), 2901.

[8] Kogut, J., Susskind, L., Phys. Rev. D11 (1975), 3594, and refeferences therein.

[9] Lowenstein, J., Swieca, J., Ann. Physics (N.Y.) 68 (1971), 172.

[10] Mattis, D., Lieb, E., J. Math. Phys. 6 (1965), 304; Coleman, S., Phys. Rev. D11 (1975), 2088; Affleck, I., Nuclear Phys. B 265 [FS15] (1986), 409; see also [8].

[11] McCartor, G., Z. Phys. C64 (1994), 349, and references therein; Vary, J. P., Fields, T. J., Pirner, H. J., Phys. Rev. D53 (1996), 7231.

[12] Pauli, H. C., Brodsky, S. J., Phys. Rev. D32 (1985), 1993.

[13] Perry, R. J., Hamiltonian Light-Front Field Theory and QCD, in: Hadron Physics 94 (Herscovitz, V. et. al., eds.) 120, World Scientific, Singapore, 1995.

[14] Schwinger, J., Phys. Rev. 128 (1972), 2425; Theoretical Physics, I.A.E.A, Vienna (1963), 89.

[15] Srivastava, P. P., Lectures on light-front quantized field theory, in: Proc. XIV Encontro Nacional de Partículas e Campos, Caxambu, MG, 154–192, Sociedade Brasileira de Física, SP, Brasil, 1993, and references therein (hep-th@xxx.lanl.gov/ 9312064).

[16] ——, On spontaneous symmetry breaking mechanism in light-front quantized field theory, Ohio State Univ. preprint 91-0481, SLAC database no. PPF-9148, November 1991; Light-front Quantization and SSB, in: Hadron Physics 94, 253; Phase Transition in Scalar Theory Quantized on the Light-Front, ibid, 257, (Herscovitz, V. et. al., eds.), World Scientific, Singapore 1995 (hep-th@xxx.lanl.gov. no. 9412204 and 205); Nuovo Cimento A107 (1994), 549; A108 (1995), 35.

[17] ——, Light-front quantization of field theory, in: Topics in Theoretical Physics, Festschrift for Paulo Leal Ferreira (V. C. Aguilera-Navarro et. al., eds.), IFT-São Paulo, SP, Brasil (1995), 206–217, (hep-th@xxx.lanl.gov/9610044).

[18] ——, Spontaneous symmetry breaking mechanism in light-front quantized field theory - *Discretized formulation*, Ohio State Univ. preprint 92-0173, SLAC PPF-9222, April 1992, Contributed to XXVI Internat. Conf. on High Energy Physics, Dallas, Texas, August 92, (AIP Conference Proceedings 272 (J. R. Sanford, ed.), 2125, hep-th@xxx.lanl.gov/ 9412193.

[19] ——, work in progress.

[20] Weinberg, S., Phys. Rev. 150 (1966), 1313.

[21] Wilson, K. G., Nucl. Phys. B (Proc. Suppl.) 17 (1990); Perry, R. J., Harindranath, A., Wilson, K. G., Phys. Rev. Lett. 65 (1990), 2959.

Clifford algebras and Witten's monopole equations

Jayme Vaz, Jr.

Abstract. In this paper Witten's monopole equations are analysed and different representations of it are obtained. First, we give a general introduction to the theory of Clifford algebras and introduce the concept of operator spinors. Witten's monopole equations are then reformulated in terms of quaternions and of those operator spinors. From these representations we obtain another one which is in Maxwell equations form. The monopole solution is then studied. Two other representations are obtained. In one of these representations the spinor is completely eliminited, while in the other we obtain a much simple spinor representation but in terms of a SU(2) connection.

1. Introduction

In 1994 Witten [1] announced a new theory of 4-manifolds. While in Donaldson theory we study the solutions of the SU(2) Yang–Mills equations, in Witten's theory we study the solutions of a system of equations based on a U(1) connection that are now called Witten's monopole equations. Our objective in this paper is to make a thoughtful study of Witten's monopole equations, in particular of the mathematical structure underlying its formulation and of different representations of it that may lead to new developments and interpretations.

We give in Section 2 a general discussion of Clifford algebras, and in particular of the Clifford algebra of the 4 dimensional space that is the structure underlying Witten's monopole equations. We introduce and discuss in some details the concept of operator spinors, that will be very usuful in our discussions.

In Section 3 we discuss the Witten's monopole equations. First we obtain a representation of it in terms of operators spinors, from which one can also obtain a representation in terms of quaternions as follows from our discussions in Section 2. Then we obtain a representation of it in a Maxwell equation form. The equation obtained in this way is a Maxwell equation including magnetic monopoles, as expected. The magnetic monopole solution is obtained, and the spinor related to it is obtained from a rather unusual but powerful and interesting approach that is what we call inversion (or recovering) theorem. This inversion theorem is also used in order to put the Maxwell-like representation of Witten's equation in a complete "electromagnetic form", that is, we eliminate the presence of the spinor from those equations and write the equations only in terms of electromagnetic quantities that are the electromagnetic field and the fields invariants. Finally, we obtain another representation which seems to be very usuful. This other representation is again a spinor representation, but the spinor has only one degree of freedom and it involves a SU(2) connection that is related to the spinor connection.

Acknoledgements. We would like to thank Professors A. P. Balachandran and W. A. Rodrigues for discussions, and FAPESP for the finnancial support.

2. Mathematical preliminaries

2.1. Clifford algebras

We start by defining a Clifford algebra [2, 3]. Let V be a vector space over $\mathbb{F} = \mathbb{R}, \mathbb{C}$ endowed with a quadratic form $Q(x)$, $x \in V$. Let A be an associative algebra with unit element 1_A. A Clifford map from V to A is a linear map φ which satisfies

$$[\varphi(x)]^2 = Q(x)1_A, \quad \forall x \in V, \tag{2.1}$$

or equivalently,

$$\varphi(x)\varphi(y) + \varphi(y)\varphi(x) = 2g(x, y)1_A, \tag{2.2}$$

where $2g(x, y) = Q(x + y) - Q(x) - Q(y)$, $x, y \in V$. A *Clifford algebra* over V is an associative algebra $\mathcal{C}\ell(V, g)$ with unit element 1 together with a Clifford map $\iota_V : V \to \mathcal{C}\ell(V, g)$ such that for every Clifford map $\varphi : V \to A$ there exists a unique homomorphism $f : \mathcal{C}\ell(V, g) \to A$ such that $\varphi = f \circ \iota_V$.

This last property characterizes what we call an *universal* Clifford algebra. We may also call it *orthogonal* Clifford algebra, in distinction to *symplectic* Clifford algebras [4]. From now on we speak only of Clifford algebras, having in mind that we are dealing with universal orthogonal Clifford algebras.

It is a well-known fact [2–6] that all Clifford algebras are semi-simple. Since all semi-simple algebras are the direct sum of two simple algebras, and from Weddenburn theorem (that is: if A is a simple algebra then it is equivalent to $\mathcal{M}(n, \mathbb{K})$, where $\mathbb{K}$ is a division algebra and n and $\mathbb{K}$ are unique (modulo isomorphisms)[3]), it follows that from the point of view of representation theory that Clifford algebras are isomorphic either to $\mathcal{M}(n, \mathbb{K})$ or $\mathcal{M}(n, \mathbb{K}) \oplus \mathcal{M}(n, \mathbb{K})$, where $\mathcal{M}(n, \mathbb{K})$ is the algebra of $n \times n$ matrices with entries in $\mathbb{K} = \mathbb{R}, \mathbb{C}, \mathbb{H}$. The classification table for all real and complex Clifford algebras can be found in the above quoted references. These matrix algebras are identified with the algebra of endomorphims of a vector space S called *space of spinors*.

From the algebraic point of view, spinors are elements of *minimal left ideals* of Clifford algebras [7]. In fact, consider the *left* regular representation $L : C\ell(V, g) \to \text{End}(C\ell(V, g))$, $L(a)b = ab$. Since the minimal left ideals transform irreducibly under this left regular representation, we arrive at the above mentioned algebraic definition. Moreover, since given a vector space we have naturally associated to it its *dual space*, then the dual space of the space of spinors is from the algebraic point of view a *minimal right ideal*. These minimal right ideals transform irreducibly under the *right* regular representation.

One of the reasons why Clifford algebras are very usuful is that they can be identified as a subalgebra of the *algebra of endomorphisms* of the *exterior (or Grassmann) algebra*. Let us see this is details. Let us consider the vector space $V = \text{span}\{v_1, v_2, \ldots, v_n\}$ and its dual $V^* = \text{span}\{\theta^1, \theta^2, \ldots, \theta^n\}$. Let $\wedge$ be the exterior product. The object $v_1 \wedge \cdots \wedge v_k$ is called a k-vector, and the vector space of k-vectors is denoted by $\bigwedge^k(V)$. Analogously, we call the object $\theta^1 \wedge \cdots \wedge \theta^k$ a k-covector, and denote the vector space by $\bigwedge^k(V^*)$. From the properties of the exterior product it follows that

$$(\theta^1 \wedge \cdots \wedge \theta^k)(v_1 \wedge \cdots \wedge v_k) = \det\{\theta^i(v_j)\}. \tag{2.3}$$

In fact, once we define the exterior product of vectors we can use the above expression in order to define the exterior product of covectors, and vice-versa. The exterior algebra of the vector space is $(\bigwedge(V), \wedge)$, where $\bigwedge(V) = \oplus_{k=0}^n \bigwedge^k(V)$, where $n = \dim(V)$ and we denoted $\bigwedge^0(V) = \mathbb{F} = \mathbb{R}, \mathbb{C}$ and $\bigwedge^1(V) = V$. We define $\bigwedge(V^*)$ in an analogous way.

Now we define [8] $\mathbf{e}^*$ $[\mathbf{e}]$: $V\,[V^*] \to \text{End}(\bigwedge(V))\,[\text{End}(\bigwedge(V^*))]$ as

$$\mathbf{e}_v^* u = v \wedge u, \quad \mathbf{e}_\theta \omega = \theta \wedge \omega, \tag{2.4}$$

and $\mathbf{i}\,[\mathbf{i}^*]$: $V\,[V^*] \to \text{End}(\bigwedge(V^*))\,[\text{End}(\bigwedge(V))]$ by

$$(\mathbf{i}_v \omega)(v_1 \wedge \cdots \wedge v_k) = \omega(\mathbf{e}_v v_1 \wedge \cdots \wedge v_k), \tag{2.5}$$

$$\omega(\mathbf{i}_\theta^*(v_1 \wedge \cdots \wedge v_k)) = (\mathbf{e}_\theta \omega)(v_1 \wedge \cdots \wedge v_k), \tag{2.6}$$

where we used the usuful notation $\mathbf{e}_v^* = \mathbf{e}^*(v)$, etc. From the definition of $\mathbf{i}$ and the properties of the exterior product it follows the following properties:

$$\mathbf{i}_v(\theta) = \theta(v), \quad \mathbf{i}_v(1) = 0, \tag{2.7}$$

$$\mathbf{i}_v(\theta_1 \wedge \cdots \wedge \theta_k) = \sum_{i=1}^k (-1)^{i+1} (\mathbf{i}_v \theta_i) \theta_1 \wedge \cdots \hat{\theta}_i \cdots \wedge \theta_k, \tag{2.8}$$

$$\mathbf{i}_{v_1 \wedge \cdots \wedge v_k} = \mathbf{i}_{v_k} \cdots \mathbf{i}_{v_1}, \tag{2.9}$$

where the hat in eq.(2.8) means omission, and with analogous expressions for $\mathbf{i}_\theta^*$.

From the above definitions one can easily find that the following commutation relations hold:

$$\mathbf{e}_{\theta_1} \mathbf{e}_{\theta_2} + \mathbf{e}_{\theta_2} \mathbf{e}_{\theta_1} = 0, \tag{2.10}$$

$$\mathbf{i}_{v_1}\mathbf{i}_{v_2} + \mathbf{i}_{v_2}\mathbf{i}_{v_1} = 0, \tag{2.11}$$

$$\mathbf{e}_\theta \mathbf{i}_v + \mathbf{i}_v \mathbf{e}_\theta = \theta(v), \tag{2.12}$$

which is an algebra like the one of fermionic creation and annihilation operators.

Now, suppose we have defined $g : V \times V \to \mathbb{F}$. Then we can define $\tau : V \to V^*$ by

$$\tau(v) = g(v, \ ) = g(\ , v). \tag{2.13}$$

If we define $\boldsymbol{\gamma}_v^*$ as

$$\gamma_v^* = \mathbf{i}^*_{\tau(v)} + \mathbf{e}_v^*, \tag{2.14}$$

then we have that $C\ell(V, g)$ can be represented as a subalgebra of $\operatorname{End}(\bigwedge(V))$. In fact, using the above definition and the commutation relations (2.10–2.12) we have that

$$\gamma_v^* \gamma_u^* + \gamma_u^* \gamma_v^* = 2g(v, u), \tag{2.15}$$

where $g(v, u) = \tau(v)(u) = \tau(u)(v)$. We see therefore that $\gamma^* = \mathbf{i}^* \circ \tau + \mathbf{e}^*$ is a Clifford map, and the Clifford algebra is $C\ell(V, g)$.

In the same way we have a representation of the Clifford algebra $C\ell(V^*, g^{-1})$ in terms of the Clifford map $\gamma = \mathbf{i} \circ \tau^{-1} + \mathbf{e}$. In fact, let $\tau^{-1}(\theta) = g^{-1}(\theta, \ ) = g^{-1}(\ , \theta)$, where $g^{-1} : V^* \times V^* \to \mathbb{R}$, and define

$$\gamma_\theta = \mathbf{i}_{\tau^{-1}(\theta)} + \mathbf{e}_\theta. \tag{2.16}$$

Then we get that

$$\gamma_\theta \gamma_\omega + \gamma_\omega \gamma_\theta = 2g^{-1}(\theta, \omega). \tag{2.17}$$

Let now $\{e_\mu\}$ $(\mu = 0, 1, \ldots, n-1)$ be an orthonormal basis of V and such that

$$g(e_\mu, e_\nu) = g_{\mu\nu} = \operatorname{diag}(\underbrace{1, \ldots, 1}_{p}, \underbrace{-1, \ldots, -1}_{q}). \tag{2.18}$$

Let us denote the corresponding Clifford algebra by $C\ell_{p,q}$. We also denote

$$\boldsymbol{e}_\mu = \gamma^*_{e_\mu}, \quad \boldsymbol{e}^\mu = \gamma_{e^\mu}, \tag{2.19}$$

where $e^\mu = g^{\mu\nu} e_\nu$ and $\boldsymbol{e}^\mu = g^{\mu\nu} \boldsymbol{e}_\nu$. It follows therefore that

$$\boldsymbol{e}_\mu \boldsymbol{e}_\nu + \boldsymbol{e}_\nu \boldsymbol{e}_\mu = 2g_{\mu\nu}, \tag{2.20}$$

and

$$\boldsymbol{e}_\mu \boldsymbol{e}_\nu - \boldsymbol{e}_\nu \boldsymbol{e}_\mu = 2(\boldsymbol{e}_\mu \wedge \boldsymbol{e}_\nu), \tag{2.21}$$

in such a way that

$$\boldsymbol{e}_\mu \boldsymbol{e}_\nu = g_{\mu\nu} + \boldsymbol{e}_\mu \wedge \boldsymbol{e}_\nu, \tag{2.22}$$

or

$$\boldsymbol{v}\boldsymbol{u} = g(\boldsymbol{v}, \boldsymbol{u}) + \boldsymbol{v} \wedge \boldsymbol{u}. \tag{2.23}$$

It is convenient now to introduce the following notation:

$$\boldsymbol{e}_{\mu_1 \ldots \mu_k} = \boldsymbol{e}_{\mu_1} \ldots \boldsymbol{e}_{\mu_k} \quad (\mu_1 \neq \cdots \neq \mu_k). \tag{2.24}$$

As vector spaces we have therefore that $\mathcal{C}\ell_{p,q} \simeq \bigwedge(\mathbb{R}^{p,q})$. The Clifford algebra $\mathcal{C}\ell_{p,q}$ is therefore a Z_2-graded algebra. We have that $\boldsymbol{e}_{\mu_1} \dots \boldsymbol{e}_{\mu_k}$ is a k-vector ($\mu_1 \neq \dots \neq \mu_k$). For $A_k \in \bigwedge^k(\mathbb{R}^{p,q})$ we define the graded involution as

$$\hat{A}_k = (-1)^k A_k, \quad A_k \in \bigwedge{}^k(\mathbb{R}^{p,q}), \tag{2.25}$$

and such that the Z_2-graded structure of $\mathcal{C}\ell_{p,q}$ is provided by the graded involution, i.e., $\mathcal{C}\ell_{p,q} = \mathcal{C}\ell^+_{p,q} \oplus \mathcal{C}\ell^-_{p,q}$, where $\mathcal{C}\ell^{\pm}_{p,q} = \{A \in \mathcal{C}\ell_{p,q} \mid \hat{A} = \pm A\}$. We can easily see that $\mathcal{C}\ell^+_{p,q}$ is a subalgebra of $\mathcal{C}\ell_{p,q}$ called even subalgebra. We also define an antiautomorphism which we call reversion as

$$\tilde{A}_k = (-1)^{k(k-1)/2} A_k, \quad A_k \in \bigwedge{}^k(\mathbb{R}^{p,q}). \tag{2.26}$$

The name reversion is due to the fact that $(\boldsymbol{e}_{\mu_1} \dots \boldsymbol{e}_{\mu_k})^{\sim} = \boldsymbol{e}_{\mu_k} \dots \boldsymbol{e}_{\mu_1}$ ($\mu_1 \neq \dots \neq \mu_k$), and reordering gives the factor $(-1)^{\binom{k}{2}} = (-1)^{k(k-1)/2}$.

In order to exploit the multivector structure of $\mathcal{C}\ell_{p,q}$ it is convenient to introduce the projectors $\langle\ \rangle_k : \mathcal{C}\ell_{p,q} \simeq \bigwedge(\mathbb{R}^{p,q}) \to \bigwedge^k(\mathbb{R}^{p,q})$. We have therefore that

$$\langle \boldsymbol{e}_{\mu_1 \dots \mu_k} \rangle_l = \delta_{kl} \boldsymbol{e}_{\mu_1 \dots \mu_k}. \tag{2.27}$$

The generalization of eq.(2.22) is

$$\boldsymbol{v} A = \boldsymbol{v} \cdot A + \boldsymbol{v} \wedge A, \tag{2.28}$$

where

$$\boldsymbol{v} \cdot A = \mathbf{i}^*_v(A) = \frac{1}{2}(\boldsymbol{v} A - \hat{A} \boldsymbol{v}), \tag{2.29}$$

$$\boldsymbol{v} \wedge A = \frac{1}{2}(\boldsymbol{v} A + \hat{A} \boldsymbol{v}). \tag{2.30}$$

There are some general theorems about the structure of the Clifford algebras which are very usuful. In particular, we have that

$$\begin{gathered} \mathcal{C}\ell_{p+1,q} = \mathcal{C}\ell_{q+1,p}, \quad \mathcal{C}\ell_{p+1,q+1} = \mathcal{C}\ell_{p,q} \otimes \mathcal{C}\ell_{1,1}, \quad \mathcal{C}\ell_{p,q+4} = \mathcal{C}\ell_{p,q} \otimes \mathcal{C}\ell_{0,4}, \\ \mathcal{C}\ell^+_{p,q+1} = \mathcal{C}\ell_{p,q}, \quad \mathcal{C}\ell^+_{p+1,q} = \mathcal{C}\ell_{q,p}. \end{gathered} \tag{2.31}$$

Let us consider the invertible elements of $\mathcal{C}\ell_{p,q}$. The set $\Gamma(p,q)$ of invertible elements $U \in \mathcal{C}\ell_{p,q}$ such that $U v U^{-1} \in \mathbb{R}^{p,q}$ for $v \in \mathbb{R}^{p,q}$ form the so called Clifford group. We define the norm of these invertible elements by $\mathcal{N}(U) = \tilde{U}U = U\tilde{U}$. One can show [5, 6] that the $\mathrm{Spin}_+ p, q$ group, that is the double covering of the component of the special orthogonal group $\mathrm{SO}_+(p,q)$ containing the identity, is given by $\mathrm{Spin}_+(p,q) = \{U \in \Gamma(p,q) \mid \mathcal{N}(U) = 1\} \cap \mathcal{C}\ell^+_{p,q}$. For low dimensions we have a very important result, that is, for $p + q \leq 5$ we have $\mathrm{Spin}_+(p,q) = \{U \in \mathcal{C}\ell_{p,q} \mid \hat{U} = U, \tilde{U} = U^{-1}\}$.

Spinors can be defined algebraically as elements of a minimal left ideal. In relation to this, we need the following result [9]: A minimal left ideal $\mathcal{I}_{p,q}$ of $\mathcal{C}\ell_{p,q}$ is of the type $\mathcal{I}_{p,q} = \mathcal{C}\ell_{p,q} f_{p,q}$, where $f_{p,q} = (1/2)(1+e_{A_1})\dots(1/2)(1+e_{A_k})$ is a primitive idempotent of $\mathcal{C}\ell_{p,q}$, and where $e_{A_1},\dots,e_{A_k}$ is a set of commuting elements of the canonical basis of $\mathcal{C}\ell_{p,q}$ such that $(e_{A_i})^2 = 1$, $(i = 1,\dots,k)$, that generates a group of order $k = q - r_{q-p}$ where r_i are the Radon–Hurwitz numbers.

2.2. The Clifford algebra of the 4 dimensional euclidean space

Let us consider the euclidean space $\mathbb{R}^4$. Its Clifford algebra $\mathcal{C}\ell_4$ is the algebra generated by $\{e_\mu\}$ ($\mu = 0, 1, 2, 3$) subject to the conditions

$$e_\mu e_\nu + e_\nu e_\mu = 2\delta_{\mu\nu}. \tag{2.32}$$

An arbitrary element of $\mathcal{C}\ell_4$ is of the form

$$\begin{aligned}\Phi &= a + a_0e_0 + a_1e_1 + a_2e_2 + a_3e_3 + a_{01}e_{01} + a_{02}e_{02} + a_{03}e_{03} + a_{12}e_{12} + a_{13}e_{13}\\ &\quad + a_{23}e_{23} + a_{012}e_{012} + a_{013}e_{013} + a_{023}e_{023} + a_{123}e_{123} + a_{0123}e_5,\end{aligned} \tag{2.33}$$

where we denoted $e_5 = e_{0123} = e_0e_1e_2e_3$. The algebra $\mathcal{C}\ell_4$ is isomorphic to the algebra of 2×2 quaternionic matrices. In order to obtain an explicit matrix representation let us consider the primitive idempotent $f = (1/2)(1+e_0)$. Spinors are elements of $\mathcal{C}\ell_4 f$. Let us consider the spinorial basis $|A\rangle$ ($A = 1, 2$) defined by

$$|1\rangle = f, \quad |2\rangle = e_5 f. \tag{2.34}$$

Taking $(e_1e_2)|\ \rangle = \boldsymbol{i}|\ \rangle$, $(e_3e_1)|\ \rangle = \boldsymbol{j}|\ \rangle$, $(e_2e_3)|\ \rangle = \boldsymbol{k}|\ \rangle$, where $\boldsymbol{i}, \boldsymbol{j}, \boldsymbol{k}$ are the quaternionic units, we can write the spinor $|\psi\rangle = \Phi f$ in the form

$$\begin{aligned}|\psi\rangle &= [(a + a_0) + (a_{12} + a_{012})\boldsymbol{i} + (-a_{13} - a_{013})\boldsymbol{j} + (a_{23} + a_{023})\boldsymbol{k}]\,|1\rangle\\ &\quad + [(-a_{123} + a_{0123}) + (a_3 - a_{03})\boldsymbol{i} + (a_2 - a_{02})\boldsymbol{j} + (a_1 - a_{01})\boldsymbol{k}]\,|2\rangle\,.\end{aligned} \tag{2.35}$$

We call $|\psi\rangle$ a Dirac spinor. From the above expression we see that $\mathcal{C}\ell_4 f \simeq \mathcal{C}\ell_4^+ f$, that is, there exists $\Phi'_+ \in \mathcal{C}\ell_4^+$ such that

$$\begin{aligned}|\psi\rangle = \Phi'_+ f &= [a' + a'_{12}\boldsymbol{i} + -a'_{13}\boldsymbol{j} + a'_{23}\boldsymbol{k}]\,|1\rangle\\ &\quad + [a'_{0123} - a'_{03}\boldsymbol{i} - a'_{02}\boldsymbol{j} - a'_{01}\boldsymbol{k}]\,|2\rangle\,.\end{aligned} \tag{2.36}$$

The name Dirac spinors is usually reserved for an object of $\mathbb{C}^4$ that carries the $D^{(1/2,0)} \oplus D^{(0,1/2)}$ representation of SL(2, $\mathbb{C}$). Algebraically, it is an element of a minimal left ideal of the Clifford algebra $\mathbb{C}\otimes\mathcal{C}\ell_4$. One primitive idempotent of $\mathbb{C}\otimes\mathcal{C}\ell_4$ (the one that leads to the standard representation of Dirac matrices) is $g = (1/2)(1 + e_0)(1+ie_{21})$, and one can easily show that $(\mathbb{C}\otimes\mathcal{C}\ell_4)g \simeq (\mathbb{C}\otimes\mathcal{C}\ell_4)^+ g \simeq \mathcal{C}\ell_4 g$, where $(\mathbb{C}\otimes\mathcal{C}\ell_4)^+ \simeq \mathcal{C}\ell_4$ is the even subalgebra of $(\mathbb{C}\otimes\mathcal{C}\ell_4)^+$. Since $g = f(1/2)(1+ie_{21})$ and $ig = e_{12}g$, $e_{12}f = \boldsymbol{i}f$, we have that

$$ig = \boldsymbol{i}g. \tag{2.37}$$

The conclusion is that a Dirac spinor can be defined as an element of $\mathcal{C}\ell_4 f$, as we did, and where the multiplication by $i = \sqrt{-1}$ is replaced by the multiplication by one of the quaternionic units, in our representation $\boldsymbol{i}$. Since $\mathbb{H}$ is a noncommutative ring, we must take some care concerning this replacement. In fact, we have that

$$i\Phi g = \Phi i g = \Phi \boldsymbol{i} g = \Phi g \boldsymbol{i}, \tag{2.38}$$

which means that we must replace $i\,|\psi\rangle = |\psi\rangle\, i$ by the right multiplication by $\boldsymbol{i}$, that is, $|\psi\rangle\, \boldsymbol{i}$.

Now, in relation to the matrix representation of $\mathcal{C}\ell_4$, we have therefore the following quaternionic matrix representation of $\{\boldsymbol{e}_\mu\}$:

$$\begin{aligned} \boldsymbol{e}_0 \leftrightarrow \gamma_0 = \begin{pmatrix} 1 & 0 \\ 0 & -1 \end{pmatrix}, \quad \boldsymbol{e}_1 \leftrightarrow \gamma_1 = \begin{pmatrix} 0 & -\boldsymbol{k} \\ \boldsymbol{k} & 0 \end{pmatrix}, \\ \boldsymbol{e}_2 \leftrightarrow \gamma_2 = \begin{pmatrix} 0 & -\boldsymbol{j} \\ \boldsymbol{j} & 0 \end{pmatrix}, \quad \boldsymbol{e}_3 \leftrightarrow \gamma_3 = \begin{pmatrix} 0 & -\boldsymbol{i} \\ \boldsymbol{i} & 0 \end{pmatrix}, \end{aligned} \tag{2.39}$$

where $\{\gamma_\mu\}$ are Dirac matrices, and the matrix representation of Φ is

$$\Phi \leftrightarrow \begin{pmatrix} z_1 & z_2 \\ z_3 & z_4 \end{pmatrix}, \tag{2.40}$$

$$\begin{aligned} z_1 &= (a + a_0) + (a_{12} + a_{012})\boldsymbol{i} + (-a_{13} - a_{013})\boldsymbol{j} + (a_{23} + a_{023})\boldsymbol{k}, \\ z_2 &= (a_{123} + a_{0123}) + (-a_3 - a_{03})\boldsymbol{i} + (-a_2 - a_{02})\boldsymbol{j} + (-a_1 - a_{01})\boldsymbol{k}, \\ z_3 &= (-a_{123} + a_{0123}) + (a_3 - a_{03})\boldsymbol{i} + (a_2 - a_{02})\boldsymbol{j} + (a_1 - a_{01})\boldsymbol{k}, \\ z_4 &= (a - a_0) + (a_{12} - a_{012})\boldsymbol{i} + (-a_{13} + a_{013})\boldsymbol{j} + (a_{23} - a_{023})\boldsymbol{k}. \end{aligned} \tag{2.41}$$

In terms of this matrix representation, the graded involution and reversion operations corresponds, respectively, to

$$\tilde{\Phi} \leftrightarrow \begin{pmatrix} z_1^* & z_3^* \\ z_2^* & z_4^* \end{pmatrix}, \quad \hat{\Phi} \leftrightarrow \begin{pmatrix} z_4 & z_3 \\ z_2 & z_1 \end{pmatrix}, \tag{2.42}$$

where $*$ means quaternionic conjugation.

From eq.(2.31) we have that $\mathcal{C}\ell_4^+ \simeq \mathcal{C}\ell_{0,3}$. This can be easily seen if we define

$$\boldsymbol{E}_i = \boldsymbol{e}_i \boldsymbol{e}_0, \quad (i = 1, 2, 3). \tag{2.43}$$

From eq.(2.33) and using eq.(2.43) we see that $\Phi_+ \in \mathcal{C}\ell_4^+$ can be written in the form

$$\Phi_+ = a - a_{01}\boldsymbol{E}_1 - a_{02}\boldsymbol{E}_2 - a_{03}\boldsymbol{E}_3 - a_{12}\boldsymbol{E}_{12} - a_{13}\boldsymbol{E}_{13} - a_{23}\boldsymbol{E}_{23} + a_{0123}\boldsymbol{E}_{123}, \tag{2.44}$$

and since $\boldsymbol{E}_i^2 = -1$ $(i = 1, 2, 3)$ we see that $\mathcal{C}\ell_4^+ \simeq \mathcal{C}\ell_{0,3}$. Note that $I = \boldsymbol{E}_{123} = \boldsymbol{e}_{0123} = \boldsymbol{e}_5$ commutes with all elements of $\mathcal{C}\ell_4^+ \simeq \mathcal{C}\ell_{0,3}$, and it follows therefore that $\mathcal{C}\ell_{0,3} \simeq \mathbb{H} \oplus \mathbb{H}$. The elements $F_\pm = (1/2)(1 \pm I)$ are primitive idempotents of $\mathcal{C}\ell_{0,3}$, and the ideals $\mathcal{I}_{0,3}^\pm = \mathcal{C}\ell_{0,3} F_\pm \simeq \mathbb{H}$. The elements of $\mathcal{I}_{0,3}^\pm$ are called Weyl spinors. In the usual physics' literature these corresponds to dotted and undotted Weyl spinors.

Since a Dirac spinor $|\psi\rangle \in \mathcal{C}\ell_4 f \simeq \mathcal{C}\ell_4^+ f$ and $\mathcal{C}\ell_4^+ \simeq \mathcal{C}\ell_{0,3}$, we can write

$$|\psi\rangle = \Phi_+ f = \Phi_+ \frac{1}{2}(1 + e_5) f + \Phi_+ \frac{1}{2}(1 - e_5) f, \tag{2.45}$$

where due to the above discussion we identify $\Phi_+(1/2)(1 \pm e_5)$ as Weyl spinors. In another words, a Dirac spinor can be written as a sum of two Weyl spinors, one dotted and one undotted, as well-known.

The group Spin(3) is given by the set of invertible elements of $\mathcal{C}\ell_3^+$ which have unitary norm, that is, $\tilde{U}U = U\tilde{U} = 1$, $U \in \mathcal{C}\ell_3^+$. Since $\mathrm{Spin}(3)/Z_2 \simeq \mathrm{SO}(3)$ we have that the operation $v \mapsto Uv\tilde{U}$, $U \in \mathrm{Spin}(3)$ is a rotation of the vector $v \in \mathbb{R}^{0,3}$. Moreover $\mathrm{Spin}(3) \simeq \mathrm{SU}(2)$. Elements of Spin(3) can be parametrized by the Euler angles (φ, θ, ξ), that is, we can write

$$U(\varphi, \theta, \xi) = \exp(E_{21}\varphi/2) \exp(E_{32}\theta/2) \exp(E_{21}\xi/2) \in \mathrm{Spin}(3). \tag{2.46}$$

The group Spin(4) is given by the set of invertible elements of $\mathcal{C}\ell_4^+$ which have unitary norm, that is, $\tilde{R}R = R\tilde{R} = 1$, $R \in \mathcal{C}\ell_4^+$. Since $\mathrm{Spin}(4)/Z_2 \simeq \mathrm{SO}(4)$ we have that the operation $v \mapsto Rv\tilde{R}$, $R \in \mathrm{Spin}(4)$ is a rotation of the vector $v \in \mathbb{R}^4$. It is not difficult to see that any $R \in \mathrm{Spin}(4)$ can be written in the form

$$R = \exp B = U_1(\varphi_1, \theta_1, \xi_1) U_2(I\varphi_2, I\theta_2, I\xi_2) \in \mathrm{Spin}(4), \tag{2.47}$$

where $B \in \bigwedge^2(\mathbb{R}^4)$ is a 2-vector, $I = E_{123} = e_{0123} = e_5$ and $U_1, U_2 \in \mathrm{Spin}(3)$, that is, $\mathrm{Spin}(4) \simeq \mathrm{Spin}(3) \times \mathrm{Spin}(3) \simeq \mathrm{SU}(2) \times \mathrm{SU}(2)$.

2.3. Operator spinors

The problem we want to consider now is in fact an old one, at least in the physics' liter-atute, that is the problem of representing spinor fields by completely skew-symmetric tensor fields (differential forms). It comes back to Ivanenko, Landau and Fock in 1928, and was reconsidered several times. Here we want to discuss the solution to this problem as discussed in [10, 11].

Let us consider a Dirac spinor. We saw that it can be considered as an element of $\mathbb{H}^2$, which is the representation space of the Clifford algebra $\mathcal{C}\ell_4$. The explicit translation for Dirac spinors as an element of $\mathbb{C}^4$ to an element of $\mathbb{H}^2$ is given by

$$\begin{pmatrix} a - a_{23}i \\ -a_{13} - a_{12}i \\ a_{0123} + a_{01}i \\ -a_{02} + a_{03}i \end{pmatrix} \leftrightarrow \begin{pmatrix} a + a_{12}\boldsymbol{i} - a_{13}\boldsymbol{j} + a_{23}\boldsymbol{k} \\ a_{0123} - a_{03}\boldsymbol{i} - a_{02}\boldsymbol{j} - a_{01}\boldsymbol{k} \end{pmatrix} = \begin{pmatrix} q_1 \\ q_2 \end{pmatrix}. \tag{2.48}$$

As an element of the minimal left ideal $\mathcal{C}\ell_4^+ f$ the Dirac spinor is represented by

$$|\psi\rangle = \begin{pmatrix} q_1 & 0 \\ q_2 & 0 \end{pmatrix} = \begin{pmatrix} q_1 & q_2 \\ q_2 & q_1 \end{pmatrix} \begin{pmatrix} 1 & 0 \\ 0 & 0 \end{pmatrix} = \psi f, \tag{2.49}$$

where we denoted the matrix representation of $\psi \in \mathcal{C}\ell_4^+$ and of the idempotent f by the same letter. The form of the matrix representation of ψ comes from eq.(2.40).

Clearly we have an isomorphism between $\mathcal{I}_4 = \mathcal{C}\ell_4^+ f$ and $\mathcal{C}\ell_4^+$, and this suggests us to work with a Dirac spinor as an element of $\mathcal{C}\ell_4^+$. We have explicitly

$$\begin{pmatrix} a - a_{23}i \\ -a_{13} - a_{12}i \\ a_{0123} + a_{01}i \\ -a_{02} + a_{03}i \end{pmatrix} \leftrightarrow a + a_{01}e_{01} + \cdots + a_{23}e_{23} + a_{0123}e_5. \tag{2.50}$$

We shall call an element of $\mathcal{C}\ell_4^+$ an *operator (Dirac) spinor.* One reason for the name is suggested by the fact that it is an element of $\mathcal{C}\ell_4^+$, whose elements can be interpreted as operators acting on the spinor space. There are, however, other details to be consider.

Let $\psi \in \mathcal{C}\ell_4^+$. It this case we have

$$\begin{aligned}&\psi = \langle\psi\rangle_0 + \langle\psi\rangle_2 + \langle\psi\rangle_4 \Rightarrow \\ &\psi\tilde{\psi} = [(\langle\psi\rangle_0)^2 - \langle((\langle\psi\rangle_2)^2\rangle_0] + [2(\langle\psi\rangle_0\langle\psi e_5\rangle_0) - 2\langle((\langle\psi\rangle_2)^2 e_5\rangle_0]e_5,\end{aligned} \tag{2.51}$$

where $(\langle\psi\rangle_0)^2 - \langle((\langle\psi\rangle_2)^2\rangle_0 = (\langle\psi\rangle_0)^2 + \mid\langle\psi\rangle_2\mid^2 \geq 0$. This implies that ψ can be written in the form

$$\psi = (\alpha + \beta e_5)R, \tag{2.52}$$

where $R\tilde{R} = \tilde{R}R = 1$ and $R \in \mathcal{C}\ell_4^+$, that is, $R \in \text{Spin}(4)$. We call this the polar decomposition of an operator spinor.

The above definition of an operator spinor is only partially correct. In fact, for an element A of $\mathcal{C}\ell_4$ the Spin(4) group acts like $A \mapsto UA\tilde{U}$, but for a spinor $|\psi\rangle$ we have $|\psi\rangle \mapsto U\,|\psi\rangle$. In order to obtain a more precise definition, let us remember that the physical quantities are related to the so called bilinear covariants, which are bilinear expressions of the spinors. Note first that from eq.(2.52) we have that $\psi e_3\tilde{\psi} = (\alpha^2 - \beta^2)Re_3\tilde{R}$, where $Re_3\tilde{R}$ is a rotation of the vector e_3. It follows therefore that any vector $v \in \mathbb{R}^4$ can be written in the form $v = \psi e_3\tilde{\psi}$. Let us denote $\Sigma = \{e_\mu\}$ and write explicitly ψ_Σ in order to emphasize that ψ_Σ is written in terms of the basis Σ like

$$\psi_\Sigma = a + a_{01}e_{01} + \cdots + a_{23}e_{23} + a_{0123}e_5. \tag{2.53}$$

Let us consider now another basis $\Sigma' = \{e'_\mu\}$ related to the former one by an element $U \in \text{Spin}(4)$, i.e., $e'_\mu = Ue_\mu\tilde{U}$. Let us write now

$$\psi_{\Sigma'} = a' + a'_{01}e'_{01} + \cdots + a'_{23}e'_{23} + a_{0123}e'_5. \tag{2.54}$$

We define now an equivalence relation by imposing the vectors defined by $\psi_\Sigma e_3\tilde{\psi}_\Sigma$ and $\psi_{\Sigma'}e'_3\tilde{\psi}_{\Sigma'}$ to be *the same*, that is,

$$v = \psi_\Sigma e_3\tilde{\psi}_\Sigma = \psi_{\Sigma'}e'_3\tilde{\psi}_{\Sigma'}. \tag{2.55}$$

Since $e_3' = Ue_3\tilde{U}$ we have that

$$\psi_{\Sigma'} = \psi_\Sigma \tilde{U}. \qquad (2.56)$$

This defines ψ_Σ and $\psi_{\Sigma'}$ as the representatives in terms of the basis Σ and Σ' of the *operator spinor* Ψ. Since most of the time there is no risk of confusion, we omit in general the reference to a particular basis and write $\Psi = \psi$.

Now, $v \mapsto Uv\tilde{U}$ defines a *new* vector v', and therefore a new *operator spinor* Ψ' such that in terms of the basis $\{\Sigma\}$ and $\{\Sigma'\}$ we have

$$v' = \psi'_\Sigma e_3 \tilde{\psi}'_\Sigma = \psi'_{\Sigma'} e'_3 \tilde{\psi}'_{\Sigma'}. \qquad (2.57)$$

It follows therefore that

$$\psi'_\Sigma = R\psi_\Sigma, \qquad (2.58)$$

and then $\Psi' = R\Psi$, which is the transformation property for a spinor under the action of Spin(4).

2.4. Clifford calculus, Dirac operator and Maxwell equations

In order to discuss the Clifford calculus over a 4 dimensional riemannian manifold (M^4, g) we need to introduce the Clifford bundle. The tangent space at $x \in M$ is $T_x(M^4) \simeq \mathbb{R}^4$, the tangent bundle is $T(M^4) = \cup_{x\in M^4} T_x(M^4)$, and the Clifford bundle is $C\ell(M^4) = \cup_{x\in M^4} C\ell(T_x(M^4), g)$, where $C\ell(T_x(M), g)$ is the Clifford algebra generated by the vector fields at $x \in M$.

Let $\{v_\mu\}$ be an orthonormal basis for $T_x(M^4)$, and $\{\theta^\mu\}$ the corresponding dual basis of 1-forms, that is, $\theta^\mu(v_\nu) = \delta^\mu_\nu$. Under the Clifford map we have that $\gamma^*(v_\mu) = e_\mu$ and $\gamma(\theta^\mu) = e^\mu$. Let ∇ be the Levi-Civita connection. We define the connection coefficients ${\Gamma_{\mu\nu}}^\sigma$ as

$$\nabla_\mu e^\sigma = {\Gamma_{\mu\nu}}^\sigma e^\nu, \quad \nabla_\mu e_\sigma = -{\Gamma_{\mu\sigma}}^\nu e_\nu, \qquad (2.59)$$

where $\nabla_\mu = \nabla_{e_\mu}$. Since $\{v_\mu\}$ was supposed to be an orthonormal basis, then the connection coefficients defined in terms of this basis satisty

$${\Gamma_\sigma}^{\mu\nu} = -{\Gamma_\sigma}^{\nu\mu}. \qquad (2.60)$$

We can now define

$$\Gamma_\sigma = \frac{1}{2}{\Gamma_\sigma}^{\mu\nu} e_\mu e_\nu, \qquad (2.61)$$

and write

$$\nabla_\mu e_\nu = \frac{1}{2}[\Gamma_\mu, e_\nu], \quad \nabla_\mu e^\nu = \frac{1}{2}[\Gamma_\mu, e^\nu], \qquad (2.62)$$

where [,] is the commutator, which can be verified by explicit computation. The covariant derivative of a k-vector field can be written as

$$\nabla_\mu(A^{\nu_1\ldots\nu_k}e_{\nu_1\ldots\nu_k}) = (\partial_\mu A^{\nu_1\ldots\nu_k})e_{\nu_1\ldots\nu_k} + A^{\nu_1\ldots\nu_k}\frac{1}{2}[\Gamma_\mu, e_{\nu_1\ldots\nu_k}]. \tag{2.63}$$

We call Dirac operator the operator ∂ defined by

$$\partial = e^\mu \nabla_\mu. \tag{2.64}$$

Using eq.(2.28–2.30) we can write

$$\partial A = \partial \cdot A + \partial \wedge A, \tag{2.65}$$

with

$$\partial \cdot A = \frac{1}{2}(e^\mu \nabla_\mu A - \nabla_\mu \hat{A} e^\mu), \tag{2.66}$$

$$\partial \wedge A = \frac{1}{2}(e^\mu \nabla_\mu A + \nabla_\mu \hat{A} e^\mu). \tag{2.67}$$

It is not difficult to show that these operators correspond, in terms of the differential d and codifferential δ operators on differential forms, to

$$\partial\wedge = d, \quad \partial\cdot = -\delta, \tag{2.68}$$

and therefore

$$\partial = d - \delta. \tag{2.69}$$

This last result was recognized by Kähler, and the operator ∂ is sometimes called Dirac–Kähler operator — although already discovered by Hodge and de Rham [12].

As an application of this last result, let us consider the Maxwell equations

$$dF = 0, \quad \delta F = -J, \tag{2.70}$$

where $F = (1/2)F^{\mu\nu}e_{\mu\nu}$, $J = J^\mu e_\mu$. It follows from eq.(2.70) that Maxwell equations can be written as a single equation, namely

$$\partial F = J. \tag{2.71}$$

It was shown by Rodrigues et al. [13] that if *magnetic monopoles* are present, then Maxwell equations can be modified *phenomenologically* in order to introduce a magnetic monopolar current as

$$\partial F = \mathcal{J} = J + Ke_5, \tag{2.72}$$

where $K = K^\mu e_\mu$ is the magnetic monopolar current and $Ke_5 \in \bigwedge^3(\mathbb{R}^4)$.

It is interesting to observe now that the so called Hodge duality operation $\star$ can be expressed as

$$\star A = \tilde{A}e_5. \tag{2.73}$$

Maxwell equations with magnetic monopoles can be written therefore as $\partial F = J + \star K$.

Finally, we need to consider the covariant derivative of spinor fields, and we are interested in the case of operator spinor fields. One can show [14] that the covariant derivative of an operator spinor is

$$\nabla_\mu \psi = \partial_\mu \psi + \frac{1}{2}\Gamma_\mu \psi, \tag{2.74}$$

where ψ is an even section of the Clifford bundle and the *spin* Dirac operator is $\partial = e^\mu \nabla_\mu$. Note that the action of ∇_μ over ψ is different that over A, and in this sense a notation like ∇^s_μ would be convenient, but the use of the same symbol is usual, and may not cause confusion. This expression must be properly written as $(\nabla_\mu \psi)_\Sigma$, and it is immediate that $(\nabla_\mu \psi)_\Sigma \neq \nabla_\mu(\psi_\Sigma)$ — see [11, 14]. Moreover, if $A \in \mathcal{C}\ell(M^4)$, we have that

$$\nabla_\mu(A\psi) = (\nabla_\mu A)\psi + A(\nabla_\mu \psi), \tag{2.75}$$

where $\nabla_\mu A$ is computed using eq.(2.63) and $\nabla_\mu(A\psi)$ and $\nabla_\mu \psi$ using eq.(2.74).

3. Witten's monopole equations

Witten's monopole equations are [1]

$$\not{\partial}\,|\psi\rangle = iA\,|\psi\rangle\,, \tag{3.76}$$

$$(F_\pm)_{\mu\nu} = \frac{i}{4}\,\langle\psi|\,[\gamma_\mu, \gamma_\nu]\,|\psi\rangle\,, \tag{3.77}$$

$$F = dA, \quad F_\pm = \frac{1}{2}(F \pm \star F). \tag{3.78}$$

In the above expressions $\not{\partial} = \gamma^\mu \nabla_\mu$ is the Dirac operator, γ^μ are Dirac matrices, A is the electromagnetic potential (U(1) connection) and $F = dA$ the field strenght, and $|\psi\rangle$ is a Weyl spinor, that is, $|\psi\rangle = \pm\gamma_5\,|\psi\rangle$. The 2-vector $F_\pm$ in this case is self-dual or anti-self-dual, that is, $F_+ = \star F_+$ or $F_- = -\star F_-$, respectively. We shall be restricted without loss of generality to self-dual fields F.

Let us now write this expression in terms of $\mathcal{C}\ell_4$. First of all, for a self-dual 2-vector F_+ we have

$$F_+ = \star F_+ = -e_5 F_+ = -F_+ e_5 \Rightarrow F_+ = \frac{1}{2}(1 - e_5)F_+. \tag{3.79}$$

We see that we can use the $\Pi_+ = (1/2)(1 - e_5)$ to project out a self-dual 2-vector F_+ from an arbitrary 2-vector as

$$F_+ = \Pi_+ F = \frac{1}{2}(1 - e_5)F. \tag{3.80}$$

Consider an arbitrary 2-vector F. In general F is not a simple 2-vector, that is, there does not exist vectors $\boldsymbol{v}$ and $\boldsymbol{u}$ such that $F = \boldsymbol{v} \wedge \boldsymbol{u}$. However, we can always find vectors $\{f_\mu\}$ such that we can write an arbitrary 2-vector F as

$$F = af_2 \wedge f_1 + bf_0 \wedge f_3. \tag{3.81}$$

We can always define $\{f_\mu\}$ in such a way that $a \geq 0$, and since there exists $R \in \text{Spin}(4)$ such that $f_\mu = Re_\mu\tilde{R}$ and $f_0f_3f_5 = f_2f_1$, we have that

$$F = \frac{1}{2}\phi e_2 e_1 \tilde{\phi}, \tag{3.82}$$

where ϕ is of the form

$$\phi = (\alpha + \beta e_5)R, \tag{3.83}$$

that is, ϕ is an operator spinor (see eq.(2.52). The factor $(1/2)$ was introduced for convenience only. In terms of the Dirac spinor $|\phi\rangle \in \mathbb{H}^2$ we have that $F_{\mu\nu} = (1/4)\langle\phi|[\gamma_\mu, \gamma_\nu]|\phi\rangle$, where $\{\gamma_\mu\}$ are the quaternionic matrix representation of the $\{e_\mu\}$. In terms of the complexified representation we have $\{\gamma_\mu\}$ given by the Dirac matrices and $|\phi\rangle \in \mathbb{C}^4$.

From eq.(3.80) we have therefore that

$$F_+ = \frac{1}{2}\psi e_2 e_1 \tilde{\psi}, \tag{3.84}$$

where we have used the fact that $\Pi_+^2 = \Pi_+$ and defined

$$\psi = \frac{1}{2}(1 - e_5)\phi = \phi\frac{1}{2}(1 - e_5). \tag{3.85}$$

From eq.(2.45) we see that ψ can be interpreted as an *operator* Weyl spinor, and eq.(3.84) is our version of the eq.(3.77).

In relation to eq.(3.76), we saw that $if = \boldsymbol{i}f$ and that $e_1e_2 |\ \rangle = \boldsymbol{i} |\ \rangle$. It follows therefore that eq.(3.76) can be written as

$$\partial\psi e_2 e_1 = A\psi, \tag{3.86}$$

where $A = A^\mu e_\mu$ and ∂ is the Dirac operator, that has as matrix representation the operator $\not\partial$.

Finally, eq.(3.78) is trivially translated as

$$F = \partial \wedge A, \tag{3.87}$$

as follows from eq.(2.68).

3.1. Witten's equations in Maxwell equations form

First of all, let us consider ψ as given by eq.(3.85). From eq.(3.83) we have that

$$\psi = \frac{(\alpha - \beta)}{2}R\frac{1}{2}(1 - e_5). \tag{3.88}$$

Now calculating $\psi\tilde{\psi}$ we easily get that

$$\psi\tilde{\psi} = \frac{(\alpha-\beta)^2}{4}\frac{1}{2}(1-e_5) = \sigma(1-e_5). \tag{3.89}$$

We can write therefore

$$\psi = \sqrt{2\sigma}R\frac{1}{2}(1-e_5), \tag{3.90}$$

where we define $(\alpha-\beta) = \pm 2\sqrt{2\sigma}$ with the signs chosen conveniently.

Now, let us consider eq.(3.86). If we multiply both sides of it on the right by $\tilde{\psi}$ we get that

$$\partial\psi e_2 e_1 \tilde{\psi} = A\sigma(1-e_5), \tag{3.91}$$

where we have used eq.(3.89). The term on the left hand side can be rewritten in the following way. First, we have

$$\partial(\psi e_2 e_1 \tilde{\psi}) = e^\mu[(\nabla_\mu\psi)e_2 e_1\tilde{\psi} + \psi e_2 e_1(\nabla_\mu\tilde{\psi})], \tag{3.92}$$

where ∇ now is the spin-connection (see [11, 14]). But $(\nabla_\mu\psi)e_2e_1\tilde{\psi} \in \bigwedge^0(\mathbb{R}^4) \oplus \bigwedge^2(\mathbb{R}^4) \oplus \bigwedge^4(\mathbb{R}^4)$, and

$$\psi e_2 e_1(\nabla_\mu\tilde{\psi}) = -((\nabla_\mu\psi)e_2e_1\tilde{\psi})\tilde{}. \tag{3.93}$$

Since the reversion (see eq.(2.26)) changes the sign of 2-vectors, but does not change the sign of scalars and 4-vectors, we conclude from eq.(3.92) that

$$\partial(\psi e_2 e_1\tilde{\psi}) = 2e^\mu\langle(\nabla_\mu\psi)e_2e_1\tilde{\psi}\rangle_2. \tag{3.94}$$

But we also have that

$$\langle(\nabla_\mu\psi)e_2e_1\tilde{\psi}\rangle_2 = (\nabla_\mu\psi)e_2e_1\tilde{\psi} - \langle(\nabla_\mu\psi)e_2e_1\tilde{\psi}\rangle_0 - \langle(\nabla_\mu\psi)e_2e_1\tilde{\psi}\rangle_4. \tag{3.95}$$

Using eqs.(3.94,3.95) in eq.(3.92) we have that

$$\partial\psi e_2e_1\tilde{\psi} = \frac{1}{2}\partial(\psi e_2e_1\tilde{\psi}) - e^\mu\langle(\nabla_\mu\psi)e_2e_1\tilde{\psi}\rangle_0 - e^\mu\langle(\nabla_\mu\psi)e_2e_1\tilde{\psi}\rangle_4. \tag{3.96}$$

On the other hand, we have that

$$\langle(\nabla_\mu\psi)e_2e_1\tilde{\psi}\rangle_4 = \langle(\nabla_\mu\psi)e_2e_1\tilde{\psi}e_5\rangle_0 e_5 = -\langle(\nabla_\mu\psi)e_2e_1\tilde{\psi}\rangle_0 e_5, \tag{3.97}$$

where we have used the fact that $\psi = -e_5\psi = -\psi e_5$, which follows from eq.(3.85). If we define the vector j as

$$j = e^\mu\langle(\nabla_\mu\psi)e_2e_1\tilde{\psi}\rangle_0, \tag{3.98}$$

we can write eq.(3.91) in the following form

$$\partial\left(\frac{1}{2}\psi e_2e_1\tilde{\psi}\right) = [\sigma A - j](1-e_5). \tag{3.99}$$

Finally, from eq.(3.84) we see that

$$\partial F_+ = \mathcal{J}, \tag{3.100}$$

where we defined

$$\mathcal{J} = [\sigma A - j](1 - e_5). \tag{3.101}$$

This is just Maxwell equations in the presence of magnetic monopoles — eq.(2.72) — with electric and magnetic monopolar currents $J = -K = \sigma A - j$. Since $F_+ = \star F_+ = -e_5 F_+$ and $\partial \wedge F_+ = -\partial \wedge (F_+ e_5) = -(\partial \cdot F_+)e_5$, it is equivalent to the equation

$$\partial \cdot F_+ = \sigma A - j. \tag{3.102}$$

Since $\partial \cdot \partial \cdot = 0$ we must have

$$\partial \cdot J = \partial \cdot (\sigma A - j) = 0, \tag{3.103}$$

that is the usual equation for the conservation of electric charge. Let us verify that it is indeed satisfied. We have, using eq.(3.98),

$$\partial \cdot J = (\partial \cdot A)\sigma + (A \cdot \partial)\sigma - \langle \Box \psi e_2 e_1 \tilde{\psi} \rangle_0 - \langle \nabla^\mu \psi e_2 e_1 \nabla_\mu \tilde{\psi} \rangle_0, \tag{3.104}$$

where we used $\Box = \partial^2$ that is the spinor laplacian in the above expression. The last term on the right hand side can be easily shown to vanish. In fact, we have $\langle\ \rangle_0 = \langle\ \rangle_0^{\sim}$, and then

$$\langle \nabla^\mu \psi e_2 e_1 \nabla_\mu \tilde{\psi} \rangle_0 = -\langle \nabla_\mu \psi e_2 e_1 \nabla^\mu \tilde{\psi} \rangle_0 = 0. \tag{3.105}$$

The third term of the right hand side can be computed using eq.(3.86). Then

$$\begin{aligned} \partial \cdot J &= (\partial \cdot A)\sigma + (A \cdot \partial)\sigma - \langle (\partial A)\psi\tilde{\psi} \rangle_0 - \langle e^\mu A \nabla_\mu \psi \tilde{\psi} \rangle_0 \qquad (3.106)\\ &= (\partial \cdot A)\sigma + (A \cdot \partial)\sigma - (\partial \cdot A)\langle \psi\tilde{\psi} \rangle_0 - 2\langle (A \cdot \partial)\psi\tilde{\psi} \rangle_0 - \langle A^2 \psi e_2 e_1 \tilde{\psi} \rangle_0. \end{aligned}$$

From eq.(3.89) we see that $\langle \psi\tilde{\psi} \rangle_0 = \sigma$ and that $\nabla_\mu \sigma = \langle \nabla_\mu \psi \tilde{\psi} \rangle_0 + \langle \psi \nabla_\mu \tilde{\psi} \rangle_0 = 2\langle \nabla_\mu \psi \tilde{\psi} \rangle_0$. Moreover, since $\langle \psi e_2 e_1 \tilde{\psi} \rangle_0 = 0$, we have finally that

$$\partial \cdot J = 0. \tag{3.107}$$

This result is very interesting since it shows that unlike other versions of Maxwell equations with self-interaction of the type the electromagnetic potential playing the role of electric current, here we have complete gauge invariance. For example: for Proca equation $\partial \cdot F = m^2 A$ we necessarily have $\partial \cdot A = 0$, that is, it only holds in Lorentz gauge; but for eq.(3.102) we have $\partial \cdot J = 0$ independently of our gauge choice. Another way to realize this is to look to eq.(3.86) and see that the gauge invariance of that equation requires that

$$A \mapsto A + \partial\eta, \qquad \psi \mapsto \psi \exp(e_{12}\eta), \tag{3.108}$$

and in this way $j \mapsto j + \partial\eta$, which implies that $J \mapsto J$.

3.2. The magnetic monopole solution

It is known [1] that when M^4 is flat, all non-trivial solutions of Witten's equation are non-L^2. This non-L^2 solutions may not be of interest in relation to Donaldson theory, but they may be physically relevant. Let us consider therefore the flat space Witten's equations and look for its singular solutions, in particular the one corresponding to a magnetic monopole.

Magnetic monopoles can be seen as arising from a solution of the free Maxwell equations with a singularity along the so-called Dirac string. So, let us consider the Witten's monopole equations in its Maxwell form, namely eq.(3.100). The free case corresponds to

$$\partial F = 0, \tag{3.109}$$

where from now on we shall write simply $F_+ = F$, and this implies that

$$A = \frac{1}{\sigma} j = \frac{1}{\sigma} e^\mu \langle \partial_\mu \psi e_2 e_1 \tilde{\psi} \rangle_0. \tag{3.110}$$

Now, from eq.(3.90) we have that $\psi = \sqrt{2\sigma} R(1/2)(1-e_5)$, with $R \in \text{Spin}(4)$. We can write R in the form $R = R_+ + R_-$ where $R_\pm = (1/2)(R \pm e_0 R e_0)$, and since there exists R'_+ such that $R_- = e_5 R'_+$, we can write $R = R_+ + e_5 R'_+$ with $R_+ = e_0 R_+ e_0$ and $R'_+ = e_0 R'_+ e_0$. In this way $R(1/2)(1-e_5) = (R_+ - R'_+)(1/2)(1-e_5)$. Since $R\tilde{R} = 1$ gives $U\tilde{U} = 1$ and $U = e_0 U e_0$ for $U = R_+ - R'_+$, we conclude that $U \in \text{Spin}(3)$. We have therefore that

$$R\frac{1}{2}(1-e_5) = U\frac{1}{2}(1-e_5), \quad U \in \text{Spin}(3). \tag{3.111}$$

We have seen — eq.(2.46) — that $U \in \text{Spin}(3)$ can be parametrized by Euler angles. We can write therefore

$$R\frac{1}{2}(1-e_5) = \exp(e_{12}\varphi/2)\exp(e_{23}\theta/2)\exp(e_{12}\xi/2)\frac{1}{2}(1-e_5). \tag{3.112}$$

Using the above expression we find, after a straightforward calculation, that

$$\begin{aligned} 2\partial_\mu R e_2 e_1 \tilde{R}\frac{1}{2}(1-e_5) = {} & [(\partial_\mu \xi - \partial_\mu \cos\theta) \\ & + e_3 e_2 (\partial_\mu \varphi \sin\theta \cos\varphi + \partial_\mu \theta \sin\varphi \cos\theta) \\ & + e_3 e_1 (\partial_\mu \varphi \sin\theta \sin\varphi - \partial_\mu \theta \cos\theta \cos\varphi) \\ & + e_1 e_2 (\partial_\mu \theta \sin\theta)]\frac{1}{2}(1-e_5). \end{aligned} \tag{3.113}$$

But using eq.(3.90) we see that

$$\partial_\mu \psi = \left(\frac{1}{\sqrt{2\sigma}}\partial_\mu \sigma R + \sqrt{2\sigma}\,\partial_\mu R\right)\frac{1}{2}(1-e_5). \tag{3.114}$$

Using the definition (3.98) of the vector j we arrive therefore to

$$j_\mu = \sigma \langle \partial_\mu R e_2 e_1 \tilde{R}(1 - e_5) \rangle_0. \tag{3.115}$$

Finally, using eq.(3.113) we obtain for A from eq.(3.110) that

$$A = \frac{1}{2}\left(\partial\xi + \partial\varphi \cos\theta\right). \tag{3.116}$$

Now we consider the case where the above Euler angles in parameter space are identified with angles in physical $\mathbb{R}^3$. Moreover, we consider the particular case $\xi = -\varphi$, that corresponds to a particular choice of gauge. In this way we obtain that

$$A = -\frac{1}{2}(\partial\varphi \sin\theta) \tan\frac{\theta}{2} = -\frac{1}{2}\vartheta^{\varphi} \tan\frac{\theta}{2}. \tag{3.117}$$

This is a singular electromagnetic potential along $\theta = \pi$ — the Dirac string — corresponding to a magnetic monopole solution.

It is not difficult to find the spinor ψ such that $F = (1/2)\psi e_2 e_1 \tilde{\psi}$ for $F = \partial \wedge A$ corresponding to the above monopole solution. We shall consider here a rather unusual approach, but which will be of great utility in next section. We shall obtain the spinor ψ by using the so called inversion theorem [15, 16].

It is well-known that in quantum theory the observables quantities are related to bilinear expressions on the spinor fields. This observable quantities are generally called the bilinear covariants. Since a spinor determines a set of bilinear covariant quantities, the question we can put is that if given those bilinear covariant quantities we can recover the spinor. The answer is afirmative, as shown in [15] for the case of Dirac spinors. Here we shall discuss this result (which in some sense could be called recovering theorem) according the discussion of [16], where some applications of it can be found.

First, let us consider ψ as given by

$$\psi = \sqrt{2\sigma} R \frac{1}{2}(1 - e_5) = \sqrt{2\sigma} U \frac{1}{2}(1 - e_5). \tag{3.118}$$

Since $U \in \mathrm{Spin}(3)$ we have that

$$\frac{1}{2}(\psi + e_2 e_1 \psi e_1 e_2)\frac{1}{2}(1 - e_5) = (a_0 + a_{12} e_1 e_2)\frac{1}{2}(1 - e_5), \tag{3.119}$$

which means that there exists η such that

$$\frac{1}{2}(\psi + e_2 e_1 \psi e_1 e_2) \exp(e_1 e_2 \eta)\frac{1}{2}(1 - e_5) = C\frac{1}{2}(1 - e_5), \tag{3.120}$$

where C is a scalar quantity. We can write therefore

$$\begin{aligned} \psi &= \psi\frac{1}{2}(1 - e_5) = C_1 \psi(\tilde{\psi} + e_2 e_1 \tilde{\psi} e_1 e_2) \exp(e_1 e_2 \eta)\frac{1}{2}(1 - e_5) \\ &= C_1(\sigma(1 - e_5) + 2F e_1 e_2) \exp(e_1 e_2 \eta)\frac{1}{2}(1 - e_5), \end{aligned} \tag{3.121}$$

where C_1 is a scalar quantity to be determined. We can determined C_1 by using $\psi\tilde{\psi} = \sigma(1 - e_5)$. This gives

$$\begin{aligned}\sigma(1-e_5) &= C_1^2[\sigma(1-e_5)+2Fe_1e_2][\sigma(1-e_5)-2e_2e_1F]\frac{1}{2}(1-e_5)\\ &= C_1^2[4\sigma^2+4\sigma(Fe_1e_2+e_1e_2F)-4F^2]\frac{1}{2}(1-e_5).\end{aligned} \tag{3.122}$$

But we have that

$$F^2 = -\sigma^2\frac{1}{2}(1-e_5), \tag{3.123}$$

$$(Fe_1e_2+e_1e_2F) = 4\langle Fe_1e_2\rangle_0\frac{1}{2}(1-e_5) = 4F_{21}\frac{1}{2}(1-e_5). \tag{3.124}$$

In order to get eq.(3.123) we just need to use $F = (1/2)\psi e_2e_1\tilde{\psi}$. For eq.(3.124) we use the fact that $(e_1e_2F) = (Fe_1e_2)\tilde{}$ and that $Fe_1e_2 + (Fe_1e_2)\tilde{} = 2(\langle Fe_1e_2\rangle_0 + \langle Fe_1e_2\rangle_4)$ with $F = -e_5F$. Using these results in eq.(3.122) we obtain that

$$\sigma(1-e_5) = C_1^2(8\sigma^2 + 16\sigma F_{21})\frac{1}{2}(1-e_5), \tag{3.125}$$

which gives

$$C_1 = \frac{1}{2\sqrt{\sigma+2F_{21}}}. \tag{3.126}$$

The expression for ψ is therefore

$$\psi = \psi\frac{1}{2}(1-e_5) = \frac{1}{\sqrt{\sigma+2F_{21}}}(\sigma + Fe_1e_2)\exp(e_1e_2\eta)\frac{1}{2}(1-e_5), \tag{3.127}$$

which gives ψ in terms of the electromagnetic field F and the fields invariants given by eq.(3.123). This is what we called inversion (or recovering) theorem. Note that it provides a *solution* of the equation $F = (1/2)\psi e_2e_1\tilde{\psi}$. We now shall use this expression in order to get the spinor for the magnetic monopole solution.

It will be convenient in what follows to use cilindrical coordinates (ρ, ϕ, z). In this case we have $e_0 = \gamma^*(\partial_t)$, $e_1 = \gamma^*(\partial_\rho)$, $e_2 = \gamma^*(\rho^{-1}\partial_\phi)$ and $e_3 = \gamma^*(\partial_z)$. The electromagnetic field F corresponding to our solution can be written as

$$F = \frac{1}{2}\frac{\rho e_3e_2 + ze_2e_1}{(\rho^2+z^2)^{3/2}}\frac{1}{2}(1-e_5). \tag{3.128}$$

From this expression we can calculate σ according to eq.(3.123). We have that

$$F^2 = -\frac{1}{4(\sigma^2+z^2)^2}\frac{1}{2}(1-e_5), \tag{3.129}$$

which gives

$$\sigma = \frac{1}{2(\rho^2+z^2)}. \tag{3.130}$$

Since

$$F_{21} = \frac{z}{4(\rho^2 + z^2)^{3/2}}, \tag{3.131}$$

we have that

$$\sqrt{\sigma + 2F_{21}} = \frac{1}{\sqrt{\rho^2 + z^2}} \sqrt{\frac{1 + z/\sqrt{\rho^2 + z^2}}{2}}, \tag{3.132}$$

and

$$(\sigma + F e_1 e_2)\frac{1}{2}(1 - e_5)$$
$$= \frac{1}{2}\left(\frac{1}{\rho^2 + z^2} + \frac{z}{(\rho^2 + z^2)^{3/2}} + \frac{\rho}{(\rho^2 + z^2)^{3/2}} e_1 e_3\right)\frac{1}{2}(1 - e_5). \tag{3.133}$$

It is convenient now to introduce θ as given by

$$\cos\theta = \frac{z}{\sqrt{\rho^2 + z^2}}, \quad \sin\theta = \frac{\rho}{\sqrt{\rho^2 + z^2}}. \tag{3.134}$$

Using eq.(3.132) and eq.(3.133) in eq.(3.127) and using $\cos^2\theta/2 = (1 + \cos\theta)/2$ we finally obtain that

$$\psi = \frac{1}{\sqrt{\rho^2 + z^2}} \exp(e_1 e_3 \frac{\theta}{2}) \exp(e_1 e_2 \eta) \frac{1}{2}(1 - e_5), \tag{3.135}$$

where η is an arbitrary phase angle. It is easy to verify now that $(1/2)\psi e_2 e_1 \tilde{\psi}$ gives F as in eq.(3.128).

3.3. Other representations of Witten's equations

If we look again to eq.(3.100) we see that it is not in a "complete" Maxwell-like form. In fact, the vector j is defined in terms of ψ, and in order to have an equation in a "complete" Maxwell form we must write ψ in terms of electromagnetic quantities like F or A. In order to do this we shall make use of the inversion theorem that we obtained in the last section, namely eq.(3.127). From eq.(3.127) we obtain that

$$\partial_\mu \psi = -\frac{(\partial_\mu \sigma + 2\partial_\mu F_{21})}{2(\sigma + 2F_{21})}\psi + \psi e_1 e_2 \partial_\mu \eta + \frac{(\partial_\mu \sigma + \partial_\mu F e_1 e_2)}{\sqrt{\sigma + 2F_{21}}} \exp(e_1 e_2 \eta)\frac{1}{2}(1 - e_5). \tag{3.136}$$

The last term can be written in a more convenient form if we use $\psi\tilde{\psi} = \sigma(1 - e_5)$. We obtain in this way that

$$\partial_\mu \psi = \frac{[-2\partial_\mu F_{21} + (\partial_\mu \ln\sigma) e_1 e_2 F + (\partial_\mu F) e_1 e_2 - \frac{1}{\sigma}(\partial_\mu F) F]}{2(\sigma + 2F_{21})}\psi + \psi e_1 e_2 \partial_\mu \eta. \tag{3.137}$$

But $F_{21} = \langle F e_1 e_2 \rangle_0$, and therefore

$$2\partial_\mu F_{21}\frac{1}{2}(1-e_5) = (\partial_\mu\langle F e_1 e_2\rangle_0 + \partial_\mu\langle F e_1 e_2\rangle_4)\frac{1}{2}(1-e_5) = \partial_\mu\frac{1}{2}(F e_1 e_2 + e_1 e_2 F). \tag{3.138}$$

Moreover, using eq.(3.123) we get that

$$(\partial_\mu \ln\sigma)\frac{1}{2}(1-e_5) = \frac{-1}{2\sigma^2}[(\partial_\mu F)F + F\partial_\mu F], \tag{3.139}$$

and from it

$$(\partial_\mu \ln\sigma)e_1 e_2 F = \frac{-1}{2\sigma^2}e_1 e_2 F(\partial_\mu F)F + \frac{1}{2}e_1 e_2(\partial_\mu F). \tag{3.140}$$

Using eq.(3.138) and eq.(3.140) in eq.(3.137) we obtain that

$$\partial_\mu\psi = \frac{[\sigma^2(\partial_\mu F)e_1 e_2 - e_1 e_2 F(\partial_\mu F)F - 2\sigma(\partial_\mu F)F]}{4\sigma^2(\sigma + 2F_{21})}\psi + \psi e_1 e_2 \partial_\mu \eta. \tag{3.141}$$

This expression for $\partial_\mu\psi$ can now be used in the definition of j. From eq.(3.98) and the definition of F we get

$$\begin{aligned} j_\mu &= \langle\partial_\mu e_2 e_1\tilde{\psi}\rangle_0 = \sigma\partial_\mu\eta + \frac{1}{2(\sigma+2F_{21})}[\langle(\partial_\mu F)e_1 e_2 F\rangle_0 \\ &\quad - \frac{1}{\sigma^2}\langle e_1 e_2 F(\partial_\mu F)F^2\rangle_0 - \frac{2}{\sigma}\langle(\partial_\mu F)F^2\rangle_0]. \end{aligned} \tag{3.142}$$

Eq.(3.123) can now be used, and from it we immediatly see that the last term vanishes. We can finally write that

$$j_\mu = \sigma\partial_\mu\eta + \frac{\langle F(\partial_\mu F)e_1 e_2\rangle_0}{(\sigma + 2F_{21})}. \tag{3.143}$$

Now, looking to eq.(3.100) we see that the current associated to the above term is $j(1-e_5)$. But we have that

$$\begin{aligned} \langle F\partial_\mu F e_1 e_2\rangle_0(1-e_5) &= [\langle\partial_\mu F e_1 e_2 F\rangle_0 + \langle\partial_\mu F e_1 e_2 F\rangle_4]\frac{1}{2}(1-e_5) \\ &= [(\partial_\mu F e_1 e_2 F) - \langle\partial_\mu F e_1 e_2 F\rangle_2]\frac{1}{2}(1-e_5) \\ &= [(\partial_\mu F e_1 e_2 F) - \frac{1}{2}\partial_\mu(F e_1 e_2 F)]\frac{1}{2}(1-e_5), \end{aligned} \tag{3.144}$$

which gives

$$j(1-e_5) = \sigma\partial\eta(1-e_5) + \frac{1}{\sigma+2F_{21}}\left[(\partial F)e_1 e_2 F - \frac{1}{2}\partial(F e_1 e_2 F)\right]\frac{1}{2}(1-e_5). \tag{3.145}$$

Witten's monopole equations can be written therefore as

$$\partial F = (\sigma A - j)(1-e_5), \tag{3.146}$$

where the last term is given by eq.(3.145), and now it involves only electromagnetic quantities. The term $\partial\eta$ that comes from the arbitrary phase of the spinor can be eliminated by an appropriate choice of gauge. The quaternionic representation of this equation can be easily obtained using the quaternionic representation of $C\ell_4$ as we have discussed.

The above representation has the merit of eliminating the spinor ψ. On the other hand, an evident disadvantage is that it is not much appropriated for manipulations due to the very complicated form of eq.(3.145). We shall now look for another representation that not only we hope to be easier to manipulate, but also that could show us some other features of the monopole equations.

Looking again to eq.(3.84) and eq.(3.90) we see that

$$F = R(\sigma \boldsymbol{e}_2\boldsymbol{e}_1)\tilde{R}\frac{1}{2}(1-\boldsymbol{e}_5). \tag{3.147}$$

We call the term in parenthesis an *extremal* electromagnetic field, that is

$$F_{\text{ext}} = \sigma \boldsymbol{e}_2\boldsymbol{e}_1. \tag{3.148}$$

The name extremal field comes from an analogy with some previous works [17] related to spinorial representations of Maxwell equations and its relationship with Dirac equation in the case of Minkowski spacetime, where we called an extremal field an electromagnetic field that has only electric or magnetic components and which is parallel to a given spatial direction.

From eq.(3.147) we can write

$$F_{\text{ext}}\frac{1}{2}(1-\boldsymbol{e}_5) = \tilde{R}FR. \tag{3.149}$$

In order to exploit this definition, let us write

$$\boldsymbol{e}'^{\mu} = \tilde{R}\boldsymbol{e}^{\mu}R, \quad \partial' = \boldsymbol{e}'^{\mu}\partial_\mu. \tag{3.150}$$

Then we have that

$$\begin{aligned}(\partial' \wedge A')\frac{1}{2}(1-\boldsymbol{e}_5) &= \boldsymbol{e}'^{\mu}\wedge\partial_\mu(\tilde{R}AR)\frac{1}{2}(1-\boldsymbol{e}_5)\\ &= \left[\boldsymbol{e}'^{\mu}\wedge[(\partial_\mu\tilde{R})RA' + A'\tilde{R}(\partial_\mu R)] + \tilde{R}(\boldsymbol{e}^{\mu}\wedge\partial_\mu A)R\right]\frac{1}{2}(1-\boldsymbol{e}_5)\\ &= \left[\boldsymbol{e}'^{\mu}\wedge\frac{1}{2}[\omega_\mu, A'] + F_{\text{ext}}\right]\frac{1}{2}(1-\boldsymbol{e}_5),\end{aligned} \tag{3.151}$$

that is,

$$F_{\text{ext}}\frac{1}{2}(1-\boldsymbol{e}_5) = \left(\partial'\wedge A' - \boldsymbol{e}'^{\mu}\wedge\frac{1}{2}[\omega_\mu, A']\right)\frac{1}{2}(1-\boldsymbol{e}_5) = (D'\wedge A')\frac{1}{2}(1-\boldsymbol{e}_5), \tag{3.152}$$

where we defined

$$D' = \boldsymbol{e}'^{\mu}D_\mu, \tag{3.153}$$

$$D_\mu = \partial_\mu - \frac{1}{2}[\omega_\mu, \ ], \tag{3.154}$$

$$\omega_\mu = 2(\partial_\mu \tilde{R})R = -2\tilde{R}\partial_\mu R, \tag{3.155}$$

where in the last equality we used the fact that $R\tilde{R} = \tilde{R}R = 1$.

In an analogous way we get that

$$\partial' F_{\text{ext}} \frac{1}{2}(1 - e_5) = \left(e'^{\mu} \frac{1}{2}[\omega_\mu, F_{\text{ext}}] + \tilde{R}(\partial F)R \right) \frac{1}{2}(1 - e_5), \tag{3.156}$$

that is,

$$D' F_{\text{ext}} \frac{1}{2}(1 - e_5) = \tilde{R}(\partial F)R. \tag{3.157}$$

Using Witten's equation in the form (3.100) we have from eq.(3.156) that

$$D' F_{\text{ext}} \frac{1}{2}(1 - e_5) = (A'\sigma - j')(1 - e_5). \tag{3.158}$$

Let us now use eq.(3.90) in order to rewrite the term $j'(1 - e_5)$. Note that we can write

$$\partial_\mu \psi = \frac{1}{2}(\partial \ln \sigma + \Omega_\mu)\psi, \tag{3.159}$$

where we defined

$$\Omega_\mu = 2(\partial_\mu R)\tilde{R}. \tag{3.160}$$

Then we have that

$$\langle \partial_\mu \psi e_2 e_1 \psi \rangle_0 = \langle \Omega_\mu F \rangle_0, \tag{3.161}$$

where we used eq.(3.84) and the fact that $\langle (\partial_\mu \ln \sigma F \rangle_0 = 0$. But using eq.(3.149) we have that

$$\langle \Omega_\mu F \rangle_0 = -\langle \omega_\mu F_{\text{ext}} \frac{1}{2}(1 - e_5) \rangle_0, \tag{3.162}$$

where we used $\langle AB \rangle_0 = \langle BA \rangle_0$ and

$$\omega_\mu = -\tilde{R}\Omega_\mu R, \tag{3.163}$$

which follows from eq.(3.155) and eq.(3.160).

Using the above results we can write

$$j'(1 - e_5) = -e'^{\mu} \langle \omega_\mu F_{\text{ext}} \frac{1}{2}(1 - e_5) \rangle_{0\oplus 4} \frac{1}{2}(1 - e_5), \tag{3.164}$$

where we defined $\langle \ \rangle_{0\oplus 4} = \langle \ \rangle_0 + \langle \ \rangle_4$.

Now, we can use the above expression in eq.(3.158). We get that

$$\begin{aligned} &e'^{\mu} \partial_\mu F_{\text{ext}} \frac{1}{2}(1 - e_5) - e'^{\mu} \frac{1}{2}[\omega_\mu, F_{\text{ext}}] \frac{1}{2}(1 - e_5) \\ &\quad - e^{\mu} \langle \omega_\mu F_{\text{ext}} \frac{1}{2}(1 - e_5) \rangle_{0\oplus 4} = A'\sigma(1 - e_5), \end{aligned} \tag{3.165}$$

where we used eqs.(3.153,3.154). But if now we used the fact that $B_2 \in \bigwedge^2(V)$

$$\langle B_2 A\rangle_2 = \frac{1}{2}[B_2, A], \quad \forall A \in \mathcal{C}\ell(V), \tag{3.166}$$

we can write, since $\omega_\mu \in \bigwedge^2(\mathbb{R}^4)$,

$$\begin{aligned} &\frac{1}{2}[\omega_\mu, F_{\text{ext}}]\frac{1}{2}(1-e_5) + \langle \omega_\mu F_{\text{ext}}\frac{1}{2}(1-e_5)\rangle_{0\oplus 4} \\ &\qquad = \langle \omega_\mu F_{\text{ext}}\frac{1}{2}(1-e_5)\rangle_{0\oplus 2\oplus 4} = \omega F_{\text{ext}}\frac{1}{2}(1-e_5). \end{aligned} \tag{3.167}$$

Eq.(3.165) can now be written as

$$e'^\mu \left(\partial_\mu F_{\text{ext}} - \omega_\mu F_{\text{ext}}\right)\frac{1}{2}(1-e_5) = A'\sigma(1-e_5), \tag{3.168}$$

or using eq.(3.148),

$$e'^\mu \left(\partial_\mu \sigma - \omega_\mu \sigma\right) e_2 e_1 \frac{1}{2}(1-e_5) = 2A'\sigma\frac{1}{2}(1-e_5). \tag{3.169}$$

The above expression is another representation of Witten's monopole equation. It has the very interesting feature that the term in parenthesis on its left hand side has the form of a spinor covariant derivative. In fact, it can be put in a more convenient form, namely,

$$e'^\mu \left(\partial_\mu - \frac{1}{2}\omega_\mu\right)\sqrt{\sigma} e_2 e_1 \frac{1}{2}(1-e_5) = A'\sqrt{\sigma}\frac{1}{2}(1-e_5), \tag{3.170}$$

or also

$$D'\Phi e_2 e_1 = A'\Phi, \tag{3.171}$$

where $D' = e'^\mu D_\mu$, with D_μ being the *spinor* covariant derivative

$$D_\mu = \partial_\mu - \frac{1}{2}\omega_\mu \tag{3.172}$$

acting on the *spinor* Φ,

$$\Phi = \sqrt{\sigma}\frac{1}{2}(1-e_5). \tag{3.173}$$

The connection $-\omega_\mu$ is the *spinor connection*. Moreover, since $R = R(1/2)(1-e_5) = U(1/2)(1-e_5)$ with $U \in \text{Spin}(3) \simeq \text{SU}(2)$, we have a connection ω_μ with values in the Lie algebra of SU(2). The above representation of Witten's monopole equation could therefore be very usuful in relation to Donaldson's theory of 4-manifolds.

References

[1] E. Witten, Math. Res. Lett. 1 (1994), 769.

[2] I. Porteous, Clifford Algebras and the Classical Groups, Cambridge Univ. Press, 1995.

[3] W. Greub, Multilinear Algebra, 2nd. ed., Springer-Verlag, 1978.

[4] A. Crumeyrolle, Orthogonal and Symplectic Clifford Algebras, Kluwer Acad. Publ., 1990.

[5] I. Benn and R. Tucker, An Introduction to Spinors and Geometry, Adam Hilger, 1987.

[6] V. L. Figueiredo, E. C. Oliveira and W. A. Rodrigues, Jr., Internat. J. Theoret. Phys. 29 (1990), 371.

[7] C. Chevalley, The Algebraic Theory of Spinors, Columbia Univ. Press, 1954.

[8] Z. Oziewicz, From Grassmann to Clifford, in: Clifford Algebras and their Applications in Mathematical Physics (J. Chisholm and A. Common, eds.), D. Reidel Publ. Co., 1986, 245.

[9] P. Lounesto and G. P. Wene, Acta Applic. Math. 9 (1987), 165.

[10] J. Vaz, Jr., PhD Thesis, Unicamp, 1993.

[11] W. A. Rodrigues, Jr., Q. A. G. Souza and J. Vaz, Jr., Spinor Fields and Superfields and Equivalence Classes of Exterior Algebra Fields, in: Clifford Algebras and Spinor Structures (R. Ablamowicz and P. Lounesto, eds.), Kluwer Acad. Publ. 1995, 177.

[12] G.de Rham, Variétés Différentiables, Formes, Courants, Formes Harmoniques, Hermann, Paris 1955.

[13] W. A. Rodrigues, Jr., M. A. Faria-Rosa, A. Maia, Jr. and E. Recami, Hadronic J. 12 (1989), 187.

[14] W. A. Rodrigues, Q. A. Souza, J. Vaz, Jr. and P. Lounesto, Dirac–Hestenes Spinor Fields over Riemann–Cartan Manifolds, Internat. J. Theoret. Phys. 35 (1996), 1849–1900.

[15] J. Crawford, J. Math. Phys. 26 (1985), 1439.

[16] J. Vaz, Jr., Construction of Monopoles and Instantons by using Spinors and the Inversion Theorem, to appear in: Clifford Algebras and their Applications in Mathematical Physics, Proceedings of the 4th International Conference, Aachen 1996.

[17] J. Vaz, Jr. and W. A. Rodrigues, Jr., Maxwell and Dirac Theories as an Already Unified Theory, Adv. Applied Clifford Algebras 7 (Supplement) (1997), 369–386.

Sobolev classes and quasi-conformal mappings on Carnot–Carathéodory spaces

Serguei Vodop'yanov†

Abstract. The theory of quasi-conformal mappings and the theory of Sobolev function classes in the geometry of vector fields satisfying Hörmander's hypoellipticity condition have been used to establish a connection between Sobolev classes and quasi-conformal mappings on Carnot–Carathéodory spaces: a homeomorphism of a compact domain of a Carnot–Carathéodory space induces an isomorphism of the Sobolev classes L_p^1, $p \geq 1$, according to the composition rule iff it is quasi-conformal when p equals the Hausdorff dimension of the space, and quasi-isometric otherwise.

1991 Mathematics Subject Classification: 31, 45.

Key words: Carnot–Carathéodory spaces, Sobolev function classes, quasi-conformal mappings, embedding theorems.

The theory of quasi-conformal mappings and the theory of Sobolev function spaces have been used to establish metric properties of homeomorphisms inducing isomorphisms of Sobolev classes in the Euclidean space [VG1, VG2, V3] and in Carnot groups [V1, V2] (see also [HK, Ma, R, V4] where different aspects of this problem in the Euclidean space have been studied). The goal of the present paper is to describe homeomorphisms of domains in Carnot–Carathéodory spaces inducing isomorphisms of Sobolev classes defined in the geometry of vector fields satisfying some conditions. The main properties of quasi-conformal mappings in such spaces were studied by Margulis and Mostow in [MM]. Embedding theorems of Sobolev spaces and pointwise estimates we use to prove our results are exposed in [CV, GN, FLW, H, L, MS].

Let M be a smooth connected manifold, T its tangent bundle and H a subbundle of T. Given $x \in M$, let T_x^i be the subspace of T_x spanned by a local basis of smooth vector fields $X_1, \ldots, X_n$ about x for the subbundle H, together with all commutators of these vector fields of order $\leq i$. The pair $(M; H)$ is called [G] a Carnot–Carathéodory space of depth m if $\dim T_x^i$ is independent of x for all $x \in M$, and $m = \inf\{k : T_x^k = T_x\}$.

† The research is financially supported by the Russian Foundation for Basic Research (Grant 96–01–01769)

We suppose that M is endowed with a metric tensor. A piecewise-smooth curve $\gamma\colon [a,b] \to M$ is called *horizontal* if $\frac{d\gamma}{dt} \in H_{\gamma(t)}$ for almost every $t \in [a,b]$. The length of such horizontal curve is defined as $\|\gamma\| = \int_a^b \|\dot\gamma(t)\|\,dt$. The Carnot–Carathéodory distance between two points $x_1, x_2 \in M$ is

$$d(x_1,x_2) = \inf\{\|\gamma\| : \gamma \text{ is a horizontal curve with endpoints } x_1 \text{ and } x_2\}.$$

The topology determined by the metric d coincides with the topology of the differentiable manifold M.

According to [Mi], the Hausdorff dimension of M with respect to the metric d is

$$\nu = \sum_{k=1}^{m} k(\dim T_x^k - \dim T_x^{k-1}).$$

It is proved in [Mi] that for every compact set $F \subset M$ and for all $x \in F$ the Lebesgue measure of the ball $B(x,r) = \{y \in M : d(x,y) < r\}$, $r > 0$, $x \in M$, satisfies the condition $c_1 r^\nu \le |B(x,r)| \le c_2 r^\nu$ for r small enough, where c_1 and c_2 depend only on the compact set F (the *regularity* property of the measure). If μ is the Hausdorff ν-measure relative to the metric d then on any compact subset of M μ is comparable with a Lebesgue measure.

If $X \in H_x$ and $g\colon M \to \mathbb{R}$ is a smooth function then the linear functional Xg on H_x has the representation $Xg = \langle X, a\rangle_x$. The vector $a \in H_x$ is called the *subgradient* of g at $x \in M$, denoted by $\nabla_{\mathcal{L}} g(x)$. For $g \in C^\infty(D)$, $D \subset M$, set

$$\|g\|_{1,p} = \left(\int_D |\nabla_{\mathcal{L}} g|^p\,d\mu\right)^{1/p}, \quad p \in [1,\infty],$$

where $|\cdot|(x) = \langle\cdot,\cdot\rangle_x^{1/2}$. Define the Sobolev space $L_p^1(D) \subset L_{p,\mathrm{loc}}(D)$ as the closure of $\{g \in C^\infty(D) : \|g\|_{1,p} < \infty\}$ with respect to the norm $\|g\|_{1,p}$. In other words, u belongs to $L_p^1(D)$ iff u is locally integrable ($u \in L_{p,\mathrm{loc}}(D)$) and there is a cross-section $v\colon D \to H_D$, defined almost everywhere on D, such that $|v| \in L_p(D)$ and

$$\int_F |g_i - u|^p\,d\mu \to 0 \quad \text{and} \quad \int_D |\nabla_{\mathcal{L}} g_i - v|^p\,d\mu \to 0 \quad \text{as} \quad i \to \infty$$

for some sequence $g_i \in C^\infty(D)$ depending on the choice of the compact domain $F \Subset M$. The vector-valued function v is called the *subgradient* of u in the space $L_p^1(D)$, denoted by $\nabla_{\mathcal{L}} u$. The sugradient of a function in $L_p^1(D)$ is well defined.

Let M, M' be two Carnot–Carathéodory spaces with metrics d, d', and let $\Omega \subset M$, $\Omega' \subset M'$ be domains in these spaces. A mapping $\varphi\colon \Omega \to M'$ is said to satisfy the *Lipschitz condition* if the quantity

$$\overline{\lim_{y\to x}} \frac{d'(\varphi(x),\varphi(x))}{d(x,y)} = L(x;\varphi)$$

is bounded on Ω. A homeomorphism $\varphi\colon \Omega \to \Omega'$ is called *quasi-isometric* if both mappings φ and φ^{-1} satisfy the Lipschitz condition. The number $L_\varphi = \max(\|L(x;\varphi) \mid L_\infty(\Omega)\|, \|L(y,\varphi^{-1}) \mid L_\infty(\Omega')\|)$ is called the *quasi-isometry coefficient* of φ. A homeomorphism $\varphi\colon \Omega \to \Omega'$ is called *quasi-conformal* if

$$H_\varphi(x) = \overline{\lim_{r\to 0}} \frac{\max\limits_{d(x,y)=r} d(\varphi(x),\varphi(y))}{\min\limits_{d(x,y)=r} d(\varphi(x),\varphi(y))}$$

is bounded in Ω. The number $H_\varphi = \|H_\varphi(x) \mid L_\infty(\Omega)\|$ is called the *quasi-conformality coefficient* of φ. It is evident that every quasi-isometric mapping is quasi-conformal with the quasi-conformality coefficient $H_\varphi \le L_\varphi^2$. The reader can consult about the properties of quasi-conformal mappings on Carnot groups and Carnot–Carathéodory spaces in [He, KR, MM, P, V1, V2] where references to earlier papers can be found.

The main results of the present article are the following theorems.

Theorem 1. *Let $\Omega \Subset M$ and $\Omega' \Subset M'$ be compact domains in Carnot–Carathéodory spaces M and M' respectively. If a homeomorphism $\varphi\colon \Omega \to \Omega'$ induces a bounded operator $\varphi^*\colon L^1_{p'}(\Omega') \to L^1_p(\Omega)$ of Sobolev spaces with $p = \nu$, $p' = \nu'$ (a bounded isomorphism with $1 \le p \le \infty$, $p \ne \nu$, $1 \le p' \le \infty$, $p' \ne \nu'$) by the composition rule $\varphi^*(u) = u \circ \varphi$ then φ is a quasi-conformal (quasi-isometric) mapping. Moreover, the Hausdorff dimensions of M and M' coincide, and $p = p'$.*

Theorem 2. *Let $\Omega \Subset M$ and $\Omega' \Subset M'$ be compact domains in Carnot–Carathéodory spaces M and M' of the same Hausdorff dimension. Then every quasi-conformal (quasi-isometric) homeomorphism $\varphi\colon \Omega \to \Omega'$ induces a bounded isomorphism $\varphi^*\colon L^1_p(\Omega') \to L^1_p(\Omega)$ of Sobolev spaces if $p = \nu$ $(1 \le p \le \infty)$.*

Remark 1. Theorem 1 in the Euclidean case and $n = n'$, $p = p'$ is essentially contained in [Ge]. In the case of Carnot groups Theorems 1–2 are proved in [V1, V2, V3]. Our proof is new when $p < \nu$ and its idea may be applicable to different situation. We formulate here the following assertion which one can prove adapting the method presented below.

Let $D \subset M$ be compact. Define the Sobolev space $\overset{\circ}{L}{}^1_p(D) \subset L^1_p(D)$ as the closure of $\{g \in C_0^\infty(D) : \|g\|_{1,p} < \infty\}$ with respect to the norm $\|g\|_{1,p}$.

Theorem 3. *Let $\Omega \Subset M$ and $\Omega' \Subset M$ be compact domains in Carnot–Carathéodory space M of the same Hausdorff dimension ν. A measurable mapping $\varphi\colon \Omega \to \Omega'$ induces a bounded isomorphism $\varphi^*\colon \overset{\circ}{L}{}^1_p(\Omega') \to \overset{\circ}{L}{}^1_p(\Omega)$ of Sobolev spaces with $1 \le p < \nu$ by the composition rule $\varphi^*(u) = u \circ \varphi$ iff φ coincides almost everywhere with a* *quasi-isometric mapping* $\widetilde{\varphi}\colon \Omega \to M$. *Moreover, in the case* $1 < p < \nu$ *the capacity* $\operatorname{cap}((\Omega' \setminus \widetilde{\varphi}(\Omega)) \cup (\widetilde{\varphi}(\Omega) \setminus \Omega');\, W_p^1(M))$ *equals zero.*

Our method is applicable to Sobolev spaces on a metric space in the sense of P. Hajłasz when $p > s$ [H].

The proof of Theorem 1 is contained in the following lemmas. To state the first of them we need the concept of a maximal function: given a locally summable function g, put

$$M_\delta g(x) = \sup\left\{\frac{1}{|B(x,r)|}\int\limits_{B(x,r)} |g|\,d\mu(x) : r \le \delta\right\}.$$

Applying the pointwise estimate of [FLW, Proposition 2.12] and the arguments of [V2, Theorem 1] we obtain the following two results.

Lemma 1. *Let* $D \Subset M$ *be a compact domain,* $1 < p \le \infty$. *There are constants* $\gamma > 2$, c *and* C *such that for every function* $f \in L_p^1(D)$ *and every ball* $B = B(a,r)$, $\gamma B = B(a,\gamma r) \subset D$ *and* $r < r_0$, *with* r_0 *a positive number depending on the geometry of vector fields, the inequality*

$$|f(x) - f(y)| \le C d(x,y)\big(M_{cd(x,y)}(|\nabla_{\mathcal{L}} f|)(x) + M_{cd(x,y)}(|\nabla_{\mathcal{L}} f|)(y)\big) \qquad (1)$$

holds for all points $x, y \in B \setminus S$, *where* $S \subset B$ *is some set of measure zero.*

Proof. For a locally integrable function and an arbitrary ball B, consider the mean $f_B = \fint\limits_B f\,dx$ of f on B. If $f \in L_p^1(D)$, $p \in (1,\infty]$, then the following pointwise estimate is valid for the potential [FLW, Proposition 2.12]:

$$|f(x) - f_B| \le \widetilde{C}\int\limits_{\tau B} \frac{|\nabla_{\mathcal{L}} f(z)|}{d(x,z)^{\nu-1}}\,d\mu(z) \qquad (2)$$

for almost every $x \in B$, where B is a ball whose radius is such that $\tau B \subset D$ for some constant $\tau > 1$ independent of the choice of B (here and in the sequel, given $B = B(x,t)$, kB is the ball $B(x,kt)$). For points $x, y \in D$ close enough to one another, find a ball B of least radius with center a such that $x, y \in B$ and $d(x,a) = d(y,a)$. Applying (2), we obtain

$$\begin{aligned}|f(x) - f(y)| &\le |f(x) - f_B| + |f(y) - f_B|\\ &\le \widetilde{C}\Big(\int\limits_{\tau B} \frac{|\nabla_{\mathcal{L}} f(z)|}{d(x,z)^{\nu-1}}\,\mu(z) + \int\limits_{\tau B} \frac{|\nabla_{\mathcal{L}} f(z)|}{d(y,z)^{\nu-1}}\,\mu(z)\Big) = C(I_1 + I_2).\end{aligned}$$

Estimate the first integral in the sum on the right-hand side; the second is estimated in a similar fashion. It is obvious that $\tau B \subset B_1 = B(x,t)$, where $t = (\tau+1)d(x,a)$.

Hence,

$$I_1 \le \int_{B_1} \frac{|\nabla_{\mathcal{L}} f(z)|}{d(x,z)^{\nu-1}}\,\mu(z) = \sum_{j=1}^{\infty} \int_{\frac{1}{2^{j-1}}B_1 \setminus \frac{1}{2^j}B_1} \frac{|\nabla_{\mathcal{L}} f(z)|}{d(x,z)^{\nu-1}}\,\mu(z)$$

$$\le t \sum_{j=1}^{\infty} \frac{2^\nu}{2^j} \frac{1}{|\frac{1}{2^{j-1}}B_1|} \int_{\frac{1}{2^{j-1}}B_1} |\nabla_{\mathcal{L}} f(z)|\,\mu(z) \le 2^\nu t M_t(|\nabla_{\mathcal{L}} f|)(x).$$

Finally, we arrive at the following result: if $f \in L_p^1(D)$, $p \in (1, \infty]$, then there is a set S of measure zero such that the following inequality holds for $x, y \in B \setminus S$:

$$|f(x) - f(y)| \le C d(x,y)(M_{cd(x,y)}(|\nabla_{\mathcal{L}} f|)(x) + M_{cd(x,y)}(|\nabla_{\mathcal{L}} f|)(y))$$

with $C = \frac{\tau+1}{2} 2^\nu \widetilde{C}$. Thus, for all points $x, y \in D$ close enough to one another, (1) holds with the constants $C = \frac{\tau+1}{2} 2^\nu \widetilde{C}$, $c = \frac{\tau+1}{2}$ and $\gamma = \tau + 2$, where the last value is defined from the condition $B(x, 2cr) \subset B(a, \gamma r)$, $x \in B(a, r)$. □

Remark 2. Inequality (1) is also valid in a more general situation. Since (2) is available in the geometry of arbitrary vector fields satisfying the Hörmander hypoellipticity condition, (1) is valid in this more general case (see similar arguments in the proof of Theorem 6.14 in [CV, VC]).

Denote by $d_D(x, y)$ the intrinsic metric in the domain D: $d_D(x, y) = \inf\{l(\gamma) \mid \gamma\colon [0,1] \to D$ is a horizontal curve with endpoints x and $y\}$.

Corollary 1. *Let $D \Subset M$ be a compact domain and let $f \in L_p^1(D)$, $1 \le p \le \infty$, be a function such that $|\nabla_{\mathcal{L}} f|(x) \le M$ almost everywhere on D. Then there exists a positive constant C such that*

$$|f(x) - f(y)| \le C M d_D(x, y).$$

Lemma 2. *Let $D \Subset M$ be a compact domain, $\nu < p \le \infty$, and $\gamma > 2$ the constant of Lemma 1. Then there is a positive constant C such that for every function $f \in L_p^1(D)$ and every ball $\widetilde{B} = B(z, r)$ such that $2\gamma \widetilde{B} = B(z, 2\gamma r) \subset D$, $r < r_0$ with r_0 a positive number depending on the geometry of vector fields, the estimate*

$$\left(\frac{1}{\mu(B)} \int_B \left| f(x) - \frac{1}{\mu(B)} \int_B f(y)\,d\mu(y)\right|^p d\mu(x)\right)^{1/p}$$

$$\le C\mu(B)^{\frac{p-\nu}{\nu p}} \left(\int_{2\gamma\widetilde{B}} |\nabla_{\mathcal{L}} f|(x)^p\,d\mu(x)\right)^{1/p}, \quad (3)$$

holds for every ball $B = B(\xi, \tau)$ with center $\xi \in \widetilde{B}$ and radius $\tau \le r$.

Proof. Put $g_t(x) = M_t(|\nabla_{\mathcal{L}} f|)(x)$, $x \in B(z,r)$, $t = 2cr$, with c the constant of Lemma 1. By the maximal function theorem [ST] g_t belongs to $L_p(\widetilde{B})$ and $\|g_t \mid L_p(\widetilde{B})\| \le M\|\nabla_{\mathcal{L}} f \mid L_p(2\gamma\widetilde{B})\|$, $1 < p \le \infty$. Let $B = B(\xi,\tau)$, $\xi \in \widetilde{B}$, be an arbitrary ball of radius $\tau \le r$ containing points $x, y \in \widetilde{B}$. Integrating the left- and right-hand sides of (1) over B with respect to y, we obtain

$$\left|f(x) - \frac{1}{\mu(B)}\int\limits_B f(y)\,d\mu(y)\right| \le 2Cr\left(g_t(x) + \frac{1}{\mu(B)}\int\limits_B g_t(y)\,d\mu(y)\right).$$

Raising the left- and right-hand sides of this relation to the power p and integrating with respect to x, we arrive at the inequality

$$\left(\frac{1}{\mu(B)}\int\limits_B\left|f(x) - \frac{1}{\mu(B)}\int\limits_B f(y)\,d\mu(y)\right|^p d\mu(x)\right)^{1/p}$$
$$\le 2C\tau\mu(B)^{-1/p}\left(\int\limits_B g_t(y)^p\,d\mu(y)\right)^{1/p}. \quad (4)$$

As was mentioned, the measure μ possesses the extra property

$$\mu(B(\xi,\tau)) \ge \kappa\tau^\nu, \quad \tau \le r, \quad (5)$$

for some positive constant κ independent of ξ. To finish the proof of (3) it suffices to replace τ on the right-hand side of (4) with the larger value $\kappa^{-1/\nu}\mu(B(\xi,\tau))^{1/(\nu)}$ (see (5)) and apply the estimate for the maximal function. □

The main result of [MS] asserts that every function satisfying condition (3), if redefined on a set of measure zero, becomes continuous and moreover satisfies the inequality

$$|f(x) - f(y)| \le C(\mu(B))^{\frac{p-\nu}{\nu p}}\left(\int\limits_{2\gamma B} |\nabla_{\mathcal{L}} f|(z)^p\,d\mu(z)\right)^{1/p} \quad (6)$$

for every ball B containing the points x and y, $2\gamma B \subset D$, with γ the constant of Lemma 1 (the constant C is independent of the choice of f and B). Taking the regularity property of the measure into account, we infer

Lemma 3. *Let $D \Subset M$ be a compact domain, $\nu < p \le \infty$, and $\gamma > 2$ the constant of Lemma* 1. *If $f \in L^1_p(D)$ and $2\gamma B \subset D$,* rad $B < r_0$, *where r_0 is a positive number depending on the geometry of vector fields, then*

$$|f(x) - f(y)| \le Cd(x,y)^{1-\frac{\nu}{p}}\left(\int\limits_{2\gamma B} |\nabla_{\mathcal{L}} f|(z)^p\,d\mu(z)\right)^{1/p}$$

for all points $x, y \in B$ (the constant C is independent of the choice of B and f).

Remark 3. Lemma 3 may also derived straightforward from (1) by using the embedding theorem of [H].

Let $D \Subset M$ be an open connected set and let F_0, F_1 two closed sets in the closure of D. The *capacity* $\operatorname{cap}(F_0, F_1; L^1_p(D))$ of the pair F_0, F_1 in the space $L^1_p(D)$ is defined to be $\inf \|f \mid L^1_p(D)\|^p$, where the infimum is taken over all continuous functions $f \in L^1_p(D)$ such that f equals 1 (0) on some neighborhood of F_1 (F_0). If $F_0 = \partial D$ then instead of $\operatorname{cap}(\partial D, F_1; L^1_p(D))$ we simply write $\operatorname{cap}(F_1; \overset{\circ}{L}{}^1_p(D))$.

In the case of non-zero capacity and $1 < p < \infty$, it is well known, see for instance [CV, VC], that there exists a unique $u \in L^1_p(D)$ (called *extremal*) such that $\|u \mid L^1_p(D)\|^p = \operatorname{cap}(F_1; \overset{\circ}{L}{}^1_p(D))$.

Lemma 4. *Let $D \Subset M$ be a compact domain, $1 \le p$, $B(x, R) \subset D$ and $0 < r < R \le r_0$, where r_0 is a positive number depending on the geometry of vector fields. Then*

$$\operatorname{cap}(B(x,r); \overset{\circ}{L}{}^1_p(B(x,R))) \sim \begin{cases} \left[\dfrac{R^{\frac{p-\nu}{p-1}} - r^{\frac{p-\nu}{p-1}}}{(p-\nu)/(p-1)}\right]^{1-p} & \text{if } 1 < p \neq \nu, \\ \left(\ln \frac{R}{r}\right)^{1-\nu} & \text{if } 1 < p = \nu, \end{cases}$$

$$\operatorname{cap}(B(x,r); \overset{\circ}{L}{}^1_1(B(x,2r))) \le c r^{\nu-1}$$

where the constants in the equivalence (inequality) depend only on D.

Proof. The claim of the lemma in the case of $p > 1$ follows from [CV, VC, Theorems 6.6 and 6.9]. The estimate in the case of $p = 1$ is obtained by inserting $u(z) = \min(1, \operatorname{dist}(z, S(x, 2r))/r)$ in the relation $\operatorname{cap}(B(x,r); \overset{\circ}{L}{}^1_1(B(x,2r))) \le \|u \mid L^1_1(B(x,2r))\|$. □

Lemma 3 implies

Lemma 5. *If $x, y \in D$ and $B(x, \gamma d(x,y)) \subset D$, with γ depending on the geometry of vector fields, then the estimate*

$$\operatorname{cap}(\{x\}, \{y\}; L^1_p(B(x, \gamma d(x,y)))) \sim \frac{1}{d(x,y)^{p-\nu}}$$

holds, provided that $p > \nu$, where the constants in the equivalence depend on D as well as on p, γ, and ν.

Lemma 6. *Let $\Omega \Subset M$ and $\Omega' \Subset M'$ be two compact domains in Carnot–Carathéodory spaces M and M' with metrics d and d' and the Hausdorff dimensions ν and ν'. If a homeomorphism $\varphi\colon \Omega \to \Omega'$ induces a bounded operator $\varphi^* : L^1_{p'}(\Omega') \to L^1_p(\Omega)$,*

$\nu < p$, $\nu' < p'$, and $\|\varphi^*\|$ is its norm then

$$d'(\varphi(x),\varphi(y))^{\frac{p'-\nu'}{p'}} \le C\|\varphi^*\|d(x,y)^{\frac{p-\nu}{p}} \tag{7}$$

for all points $x, y \in \Omega$ close enough to one another. Moreover, $\frac{p-\nu}{p} \le \frac{p'-\nu'}{p'}$.

Proof. Let $x, y \in \Omega$ be such that the assertion of Lemma 5 is valid for the pair of the one-point sets $\{x\}$, $\{y\}$, and the assertion of Lemma 4 holds for the pair of the sets $S(\varphi(x), d'(\varphi(x),\varphi(y)))$, $\{\varphi(x)\} \subset \Omega'$. Denoting $B = B(\varphi(x), d'(\varphi(x),\varphi(y)))$, we have $\partial B = S$. Therefore,

$$\begin{aligned}\frac{1}{d(x,y)^{\frac{p-\nu}{p}}} &\le c_1 \operatorname{cap}\bigl(\{y\},\{x\}; L^1_p(B(x,\gamma d(x,y)))\bigr)^{1/p} \\ &\le \operatorname{cap}\bigl(\varphi^{-1}(\partial B),\{x\}; L^1_p(\varphi^{-1}(B))\bigr)^{1/p} = c_1 \operatorname{cap}\bigl(\{x\}; \overset{\circ}{L}{}^1_p(\varphi^{-1}(B))\bigr)^{1/p} \\ &\le c_1\|\varphi^*\| \operatorname{cap}\bigl(\{\varphi(x)\}; \overset{\circ}{L}{}^1_{p'}(B)\bigr)^{1/p'} \le \frac{c_1c_2\|\varphi^*\|}{d'(\varphi(x),\varphi(y))^{\frac{p'-\nu'}{p'}}}.\end{aligned}$$

This implies (7). Were $\frac{p-\nu}{p} > \frac{p'-\nu'}{p'}$, we would have $|\nabla_{\mathcal{L}}(f\circ\varphi)|(x) = 0$ almost everywhere on Ω for every non-constant function $f\colon \Omega' \to \infty$ satisfying the Lipschitz condition. By Corollary 1, $f\circ\varphi$ would identically equal some constant, which is impossible since φ is a homeomorphism. □

Proof of Theorem 1. Theorem 1 in the case of $\nu < p$, $\nu' < p'$ follows from Lemma 6. Indeed, if φ^* is an isomorphism then $\frac{p-\nu}{p} = \frac{p'-\nu'}{p'}$ and φ is quasi-isometric and hence quasi-conformal. From [MM] we have $\nu = \nu'$, implying $p = p'$. □

Lemma 7. *Let $D \Subset M$ be a compact domain, $1 \le p < \nu$. Then there is a constant C such that for every measurable set $A \subset B$, $B \subset D$,* rad B *is small enough, the inequality $\mu(A)^{\frac{\nu-p}{\nu p}} \le C\operatorname{cap}\bigl(A; \overset{\circ}{L}{}^1_p(B)\bigr)^{1/p}$ holds.*

Proof. Assume the size of the ball B small so that it possible to apply the Sobolev inequality [GN, H, L]: $\|f \mid L_q(B)\| \le C\bigl\|f \mid \overset{\circ}{L}{}^1_p(B)\bigr\|$, where $q = \nu p/(\nu - p)$. Involving the inequality $\mu(A)^{1/q} \le \|f \mid L_q(B)\|$ valid for all measurable functions $f \in L_q(B)$ such that $f \ge 1$ almost everywhere on A, we have $\mu(A)^{\frac{\nu-p}{\nu p}} \le C\bigl\|f \mid \overset{\circ}{L}{}^1_p(B)\bigr\|$ with $f \ge 1$ almost everywhere on A. □

Lemma 8. *Let $\Omega \Subset M$ and $\Omega' \Subset M'$ be two compact domains in Carnot–Carathéodory spaces M and M' with Hausdorff measures μ and μ' and Hausdorff dimensions ν and ν'. If a homeomorphism $\varphi\colon \Omega \to \Omega'$ induces a bounded operator $\varphi^*\colon L^1_{p'}(\Omega') \to L^1_p(\Omega)$, $1 \le p < \nu$, $1 \le p' < \nu'$, then there is a constant c*

such that $\mu(\varphi^{-1}(B(x,r))^{\frac{\nu-p}{\nu p}} \le c\mu'(B(x,r))^{\frac{\nu'-p'}{\nu' p'}}$ *for every ball* $B(x,r) \subset \Omega'$ *of a sufficiently small radius. Moreover,* $\frac{\nu-p}{\nu p} \ge \frac{\nu'-p'}{\nu' p'}$.

If $\varphi^*\colon L^1_{p'}(\Omega') \to L^1_p(\Omega)$ *is a bounded isomorphism then* $\frac{\nu-p}{\nu p} = \frac{\nu'-p'}{\nu' p'}$. *In this event, the set function* $A \to \mu'(\varphi(A))$, $A \subset \Omega$ ($A \to \mu(\varphi^{-1}(A))$, $A \subset \Omega'$) *is absolutely continuous and there is a positive constant* M *such that*

$$M^{-1}\|\varphi^*\|^{-\frac{\nu p}{\nu-p}} \le \mathcal{J}(x;\varphi) = \lim_{r\to 0}\frac{\mu'(\varphi(B(x,r)))}{\mu(B(x,r))} \le M\|\varphi^{*-1}\|^{\frac{\nu p}{\nu-p}} \tag{8}$$

for all $x \in \Omega$.

Proof. Assume the radius of $B = B(x,r)$ be small so that $2B \subset \Omega'$, Lemma 4 applies to $2B$ and $\varphi^{-1}(B)$ is contained in some ball Q wherein Lemma 7 is effective. Then we have

$$\begin{aligned}\mu(\varphi^{-1}(B))^{\frac{\nu-p}{\nu p}} &\le C\operatorname{cap}\big(\varphi^{-1}(B); \overset{\circ}{L}{}^1_p(\varphi^{-1}(2B))\big)^{1/p}\\ &\le C\|\varphi^*\|\operatorname{cap}\big(B; \overset{\circ}{L}{}^1_p(2B)\big)^{1/p'} \le Cc\|\varphi^*\| r^{\frac{\nu'-p'}{p'}} \le Cc\|\varphi^*\|\mu'(B)^{\frac{\nu'-p'}{\nu' p'}}.\end{aligned}$$

If $\frac{\nu-p}{\nu p} < \frac{\nu'-p'}{\nu' p'}$ then the set function $F \to \mu(\varphi^{-1}(F))$ defined on the σ-algebra of the Borel sets $F \subset \Omega'$ is absolutely continuous and its Radon–Nikodim derivative vanishes by the Lebesgue differentiation theorem. Therefore $\mu(\varphi^{-1}(F)) = 0$ for every Borel set $F \subset \Omega'$. This contradicts the homeomorphy of φ.

If $\varphi^*\colon L^1_{p'}(\Omega') \to L^1_p(\Omega)$ is a bounded isomorphism then $\frac{\nu-p}{\nu p} = \frac{\nu'-p'}{\nu' p'}$ by symmetry. This yields the remaining assertions of Lemma 8. □

Lemma 9. *Under the condition of the previous lemma, let* $\varphi^*\colon L^1_{p'}(\Omega') \to L^1_p(\Omega)$ *be a bounded isomorphism. Then* $p = p'$ *and* $\nu = \nu'$. *Moreover, there is a constant* C *such that, for every function* $u \in L^1_{p'}(\Omega')$ *with* $|\nabla_{\mathcal{L}} u|(x) \le 1$ *at almost every* $x \in \Omega'$, *the composite function* $u \circ \varphi \in L^1_p(\Omega)$ *satisfies the inequality* $|\nabla_{\mathcal{L}}(u \circ \varphi)|(x) \le C$ *almost everywhere on* Ω.

Proof. Let $u\colon \Omega' \to \mathbb{R}$ be a continuous function belonging to $L^1_\infty(\Omega')$ and such that $\|u \mid L^1_\infty(\Omega')\| \le 1$. Fix a point $x \in \Omega'$ and put $\eta(z) = \min(1, \operatorname{dist}(z, S(x,2r))/r)$ where r is such that $2B \subset \Omega'$. From [CV, VC, Lemma 1.1] (see also Lemmas 11 and 12 below) we infer that η belongs to $L^1_\infty(\Omega')$. The following inequalities hold:

$$\begin{aligned}&\Big(\int\limits_{\varphi^{-1}(B)} |\nabla_{\mathcal{L}}(u\circ\varphi)|^p(z)\,d\mu(z)\Big)^{1/p}\\ &\qquad\le \|\varphi^*\|\Big(\int\limits_{2B} |\nabla_{\mathcal{L}}((u-u(x))\eta)|^{p'}\,d\mu'(y)\Big)^{1/p'} \le\end{aligned}$$

$$\le \|\varphi^*\|\Big(\int\limits_{2B} |\nabla_{\mathcal{L}} u|^{p'}(y)\eta^{p'}(y)\,d\mu'(y)\Big)^{1/p'}$$

$$+ \|\varphi^*\|\Big(\int\limits_{2B} |\nabla_{\mathcal{L}}\eta)|^{p'}(y)|u(y)-u(x)|^{p'}(y)\,d\mu'(y)\Big)^{1/p'}$$

$$\le \|\varphi^*\|\mu'(2B)^{1/p'} + K(c_1 r^{-1}c_2 r\mu'(2B))^{1/p'} = C\|\varphi^*\|\mu'(B)^{1/p'}.$$

In view of (8), it follows that

$$\Big(\int\limits_{B} |\nabla_{\mathcal{L}}(u\circ\varphi)|^{p}(\varphi^{-1}(y))\,d\mu'(y)\Big)^{1/p}$$

$$= \Big(\int\limits_{\varphi^{-1}(B)} |\nabla_{\mathcal{L}}(u\circ\varphi)|^{p}(z)\mathcal{J}(z;\varphi)\,d\mu(z)\Big)^{1/p}$$

$$\le \Big(M\|\varphi^{*-1}\|^{\frac{\nu p}{\nu-p}} \int\limits_{\varphi^{-1}(B)} |\nabla_{\mathcal{L}}(u\circ\varphi)|^{p}(z)\,d\mu(z)\Big)^{1/p}$$

$$\le CM^{1/p}\|\varphi^*\|\|\varphi^{*-1}\|^{\frac{\nu}{\nu-p}}\mu'(B)^{1/p'}.$$

If $1/p' > 1/p$ then, applying the Lebesgue differentiation theorem, we obtain

$$|\nabla_{\mathcal{L}}(u\circ\varphi)|^{p}(\varphi^{-1}(y)) = 0 \text{ for almost every } y\in\Omega'.$$

Hence, $u\circ\varphi$ is a constant by Corollary 1. This contradicts the homeomorphy of φ. Thus, $1/p' \le 1/p$ and, by symmetry, $1/p' \ge 1/p$; whence it follows that $p = p'$ and $\nu = \nu'$.

Lemmas 8 and 9 imply Theorem 1 in case $1 \le p < \nu$, $1 \le p' < \nu'$. First and foremost observe that $p = p'$, $\nu = \nu'$. By the Lebesgue differentiation theorem, these relations imply the estimate

$$|\nabla_{\mathcal{L}}(u\circ\varphi)|^{p}(\varphi^{-1}(y)) \le C^{p}M\|\varphi^*\|^{p}\|\varphi^{*-1}\|^{\frac{\nu p}{\nu-p}}.$$

By Lemma 1, this implies $|u\circ\varphi(x) - u\circ\varphi(y)| \le cd(x,y)$ for the points $x, y\in\Omega$ close enough to one another, where $c = \widetilde{C}^{p}M\|\varphi^*\|^{p}\|\varphi^{*-1}\|^{\frac{\nu p}{\nu-p}}$. On putting $u(z) = d'(\varphi(x), z)$ in this relation, we have $d'(\varphi(x),\varphi(y)) \le cd(x,y)$, implying that φ is Lipschitzian (by Lemmas 11 and 12, u belongs to a Sobolev class). By symmetry, φ^{-1} is Lipschitzian as well. □

The quasi-conformality of φ in the case $p = \nu$ can be proved in the same way as in [V1, Theorem 6] wherein the case of the Carnot groups was settled. We should only involve Lemma 4 with $p = \nu$ (the estimate from above) and the following result.

Lemma 10 ([CV], [VC, Corollary 6.12]). *Let $D \Subset M$ be a compact domain and $p = \nu$. Then there exist constants $\overline{\gamma}$, r_0 and $c > 0$ such that, for every ball $B(x, r)$, $r \le r_0$, $\overline{\gamma}B(x,r) \subset D$, the relation*

$$\inf_{F_0, F_1} \operatorname{cap}\big(F_0, F_1; L^1_\nu(B(x, 2r))\big) \ge c > 0$$

holds, where the infimum is taken over all connected compact sets F_0 and F_1 such that F_0 meets both $\overline{B}(x,r)$ and $D \setminus B(x, 2r)$, and F_1 contains the point x and meets $D \setminus B(x, r)$.

Remark 4. A result related to the assertion of Lemma 10, is obtained independently in [HeK]).

To prove Theorem 2 we need other descriptions for of class $L^1_p(\Omega)$. They are of interest in their own right and we formulate them below.

Let $\mathcal{U} \subset \Omega$ be an open connected set with the following properties: $\overline{\mathcal{U}} \subset \Omega$ and there is a non-vanishing horizontal vector field X on $\mathcal{U}$ such that for some manifold S of codimension 1 transversal to X (in local coordinates S can be considered as part of a hyperplane) $g\colon (s,t) \to \exp tX(s)$, $-a < t < a$, is a diffeomorphism from $S \times (-a, a)$ onto $\mathcal{U}$. Set $\gamma_s = g(\{s\} \times (-a, a))$ for $s \in S$ and let $\pi : \mathcal{U} \to S$ be the fibration $\mathrm{pr}_1 \circ g^{-1}$, $\pi^{-1}(s) = \gamma_s$, $s \in S$.

A locally summable function v is said to be a *weak derivative* of $f \in L_{p,\mathrm{loc}}(\mathcal{U})$ along the vector field X if the relation

$$\int\limits_{\mathcal{U}} v\psi \, d\mu(x) = \int\limits_{\mathcal{U}} f X^*\psi \, d\mu(x)$$

holds for any test function $\psi \in C_0^\infty(\mathcal{U})$, where $X^* = -\sum\limits_{i=1}^{k} \frac{\partial}{\partial x_i}(a_i *)$ denotes the formal adjoint of $X = \sum\limits_{i=1}^{k} a_i \frac{\partial}{\partial x_i}$ with k the topological dimension of M.

A function $f \in L_{p,\mathrm{loc}}(\mathcal{U})$ is said to belong to ACL_X (absolutely continuous along the vector field X) if f is absolutely continuous on any compact subset of γ_s for almost every $s \in S$ relative the surface measure ds on S. If $f \in L_{p,\mathrm{loc}}(\mathcal{U})$ is absolutely continuous then the derivative $\frac{\partial}{\partial t}(f \circ \gamma_s) = Xf$ along the vector field X is well defined for almost every $s \in S$ and almost every $t \in (-a, a)$ (and so almost everywhere on $\mathcal{U}$ since the Hausdorff ν-measure is comparable on $\mathcal{U}$ with the product of the arc length dt on the fiber $\{\pi^{-1}(s) : s \in S\}$ and the quotient measure ds on S).

Lemma 11. *A function $u \in L_{p,\mathrm{loc}}(\mathcal{U})$, where $\overline{\mathcal{U}}$ is a compact subset of M, has a weak derivative $v \in L_p(\mathcal{U})$ along the vector field X iff u is absolutely continuous along X and Xu belongs to $L_p(\mathcal{U})$. In this event, Xu coincides with the weak derivative v almost everywhere.*

Proof. If $u \in L_p(\mathcal{U})$ has a weak derivative $v \in L_p^1(\mathcal{U})$ then, for every domain ω, $\overline{\omega} \subset \mathcal{U}$, there is [GN] a sequence $u_n \in C^\infty(\omega)$ such that $\|u - u_n \mid L_p(\omega)\| \to 0$ and $\|v - Xu_n \mid L_p(\omega)\| \to 0$ as $n \to \infty$. Then

$$\int\limits_S ds \int\limits_{\gamma_s \cap \omega} |v(s,t) - Xu_n(s,t)|^p \, dt \to 0.$$

Dropping to a subsequence, we can assume that, for almost every $s \in S$,

$$\|u(s,t) - u_n(s,t) \mid L_p(\gamma_s \cap \omega)\| \to 0$$

and

$$\|v(s,t) - Xu_n(s,t) \mid L_p(\gamma_s \cap \omega)\| \to 0$$

simultaneously as $n \to \infty$. Take $s \in S$ such that the convergences take place. Then it is easy to verify that

$$u(s,\tau) = u(s,\alpha) + \int\limits_\alpha^\tau v(s,t)\,dt, \qquad \tau \in [\alpha, \beta],$$

for every segment $[\alpha, \beta] \subset \gamma_s$. Hence, u is absolutely continuous along the vector field X and $Xu = v$ almost everywhere on $\mathcal{U}$.

Now, assume $u \in ACL_X$ and $Xu \in L_p(\mathcal{U})$. Fix $\psi \in C_0^\infty(\mathcal{U})$ and consider

$$M_\varepsilon u(s,t) = \int\limits_S K_1(s + \varepsilon_1 h)\,dh \int\limits_{\gamma_s} K_2(t + \varepsilon_2 \tau) u(h,\tau)\,d\tau$$

where $K_1(h)$ and $K_2(\tau)$ are Sobolev mollifiers, i.e., $K_1 \in C_0^\infty(B)$, $B \subset S$, $\int_B K_1(h)\,dh = 1$, $K_2 \in C_0^\infty([-1,1])$, $\int_{-1}^1 K_2(\tau)\,d\tau = 1$. These properties imply that $M_\varepsilon u$ is defined on supp ψ for $\varepsilon_1, \varepsilon_2$ small enough and $\|u - M_\varepsilon u \mid L_p(\omega)\| \to 0$, $XM_\varepsilon u = M_\varepsilon Xu$ on ω and $\|Xu - M_\varepsilon Xu \mid L_p(\omega)\| \to 0$. Since

$$\int\limits_\omega XM_\varepsilon u\psi = \int\limits_\omega M_\varepsilon Xu\psi = \int\limits_\omega M_\varepsilon u X^*\psi,$$

passing to the limit as $\varepsilon \to 0$ we infer that Xu is a weak derivative of u. □

Lemma 12. *Let $\mathcal{U} \subset M$ be an open subset of M such that $\overline{\mathcal{U}}$ is compact and there is a collection of vector fields X_i, $i = 1, \dots, n$, forming a basis for H_x at every point $x \in \mathcal{U}$. Then the following conditions are equivalent:*

(1) $u \in L_p^1(\mathcal{U})$;

(2) *$u \in L_{p,\mathrm{loc}}(\mathcal{U})$ and there are weak derivatives along the vector fields $X_1, \dots, X_n$ which are summable in power p on $\mathcal{U}$;*

(3) $u \in L_{p,\mathrm{loc}}(\mathcal{U})$ *and* $u \in ACL_{X_i}$, $X_i u \in L_p(\mathcal{U})$, $i = 1, \dots, n$.

Proof. The equivalence of 1 and 2, on account of the relation

$$0 < c_1 |\nabla_{\mathcal{L}} f| \le \left\{ \sum_{i=1}^{n} (X_i f) \right\}^{1/2} \le c_2 |\nabla_{\mathcal{L}} f| < \infty, \tag{9}$$

is proved in [GN]. The equivalence of 2 and 3 follows from Lemma 11. □

Proof of Theorem 2. If $\varphi\colon \Omega \to \Omega'$ is quasi-isometric then it is easy to verify that the composite function $u \circ \varphi$ is locally Lipschitzian on Ω for every smooth function $u\colon \Omega' \to \mathbb{R}$. Therefore, in view of Lemma 12, $u \circ \varphi$ belongs to $L^1_p(\Omega)$ for all $p \ge 1$. Moreover,

$$|\nabla_{\mathcal{L}}(u \circ \varphi)| \le (|\nabla_{\mathcal{L}} u| \circ \varphi) L(x; \varphi) \text{ where } L(x; \varphi) = \sup_{x \in \Omega} \overline{\lim_{y \to x}} \frac{d(\varphi(y), \varphi(x)}{d(x, y)}.$$

If $\varphi\colon \Omega \to \Omega'$ is quasi-conformal then by virtue of the results of [MM] the composite $u \circ \varphi$ with a smooth function $u\colon \Omega' \to \mathbb{R}$ belongs to ACL_X for every non-vanishing horizontal field X. Also, $|X(u \circ \varphi)| \le (|Xu| \circ \varphi) \| D\varphi \|$, where $\| D\varphi \|$ is the norm of the horizontal differential $D\varphi\colon H \to H$ which is defined almost everywhere on Ω. (The differentiability of quasi-conformal mappings and its properties on Carnot–Carathéodory spaces are established in [MM].)

Taking into account the relation $L(x; \varphi)^\nu \le c H^\nu_\varphi \mathcal{J}(x, \varphi)$ well-known in the theory of quasi-conformal mappings (the constant c is independent of the choice of $x \in \Omega$) and the absolute continuity of the set function $A \to \mu'(\varphi(A))$ [MM], we obtain

$$\begin{aligned} \int_\Omega |\nabla_{\mathcal{L}} u \circ \varphi|^p(x)\, d\mu(x) &\le \int_\Omega |\nabla_{\mathcal{L}} u|^p(\varphi(x)) L(x; \varphi)^\nu L(x; \varphi)^{p-\nu}\, d\mu(x) \\ &\le M \int_\Omega |\nabla_{\mathcal{L}} u|^p(\varphi(x)) J(x; \varphi)\, d\mu(x) = M \int_{\Omega'} |\nabla_{\mathcal{L}} u|^p(y)\, d\mu'(y) \end{aligned}$$

where $M = c H^\nu_\varphi$ at $p = \nu$ and $M = c L^{\nu+p}_\varphi$ at $p \ne \nu$. Thus, the estimate

$$\| u \circ \varphi \mid L^1_p(\Omega) \| \le M^{1/p} \| u \mid L^1_p(\Omega') \|$$

is proved for every smooth function u. Extending the operator to the whole function space, we obtain boundedness of $\varphi^*\colon L^1_p(\Omega') \to L^1_p(\Omega)$. The above-obtained estimates imply that $\| \varphi^* \| \le c^{1/\nu} H_\varphi$ if $p = \nu$, and $\| \varphi^* \| \le c^{1/p} L^{1+\nu/p}_\varphi$ otherwise. Since the converse of a quasi-conformal (quasi-isometric) mapping is itself quasi-conformal (quasi-isometric) [MM], φ^* is a bounded isomorphism. □

References

[CV] V. M. Chernikov and S. K. Vodop'yanov, Sobolev Spaces and Hypoelliptic Equations. II, Siberian Adv. Math 6:4 (1996), 64–96.

[FLW] B. Franchi, G. Lu and R. L. Wheeden, Representation Formulas and Weighted Poincaré Inequalities for Hörmander Vector Fields, Ann. Inst. Fourier, Grenoble. 45:2 (1995), 577–604.

[GN] N. Garofalo and Duy-Minh Nhieu, Isoperimetric and Sobolev inequalities for Carnot–Caratheodory spaces and existence of minimal surfaces, Comm. Pure Applied Math. 49 (1996), 1081–1144.

[G] M. Gromov, Carnot–Caratheodory spaces seen from within, preprint, IHES, Bures-sur-Yvette, 1994.

[Ge] F. W. Gehring, Lipschitz Mappings and the p-capacity of Rings in n-space, Advances in the Theory of Riemann surfaces, Ann. of Math. Stud. 66 (1971), 175–193.

[H] P. Hajłasz, Sobolev spaces on an arbitrary metric space, Potential Analysis 5:4 (1996), 403–415.

[He] J. Heinonen, Calculus on Carnot groups, Fall school in analysis (Javäskylä, 1994), Javäskylä, University of Javäskylä, 1994, 1–32.

[HeK] J. Heinonen and P. Koskela, From local to global in quasiconformal structures, Proc. Nat. Acad. Sci. USA 93 (1996), 554–556.

[HK] D. A. Herron and P. Koskela, Poincaré domains and quasiconformal mappings, Sibirsk. Mat. Zh. 35:6 (1995), 1417–1434.

[KR] A. Koranyi and H. M. Reimann, Foundations for the theory of quasiconformal mappings on the Heisenberg group, Adv. in Math. 111 (1995), 1–87.

[L] G. Lu, The sharp Poincaré inequality for free vector fields: An endpoint result, Revista Mat. Iberoamericana 10:2 (1994).

[Ma] I. G. Markina, On change of variable in spaces of differentiable functions, Sibirsk. Mat. Zh. 31:3 (1990), 73–84; English translation: Siberian Math. J. 31:3 (1990), 422–432.

[Mi] J. Mitchell, On Carnot–Caratheodory metrics, J. Differential Geom. 21 (1985), 35–45.

[MS] R. Macías and C. Segovia, Lipschitz functions on spaces of homogeneous type, Adv. in Math. 33:3 (1979), 257–270.

[MM] G. A. Margulis and G. D. Mostow, The differential of a quasi-conformal mapping of a Carnot–Caratheodory space, Geometric and Functional Analysis. 5:2 (1995), 402–433.

[P] P. Pansu, Métriques de Carnot–Carathéodory et quasiisométries des espaces symétriques de rang un, Ann. of Math. 129 (1989), 1–60.

[ST] J, D. Strömberg and A. Torchinsky, Weighted Hardy spaces, Lecture Notes in Math. 1381, Springer-Verlag, Berlin 1989.

[R] A. S. Romanov, On change-of-variable in the Bessel and Riesz potential spaces, Functional analysis and mathematical physics, Institute of Mathematics, Novosibirsk 1985, 117–133 (in Russian).

[V1] S. K. Vodop'yanov, Quasiconformal mappings on Carnot groups, Dokl. Russian Acad. Sci. 347:4 (1996), 439–442.

[V2] S. K. Vodop'yanov, Monotone functions and quasi-conformal mappings on Carnot groups, Sibirsk. Mat. Zh. 36:6 (1996), 1269–1295; English translation: Siberian Math. J. 36:6 (1996), 1113–1136.

[V3] S. K. Vodop'yanov, L_p-potential theory and quasiconformal mappings on homogeneous groups, Contemporary Problems on Geometry and Analysis [in Russian], Nauka, Novosibirsk 1989, 45–89.

[V4] S. K. Vodop'yanov, Mappings of homogeneous groups and embeddings of function spaces, Sibirsk. Mat. Zh. 30:5 (1989), 25–41; English translation: Siberian Math. J. 30:5 (1989), 191–202.

[VC] S. K. Vodop'yanov and V. M. Chernikov, Sobolev Spaces and Hypoelliptic Equations, Trudy Sobolev Institute Mat. 29 (1995), 7–62; English translation: Siberian Adv. Math. 6:3 (1996), 27–67; 6:4 (1966), 64–96).

[VG1] S. K. Vodop'yanov and V. M. Gol'dshteĭn, Lattice isomorphisms of the spaces W_n^1 and quasi-conformal mappings, Sibirsk. Mat. Zh. 16:2 (1975), 224–246; English translation: Siberian Math. J. 16:2 (1975), 174–189.

[VG2] S. K. Vodop'yanov and V. M. Gol'dshteĭn, Functional characteristics of quasi-isometric mappings, Sibirsk. Mat. Zh. 17:4 (1976), 768–773; English translation: Siberian Math. J. 17:4 (1976), 580–584.

Harmonic maps and morphisms in 4 dimensions

John C. Wood

Abstract. We discuss (i) harmonic maps from surfaces to 4-manifolds, especially, twistor constructions and a recent application to the study of the space of harmonic maps from S^2 to $\mathbb{C}P^2$, (ii) harmonic morphisms from 4-manifolds to surfaces, especially relationships with Hermitian structures, shear-free ray congruences and a twistorial construction of foliations by superminimal surfaces of hyperbolic 4-space with given boundary values at infinity.

1991 Mathematics Subject Classification: 58E2O, 53A10, 83C60.

Key words: harmonic map, harmonic morphism, minimal surface, superminimal, ray congruence.

1. Introduction

Harmonic maps are maps between Riemannian manifolds which extremize a natural energy integral; there are many thousands of papers on this subject, see the bibliography [BLRww]. Less well-known are *harmonic morphisms*, which are maps between Riemannian manifolds which pull back germs of harmonic functions to germs of harmonic functions (Definition 4.1 below), see [Gudww] for a frequently updated bibliography. We discuss harmonic maps *from* a surface *to* a 4-manifold and harmonic morphisms *from* a 4-manifold *to* a surface. These subjects are dual in many aspects. Our aim is to point out some of these aspects whilst concentrating on some recent advances in both subjects.

After reminding the reader in §2 about twistor constructions for harmonic maps from surfaces to 4-manifolds, where we take the opportunity to set out the relationship between *real isotropic*, *superminimal* and *complex isotropic*, we indicate in §3 how the well-known twistor construction has been recently used to study the topology [Crax1] and the smooth structure [LW96] of the space of harmonic maps from the 2-sphere to complex projective 2-space. In §4, we turn to the study of harmonic morphisms, especially those with values in a surface, giving some basic properties including minimality of the fibres (Proposition 4.2). Then, in §5, we describe the twistor construction [Woo92] of harmonic morphisms from certain 4-manifolds, in particular discussing the connection with Hermitian structures and superminimality.

Analogously, we show [BWx2] that harmonic morphisms from Minkowski 4-space are essentially equivalent to the shear-free ray congruences of Mathematical Physics. Finally, we discuss harmonic morphisms from hyperbolic 4-space $\mathbb{H}^4$. By [Woo92] these have superminimal fibres and, in [BWx2], their boundary values at infinity are characterized; we can thus construct foliations by superminimal surfaces of $\mathbb{H}^4$ with given boundary values at infinity (Corollary 5.9).

The author would like to thank the organizers for their hospitality and the opportunity of presenting this lecture at the first Brazilian–USA Workshop on Geometry, Topology and Physics in Campinas, July, 1996.

2. Harmonic maps and twistors

Let $M = (M^m, g)$ and $N = (N^n, h)$ be smooth (C^∞) Riemannian manifolds. Recall that a *harmonic map* $\phi : M \to N$ is a smooth map which extremizes the Dirichlet or energy integral

$$\frac{1}{2}\int_D |d\phi|^2 \, v_M$$

over any compact domain D, equivalently, it is a solution to the harmonic equation

$$\tau(\phi) \equiv \operatorname{trace} \nabla d\phi = 0 \tag{1}$$

where ∇ is the connection on $T^*M \otimes \phi^{-1}TN$ induced from the Levi-Civita connections D^M, D^N on M, N; more explicitly,

$$\tau(\phi) = \sum \left(D_{e_i}\left(d\phi(e_i)\right) - d\phi(D^M_{e_i} e_i)\right)$$

where we sum over an orthonormal basis $\{e_i\}$ of TM, and D is the pull-back of D^N to $\phi^{-1}TN$. For general facts on harmonic maps, see [EL78, EL88, Xin96].

Suppose, now, that the domain M is of dimension 2, and let (x, y) be isothermal coordinates. Writing $z = x + \mathrm{i}y$, $\frac{\partial}{\partial z} = \frac{1}{2}\left(\frac{\partial}{\partial x} - \mathrm{i}\frac{\partial}{\partial y}\right)$, $\frac{\partial}{\partial \bar z} = \frac{1}{2}\left(\frac{\partial}{\partial x} + \mathrm{i}\frac{\partial}{\partial y}\right)$, $\frac{\partial^c \phi}{\partial z} = \frac{1}{2}\left(\frac{\partial \phi}{\partial x} - \mathrm{i}\frac{\partial \phi}{\partial y}\right)$, $\frac{\partial^c \phi}{\partial \bar z} = \frac{1}{2}\left(\frac{\partial \phi}{\partial x} + \mathrm{i}\frac{\partial \phi}{\partial y}\right)$, $\frac{D}{\partial z} = D_{\partial/\partial z}$, $\frac{D}{\partial \bar z} = D_{\partial/\partial \bar z}$, Eqn. (1) reads

$$\frac{D}{\partial \bar z}\frac{\partial^c \phi}{\partial z} \equiv \frac{D}{\partial z}\frac{\partial^c \phi}{\partial \bar z} = 0\,. \tag{2}$$

In particular, we have the following *conformal invariance in the domain*:

Harmonicity depends only on the conformal structure of M^2 so that the concept of harmonic map from a Riemann surface *is well-defined.* (3)

Recall that a map $\phi : M^2 \to N^n$ is called *weakly conformal* if, for each $p \in M^2$, either $d\phi_p = 0$, in which case p is called a *branch point* or *ramification point* of ϕ, or

$d\phi_p$ maps T_pM^2 conformally onto its image, in which case p is called a *regular point* of ϕ.

For example, a smooth map $\phi : A \to \mathbb{C}$ from an open subset of $\mathbb{C}$ or a Riemann surface to $\mathbb{R}^n$ is weakly conformal if and only if

$$\sum_{\alpha=1}^{n} \left(\frac{d\phi_\alpha}{dz} \right)^2 = 0. \tag{4}$$

As a consequence of the above conformal invariance we have:

The composition of a (weakly conformal) harmonic map $\phi : M^2 \to N^n$ from a surface with a weakly conformal map $\tilde{M}^2 \to M^2$ between surfaces is a (weakly conformal) harmonic map. (5)

In isothermal coordinates (x, y), weak conformality at a point is expressed by

$$\left\langle \frac{\partial \phi}{\partial x}, \frac{\partial \phi}{\partial x} \right\rangle = \left\langle \frac{\partial \phi}{\partial y}, \frac{\partial \phi}{\partial y} \right\rangle \text{ and } \left\langle \frac{\partial \phi}{\partial x}, \frac{\partial \phi}{\partial y} \right\rangle = 0$$

where $\langle\ ,\ \rangle$ denotes the inner product on TN (extended to T^cN by bilinearity), equivalently,

$$\eta_{1,1} \equiv \left\langle \frac{\partial^c \phi}{\partial z}, \frac{\partial^c \phi}{\partial z} \right\rangle = 0. \tag{6}$$

Given an immersion $\phi : M^2 \to N^4$, we have an orthogonal decomposition of the pull-back $\phi^{-1}TN$ into tangent and normal bundles, viz.

$$\phi^{-1}TN = TM \oplus \nu M. \tag{7}$$

Then recall that the *second fundamental form* of ϕ is defined for each $p \in M, X, Y \in T_pM$ by $B(X, Y) =$ the normal component of D_XY (where Y is extended arbitrarily to a section of TM). Since, for a weakly conformal map,

$$\left\langle \frac{D}{\partial \bar{z}} \frac{\partial^c \phi}{\partial z}, \frac{\partial^c \phi}{\partial z} \right\rangle = \frac{1}{2} \frac{\partial}{\partial \bar{z}} \left\langle \frac{\partial^c \phi}{\partial z}, \frac{\partial^c \phi}{\partial z} \right\rangle = 0,$$

if follows that, at regular points, $\frac{D}{\partial \bar{z}} \frac{\partial^c \phi}{\partial z}$ is normal and coincides with the second fundamental form $B\left(\frac{\partial}{\partial \bar{z}}, \frac{\partial}{\partial z}\right) = \frac{1}{4}\left(B\left(\frac{\partial}{\partial x}, \frac{\partial}{\partial x}\right) + B\left(\frac{\partial}{\partial y}, \frac{\partial}{\partial y}\right)\right)$, hence the tension field of ϕ is a non-zero multiple of the mean curvature vector H. Thus:

A weakly conformal map is minimal at regular points if and only if it is harmonic. (8)

Such a map is called a *branched minimal immersion* and has isolated branch points, the form of which are given by [GOR73], see also [ET88]. More generally, for any integers $r, s \geq 1$, set

$$\eta_{r,s} = \left\langle \frac{D^{r-1}}{\partial z^{r-1}} \frac{\partial^c \phi}{\partial z}, \frac{D^{s-1}}{\partial z^{s-1}} \frac{\partial^c \phi}{\partial z} \right\rangle. \tag{9}$$

We say that ϕ is *real isotropic* if $\eta_{r,s}$ is identically zero for all r, s. For general information, see [Woo84]. For now, we take the codomain to be an oriented 4-dimensional Riemannian manifold N^4. Given an immersion $\phi : M^2 \to N^4$ from a Riemann surface to such a manifold, there are two *adapted almost Hermitian structures* $J^{\pm}$ along ϕ, namely the sections of $\mathrm{End}(\phi^{-1}TN)$ defined by rotation through $+\pi/2$ on TM together with rotation through $\pm\pi/2$ on the normal space νM.

Definition 2.1. An immersion $\phi : M^2 \to N^4$ is said to be *superminimal with positive (resp. negative) spin* if its second fundamental form B satisfies

$$B(X, J^{\pm}Y) = J^{\pm}B(X, Y) \quad \text{for all } p \in M,\ X, Y \in T_pM^2 . \tag{10}$$

Equivalently, $J^{\pm}$ is parallel along the immersion, viz.

$$\nabla_X J^{\pm} = 0 \quad \text{for all } X \in TM ,$$

where ∇ denotes the connection on $\mathrm{End}(\phi^{-1}TN^4)$ induced from the Levi-Civita connection on N^4. We say that ϕ is *superminimal* if, at each point, it is superminimal with positive or negative spin (the sign might be different at different points). Sometimes the terminology *right (resp. left) superminimal* is used for *superminimal with positive (resp. negative) spin*. For some history and geometric interpretations, see [Fri84, Frix1].

Clearly, a superminimal immersion is minimal. Further, a minimal immersion is superminimal if and only if the quartic form $\langle B(Z, Z), B(Z, Z)\rangle$ is zero for $Z = \partial/\partial z$. The concept of *superminimal* can be extended to any map which is an immersion except at a discrete set of points at which the decomposition (7) extends smoothly. Such maps are called *pseudo-immersions* in [Gau86]; by a result in [GOR73] they include branched minimal immersions (see [ET88, Gau86]). We have ([ES85] §6):

Proposition 2.2. *Let $\phi : M^2 \to N^4$ be a weakly conformal harmonic map. Then ϕ is superminimal if and only if it is real isotropic.*

Proof. Note first that

$$\langle B(Z, Z), B(Z, Z)\rangle = \eta_{2,2}$$

so that ϕ is superminimal if and only if $\eta_{2,2} = 0$. We show that this condition implies that $\eta_{r,s} = 0$ for all $r, s \geq 1$. Firstly, differentiating $\eta_{1,1} = 0$ gives $\eta_{2,1} = -\eta_{1,2} = 0$. Differentiating this shows that $\eta_{3,1} = -\eta_{2,2} = 0$. By similar differentiations, it suffices to show that $\eta_{r,1} = 0$ at all points; by continuity, it suffices to show this at regular points. Let $p \in M$ be a regular point. Suppose that (i) $\frac{D}{\partial z}\frac{\partial^c\phi}{\partial z}$ and $\frac{\partial^c\phi}{\partial z}$ are linearly independent at p. Then they remain linearly independent on some neighbourhood of p and so span a maximal isotropic subbundle W of $\phi^{-1}TN$. Differentiating $\eta_{2,2} = 0$ shows that $\eta_{3,2} = 0$ which, together with $\eta_{3,1} = 0$, implies that $\frac{D^2}{\partial z^2}\frac{\partial^c\phi}{\partial z}$ lies in W, whence, by differentiating, $\frac{D^{r-1}}{\partial z^{r-1}}\frac{\partial^c\phi}{\partial z}$ lies in W for all r so that $\eta_{r,1} = 0$. Suppose, instead, that (ii) $\frac{D}{\partial z}\frac{\partial^c\phi}{\partial z}$ and $\frac{\partial^c\phi}{\partial z}$ are linearly dependent at p. Then, either p is the limit point of points of type (i) in which case $\eta_{r,1} = 0$ for all r by continuity, or p is contained in an open subset of regular points of type (ii), so that, on this open set, $\frac{D}{\partial z}\frac{\partial^c\phi}{\partial z}$ is linearly dependent on $\frac{\partial^c\phi}{\partial z}$, and, by differentiation, so is $\frac{D^{r-1}}{\partial z^{r-1}}\frac{\partial^c\phi}{\partial z}$, so that again $\eta_{r,1} = 0$ for all r. □

Superminimality can be phrased in terms of twistor spaces: Let Z^6_+ (resp. Z^6_-) be the positive (resp. negative) twistor space of N^4. Thus $Z^6_\pm$ is the total space of a fibre bundle $\pi : Z^6_\pm \to N^4$ with each fibre $(Z^6_\pm)_p = \pi^{-1}(p)$ the set of positive (resp. negative) almost Hermitian structures at p. Recall that for each $p \in N^4$, $J \in (Z^6_\pm)_p$, the Levi-Civita connection on M^4 defines a *horizontal (sub)space* $H_J(Z^6_\pm)$ of the tangent space to $Z^6_\pm$ at J and so a decomposition $T_J Z^6_\pm = T_J(Z^6_\pm)_p \oplus H_J(Z^6_\pm)$ with the first factor the tangent space to the fibre of π. Then an almost Hermitian structure J at p can be lifted to one on the horizontal space at J; together with the canonical Hermitian structure on each fibre $(Z^6_\pm)_p \cong SO(4)/U(2) \cong S^2$ this defines the canonical almost Hermitian structure $\mathcal{J}$ on Z^6_+ (resp. Z^6_-); recall that this is integrable if and only if M^4 is anti-self-dual (resp. self-dual) [AHS78].

Given a pseudo-immersion $\phi : M^2 \to N^4$ we have canonical lifts $\tilde{\phi}^\pm : M^2 \to Z^6_\pm$ given by $\tilde{\phi}^\pm(p) =$ the adapted almost Hermitian structure $J^\pm_p$ at p defined above. Then, by the definition of the almost Hermitian structure and the horizontal subbundle we quickly have the well-known (e.g. [Gau86])

Proposition 2.3. *A pseudo-immersion $\phi : M^2 \to N^4$ is weakly conformal and superminimal with positive (resp. negative) spin if and only if its canonical lift into Z^6_+ (resp. Z^6_-) is holomorphic and horizontal, i.e. for each $p \in M$, $d\tilde{\phi}^\pm_p$ maps T_pM into a horizontal space of Z^6_+ (resp. Z^6_-).*

In fact this gives bijections between non-constant weakly conformal superminimal maps from M^2 to N^4 of positive (resp. negative) spin and non-constant horizontal holomorphic maps from M^2 to Z^6_+ (resp. Z^6_-).

When N is an almost Hermitian manifold, we have the decomposition of the complexified tangent bundle: $T^cN = T'N \oplus T''N$ into $(1,0)$ and $(0,1)$ parts and corresponding decompositions: $\frac{\partial^c\phi}{\partial z} = \frac{\partial\phi}{\partial z} + \frac{\partial\bar{\phi}}{\partial z}$, $\frac{\partial^c\phi}{\partial \bar{z}} = \frac{\partial\phi}{\partial \bar{z}} + \frac{\partial\bar{\phi}}{\partial \bar{z}}$ with $\frac{\partial\bar{\phi}}{\partial z} = \overline{\frac{\partial\phi}{\partial \bar{z}}}$ and $\frac{\partial\bar{\phi}}{\partial \bar{z}} = \overline{\frac{\partial\phi}{\partial z}}$. If M is Kähler, D^N_X maps $T'N$ to itself for all $X \in T^cN$. More generally, an almost Hermitian manifold with this property is called $(1,2)$-*symplectic*. For such a manifold, the harmonic condition (1) is equivalent to

$$\frac{D}{\partial \bar{z}}\frac{\partial\phi}{\partial z} = 0, \quad \text{equivalently} \quad \frac{D}{\partial z}\frac{\partial\phi}{\partial \bar{z}} = 0.$$

In particular we recover the well-known fact:

A ±-holomorphic map from a Riemann surface to a Kähler or (1, 2)-symplectic manifold is a weakly conformal harmonic map. (11)

Here by *±-holomorphic* we mean *holomorphic or antiholomorphic*. (Harmonicity remains true for a ±–holomorphic map from a Kähler manifold [ESa64], or, more generally, a *cosymplectic* manifold [Lic69, Sal85] of arbitrary dimension, though such a map is not, in general, weakly conformal.)

Recall that a map $\phi : M^2 \to N^n$ from a Riemann surface to a Kähler manifold is called *complex isotropic* [EW83] if

$$\eta^c_{r,s} = \left\langle \frac{D^{r-1}}{\partial z^{r-1}}\frac{\partial\phi}{\partial z}, \frac{D^{s-1}}{\partial \bar{z}^{s-1}}\frac{\partial\phi}{\partial \bar{z}} \right\rangle^{\text{Herm}} = 0 \quad \text{for all integers } r, s \geq 1\,. \tag{12}$$

where $\langle\ ,\ \rangle^{\text{Herm}}$ denotes the Hermitian extension of the inner product on TN. For a 4 (real-)dimensional Kähler manifold, the negative twistor space can be identified with the projectivised (1, 0)-tangent bundle $P(T'N)$ and then the twistor lift $\tilde{\phi}^-$ is given by the projective class of $\partial\phi/\partial z$ (see for example, [Sal85, ES85, Woo87]). Then a similar argument to that for Proposition 2.2 gives

Proposition 2.4 ([ES85] Corollary 11.3). *A weakly conformal harmonic map from a Riemann surface to a 4-dimensional Kähler manifold is complex isotropic if and only if it is superminimal with negative spin or it is ±-holomorphic.*

In particular, complex isotropic maps from Riemann surfaces to 4-dimensional Kähler manifolds are given twistorially as the projections of horizontal holomorphic maps into the negative twistor space. See [Bur87, Woo87, Raw85] for more information on twistor methods.

3. The space of harmonic maps from S^2 to $\mathbb{C}P^2$ as a manifold

We study the space $\text{Harm}(S^2, \mathbb{C}P^2)$ of all harmonic maps from S^2 to $\mathbb{C}P^2$. We recall the construction of such harmonic maps, for details and history, see, [EW83]; we take a twistorial approach as in [ES85]. First note that any holomorphic or antiholomorphic map is harmonic, we exclude these from the subsequent discussion. An easy inductive argument shows that any harmonic map $S^2 \to \mathbb{C}P^2$ is complex isotropic and so, by Proposition 2.4, any non ±-holomorphic harmonic map is superminimal with negative spin. It is therefore given by the projection of a horizontal holomorphic map from S^2 to the negative twistor space $(Z^6_-, \mathcal{J})$ This twistor space may be identified with the flag manifold $F_{1,2} = \{(V, X) \in G_1(\mathbb{C}^3) \times G_2(\mathbb{C}^3) : V \subset X\}$ with the complex structure induced from the inclusion in $G_1(\mathbb{C}^3) \times G_2(\mathbb{C}^3)$, and projection π given by $\pi(V, X) =$

$V^\perp \cap X$ (see, for example [ES85, Gau86, Woo87]. Horizontal holomorphic maps are given [ErW83, Proposition 3.3] by $(V, X) = (f, f_{(1)})$ where $f : S^2 \to \mathbb{C}P^2$ is a holomorphic map which is *full*, i.e. does not have image in a complex projective line and $f_{(1)}$ is its first associated curve defined as follows:

On a local coordinate chart (U, z), write $f(z) = [F_0(z), F_1(z), F_2(z)]$ where $F = (F_0, F_1, F_2) : U \to \mathbb{C}^3 \setminus \{0\}$ is holomorphic. Define $f_{(1)} : S^2 \to G_2(\mathbb{C}^3)$ by $f_{(1)}(z) = \text{span}\,(F(z), F'(z))$ at points where $F(z)$ and $F'(z) = dF/dz$ are linearly independent. Identifying $G_2(\mathbb{C}^3)$ with $\mathbb{C}P^1$ we can write $f_{(1)} = [F \wedge F']$. Now $F \wedge F'$ is zero precisely at the branch (ramification) points of f and, in the neighbourhood of such a point z_0, we can write $F(z) \wedge F'(z) = (z - z_0)^\rho \Psi(z)$ for some $\rho \in \mathbb{N}$ called the *ramification index of f at z_0*, and some smooth function Ψ with $\Psi(z_0) \neq 0$. We then set $f_{(1)}(z_0) = [\Psi(z_0)]$ thus extending $f_{(1)}$ to a smooth map on S^2. Note that the ramification index of f at z_0 is just the order of the zero of df at that point; the sum of the ramification indices over all ramification points is called the *(total) ramification index of f*. This process leads to a bijection, see, [EW83]:

Theorem 3.1. *The map G' : $\text{Hol}^*(S^2, \mathbb{C}P^2) \to \text{Harm}_{NH}(S^2, \mathbb{C}P^2)$ defined by $G'(f) = f^\perp \cap f_{(1)}$ is a bijection from the set of full holomorphic maps from S^2 to $\mathbb{C}P^2$ to the set of all non $\pm$-holomorphic harmonic maps from S^2 to $\mathbb{C}P^2$.*

To study the latter space, we first study $\text{Hol}^*(S^2, \mathbb{C}P^2)$. Giving this space the uniform topology, the components are the spaces $\text{Hol}^*_k(S^2, \mathbb{C}P^2)$ of full holomorphic maps of degree k, $(k \in \mathbb{N})$. These are complex manifolds of dimension $3k + 2$. We should like to transfer this structure to the space of harmonic maps. However, the map G' may not be continuous on a curve f_t along which the total ramification index changes, see for example [LW96, Example 2.4]. To fix this, for each $r \in \{0, 1, 2, \dots\}$ we consider the space

$$\text{Hol}^*_{k,r}(S^2, \mathbb{C}P^2) = \{f \in \text{Hol}^*_k(S^2, \mathbb{C}P^2) : \text{total ramification index of } f = r\}.$$

Note that this space is non-empty precisely for $0 \leq r \leq \frac{3}{2}k - 3$. We have

Theorem 3.2 ([Crax1]). *The space $\text{Hol}^*_{k,r}(S^2, \mathbb{C}P^2)$ is a connected complex manifold of dimension $3k - r + 2$.*

Now the image of $\text{Hol}^*_{k,r}(S^2, \mathbb{C}P^2)$ under G' is the space $\text{Harm}_{d,E}(S^2, \mathbb{C}P^2)$ of harmonic maps of degree d and energy $4\pi E$ where $d = k - r - 2$ and $E = 3k - r - 2$. In particular, these spaces are connected, and together with the spaces of $\pm$-holomorphic maps of a given degree, give the components of $\text{Harm}(S^2, \mathbb{C}P^2)$. Crawford [Crax1] shows that G' is a homeomorphism. In [LW96] we take this further and study G' as a map into the C^∞ manifold $C^j(S^2, \mathbb{C}P^2)$ of C^j maps from S^2 to $\mathbb{C}P^2$ establishing:

Theorem 3.3 ([LW96]). *The map*

$$G' : \mathrm{Hol}^*_{k,r}(S^2, \mathbb{C}P^2) \to C^j(S^2, \mathbb{C}P^2)$$

is a smooth embedding onto $\mathrm{Harm}_{k-r-2,3k-r-2}(S^2, \mathbb{C}P^2)$ *for any* $j \geq 2$.

Each component $\mathrm{Harm}_{d,E}(S^2, \mathbb{C}P^2)$ *of* $\mathrm{Harm}(S^2, \mathbb{C}P^2)$ *is a closed smooth submanifold of* $C^j(S^2, \mathbb{C}P^2)$ *of dimension* $6E + 4$ *if* $E = |d|$ *(in which case it consists of holomorphic or antiholomorphic maps) and of dimension* $2E + 8$ *(otherwise).*

The methods used will not readily extend to higher dimensional projective spaces or other symmetric space targets, and it remains an open question as to what spaces of harmonic maps to these targets are manifolds. For a related question on Jacobi fields along harmonic maps, see [Quest].

4. Harmonic morphisms

We now turn to a subject which is, in many ways, dual to the theory of weakly conformal harmonic maps:

Definition 4.1. A *harmonic morphism* $\phi : (M^m, g) \to (N^n, h)$ between Riemannian manifolds is a smooth map such that, if $f : V \to \mathbb{R}$ is a harmonic function on an open subset V of N, then, if $U = \phi^{-1}(V) \subset M$ is non-empty, $f \circ \phi$ is a harmonic function on U.

Harmonic morphisms can be characterized as follows: Say that a smooth map $\phi : M^m \to N^n$ is *horizontally weakly conformal* if, at each point $p \in M$, either the differential $d\phi_p = 0$, in which case we call p a *critical point*, or the restriction of $d\phi_p$ to the horizontal space $(\ker d\phi_p)^\perp$ is conformal and surjective, in which case we call p a *regular point*. Note the duality with the definition of weakly conformal in §2. For example, a smooth map ϕ from an open subset of $\mathbb{R}^m$ to $\mathbb{C}$ is horizontally weakly conformal if and only if

$$\sum_{i=1}^{m} \left(\frac{\partial \phi}{\partial x_i} \right)^2 = 0, \tag{13}$$

cf. (4). Then we have [Fug78, Ish79]:

A smooth map $M^m \to N^n$ *is a harmonic morphism if and only if it is a horizontally weakly conformal harmonic map.* (14)

Note also that the set of critical points of a harmonic morphism is polar, in particular, in the real-analytic case, is a real-analytic set of codimension at least 2, and that any harmonic morphism is an open mapping [Fug78]. Lastly, a foliation is called *conformal* if Lie transport along the leaves of vectors normal to the leaves is conformal, see [Vai79, Woo86]; clearly, the fibres of a horizontally conformal submersion form a conformal foliation, in particular, at regular points, the fibres of a harmonic morphism form such a foliation.

Now let the codomain N be of dimension 2. Then dually to (5) we have

The composition of a harmonic morphism $\phi : M^m \to N^2$ to a surface with a weakly conformal map $N^2 \to \tilde{N}^2$ between surfaces is a harmonic morphism. (15)

In particular, dually to (3) we have

The concept of harmonic morphism to a Riemann surface *is well-defined.* (16)

Dually to (8) we have a strengthened form [Gudx1] of a result of Baird and Eells [BE81] which we further extend slightly to allow critical points:

Proposition 4.1. *A non-constant horizontally weakly conformal map $\phi : M^m \to N^2$ to a surface is a harmonic morphism if and only if the fibres are minimal at regular points.*

Proof. By [BE81, Gudx1], the tension field of ϕ at p is a non-zero multiple of $d\phi_p(H)$ where H is the mean curvature of the fibre through p, so that the "only if" part is immediate. For the "if" part, note that, if the fibres are minimal at regular points, then ϕ is harmonic at those points. Now let p be a critical point. Then (i) *either* p is the limit point of regular points, in which case ϕ is harmonic at p by continuity, *or* (ii) p is contained in an open set U of critical points, in which case ϕ is constant on U and hence harmonic. By [Fug78, Ish79], ϕ is a harmonic morphism. □

Note that, in fact, case (ii) cannot occur by the unique continuation property of harmonic maps, but we do know this *a priori*.

We have dually to (11):

A holomorphic map from a Kähler manifold to a Riemann surface is a horizontally weakly conformal harmonic map, i.e. a harmonic morphism. (17)

To see this, merely note that the Cauchy–Riemann equations imply both horizontal weak conformality, and, as in [ESa64], harmonicity. In fact, by [Lic69], the result remains true when the domain is a cosymplectic almost Hermitian manifold.

Finally note that, (i) the composition of harmonic morphisms is a harmonic morphism, (ii) in stochastic processes, harmonic morphisms are "Brownian path-preserving transformations", see, for example, [Duh97], (iii) any stable harmonic map from a Riemannian manifold to the 2-sphere is a harmonic morphim [Che93].

See [GLMM] for an updated list of the known harmonic morphisms.

5. Harmonic morphisms from 4-dimensional manifolds

Just as harmonic maps *from* surfaces *to* 4-manifolds have some special properties, so do harmonic morphisms *from* 4-manifolds *to* surfaces. Given a submersion $M^4 \to N^2$ from an oriented 4-dimensional Riemannian manifold to a Riemann surface, there are two adapted almost Hermitian structures $J^{\pm}$ on M^4 defined as rotation through $+\pi/2$ on the horizontal space together with rotation through $\pm\pi/2$ on the vertical space. Then we have

Theorem 5.1 ([Woo92]). *Let $\phi : M^4 \to N^2$ be a submersive harmonic morphism from an oriented Einstein 4-manifold to a Riemann surface. Then*

1) *one of the adapted almost Hermitian structures $J^{\pm}$ is integrable,*
2) *correspondingly, the fibres of ϕ are superminimal with positive (resp. negative) spin.*

Note that a holomorphic map from a non-Kähler Hermitian manifold is not, in general, a harmonic morphism, for conditions under which this holds see [GW97] and, for maps from open subsets of $\mathbb{R}^m$ to $\mathbb{C}$, [BW95].

The idea of the proof is that, since the fibres of ϕ form a conformal foliation by minimal surfaces, a horizontal basic vector field X satisfies the Jacobi equation for minimal surfaces as does $J^{\pm}X$. This implies that the term involving the second fundamental form must be $J^{\pm}$-equivariant which is equivalent to superminimality. This, in turn, is equivalent to integrability of J^+ or J^-. Note that neither of $J^{\pm}$ is, in general, Kähler.

Say that the harmonic morphism ϕ is of *positive (resp. negative) spin* according as J^+ (resp. J^-) is integrable. (In [AGx1], the term *right* (resp. *left*) harmonic morphism is used.)

Changing the orientation of M^4 clearly changes the sign of the spin, so, without loss of generality, we shall henceforth take it to be positive. Then by Proposition 2.3, the superminimal fibres of ϕ are given by the projections of horizontal holomorphic maps, and since these form a conformal foliation, they vary holomorphically with respect to J^+. This leads to

Corollary 5.2 ([Woo92]). *Let $\phi : M^4 \to N^2$ be a submersive harmonic morphism of positive spin from an oriented Einstein 4-manifold to a Riemann surface. Then, the fibres of ϕ are given locally by $\xi \mapsto \pi \circ H(z, \xi)$ where H is a holomorphic function of two complex variables with values in $(Z^6_+, \mathcal{J})$ such that, for each z, the map $\xi \mapsto H(z, \xi)$ is horizontal and $h = \pi \circ H$ is a diffeomorphism of an open subset of $\mathbb{C}^2$ onto an open subset of M^4.*

Conversely, any such function H defines a harmonic morphism ϕ on an open subset of M^4 as the inverse of $h = \pi \circ H$ followed by natural projection $(z, \xi) \mapsto z$.

We shall call such a function H *a holomorphic parametrization (of ϕ).*

If M^4 is anti-self-dual, the almost complex structure $\mathcal{J}$ on Z^6_+ is integrable [AHS78], and, when M^4 is Einstein as well, the horizontal bundle of type $(1,0)$ of π is a holomorphic subbundle of $T'Z^6_+$ [ES85], hence, in this case, there is an abundance of functions H of the above type. For example, if $N^4 = S^4$, R. Bryant [Bry82] showed that essentially all horizontal holomorphic maps from an open subset of $\mathbb{C}$ to $Z^6_+ = \mathbb{C}P^3$ are locally of the form

$$\xi \mapsto \left[f^1(\xi) - \frac{1}{2} f^2(\xi) \frac{df^1}{df^2}, f^2(\xi), \frac{1}{2} \frac{df^1}{df^2} \right]$$

where f^1 and f^2 are holomorphic. Letting these functions depend holomorphically on another complex variable z gives holomorphic parametrizations of harmonic morphisms from open subsets of S^4 to $\mathbb{C}$, or, via a chart, to a Riemann surface; for example, setting $f^1(z, \xi) = z$, $f^2(z, \xi) = \xi$ gives a harmonic morphism which is the composition $S^4 \setminus (0, \pm 1) \to S^3$ of the orthogonal projection along great circles to the equatorial sphere, the Hopf map $S^3 \to S^2$ and stereographic projection $S^2 \to \mathbb{C} \cup \{\infty\}$. See [Woo92] for details and similar constructions of harmonic morphisms from $\mathbb{C}P^2$, and see [Bai92] for $\mathbb{H}^4$ and more discussion. For $\mathbb{R}^4$, we obtain Weierstrass-like formulae [Woo92]; for generalizations to higher dimensions, see [BWx1].

The result can be phrased equivalently as a correspondence between harmonic morphisms and Hermitian structures, see [Woo92, Proposition 5.2], where it is shown that a nowhere Kähler Hermitian structure J on an anti-self-dual Einstein manifold defines local harmonic morphisms of positive spin. Indeed, the anti-self-duality ensures the integrability of the distribution $\{v \in TM^4 : \nabla^M_v J = 0\}$, then its integral surfaces form a conformal foliation by superminimal surfaces and are thus locally the fibres of a harmonic morphism. Recalling that the condition for anti-self-duality is the vanishing of the self-dual part W^+ of the Weyl curvature tensor, Apostolov and Gauduchon have improved this result to the following:

Theorem 5.3 ([AGx1]). *Let M^4 be an oriented Einstein 4-manifold. The following three conditions are equivalent:*

1) *the self-dual part W^+ of the Weyl tensor is degenerate everywhere, i.e. has at most two distinct eigenvalues,*
2) *for each point $p \in M^4$, there exists a positive Hermitian structure on a neighbourhood of p,*
3) *there exists a submersive harmonic morphism of positive spin on an open subset of M^4.*

In the case that M^4 = an open subset A of $\mathbb{R}^4$, an almost Hermitian structure J is given by a map $J : A \to SO(4)/U(2) \cong S^2$. It is easy to see that this is a harmonic morphism [Woo92]; conversely, Theorem 5.1 implies

Corollary 5.4 ([Woo92]). *Let $\phi : A \to N^2$ be a submersive harmonic morphism from an open subset of $\mathbb{R}^4$ to a Riemann surface. Then, for any $p \in A$, there is a neighbourhood A' of p and a holomorphic map $\zeta : V \to S^2$ on an open subset of N^2 such that $J = \zeta \circ \phi$ defines a Hermitian structure on A'.*

Thus the problems of finding submersive harmonic morphisms and Hermitian structures are essentially equivalent. The above results should help us to understand the *space* of all harmonic morphisms in the various cases.

There is a version of this for Minkowski 4-space $\mathbb{M}^4 = (\{(t, x_1, x_2, x_3) \in \mathbb{R}^4\}, g^M)$ where $g^M = -dt^2 + dx_1^2 + dx_2^2 + dx_3^2$. First note that the concepts of *harmonic map* and *harmonic morphism* make sense for maps $\phi : M^m \to N^n$ between *semi-Riemannian manifolds* and the characterization (14) still holds [Fug96]. However, there are now three sorts of points $p \in M$: (i) points where $d\phi_p = 0$, called *critical points* as before; (ii) points where $d\phi_p \neq 0$ and the kernel of $d\phi_p$ is non-degenerate, horizontal conformality is described as before; (iii) points where $d\phi_p \neq 0$ and the kernel of $d\phi_p$ is degenerate. In case (iii) the horizontal space is contained in the kernel and has dimension less than $m - n$ so that $d\phi_p$ is not surjective. A map ϕ with such points may not be an open mapping; a simple example is the harmonic morphism $\mathbb{M}^4 \to \mathbb{C}$ defined by $(t, x_1, x_2, x_3) \mapsto t - x_1$, see [Fug96] and [BWx2] for further examples.

The analogue of Hermitian structure for $\mathbb{M}^4$ is *shear-free ray congruence*: this is a foliation ℓ of an open subset A of $\mathbb{M}^4$ by null lines which enjoys the following conformality property: Let V be its tangent vector field. Then, at any point $p \in A$, $V_p^{\perp}$ is 3-dimensional and contains V_p, the quotient $V_p^{\perp}/V_p$ (or any complement of V_p in $V_p^{\perp}$), called the *screen space* at p, is two-dimensional and inherits a canonical positive definite inner product. Then ℓ is called *shear-free* if Lie transport along V of vectors in the screen space is conformal, see [BWx2] for more details and [PR86, Ben94] for other descriptions.

Writing the tangent field V of a shear-free ray congruence on A as $V = \partial/\partial t + U$ where $U : A \to S^2$, then U defines a harmonic morphism [BWx2]; conversely we have

Theorem 5.5 ([BWx2]). *Let $\phi : A \to N^2$ be a real-analytic harmonic morphism without critical points from an open subset A of Minkowski 4-space to a Riemann surface. Then there is a shear-free ray congruence ℓ on A such that each connected component of a fibre of ϕ is the union of parallel null lines of ℓ.*

Corollary 5.6 ([BWx2]). *Let $\phi : A \to N^2$ be a real-analytic harmonic morphism without critical points from an open subset of Minkowski 4-space to a Riemann surface. Then, for any $p \in A$, there is a neighbourhood A' of p and a holomorphic map $\rho : V \to S^2$ from an open subset of N^2 such that setting $U = \rho \circ \phi$, $V = \partial/\partial t + U$ is the tangent vector field to a shear-free ray congruence.*

Shear-free ray congruences and Hermitian structures both complexify to *holomorphic foliations of open subsets of* $\mathbb{C}^4$ *by null planes*; there is a version of Theorem 5.5 for these involving *complex harmonic morphisms* [BWx2].

Lastly we consider harmonic morphisms from hyperbolic 4-space $\mathbb{H}^4$ and the construction of families of superminimal surfaces in $\mathbb{H}^4$. First note that $\mathbb{H}^4$ is a self-dual and anti-self-dual Einstein manifold, so the above theory applies. We have a new fact: the boundary values at infinity of a hyperbolic harmonic morphism form a horizontally conformal map. More precisely, equip $\breve{\mathbb{R}}^4 = \mathbb{R}^4 \setminus \mathbb{R}^3 = \{(x_0, x_1, x_2, x_3) \in \mathbb{R}^4 : x_0 \neq 0\}$ with the *hyperbolic metric* $g^H = \left(\sum_{i=0}^3 dx_i^2\right)/x_0^2$ so that each component of $\breve{\mathbb{R}}^4$ is isometric to hyperbolic 4-space $\mathbb{H}^4$. Then

Proposition 5.7 ([BWx2]). *Let $\phi : A^4 \to \mathbb{C}$ be a smooth map which is a harmonic morphism on $A^4 \setminus \mathbb{R}^3$ with respect to the hyperbolic metric. Then the restriction of ϕ to $A^3 = A^4 \cap \mathbb{R}^3$ is horizontally weakly conformal.*

Proof. The equation for harmonicity of ϕ reads:

$$x_0 \sum_{i=0}^{4} \frac{\partial^2 \phi}{\partial x_i^2} - 2\frac{\partial \phi}{\partial x_0} = 0 .$$

This implies that $\partial\phi/\partial x_0 = 0$ on $\mathbb{R}^3 = \{x_0 = 0\}$, which reduces the horizontal weak conformality equation (13) for $m = 4$ to that for $m = 3$. □

Using a twistorial technique, we have conversely,

Theorem 5.8 ([BWx2]). *Let $f : A^3 \to \mathbb{C}$ be a real-analytic horizontally conformal submersion on an open subset of $\mathbb{R}^3$. Then there is an open subset A^4 of $\mathbb{R}^4$ with $A^4 \cap \mathbb{R}^3 = A^3$ and a real-analytic submersion $\phi : A^4 \to \mathbb{C}$ with $\phi|A^3 = f$ such that $\phi|A^4 \setminus \mathbb{R}^3$ is a harmonic morphism with respect to the hyperbolic metric.*

Since the fibres of a harmonic morphism from $\mathbb{H}^4$ are minimal (in fact, superminimal) this can be used to find minimal surfaces as follows:

Corollary 5.9 ([BWx2]).

1) *Let C be a real-analytic conformal foliation by curves of an open subset of $\mathbb{R}^3$. Then there is a real-analytic foliation of an open subset A^4 of $\mathbb{R}^4$ by surfaces which are minimal in $A^4 \setminus \mathbb{R}^4$ with respect to the hyperbolic metric and intersect $\mathbb{R}^3$ in parts of leaves of C.*
2) *Let c be an embedded real-analytic curve in $\mathbb{R}^3$. Then there is an embedded real-analytic surface s in $\mathbb{R}^4$ which is minimal in $\mathbb{R}^4 \setminus \mathbb{R}^3$ with respect to the hyperbolic metric and intersects $\mathbb{R}^3$ in c.*

Proof. 1) Representing the leaves of $\mathcal{C}$ as the level curves of a real-analytic horizontally conformal submersion $f : A^3 \to \mathbb{C}$ on an open subset of $\mathbb{R}^3$, construct a hyperbolic harmonic morphism ϕ as in the theorem: its fibres give the desired real-analytic foliation.

2) We can embed c in a real-analytic conformal foliation by curves of an open subset of $\mathbb{R}^3$. Then apply 1). □

For examples of harmonic morphisms on hyperbolic spaces and other symmetric spaces, see also [Bai92,Gudx1], and for related methods of finding minimal surfaces, [BG92].

References

[AHS78] M. F. Atiyah, N. J. Hitchin and I. M. Singer, Self-duality in four-dimensional Riemannian geometry, Proc. Roy. Soc. London Ser. A 362 (1978), 425–461.

[AGx1] V. Apostolov and P. Gauduchon, The Riemannian Goldberg–Sachs theorem, Internat. J. Math., to appear.

[Bai92] P. Baird, Riemannian twistors and Hermitian structures on low-dimensional space forms, J. Math. Phys. 33 (1992), 3340–3355.

[BE81] P. Baird and J. Eells, A conservation law for harmonic maps, in: Geometry Symposium, Utrecht 1980 (E. Looijenga, D. Siersma and F. Takens, eds.), Lecture Notes in Math. 894, Springer-Verlag, Berlin–Heidelberg–New York 1981, 1–25.

[BG92] P. Baird and S. Gudmundsson, *p*-Harmonic maps and minimal submanifolds, Math. Ann. 294 (1992), 611–624.

[BW95] P. Baird and J.C. Wood, Hermitian structures and harmonic morphisms on higher dimensional Euclidean spaces, Internat. J. Math. 6 (1995), 161–192.

[BWx1] P. Baird and J. C. Wood, Weierstrass representations for harmonic morphisms on Euclidean spaces and spheres, Math. Scand., to appear.

[BWx2] P. Baird and J. C. Wood, Harmonic morphisms, conformal foliations and shear-free ray congruences, Bull. Belg. Math. Soc., to appear.

[Ben94] I. M. Benn, A unified description of null and non-null shear-free congruences, J. Math. Phys. 35 (1994), 1796–1802.

[Bry82] R. Bryant, Conformal and minimal immersions of compact surfaces into the 4-sphere, J. Differential Geom. 17 (1982), 455–473.

[Bry85] R. Bryant, Lie groups and twistor spaces, Duke Math. J. 52 (1985), 223–261.

[Bur87] F. E. Burstall, Twistor methods for harmonic maps, in: Differential Geometry (V. L. Hansen, ed.), Lecture Notes in Math. 1263, Springer-Verlag, Berlin–Heidelberg–New York 1987, 55–96.

[BLRww] F. E. Burstall, L. Lemaire and J. Rawnsley, Harmonic maps bibliography. http://www.bath.ac.uk/~masfeb/harmonic.html

[BR90] F. E. Burstall and J. Rawnsley, Twistor theory for Riemannian symmetric spaces, Lecture Notes in Math. 1424, Springer-Verlag, Berlin–Heidelberg–New York 1990.

[Che93] J. Y. Chen, Stable harmonic maps into S^2, in: Geometry and Global Analysis (T. Kotake, S. Nishikawa and R. Schoen, eds.), Tohoku Univ. Sendai 1993, 431–436.

[Crax1] T. A. Crawford, The space of harmonic maps from the 2-sphere to the complex projective plane, Canadian Math. Bull., to appear.

[DS93] J. Davidov and A. G. Sergeev, Twistor spaces and harmonic maps, Uspekhi Mat. Nauk 48 (3) (1993), 3–96; English transl. in Russian Math. Surveys 48 (3) (1993), 1–91.

[Duh97] F. Duheille, Une preuve probabiliste élémentaire d'un résultat de P. Baird et J. C. Wood, Ann. Inst. H. Poincaré Probab. Statist. 33 (1997), 283–291.

[EL78] J. Eells and L. Lemaire, A report on harmonic maps, Bull. London Math. Soc. 10 (1978) 1–68.

[EL88] J. Eells and L. Lemaire, Another report on harmonic maps, Bull. London Math. Soc. 20 (1988), 385–524.

[ES85] J. Eells and S. M. Salamon, Twistorial construction of harmonic maps of surfaces into four-manifolds, Ann. Scuola Norm. Sup. Pisa Cl. Sci. 12 (1985), 589–640.

[ESa64] J. Eells and J. H. Sampson, Harmonic mappings of Riemannian manifolds, Amer. J. Math. 86 (1964), 109–160.

[EW83] J. Eells and J. C. Wood, Harmonic maps from surfaces to complex projective spaces, Adv. in Math. 49 (1983), 217–263.

[ErW83] S. Erdem and J. C. Wood, On the construction of harmonic maps into a Grassmannian, J. London Math. Soc. 28 (1983), 161–174.

[ET88] J.-H. Eschenburg and R. Tribuzy, Branch points of conformal mappings of surfaces, Math. Ann. 279 (1988), 621–633.

[Fri84] T. Friedrich, On surfaces in four-spaces, Anal. Global Anal. Geom. 2 (1984), 257–287.

[Frix1] T. Friedrich, On superminimal surfaces, preprint, SFB 288, T.U., Berlin 1996.

[Fug78] B. Fuglede, Harmonic morphisms between Riemannian manifolds, Ann. Inst. Fourier (Grenoble) 28 (1978), 107–144.

[Fug96] B. Fuglede, Harmonic morphisms between semi-Riemannanian manifolds, Ann. Acad. Sci. Fennicae 21 (1996), 31–50.

[Gau86] P. Gauduchon, Pseudo-immersions superminimales d'une surface de Riemann dans une variété riemannienne de dimension 4, Bull. Soc. Math. France 114 (1986), 447–508.

[Gudx1] S. Gudmundsson, On the existence of harmonic morphisms from symmetric spaces of rank one, Manuscripta Math., to appear.

[Gudww] S. Gudmundsson, The bibliography of harmonic morphisms. http://www.maths.lth.se/matematiklu/personal/sigma/harmonic /bibliography. html

[GLMM] S. Gudmundsson, L. Loubeau, S. Montaldo and M. T. Mustafa, The atlas of harmonic morphisms. http://www.maths.lth.se/matematiklu/personal/sigma/harmonic/atlas.html

[GW97] S. Gudmundsson and J. C. Wood, Harmonic morphisms between almost Hermitian manifolds, Boll. Un. Mat. Italiana (7) 11-B, Suppl. Fasc. 2 , (1997), 185–197.

[GOR73] R. Gulliver, R. Osserman and H. L. Royden, A theory of branched immersions of surfaces, Amer. J. Math. 95 (1973), 750–812.

[Ish79] T. Ishihara, A mapping of Riemannian manifolds which preserves harmonic functions, J. Math. Kyoto Univ. 19 (1979), 215–229.

[LW96] L. Lemaire and J. C. Wood, On the space of harmonic 2-spheres in $\mathbf{C}P^2$, Internat. J. Math. 7 (1996), 211–225.

[Lic69] A. Lichnerowicz, Applications harmoniques et variétés kählériennes, Rend. Sem. Mat. Fis. Milano 39 (1969), 186–195.

[PR86] R. Penrose and W. Rindler, Spinors and space-time (two volumes), Cambridge Monographs on Mathematical Physics, Cambridge Univ. Press 1986.

[Raw85] J. Rawnsley, f-Structures, f-twistor spaces and harmonic maps, in: Geometry Seminar "Luigi Bianchi" II—1984 (E. Vesentini, ed.) Lecture Notes in Math. 1164, Springer-Verlag, Berlin–Heidelberg–New York 1985, 85–159.

[Sal85] S. M. Salamon, Harmonic and holomorphic maps, in: Geometry Seminar "Luigi Bianchi" II—1984 (E. Vesentini, ed.), Lecture Notes in Math. 1164, Springer-Verlag, Berlin–Heidelberg–New York 1985, 161–224.

[Vai79] I. Vaisman, Conformal foliations, Kodai Math. J. 2 (1979), 26–37.

[Woo84] J. C. Wood, Holomorphic differentials and classification theorems for harmonic maps and minimal immersions, in: Global Riemannian Geometry (T. J. Willmore and N. J. Hitchin, eds.), Ellis Horwood Ser. Math. Appl., E. Horwood Chichester 1984, 168–175.

[Woo86] J. C. Wood, Harmonic morphisms, foliations and Gauss maps, in: Complex differential geometry and nonlinear partial differential equations (Y.T. Siu, ed.) Contemp. Math. 49, Amer. Math. Soc., Providence, R. I. 1986, 145–184.

[Woo87] J. C. Wood, Twistor constructions for harmonic maps, in: Differential Geometry and Differential Equations (C. H. Gu, M. Berger and R. Bryant, eds.), Lecture Notes in Math. 1255, Springer-Verlag, Berlin–Heidelberg–New York 1987, 130–150.

[Woo92] J. C. Wood, Harmonic morphisms and Hermitian structures on Einstein 4-manifolds, Internat. J. Math. 3 (1992), 415–439.

[Xin96] Y.-L. Xin, Geometry of Harmonic Maps, Progress in Nonlinear Differential Equations and their Applications 23, Birkhäuser Boston 1996.

[Quest] Problem 15, in: Open Problems Session, Geometry, Topology and Physics, Proceedings of the First Brazil–USA Workshop, Campinas, Brazil, 1996, Walter de Gruyter, Berlin–New York 1997, 338 (this volume).

Open Problems Session

1. (Dmitrii Alekseevsky) Describe compact 4-manifolds which admit doubly-Hermitian metric, that is a Riemannian metric g together with two g-orthogonal complex structures $J, J' \neq \pm J$ of the same orientation. Besides hyper-Kähler metrics on K3 surfaces, some non hyper-Kähler examples are known, [K]. The classification of compact 4-manifolds which admit antiselfdual doubly-Hermitian metric was obtained recently by M. Pontecorvo, 1996.

2. (Boris Apanasov) What are building blocks of smooth 4-manifolds from the point of view of appropriate generalizations of the Thurston geometrization (see [Th, Sc]) to 4-manifolds? Are homogeneous geometries and splitting along homology spheres (additional to the splittings along 3-spheres and almost nilpotent 3-manifolds representing finite volume cusps of locally homogeneous manifolds, see [Ap1]) relevant for a geometrization of smooth 4-manifolds? There are known limitations for that in dimension 4, see [Ap1, Ap2, H]. In particular, due to [H], a closed 4-manifold which admits a finite decomposition (along cusp-surfaces) into finite volume geometric pieces is essentially either geometric itself or is aspherical. Does the so called "fiber connect sum" [FS] along submanifolds of dimension 3 which are fibered over a circle with the fiber a Riemann surface provide a splitting of 4-manifolds into geometric pieces (maybe of infinite volume)?

3. (Boris Apanasov) Are there complex hyperbolic surfaces (of infinite volume, see [A1]) with finitely generated fundamental group and with at least two boundary components at infinity? We note that since all these boundary components of such a complex surface must be non-compact due to D. Burns theorem [BM, NR1, NR2], none of them can be a homology sphere This problem has links with the above Problem 2 and with Problems 3.14 and 6.7 in Apanasov's paper [Ap1] in this volume related to geometrical finiteness of complex hyperbolic surfaces [AX].

4. (David Auckly) Use the Seiberg–Witten equations to reprove C. Taubes' theorem on manifolds with periodic ends [Ta]. How does the condition that the end is simply connected appear in this new argument? Generalizations of this theorem could be interesting topologically. Z. Bizaca and J. Ethnyre [BE] have shown that any open 4-manifold with at least one collarable end has infinitely many differential structures. A version of Furuta's 10/8-ths theorem [Fu] for a manifold with periodic ends would prove that any 4-manifold as above has uncountably many distinct differential structures.

5. (Peter Gilkey) In the Gromov–Lawson Conjecture, it is stated that the vanishing of the $\hat{A}$-genus (or more generally the vanishing of the α invariant defined by Rosenberg) is the only obstacle to a closed spin manifold of dimension $n \geq 5$ admitting a metric of positive scalar curvature. The Seiberg–Witten equations can be used to show that the situation is very different in dimension $n = 4$. One would want to know if there are equations of this type which can be used in higher dimensions to get information about this conjecture in the non-simply connected setting.

There are some interesting test cases. Let π be an elementary Abelian p group of rank 5 where p is an odd prime. Do surgery on the 5-torus to construct a spin manifold M_p^5 with fundamental group π which is π equivariant spin bordant to the original torus. Current techniques do not suffice to settle the Gromov–Lawson conjecture for M_p^5 and it remains a potential counter example to the Gromov–Lawson conjecture.

6. (Bill Goldman) Which closed 3-manifolds admit geometric structures? In particular which closed 3-manifolds admit flat affine or real-projective structures?

For example, the only closed 3-manifolds known NOT to admit flat affine structures are 3-manifolds with finite fundamental group, connected sums of 3-manifolds with finite fundamental group (John Smillie), some Seifert fibered 3-manifolds, and several hyperbolic 3-manifolds obtained by Dehn surgery on the complement of the figure-eight knot (recent work of David Fried). I think all these manifolds are rational homology 3-spheres. For example, can a hyperbolic 3-manifold admit an affine structure?

The only closed 3-manifolds known TO admit affine structures are those with infinite solvable fundamental group (except for the infinite dihedral group) and some Seifert-fibered 3-manifolds covered by products with the circle. 3-manifolds which admit flat projective structures include, in addition to these classes described above, more Seifert-fibered 3-manifolds (some with nonzero Euler class over hyperbolic base), hyperbolic 3-manifolds, and elliptic 3-manifolds.

7. (Bill Goldman) Symplectic automorphisms of the moduli space.

Let Σ be a closed surface of genus $g > 1$, $\pi = \pi_1(\Sigma)$ be its fundamental group and $\Gamma = \mathrm{Out}(\pi)$ its mapping class group. Let G be a Lie group with bi-invariant pseudo-Riemannian metric, e.g. a real reductive group). Then the moduli space $X = \mathrm{Hom}(\pi, G)/G$ inherits a natural symplectic structure ω preserved by Γ.

Does the image of the resulting homomorphism

$$\Phi : \Gamma \longrightarrow \mathrm{Aut}(X, \omega)$$

meet every component of the symplecomorphism group of (X, ω)? Characterize the image of Φ in terms of the symplectic geometry of (X, ω).

One might try to find a variational characterization of the image of Φ, in terms of some kind of symplectic invariants of the moduli space. When G is a compact Lie group with 1- and 3- dimensional simple factors, then the Γ-action is ergodic. For G nonamenable, this moduli space contains open subsets upon which Γ acts properly.

For an arbitrary Lie group G determine the ergodic decomposition of the Γ-action on X. This question is already quite interesting for $G = \mathrm{PSL}(2, R)$, where there are open subsets (corresponding to Teichmüller space upon which the group acts properly) but other subsets where the action is chaotic.

8. (Valery Marenich) Let $f : \Sigma \to E^4$ be an immersion of a closed surface of genus $g > 0$ into 4-dimensional manifold E, and ω_f be the holonomy form of this immersion. What is m – the infimum of

$$m_{\tilde{f}} = 2\pi \left| \int \omega_{\tilde{f}} d\Sigma \right|$$

taken over all immersions $\tilde{f}$ homotopic to f?

(1) Is it true that $m = m_{f^*}$, where f^* is the harmonic map (if it exists) obtained by the heat-flow with the initial value f (i.e., $f^* = \lim_{t\to\infty} f_t$ where $\frac{\partial}{\partial t} f_t = \Delta f_t$, $f_0 = f$)?

(2) Is it true, that

$$m_{f^*} \le m_f - 2\pi \sum_{p\in\Sigma} (b_p - 1),$$

where b_p is a branching number of a point p on Σ?

9. (Dan Pollack) Is a complete, embedded, constant mean curvature surface in $\mathbf{R}^3$, all of whose ends are asymptotically cylindrical, necessarily a cylinder? Analogously, is a complete conformally flat manifold with constant positive scalar curvature and asymptotically cylindrical ends necessarily a cylinder? In either case if you can show that there are only two ends then you would necessarily have a cylinder. It may be helpful to assume that the surface has genus zero (or, respectively, that the manifold is simply connected).

10. (Lorenzo A. Sadun) What do Donaldson invariants (or Seiberg–Witten invariants) tell us about the structure of ASD moduli spaces? In particular, the simple type condition, generally believed to hold in all manifolds with $b^+ > 1$, suggests a similarity in structure between $\mathcal{M}_k$ and $\mathcal{M}_{k+1}$. (This similarity is not merely due to bubbling.) Just how are $\mathcal{M}_k$ and $\mathcal{M}_{k+1}$ related? Can one embed $\mathcal{M}_k$ into the interior of $\mathcal{M}_{k+1}$? If so, how? By "interior" we mean those points of $\mathcal{M}_{k+1}$ that cannot be viewed as a charge-k background together with a charge-1 bubble.

11. (Nikolai Saveliev) Suppose that an integral homology sphere Σ bounds a smooth acyclic 4-manifold W. Is that true that all the relative Donaldson polynomials of the pair (W, Σ) are trivial?

12. (Nikolai Saveliev) Let $\Sigma(p, q, r)$ be the Brieskorn homology sphere, which is the link of the singularity at 0 of the hypersurface $x^p + y^q + z^r = 0$ in $\mathbb{C}^3$. The following question is attributed in [NW] to M. Atiyah: Is there a Milnor fiber description of the

Floer homology of $\Sigma(p,q,r)$? Is it related to the action of complex conjugation on the homology of the Milnor fiber?

Conjecture. The ranks of Floer homology groups of the Brieskorn homology sphere $\Sigma(p,q,r)$ are equal to $\frac{1}{2}(\lambda(\Sigma(p,q,r)) \pm \bar{\mu}(\Sigma(p,q,r))$ where λ is the Casson invariant and $\bar{\mu}$ is the invariant of Neumann [N] and Siebenmann [Sb].

While this volume was in preparation, the conjecture was proved to be correct, see [S1] and [S2]. The question remains if the same formula holds for all graph homology 3-spheres.

13. (Karen Uhlenbeck) Give a comprehensible proof of the theorem of Caffarelli, Gidas and Spruck. One wants a local strong estimate on solutions of the Yamabe equation with an isolated singularity which shows that the end is either removable or has an asymptotically periodic structure.

Remark added in proof. While this volume was in preparation, such an estimate has recently been provided in a preprint [KMPS] by Korevaar, Mazzeo, Pacard and Schoen. They actually provide a stronger estimate which not only establishes exponential decay to a periodic solution but also identifies the precise asymptotic Delaunay solution (as a complete metric on the complement of two points in the n-sphere) by analyzing the second term in the asymptotic expansion.

14. (Karen Uhlenbeck) Develop gauge theory with projective or affine connections. Are there interesting constructions in gauge theory using affine or projective constants?

15. (John Wood) Let $\phi : M \to N$ be a harmonic map between compact C^∞ Riemannian manifolds and let v be a Jacobi field (infinitesimal deformation) for the energy along ϕ (see, for example [GW89]). Say that v is *integrable* if it arises as the variation vector field $v = d\phi/dt(0)$ of a family $\phi_t : M \to N$ of harmonic maps.

Question: For what (M, N, ϕ) is any Jacobi field along ϕ integrable?

Note that:

(1) This is not always the case, e.g. for the normal Jacobi field to a closed geodesic around the flat waist of a surface of revolution.

(2) It *is* the case when

  (a) N is real-analytic and of non-positive sectional curvature [SY79],

  (b) $M = N = S^2$ [GW89],

  (c) $M = S^2$, $N = \mathbb{C}P^n$ and ϕ is holomorphic or antiholomorphic [LWxx].

In the case $M = S^m$, the question has important implications for the behaviour of blow-ups near singularities (see, for example, [AS88, GW89, Har97, Sim91]). In the case $M = S^2$, $N = \mathbb{C}P^2$, the components of the space of harmonic maps are smooth submanifolds of, say, $C^2(S^2, \mathbb{C}P^2)$ (see [LW96, Woo96]) and one might hope for a positive answer.

16. (Lyndon Woodward) What are the harmonic maps of S^2 into a compact Lie group G for which the induced metric is of constant curvature?

A related problem, which is germane to this question, is to find a congruence theorem for harmonic maps of S^2 into G. What other invariants are needed besides the induced metric?

References

[AS88] D. Adams and L. Simon, Rates of asymptotic convergence near isolated singularities of geometric extrema, Indiana Univ. Math. J. 37 (1988), 225–254.

[Ap1] Boris Apanasov, Quasiconformality and geometrical finiteness in Carnot–Carathéodory and negatively curved spaces, this volume.

[Ap2] ——, Geometry and topology of complex hyperbolic CR-manifolds, Russian Math. Surveys, to appear (Preprint 1997-007, Math. Sci. Res. Inst., Berkeley, 1997).

[AX] B. Apanasov and X. Xie, Geometrically finite complex hyperbolic manifolds, Intern. J. Math, to appear.

[BE] Z. Bizaca, J. Etnyre, Smooth structures on collarable ends of 4-manifolds, preprint.

[BM] Dan Burns and Rafe Mazzeo, On the geometry of cusps for $SU(n, 1)$, Preprint, Univ. of Michigan, 1994.

[FS] Ronald Fintushel and Ronald Stern, Surgery in cusp neighborhoods and the geography of irreducible 4-manifolds, Invent. Math. 117 (1994), 455–523.

[Fu] M. Furuta, Monopole equation and the 11/8 conjecture, preprint.

[GW89] R. Gulliver and B. White, The rate of convergence of a harmonic map at a singular point, Math. Ann. 283 (1989), 539–549.

[Har97] R. M. Hardt, Singularities of harmonic maps, Bull. Amer. Math. Soc. (N.S.) 34 (1997), 15–34.

[H] J. A. Hillman, On 4-manifolds which admit geometric decompositions, preprint, University of Sydney, 1996.

[K] P. Kobak, Explicit doubly-Hermitian metrics, preprint ESI, 197, 1995.

[KMPS] N. Korevaar, R. Mazzeo, F. Pacard and R. Schoen, Refined asymptotics for constant scalar curvature metrics with isolated singularities, preprint, 1997.

[LW96] L. Lemaire and J. C. Wood, On the space of harmonic 2-spheres in $\mathbf{C}P^2$, Internat. J. Math. 7 (1996), 211–225.

[LWxx] L. Lemaire and J. C. Wood, in preparation.

[NR1] T. Napier and M. Ramachandran, Structure theorems for complete Kähler manifolds and applications to Lefschetz type theorems, Geom. Funct. Anal. 5 (1995), 807–851.

[NR2] T. Napier and M. Ramachandran, The L^2 $\bar{\partial}$-method, weak Lefschetz theorems, and topology of Kähler manifolds, Preprint, 1996.

[N] W. Neumann, An invariant of plumbed homology spheres, Lecture Notes in Math. 788, Springer-Verlag, Berlin and New-York, 1980, 125–144.

[NW] W. Neumann, J. Wahl, Casson invariant of link of singularities, Comm. Math. Helv. 65 (1990), 58–78.

[P] M. Pontecorvo, 1996.

[Sa1] Nikolai Saveliev, Floer homology of Brieskorn homology spheres: solution to Atiyah's problem. MSRI Preprint No. 1997–006

[Sa2] Nikolai Saveliev, On the homology cobordism group of homology 3-spheres, this volume

[SY79] R. Schoen and S. T. Yau, Compact group actions and the topology of manifolds with non-positive curvature, Topology 18 (1979), 361–380.

[Sc] Peter Scott, The geometries of 3-manifolds, Bull. London Math. Soc. 15 (1983), 407–487.

[Sb] L. Siebenmann, On vanishing of the Rohlin invariant and nonfinitely amphicheiral homology 3-spheres, Lecture Notes in Math. 788, Springer-Verlag, Berlin and New-York, 1980, 172–222.

[Sim91] L. Simon, The singular set of minimal submanifolds and harmonic maps, in: Progress in Partial Differential Equations: Calculus of Variations, Applications (Pont-a-Mousson, 1991), Pitman Res. Notes Math. 267, Longman Sci. Tech., Harlow, 1992, 99–108.

[Ta] C. Taubes, Gauge theory on asymptotically periodic 4-manifolds, J. Diff. Geom. 25 (1987), 363–430.

[Th] W. P. Thurston, The geometry and topology of three-manifolds, Lect. Notes, Princeton Univ., 1979.

[Woo96] J. C. Wood, Harmonic maps and morphisms in 4-dimensions, this volume.

List of Contributors

D. V. Alekseevsky, ESI, Boltzmanngasse 9, A-1090 Austria
daleksee@esi.ac.at

Boris Apanasov, University of Oklahoma, Department of Mathematics, Norman, OK 73019, U.S.A.
apanasov@ou.edu

John Bolton, Department of Mathematical Sciences, University of Durham, Durham DH1 3LE, England, UK
John.Bolton@durham.ac.uk

Steven Bradlow, Department of Mathematics, University of Illinois, Urbana, IL 61801, U.S.A.
bradlow@math.uiuc.edu

Suhyoung Choi, Department of Mathematics, Seoul National University, 151-742 Seoul, Korea
shchoi@math.snu.ac.kr

Georgios D. Daskalopoulos, Department of Mathematics, Brown University, Providence, RI 02912, U.S.A.
daskal@math.brown.edu

Peter B. Gilkey, Department of Mathematics, University of Oregon, Eugene, OR 97403-1222, U.S.A.
gilkey@bright.uoregon.edu

James Glazebrook, Department of Mathematics, University of Illinois, Urbana, IL 61801, U.S.A.; and Department of Mathematics, Eastern Illinois University, Charleston, IL 61920, U.S.A.
glazebro@math.uiuc.edu

Franz Kamber, Department of Mathematics, University of Illinois, Urbana, IL 61801, U.S.A.
kamber@math.uiuc.edu

Yoshinobu Kamishima, Department of Mathematics, Kumamoto University, Kumamoto 860, Japan
yoshi@kumamoto-u.ac.jp

Antonella Marini, Dipartimento di Matematica Pura ed Applicata, Università degli Studi di L'Aquila, 67100 L'Aquila, Italy
marini@smaq20.univaq.it

Caio J. C. Negreiros, Universidade Estadual de Campinas, IMECC, C..P.: 6065, 13083-970 Campinas - SP, Brazil
caione@ime.unicamp.br

Mauricio M. Peixoto, Instituto de Matemática Pura e Aplicada, 110 Estrada Dona Castorina, Rio de Janeiro, 22460-320, Brazil
peixoto@impa.br

F. Podestà, Dipartimento di Matematica "U. Dini", Viale Morgagni 67/A, 50100 Firenze, Italy
podesta@udini.math.unifi.it

Lorenzo Sadun, Department of Mathematics, University of Texas, Austin, TX 78712, U.S.A.
sadun@math.utexas.edu

Nikolai Saveliev, Department of Mathematics, University of Michigan, Ann Arbor, MI 48109, U.S.A.
saveliev@math.lsa.umich.edu

P. Srivastava, DFNAE, Instituto de Fisica, Universidade do Estado de Rio de Janeiro-UERJ, Rua San Francisco Xavier, 524, 20550-013 Rio de Janeiro, Brazil
prem@lafexsu1.lafex.cbpf.br

Jayme Vaz, Jr., Department of Physics, Syracuse University, Syracuse, NY 13244-1130, U.S.A.
vaz@suhep.phy.syr.edu

Serguei Vodop'yanov, Sobolev Institute of Mathematics, 630090 Novosibirsk, Russia
vodopis@math.nsc.ru

Richard A. Wentworth, Department of Mathematics, University of California, Irvine, 420 Physical Sciences I, Irvine, CA 92697-3875, U.S.A.
rwentwor@math.uci.edu

John C. Wood, School of Mathematics, University of Leeds, Leeds LS2 9JT, England, UK
J.C.Wood@Leeds.ac.uk

Lyndon M. Woodward, Department of Mathematical Sciences, University of Durham, Durham DH1 3LE, England, UK
L.M.Woodward@uc.ac.durham

List of Participants

Dmitry Alekseevsky, ESI, Boltzmanngasse 9, A-1090 Austria, daleksee@esi.ac.at

Luis Carlos G. de Andrade, UFRJ, 524 Rua Sao Francisco Xavier, Maracana - Rio de Janeiro, RJ CEP-20559-900, Brazil

Boris Apanasov, University of Oklahoma, Department of Mathematics, Norman, OK 73019, U.S.A., apanasov@ou.edu

José Ricardo Arteaga, Universidade de Los Andes, Apt Aereo 4976, Bogota, Colombia, jarteaga@uniandes.edu.co

Dave Auckly, University of California at Berkeley, Berkeley, CA 94720, U.S.A., dav@math.berkeley.edu

Paul Baird, Université de Bretagne, 6 avenue V. Legorgeu, 29287 Brest, France, Paul.Baird@univ-brest.fr

John Bolton, Department of Mathematical Sciences, University of Durham, Durham DH1 3LE, England, UK, John.Bolton@durham.ac.uk

Steven Bradlow, Department of Mathematics, University of Illinois, Urbana, IL 61801, U.S.A., bradlow@math.uiuc.edu

Aldir C. Brasil Jr., Univ. de Sao Paulo, 2265 Rua Cristovao Colombo, Sao Jose do Rio Preto, SP CEP-15100, Brazil

Francis Burstall, School of Mathematics, University of Bath, Claverton Down, Bath BA2 7AY, England, UK, F.E.Burstall@maths.bath.ac.uk

Suhyoung Choi, Department of Mathematics, Seoul National University, 151-742 Seoul, Korea, shchoi@math.snu.ac.kr

Antonio Conde, Univ. de Sao Paulo, Cx. P. 668, Sao Paulo, CEP - 13560, Brazil conde@icmsc.sc.usp.br

Luciano Costa, Univ. de Sao Paulo, Cx.P.66281, Sao Paulo, SP CEP-05389-970, Brazil, luciano@snfma1.if.usp.br

Sérgio Santa Cruz, Univ. de Pernambuco, RE Braga 584, Iputinga - Recife, CEP - 50760-380, Brazil, ssc@dmat.ufpe.br

Georgios D. Daskalopoulos, Department of Mathematics, Brown University, Providence, RI 02912, U.S.A., daskal@math.brown.edu

Joseph Datskovsky, Pontifica Univ. Catolica, 114/1001 Rua Republica do Peru, Copacabana - Rio de Janeiro, RJ CEP-22021-040, Brazil

Ron Donagi, University of Pennsylvania, 209 S 33rd Str, Philadelphia, PA 19104-6395, U.S.A., donagi@math.upenn.edu

Celso M. Doria, Univ. Fed. de Santa Catarina, Trindade - Florianopolis - SC, Brazil, cmdoria@mtm.ufsc.br

Francisco Dutenhefner, Univ. de Sao Paulo, Cx. P. 668, Sao Paulo, CEP - 13560, Brazil, chico@icmsc.sc.usp.br

Cecilia Ferreira, Univ. of Lisbon, 2 Ave. Prof. Gama Pinto, Lisboa, Portugal 1699, cecilia@ptmat.lmc.fc.ul.pt

Ronald Fintushel, Michigan State University, East Lansing, MI 48824, U.S.A., ronfint@math.msu.edu

Dmitry Galtsov, Department of Theor. Physics, Moscow State Univ., 17234 Moscow, Russia

Oscar Garcia-Prada, Universidad Autonoma de Madrid, 28049 Madrid, Spain, ogprada@ccuam3.sdi.uam.es

Peter B. Gilkey, Department of Mathematics, University of Oregon, Eugene, OR 97403-1222, U.S.A., gilkey@bright.uoregon.edu

William Goldman, University of Maryland, College Park, MD 20742-0001, U.S.A., wmg@math.umd.edu

Guillermo Gonzalez, Univ. of Campinas, IMECC, Cx.P.6065, Campinas, SP CEP-13081-970, Brazil, guille@ime.unicamp.br

Nikolai Goussevskii, Univ. Fed. de Minas Gerais, 6627 Avenida Antonio Carlos, Campus Universitario-Pampulha, Cx. P.1621, Belo Horizonte, MG CEP-31270-901, Brazil, nikolay@mat.ufmg.br

Irwen Guadalupe, Univ. of Campinas, IMECC, Cx.P.6065, Campinas, SP CEP-13081-970, Brazil

Derek Hacon, Pontifica Universidade Catolica, 225 Rua Marques de Sao Vicente, Gavea - Rio de Janeiro, RJ CEP-22453-900, Brazil, hacon@lncc.bitnet and derec@saci.mat.puc-rio.br

Rafael Herrera, Oxford University, Institute of Mathematics, 24–29 St. Giles, Oxford OX1 3LB, England, UK, herrera@maths.ox.ac.uk

Yoshinobu Kamishima, Department of Mathematics, Kumamoto University, Kumamoto 860, Japan, yoshi@kumamoto-u.ac.jp

Takashi Kimura, Boston University, 111 Cummington St., Boston, MA 02215, U.S.A., kimura@math.bu.edu

Eduardo Lima, Univ. de Sao Paulo, 2265 Rua Cristovao Colombo, Sao Jose do Rio Preto, SP CEP-15100, Brazil, lima@nimitz.ibilce.unesp.br

Armando Machado, Univ. of Lisbon, 2 Av. Prof. Gama Pinto, Lisboa, Portugal 1699, armac@ptmat.lmc.fc.ul.pt

Ton Marar, Univ. de Sao Paulo, USP-ICMSC, Cx.P.668, SP CEP-13560, Brazil, walmarar@cmsc.sc.usp.br

Valery Marenich, Univ. of Campinas, IMECC, Cx.P.6065, Campinas, SP CEP-13081-970, Brazil, marenich@ime.unicamp.br

Antonella Marini, Dipartimento di Matematica Pura ed Applicata, Università degli Studi di L'Aquila, 67100 L'Aquila, Italy, marini@smaq20.univaq.it

Antonio Carlos Gilli Martins, Univ. of Campinas, IMECC, Cx. P.6065, Campinas, SP CEP-13081-970, Brazil, gilli@ime.unicamp.br

Miriam Percia Mendes, Univ. de Sao Paulo, USP-ICMSC, Cx.P.668, SP CEP-13560, Brazil, mpmendes@icmsc.sc.usp.br

Ramon Mendoza, Univ. Fed. de Pernambuco, RE Braga 584, Recife, PE CEP-50760-380, Brazil, ramon@dmat.ufpe.br

Fernando Moraes, Univ. Fed. de Pernambuco, Recife, PE CEP-50740-540, Brazil, moraes@lftc.ufpe.br

Joao Batista do Nascimento, Univ. de Sao Paulo, Cx.P.66281, Sao Paulo, SP CEP-05389-970, Brazil

Caio J. C. Negreiros, Universidade Estadual de Campinas, IMECC, C..P.: 6065, 13083-970 Campinas - SP, Brazil, caione@ime.unicamp.br

Seme Gebara Neto, UFMG, 6627 Avenida Antonio Carlos, Cx. P.1621, Belo Horizonte, MG CEP-31270-901, Brazil

Edmundo Capelas de Oliveira, Univ. of Campinas, IMECC, Cx.P.6065, Campinas, SP CEP-13081-970, Brazil, capelas@ime.unicamp.br

Burak Ozbagchi, Univ. of California at Irvine, Irvine, CA 92697-3875, U.S.A., bosbagci@math.uci.edu

Frederico Palmeira, Pontifica Univ. Catolica, 225 Rua Marques de Sao Vicente, Gavea - Rio de Janeiro, RJ CEP-22453-900, Brazil, fredpalm@saci.mat.puc-rio.br

Marlio Paredes, Univ. of Campinas, IMECC, Cx.P.6065, Campinas, SP CEP-13081-970, Brazil, marlio@ime.unicamp.br

M. Peixoto, Instituto de Matemática Pura e Aplicada, 110 Estrada Dona Castorina, Rio de Janeiro, 22460-320, Brazil, peixoto@impa.br

Daniel Pollack, University of Washington, Seattle, WA 98185, U.S.A., pollack@math.washington.edu

Jonathan Poritz, University of Maryland, College Park, MD 20742-4015, U.S.A., kaleb@math.umd.edu

Rohan A. dos Reis, Univ. de Sao Paulo, Cx.P.66281, Sao Paulo, SP CEP-05389-970, Brazil

Lorenzo Sadun, Department of Mathematics, University of Texas, Austin, TX 78712, U.S.A., sadun@math.utexas.edu

Luiz A. Saeger, Univ. de Sao Paulo-IFT, 145 Rua Pamplona145, Sao Paulo, SP CEP-01405-900, Brazil, saeger@axp.ift.unesp.br

Nikolai Saveliev, Department of Mathematics, University of Michigan, Ann Arbor, MI 48109, U.S.A., saveliev@math.lsa.umich.edu

Paul Schweitzer, Pontifica Univ. Catolica, 225 Rua Marques de Sao Vicente, Gavea - Rio de Janeiro, RJ CEP-22453-900, Brazil, paul@mat.puc-rio.br

Antonio Roberto da Silva, UFRJ-IM, Cx.P.68519, Rio de Janeiro, RJ CEP-21945-970, Brazil, beto@lpim.ufrg.br

P. Srivastava, DFNAE, Instituto de Fisica, Universidade do Estado de Rio de Janeiro-UERJ, Rua San Francisco Xavier, 524, 20550-013 Rio de Janeiro, Brazil, prem@lafexsu1.lafex.cbpf.br

Alexandre Luiz Trovon, Univ. Fed. do Parana, Cx.P.19081, Curitiba, PR CEP-81531-990, Brazil, trovon@mat.ufpr.br

Karen Uhlenbeck, University of Texas at Austin, Austin, TX 78712-1082, U.S.A., uhlen@math.utexas.edu

Jayme Vaz, Jr., Department of Physics, Syracuse University, Syracuse, NY 13244-1130, U.S.A., vaz@suhep.phy.syr.edu

Serguei Vodop'yanov, Sobolev Institute of Mathematics, 630090 Novosibirsk, Russia, vodopis@math.nsc.ru

Marcus Wanderley, Univ. Fed. de Pernambuco, RE Braga 584, Recife, PE CEP-50760-380, Brazil, marcus@dmat.ufpe.br

John C. Wood, School of Mathematics, University of Leeds, Leeds LS2 9JT, England, UK, J.C.Wood@Leeds.ac.uk

Lyndon M. Woodward, Department of Mathematical Sciences, University of Durham, Durham DH1 3LE, England, UK, L.M.Woodward@uc.ac.durham

Yu Yan-Lin. Inst. of Math., Chinese Acad. Science, Beijing 100080, Peoples Rep. of China

Jose Ricardo R. Zeni, Pontifica Univ. Catolica, 225 Rua Marques de Sao Vicente, Gavea - Rio de Janeiro, RJ CEP-22453-900, Brazil, zeni@saci.mat.puc-rio.br